AF594966

'Eyes', 'Brain', 'Feet' and 'Hands' of Efficient Harvesting Machinery

'Eyes', 'Brain', 'Feet' and 'Hands' of Efficient Harvesting Machinery

Editors

Cheng Shen
Zhong Tang
Maohua Xiao

Basel • Beijing • Wuhan • Barcelona • Belgrade • Novi Sad • Cluj • Manchester

Editors
Cheng Shen
Nanjing Institute of Agricultural Mechanization, Ministry of Agriculture and Rural Affairs
Nanjing, China

Zhong Tang
Jiangsu University
Zhenjiang, China

Maohua Xiao
Nanjing Agricultural University
Nanjing, China

Editorial Office
MDPI
St. Alban-Anlage 66
4052 Basel, Switzerland

This is a reprint of articles from the Special Issue published online in the open access journal *Agriculture* (ISSN 2077-0472) (available at: https://www.mdpi.com/journal/agriculture/special_issues/I00178P8H3).

For citation purposes, cite each article independently as indicated on the article page online and as indicated below:

Lastname, A.A.; Lastname, B.B. Article Title. *Journal Name* **Year**, *Volume Number*, Page Range.

ISBN 978-3-0365-9212-1 (Hbk)
ISBN 978-3-0365-9213-8 (PDF)
doi.org/10.3390/books978-3-0365-9213-8

Cover image courtesy of Cheng Shen

Contents

About the Editors

Cheng Shen

Cheng Shen, born in October 1989, is an Associate Professor of Nanjing Institute of Agricultural Mechanization, Ministry of Agriculture and Rural Affairs, mainly engaged in crop harvesting machinery and agricultural machinery intelligence. He is also the Deputy Secretary General of the Youth Working Committee of the Chinese Society for Agricultural Machinery, the Deputy Director and Secretary General of the Youth Working Committee of the Jiangsu Society of Agricultural Engineering, and an expert in the evaluation of innovation and application projects of the Ministry of Innovation and Development of the Republic of Uzbekistan. He was awarded the title of "Jiangsu Youth Science and Technology Innovation 'U35 Innovator'" by multiple departments of Jiangsu Province, "Young Outstanding Talent" and the first "Top Ten Youth" at the Institute level.

In recent years, as a PI, he has undertaken 7 projects including the National Natural Science Foundation of China; as one of the drafters, he participated in the preparation of 3 national-level plans/outlines; as the first author, he submitted a report to the Scientists' Suggestions, the internal reference published by the China Association for Science and Technology; as the first or corresponding author, he published 18 papers indexed by SCI/EI; as the first inventor, he has authorized 7 China Invention Patents; he completed more than 200 reviews for more than 20 journals, and made 3 consecutive reports at the CIGR World Congress. He won the Excellent Product Award of the China High-tech Fair, first prize for the "Science and Technology" special article of the young and middle-aged cadres of the Ministry of Agriculture and Rural Affairs, the Excellent Paper Award of the Chinese Society of Agricultural Engineering, and the Outstanding Reviewer Award of IJABE.

Zhong Tang

Zhong Tang was born in April 1982, professor and doctoral supervisor of School of agricultural engineering, Jiangsu University, mainly engaged in the development and research of key technologies and products of agricultural equipment, such as combine harvesters. He has conducted theoretical and experimental research on cutting, threshing separation, cleaning mechanisms and intelligent technology in harvesting machinery, successively obtaining the national natural science foundation project, Jiangsu province youth science foundation project, China postdoctoral foundation project, Jiangsu province postdoctoral foundation project, and Jiangsu province education department natural science foundation project.

He won first prize (fourth) for the science and technology award of China association of machinery industry in 2017, first prize (fourth) for the science and technology progress award of the ministry of education in 2017, and the Chinese patent prize in 2018. So far, he has published more than 40 SCI/EI academic papers as the first author, applied for more than 30 invention patents as the first author, and obtained 12 authorized invention patents. He has applied for 5 PCT patents, and published a textbook entitled Structure and Use Maintenance of Combine Harvester.

Maohua Xiao

Maohua Xiao, born in September 1981, is a Professor and Doctoral Supervisor of College of Engineering, Nanjing Agricultural University, mainly engaged in research on intelligent agricultural machinery power technology and precision operation equipment, intelligent control and unmanned system of agricultural machinery equipment. He has been selected into the talent program of Jiangsu Province Young Science and Technology Talent Promotion Project, Jiangsu Province "Blue Project" young and middle-aged academic leaders, and Jiangsu Province "Six talent peaks". He is also the deputy director of Jiangsu Province Electric Agricultural Equipment Science and Technology Innovation Center, the deputy director of China Machinery Industry Agricultural Machinery Power and farming machinery Reliability Technology Key Laboratory, and the secretary general of Jiangsu Province tractor industry Technology innovation Strategic alliance.

He has presided over more than 10 projects under the National Natural Science Foundation, national key research and development programs, and national key agricultural core technology research projects. More than 70 papers have been published, including nearly 70 indexed by SCI/EI; more than 60 patents have been authorized (ranked first in 11 authorized invention patents), and more than 30 software copyrights, of which two patents have been successfully converted and applied. He has won 8 provincial and ministerial awards including the first prize of China Agricultural Machinery Science and Technology Award. He has also participated in the research and development of the first multi-link high-speed ultra-precision press in China and the first large-power tractor hydraulic mechanical stepless transmission in China, and five products developed were rated as the "first (set) major equipment products" in Jiangsu Province.

Preface

Harvesting is one of the most important aspects of agricultural production. The development of harvesting tools has a long history: as early as the Stone Age, people used stone sickles and other tools to harvest grain. In the late 19th century, agricultural mechanization began to emerge, and people began to try to use mechanized equipment to replace the traditional human labor. In 1799, the earliest horse-drawn disc cutter harvester appeared in the United Kingdom; 1822 saw the cutter on top of the additional paddle device. In 1826 marked the emergence of the use of a reciprocating cutter and paddle wheel of the prototype of the modern harvester, with horse traction and through the rotation of the ground wheel drive cutter. Around 1831∼1835, similar animal-powered wheat harvester in the United States were commercialized; in 1851 saw the emergence of machines to cut down stalks into a pile with a rocking-arm harvester as well as the application of internal combustion engines to make the harvester more efficient and flexible. Years later, in 1920, due to the widespread use of tractors, the tractor with a harvester began to replace the animal-powered harvester. The types of harvests also expanded gradually from the earliest grains to a wide range of crops such as fruits, vegetables, and cotton. In the 21st century, with the continuous progress of science and technology, the degree of automation of the harvester is higher and higher. Especially in recent years, visual perception technology, satellite navigation technology and intelligent decision-making algorithms have been applied on a large scale on the harvester, which has enabled the harvesting machine intelligence to rapidly improve and gradually transition to unmanned operation.

The focus of harvester research efforts now includes mechanical components and intelligent systems, including the development and optimization of harvesting components for specific crops and the design and optimization of travel components for different terrain and soil conditions. Intelligent system research includes operating environment sensing and crop target recognition, decision-making and control of operation and traveling. It is graphically summarized as the following key technologies: precise identification and positioning systems ("eyes"), sensitive decision-making and control systems ("brain"), highly adaptable chassis and mobile platforms ("feet"), efficient end-effectors and harvesting components ("hands"), etc. In order to gather excellent papers in this field and showcase the latest ideas, we have organized a Special Issue of Agriculture entitled "'Eyes', 'Brain', 'Feet' and 'Hands' of Efficient Harvesting Machinery", and a total of 21 research papers were selected for inclusion in the Special Issue. These papers address harvesting components, traveling mechanisms, sensing systems, and decision-making algorithms for a wide range of crop harvesters, including grains, vegetables, and fruits, and provide key technical references for efficient harvesting machinery research.

Cheng Shen, Zhong Tang, and Maohua Xiao
Editors

Editorial

"Eyes", "Brain", "Feet" and "Hands" of Efficient Harvesting Machinery

Cheng Shen [1,*], Zhong Tang [2] and Maohua Xiao [3]

[1] Nanjing Institute of Agricultural Mechanization, Ministry of Agriculture and Rural Affairs, Nanjing 210014, China
[2] School of Agricultural Engineering, Jiangsu University, Zhenjiang 212013, China; tangweizhong0427@163.com
[3] College of Engineering, Nanjing Agricultural University, Nanjing 210031, China; xiaomaohua@njau.edu.cn
* Correspondence: shencheng@caas.cn

The main function of harvesting is the cutting, picking, or digging of mature crop seeds, fruits, stalks, leaves, root parts, or the whole plant. They are also necessary to complete harvesting, threshing, cleaning, transfer, and other operations. The characteristics of harvesting machinery are as follows: first, the operating object is biological; second, the working process refers to either the separation of plant tissues or the separation of a plant from the soil; and third, the working environment is unstructured. Therefore, efficient harvesting machinery is a comprehensive research area that integrates the fields of mechanical and biological material, information, and computers. With the improvement of the intelligent level of agricultural equipment, the traditional harvesting machine has been gradually developing toward harvesting robots; the focus of harvester research efforts now includes mechanical components and intelligent systems, which include the development and optimization of harvesting components for specific crops as well as the design and optimization of travel components for different terrain and soil conditions. Intelligent system research includes operating environment sensing and crop target recognition, decision making, and the control of operation and traveling. This research area can be graphically summarized as these key technologies: precise identification and positioning systems ("eyes"), sensitive decision-making and control systems ("brain"), highly adaptable chassis and mobile platforms ("feet"), efficient end-effectors and harvesting components ("hands"), etc. In order to gather excellent papers in this field, showcase the latest ideas, and provide key technical references for efficient harvesting machinery research, we have organized the Special Issue entitled "'Eyes', 'Brain', 'Feet' and 'Hands' of Efficient Harvesting Machinery" in Agriculture.

Citation: Shen, C.; Tang, Z.; Xiao, M. "Eyes", "Brain", "Feet" and "Hands" of Efficient Harvesting Machinery. *Agriculture* **2023**, *13*, 1861. https://doi.org/10.3390/agriculture13101861

Received: 12 September 2023
Revised: 18 September 2023
Accepted: 18 September 2023
Published: 22 September 2023

In total, 21 papers are published in this Special Issue, including two papers related to the physical property parameter of materials or crops [1,2]. Thirteen papers describe efficient end-effectors and harvesting components ("hands"), with the crop objects harvested including Cabbage [3], Cotton Stalk [4], Potato [5], Dandelion Seed [6], Broccoli [7], Peanuts [8], Rice [9], Supernormal Jujube Branches [10], Silage Corn [11], Small Spherical Fruit [12], Chinese Little Greens [13], Hazelnut [14], and Chinese Milk Vetch (*Astragalus sin-icus* L.) Seeds [15]. Two papers describe highly adaptable chassis and mobile platforms ("feet"), which focuses on machine–soil relationship [16,17]. Two papers described precise identification and positioning systems ("eyes"); these studies applied the improved YOLO algorithm to the recognition of tea and apples [18,19]. Two papers described sensitive decision-making and control systems ("brain") [20,21]; these studies use intelligent control algorithms, such as BP neural network algorithms, to control actuating components or for fault diagnosis. These papers address harvesting components, traveling mechanisms, sensing systems and decision-making algorithms for a wide range of crop harvesters, including grains, vegetables and fruits.

From these studies, it is clear that the research focus with relation to harvesting machinery is on equipment intelligence, in addition to design improvements for crop-specific harvesting components. The advantages of agricultural equipment intelligence include improving the operating efficiency of agricultural machinery, reducing labor costs, reducing pesticides and fertilizer use, improving farmland ecosystem protection, and promoting the sustainable development of agriculture. At the same time, the intelligentization of agricultural machinery and equipment also faces some challenges, such as the high cost of technology research and development, low acceptance by farmers, etc., which requires joint efforts from the government, enterprises and farmers to promote the development of the intelligentization of agricultural machinery and equipment.

Acknowledgments: This material is based upon work that is supported by the Jiangsu Agricultural Science and Technology Innovation Fund No. CX(21)3149. We would like to sincerely thank all authors who submitted papers to the Special Issue of *Agriculture* entitled "'Eyes', 'Brain', 'Feet' and 'Hands' of Efficient Harvesting Machinery", the reviewers of these papers for their constructive comments and thoughtful suggestions, and the editorial staff of *Agriculture*.

Conflicts of Interest: The authors declare no conflict of interest.

References

1. Hu, Y.; Xiang, W.; Duan, Y.; Yan, B.; Ma, L.; Liu, J.; Lyu, J. Calibration of ramie stalk contact parameters based on the discrete element method. *Agriculture* **2023**, *13*, 1070. [CrossRef]
2. Fu, J.; Cui, Z.; Chen, Y.; Guan, C.; Chen, M.; Ma, B. Simulation and experiment of compression molding behavior of substate block suitable for mechanical transplanting based on Discrete Element Method (DEM). *Agriculture* **2023**, *13*, 883. [CrossRef]
3. Tong, W.; Zhang, J.; Cao, G.; Song, Z.; Ning, X. Design and experiment of a low-loss harvesting test platform for cabbage. *Agriculture* **2023**, *13*, 1204.
4. Wang, Z.; Zhao, W.; Fu, J.; Xie, H.; Zhang, Y.; Chen, M. V-Shaped toothed roller cotton stalk puller: Numerical modeling and field-test validation. *Agriculture* **2023**, *13*, 1157. [CrossRef]
5. Wang, L.; Liu, F.; Wang, Q.; Zhou, J.; Fan, X.; Li, J.; Zhao, X.; Xie, S. Design of a spring-finger potato picker and an experimental study of its picking performance. *Agriculture* **2023**, *13*, 945. [CrossRef]
6. Qu, Z.; Lu, Q.; Shao, H.; Liu, L.; Wang, X.; Lv, Z. Design and testing of a self-propelled dandelion seed harvester. *Agriculture* **2023**, *13*, 917.
7. Xu, H.; Yu, G.; Niu, C.; Zhao, X.; Wang, Y.; Chen, Y. Design and experiment of an underactuated broccoli-picking manipulator. *Agriculture* **2023**, *13*, 848. [CrossRef]
8. Shen, H.; Yang, H.; Gao, Q.; Gu, F.; Hu, Z.; Wu, F.; Chen, Y.; Cao, M. Experimental research for digging and inverting of upright peanuts by digger-inverter. *Agriculture* **2023**, *13*, 847.
9. Ma, L.; Xie, F.; Liu, D.; Wang, X.; Zhang, Z. An application of artificial neural network for predicting threshing performance in a flexible threshing device. *Agriculture* **2023**, *13*, 788. [CrossRef]
10. Zhang, R.; Wang, G.; Wang, W.; Ren, D.; Gong, Y.; Yue, X.; Hou, J.; Yang, M. Numerical simulation of the picking process of supernormal jujube branches. *Agriculture* **2023**, *13*, 408. [CrossRef]
11. Wang, F.; Wang, J.; Ji, Y.; Zhao, B.; Liu, Y.; Jiang, H.; Mao, W. Research on the measurement method of feeding rate in silage harvester based on components power data. *Agriculture* **2023**, *13*, 391.
12. Zhang, F.; Chen, Z.; Wang, Y.; Bao, R.; Chen, X.; Fu, S.; Tian, M.; Zhang, Y. Research on flexible end-effectors with humanoid grasp function for small spherical fruit picking. *Agriculture* **2023**, *13*, 123. [CrossRef]
13. Wang, W.; Lv, X.; Yi, Z. Parameter optimization of reciprocating cutter for chinese little greens based on finite element simulation and experiment. *Agriculture* **2022**, *12*, 2131. [CrossRef]
14. Ren, D.; Yu, H.; Zhang, R.; Li, J.; Zhao, Y.; Liu, F.; Zhang, J.; Wang, W. Research and experiments of hazelnut harvesting machine based on CFD-DEM analysis. *Agriculture* **2022**, *12*, 2115. [CrossRef]
15. You, Z.; Gao, X.; Yan, J.; Wei, H.; Wu, H.; He, T.; Wu, J. Design and multi-parameter optimization of a combined chinese milk vetch (*Astragalus sinicus* L.) Seed Harvester. *Agriculture* **2022**, *12*, 2074. [CrossRef]
16. Zhang, J.; Xia, M.; Chen, W.; Yuan, D.; Wu, C.; Zhu, J. Simulation analysis and experiments for blade-soil-straw interaction under deep ploughing based on the discrete element method. *Agriculture* **2023**, *13*, 136. [CrossRef]
17. Cardei, P.; Constantin, N.; Muraru, V.; Persu, C.; Sfiru, R.; Vladut, N.-V.; Ungureanu, N.; Matache, M.; Muraru-Ionel, C.; Cristea, O.-D.; et al. The random vibrations of the active body of the cultivators. *Agriculture* **2023**, *13*, 1565.
18. Wang, Y.; Xiao, M.; Wang, S.; Jiang, Q.; Wang, X.; Zhang, Y. Detection of famous tea buds based on improved YOLOv7 network. *Agriculture* **2023**, *13*, 1190. [CrossRef]
19. Zhou, X.; Sun, G.; Xu, N.; Zhang, X.; Cai, J.; Yuan, Y.; Huang, Y. A method of modern standardized apple orchard flowering monitoring based on S-YOLO. *Agriculture* **2023**, *13*, 380.

20. Wang, P.; Chen, Y.; Xu, B.; Wu, A.; Fu, J.; Chen, M.; Ma, B. Intelligent algorithm optimization of liquid manure spreading control. *Agriculture* **2023**, *13*, 278. [CrossRef]
21. Wang, J.; Lu, Z.; Wang, G.; Hussain, G.; Zhao, S.; Zhang, H.; Xiao, M. Research on fault diagnosis of HMCVT shift hydraulic system based on optimized BPNN and CNN. *Agriculture* **2023**, *13*, 461. [CrossRef]

Article

Design and Multi-Parameter Optimization of a Combined Chinese Milk Vetch (*Astragalus sinicus* L.) Seed Harvester

Zhaoyan You [1], Xuemei Gao [1], Jianchun Yan [1], Hai Wei [1], Huichang Wu [1,*], Tieguang He [2] and Ji Wu [3]

1 Nanjing Institute of Agricultural Mechanization, Ministry of Agriculture and Rural Affairs, Nanjing 210014, China
2 Agricultural Resource and Environment Research Institute, Guangxi Academy of Agricultural Sciences, Nanning 530007, China
3 Soil and Fertilizer Institute, Anhui Academy of Agricultural Sciences, Hefei 230031, China
* Correspondence: wuhuichang@caas.cn

Abstract: In order to solve problems such as poor applicability of headers, weak separation ability of threshing mechanisms and poor impurity-removal ability of cleaning devices in the existing seed harvest methods of Chinese milk vetch (*Astragalus sinicus* L.), a combined Chinese milk vetch seed harvester was designed in this paper. The parameters of the key components, such as the flexible anti pod-dropping seedling-lifting header, the longitudinal rod-teeth-type threshing device and the air-sieve-type layered impurity-controlled cleaning device, were designed and optimized. Aiming at reducing seed loss rate, breakage rate and impurity rate of Chinese milk vetch during the mechanical harvesting process, through multi-parameter optimization, the best combination of working parameters was obtained: machine forward speed was 3 km·h^{-1}, rotation speed of the threshing drum was 550 r·min^{-1}, rotation speed of the cleaning fan was 990 r·min^{-1} and the scale sieve's opening was 35 mm. Field tests were performed under these parameters, and the results showed that the seed loss rate of Chinese milk vetch was 2.35%, the breakage rate was 0.22% and the impurity rate was 0.51%, which were better than the technical requirements of loss rate and breakage rate less than 5% and impurity rate less than 3% specified in relevant standards. The research results can solve the shortage problem of efficient seed harvest equipment in large-scale planting areas of Chinese milk vetch, and will further help to carry out seed harvest experiments on different varieties of Chinese milk vetch and other green manure varieties in paddy fields.

Keywords: Chinese milk vetch; green manure; seed harvest; parameter optimization; orthogonal test

Citation: You, Z.; Gao, X.; Yan, J.; Wei, H.; Wu, H.; He, T.; Wu, J. Design and Multi-Parameter Optimization of a Combined Chinese Milk Vetch (*Astragalus sinicus* L.) Seed Harvester. *Agriculture* **2022**, *12*, 2074. https://doi.org/10.3390/agriculture12122074

Academic Editor: Jin He

Received: 9 November 2022
Accepted: 1 December 2022
Published: 2 December 2022

1. Introduction

China is the country of origin of Chinese milk vetch (*Astragalus sinicus* L.), and also the country with the earliest utilization and cultivation of Chinese milk vetch and the largest planting area in the world. Chinese milk vetch, also known as grass seed, zi yunying and so on, is one of the main winter green manure crops in paddy fields of central and southern China [1–3]. In the 1970s, the planting area of Chinese milk vetch once exceeded 6.7 million square kilometers, mainly distributed in Sichuan, Hubei, Hunan, Jiangxi, Anhui, Jiangsu, Zhejiang and other provinces to the south of the Yangtze River. It has a strong ability of nitrogen fixation and high utilization efficiency, which can improve soil fertility and protect the ecological environment [4–6]. Currently, the harvest methods of Chinese milk vetch green manure seeds mainly include artificial harvest and mechanical harvest. Artificial harvest is time-consuming and laborious, and the yield of reserved seeds for planting is low. Generally, the seed yield of Chinese milk vetch in a paddy field is 300~650 kg·hm^{-2}. There are two common methods of mechanical harvest: the first one is segmented harvest, which uses a rice, wheat, rape or bean swather to harvest Chinese milk vetch, and then, through natural drying, a thresher is used for threshing afterwards; the second one is combine harvest, done by adjusting parameters and changing working components of a traditional grain harvester or rapeseed combine harvester to complete seed harvesting of Chinese milk vetch [7]. The segmented harvesting process is cumbersome and inefficient, and combine harvesting is of high efficiency, which is the development trend of Chinese milk vetch seed harvesting, however, the harvest quality of existing Chinese milk vetch green manure seed combine harvesters is affected by the unreasonable structure configuration of headers, weak separation ability of threshing mechanisms and poor impurity-removal function of cleaning devices. The loss rate of machine harvesting in field tests is in the range from 31.5% to 32.1% [8], which seriously affects the scale promotion and application of Chinese milk vetch.

At present, existing research on seed combine harvesters, both at home and abroad, mainly focuses on food crops such as rice, wheat, maize, as well as commercial crops such as rapeseed, soybean and flax. Wang et al. [9] designed a cutting table to be a stepless speed-adjustable telescopic structure to harvest rapeseed, and the threshing device was designed to be a longitudinal-axis drum with the same diameter and different speed. Wang et al. [10] developed a cleaning device with segmented vibrating screens, whose holes were round so that the cleaning rate and loss rate of maize grain could meet the requirements of national standards for maize grain harvesters. Zhang et al. [11] used the Plackett–Burman test method to study the impacts of vibration screen amplitude, crank revolving speed, fan revolving speed and fan dip angle on cleaning loss ratio and impurity percentage of rapeseed based on a two-roller and air-screen field mobile harvest testbed. Jin et al. [12] used a series of field trials to explore the influence of nine key working parameters on the quality of soybean harvesting operations, and figured out the optimal combination of parameters systematically. Shi et al. [13] designed a track combine harvester for hilly mountain flax, which included a crawler-type walking system, a low-damage header to prevent winding, a transverse-flow beater with the grain rod and rod teeth with small taper and narrow-grid concave plates, but the impurity rate was relatively high. Bruce et al. [14] studied the effect of threshing rotor speed and concave clearance on the threshing performance of shattering rape pods. Mekonnen et al. [15] presented the effect of a cross-flow opening on the distribution of flow along the width of a forward curved, wide centrifugal fan with two parallel outlets, and Computational Fluid Dynamics (CFD) was utilized to study the effect of the addition of a cross-flow opening on the performance of the fan using three fans of similar geometries but different in their cross-flow opening. Shreekant et al. [16] studied the effects of mechanical damage of soybean seeds during vertical bucket lifting, cleaning and grading on seed germination and seed activity.

Existing studies have provided some research on the seed combine harvester, however, there have been much fewer reports on the combined seed harvest of Chinese milk vetch. Therefore, in this paper, the Chinese milk vetch (*Astragalus sinicus* L.) seed combine harvester was developed based on the "World Group 4LZ-5.0E Ryzen Grain Harvester", the structure and movement parameters of the key components, such as the flexible anti pod-dropping seedling-lifting header, the longitudinal rod-teeth-type threshing device and the air-sieve-type layered impurity-controlled cleaning device were newly designed and optimized according to the harvest characteristics of Chinese milk vetch. The factors that had great influence on the seed harvest quality of Chinese milk vetch were chosen to be machine forward speed, rotation speed of the threshing drum, rotation speed of the fan and the scale sieve's opening. A four-factor and three-level response surface test was carried out, the effects of various factors on the evaluation indexes of seed harvest quality of Chinese milk vetch were explored and the ideal combination of parameters was obtained as well. On this basis, a combined Chinese milk vetch green manure seed harvester was developed, which could complete the work of lifting, dividing, cutting, conveying, threshing, cleaning and unloading of Chinese milk vetch at one time. The performance of the machine was verified by field test, which provides reference for the development and application of mechanized seed harvesting technology and equipment for different Chinese milk vetch varieties and other green manures.

2. Materials and Methods

Chinese milk vetch (*Astragalus sinicus* L.) is a legume plant. The growth states of Chinese milk vetch at different periods are shown in Figure 1. It is an expert at using rhizobia to fix nitrogen, as shown in Figure 1a. A large amount of nitrogen, required by rhizobia in the growth process, is converted into ammonia nitrogen through the nodules, and what it returns to the rhizobia is the carbohydrates they need for survival. In this way, every part of the whole unit enjoys mutual benefit and peaceful symbiosis. Fruit pods of Chinese milk vetch at harvest time are shown in Figure 1b. When the average black pod rate of Chinese milk vetch reaches more than 75%, it should be harvested by machine in time.

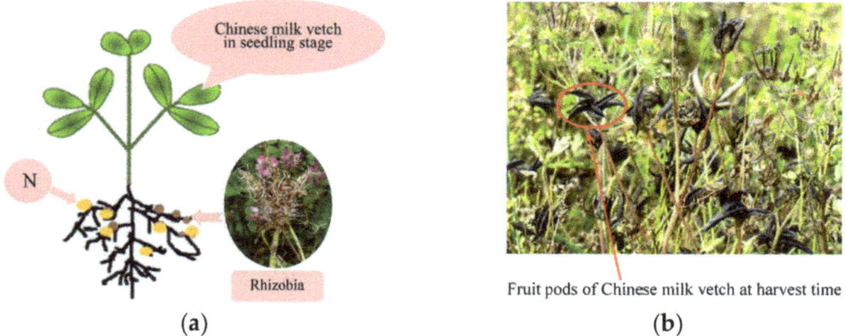

Figure 1. Growth states of Chinese milk vetch at different periods. (**a**) Nitrogen-fixation effect picture of Chinese milk vetch in seedling stage. (**b**) Chinese milk vetch during harvest.

2.1. The Overall Structure and Technical Parameters

The structure and operation process of the developed Chinese milk vetch seed combine harvester is shown in Figure 2. The structure of the Chinese milk vetch green manure seed harvester is mainly composed of a reel, divider, grain lifter, conveying device, crawler chassis, cleaning device, seed bin, threshing device and discharge device. The key technical parameters of the Chinese milk vetch green manure seed combine harvester are shown in Table 1.

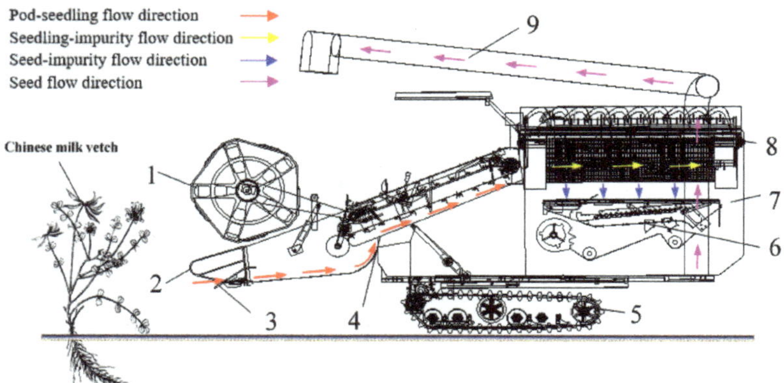

Figure 2. Structure and operation process diagram of the Chinese milk vetch green manure seed combine harvester: (1) reel; (2) divider; (3) grain lifter; (4) conveying device; (5) crawler chassis; (6) cleaning device; (7) seed bin; (8) threshing device; (9) discharge device.

Table 1. Key technical parameters of the Chinese milk vetch seed combine harvester.

Parameters	Design Values
Machine size (L × W × H)/(mm)	5100 × 2890 × 2700
Overall weight/kg	2800
Matched power/kW	62
Operation width/mm	2200
Maximum feeding quantity/(kg·s^{-1})	5.0
Work efficiency/(hm^2·h^{-1})	0.53~0.87
Loss rate/%	≤5
Breakage rate/%	≤2
Impurity rate/%	≤3

2.2. *Working Principle*

When the Chinese milk vetch seed combine harvester works in the field, the pods and stems of Chinese milk vetch enter the longitudinal rod-teeth-type threshing device under the combined action of the reel, the flexible anti pod-dropping seedling-lifting header and the conveying device. At this time, the pods of Chinese milk vetch are threshed under the rotating strike of the rod teeth and extruded by the concave sieve. Then, the threshed mixture falls into the sieve surface of the layered impurity-controlled cleaning device and the air cleaning of threshed materials under the action of a centrifugal fan is completed. The Chinese milk vetch seeds fall along the gaps between the sieving slices of the scale sieve, and fall into the horizontal seed auger through the round-hole screen. The seed auger directly transports them to the seed bin, and a small part of the unthreshed materials are not blown out of the machine; instead, seeds fall into the horizontal residual auger, and the residual auger transports this part of materials back to the threshing device for secondary threshing and cleaning so as to reduce the loss rate and impurity rate of seeds and complete the harvest operation of Chinese milk vetch seeds.

3. Design of Critical Components

3.1. *Flexible Anti Pod-Dropping Seedling-Lifting Header*

Due to the influence of weather, the harvest of Chinese milk vetch is usually creeped when it becomes mature. In order to reduce the seed loss rate and improve the harvest quality of Chinese milk vetch, the flexible anti pod-dropping seedling-lifting header was designed, as shown in Figure 3a. It is mainly composed of a reel, a divider, four lifters, cutter components, a depth-control equipment, a feeding auger and a side cutter, which can complete the operations of dividing, supporting, poking, cutting, guiding and conveying at one time.

The special lifter for the Chinese milk vetch seed harvester, as shown in Figure 3b, is mainly composed of a seedling-lifted rod head, spring rod connecting plate, spring rod and mounting plate. The seedling-lifted rod head was kept at an angle of 32~38° with the ground, and the stems and pods of Chinese milk vetch were guided to the rear by the front seedling-lifted rod head to lift the creeping or lodging Chinese milk vetch to be harvested, so as to avoid the cutter directly contacting the growth segment of Chinese milk vetch pods, and effectively reduce the header loss rate of the Chinese milk vetch seed combine harvester.

The cutter components, as shown in Figure 3c, are composed of a blade guard, cutting blade, left pushed saw-teeth, right pushed saw-teeth, transmitted connecting rod, etc. There were 26 blade guards in total, the type of blade guard was type II double finger, there was a tongue guard at the top of the tip to form a double support for the stems of Chinese milk vetch and the type of cutting blade was standard type II with a blade thickness of 2~3 mm and 6~7 teeth per centimeter of blade length.

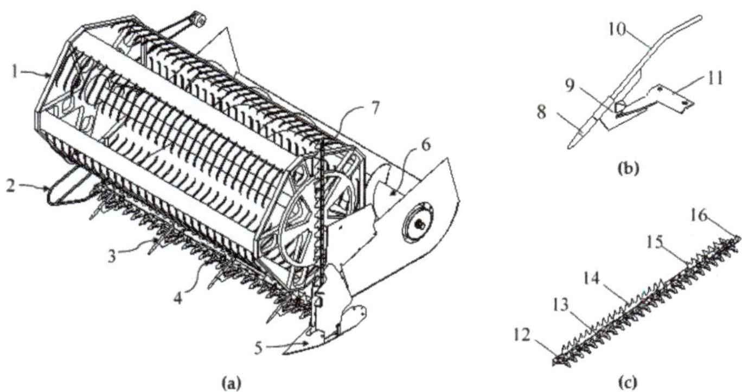

Figure 3. Schematic diagram of the flexible anti pod-dropping seedling-lifting header and its key parts: (1) reel; (2) divider; (3) lifter; (4) cutter components; (5) depth-control equipment; (6) feeding auger; (7) side cutter; (8) seedling-lifted rod head; (9) connecting plate of spring rod; (10) spring rod; (11) mounting plate; (12) blade guard; (13) cutting blade; (14) left pushed saw-teeth; (15) right pushed saw-teeth; (16) transmitted connecting rod. (**a**) Schematic diagram of the header. (**b**) Lifter for harvesting Chinese milk vetch. (**c**) Cutter components.

3.2. Longitudinal Rod-Teeth-Type Threshing Drum

In order to enhance the threshing performance of the threshing device, a longitudinal rod-teeth-type threshing drum was designed (shown in Figure 4), mainly consisting of a top cover, guide plate, spoke disc, feeding section auger, front end cap, main axle, mounting tube, rod teeth, concave sieve, bearing block, rear end cap, etc. The threshing roller is divided into the feeding section, threshing section and grass-discharging section. The total length of the threshing section was 1550 mm, the length of the feeding section was about four-fifths of the total length of the threshing section [17], that is, 315 mm, the length of grass-discharging section was 148 mm, the top cover above the roller and the concave sieve formed a cylindrical threshing chamber and the inner wall of the top cover was equipped with spiral guide plates so that stems of Chinese milk vetch will move along the axis to the grass-discharging section in the joint action of threshing rod teeth and guide plates.

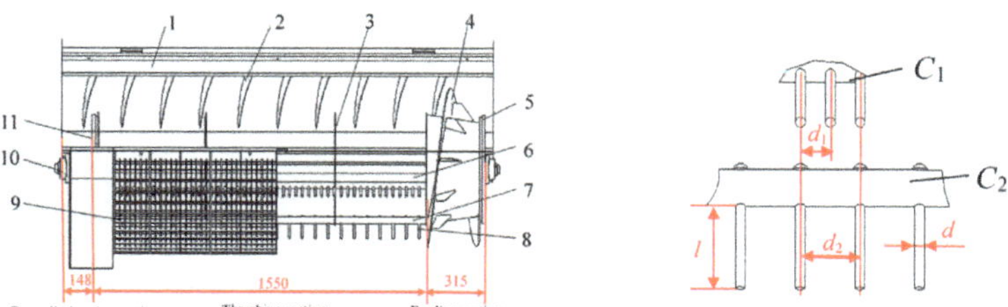

Figure 4. Schematic diagram of the longitudinal rod-teeth-type threshing device: (1) top cover; (2) deflector; (3) spoke disc; (4) auger of feeding section; (5) front end cap; (6) main axle; (7) mounting tube; (8) rod teeth; (9) concave sieve; (10) bearing block; (11) rear end cap.

The number of rod teeth on the roller is determined by the productivity of the threshing device [18],

$$z \geq (1-\beta)q/0.6q_d \tag{1}$$

where, z is the number of teeth on the roller; q is the feed quantity of the thresher, set to 5 kg/s; β refers to the weight proportion of crop pods in the feed materials, with the black pod harvest of Chinese milk vetch being 0.4; and q_d is the threshing ability of each rod tooth (threshing ability is 0.025 kg·s^{-1} for the combine harvester). In Equation (1), in the case β = 0.4, q = 5 kg·s^{-1} and q_d = 0.025 kg·s^{-1}, the number of teeth on the roller is $z \geq 200$.

In order to improve the threshing performance, the teeth rows C_1 and C_2 with different rod-tooth spacing were installed alternately along the circumference of the drum, with the specific parameters designed as follows: rod-tooth spacing of teeth row C_1, d_1 was 30 mm; rod-tooth spacing of teeth row C_2, d_2 was 60 mm; working height of rod teeth, l, was 67 mm; diameter of rod teeth, d, was 10 mm. Through arrangement calculation, under the condition of ensuring the reliable working state of the threshing drum, the rod-teeth number of teeth row C_1 was 53, the rod-teeth number of teeth row C_2 was 27 and the total number of rod teeth on the roller was 240, which meets the productivity requirements of the Chinese milk vetch threshing device.

3.3. *Air-Sieve-Type Layered Impurity-Controlled Cleaning Device*

The cleaning structure of the Chinese milk vetch seed combine harvester was designed as an air-sieve-type layered impurity-controlled cleaning device. The structure diagram is shown in Figure 5. It is mainly composed of the centrifugal fan and the layered impurity-controlled cleaning sieve. In the air-sieve cleaning devices of the existing rice or wheat combine harvesters, most of the fans are single-channel centrifugal fans; when the structure and operation parameters are determined, the velocity and direction of air flow produced by single air-duct centrifugal fans cannot meet the different demands of the whole sieve in the process of material screening [19,20]. Referring to the structure characteristics of multi-air-duct centrifugal fans both at home and abroad, the number of blades of the centrifugal fan in the cleaning unit of Chinese milk vetch green manure seed combine harvester was designed as three blades, and the impeller diameter was 385 mm. Two adjustable inclined air flow dampers were installed at the air outlet of the fan, forming the upper, middle and lower air ducts, which can meet the cleaning requirements of the front, middle and rear of the layered impurity-controlled cleaning sieve.

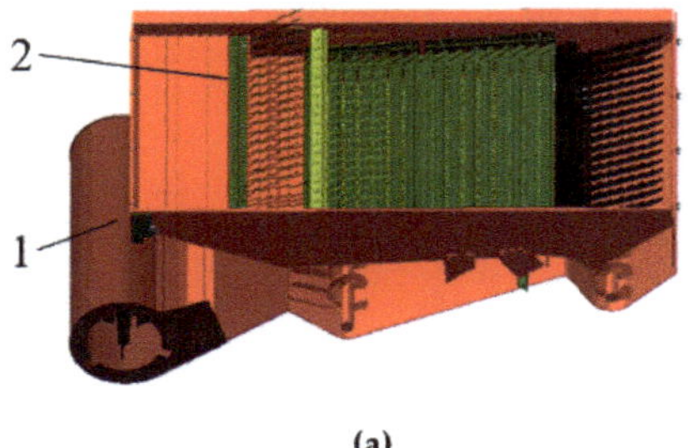

(a)

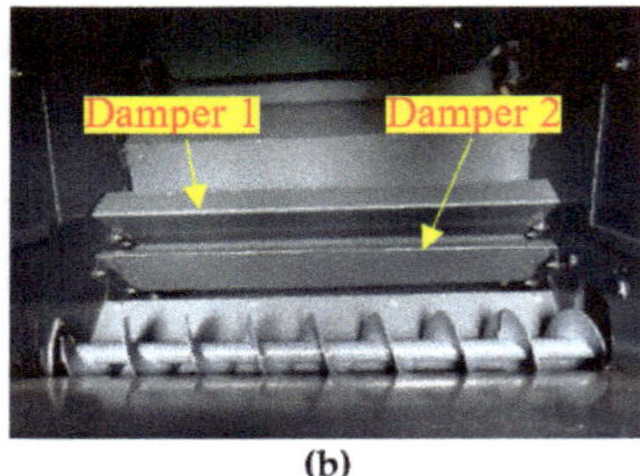

(b)

Figure 5. Three-dimensional and internal structure diagram of the air–sieve-type cleaning device: (1) centrifugal fan; (2) layered impurity-controlled sieve. (**a**) Three-dimensional diagram of the air–sieve-type cleaning device. (**b**) Internal structure diagram of the air–sieve-type cleaning device.

3.3.1. Numerical Simulation of Internal Airflow Field in a Three-Duct Centrifugal Fan

In order to verify whether the flux and distribution of the internal airflow field of the designed three-duct centrifugal fan could meet the cleaning requirements of Chinese milk vetch seeds, the whole flow channel model of the three-duct centrifugal fan was established by using Inventor, and ICEM-CFD was adopted to divide the grid of the three-duct centrifugal fan. Considering the large volume of the channel model and the complexity of its internal flow, tetrahedral unstructured meshing was used to divide the flow channel model. The contact areas between the fan inlet and the impeller, as well as the contact areas between the impeller and the spiral case were all set as the interfaces, and the meshings at the interfaces were encrypted [21]. The meshed centrifugal fan is shown in Figure 6a, and the total number of model grids is 3,617,588.

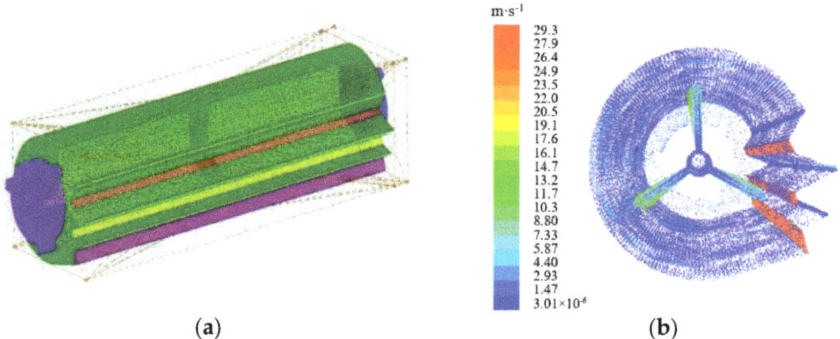

Figure 6. Mesh dividing and internal air velocity simulation of the three-channel centrifugal fan. (**a**) Mesh mode of centrifugal fan with three-channels. (**b**) Internal flow velocity vector diagram of the fan.

Then, the grid model was imported into Fluent 15.0 software, the standard k-ε turbulence model was used for numerical simulation and the SIMPLEC pressure-velocity coupling algorithm was used to calculate velocity and pressure of airflow in the fan. The spiral case of the fan was set as the static wall, and the meshing areas at the impeller of the centrifugal fan were set as the MRF rotation areas. Referring to the structure and working parameter design of existing centrifugal fans, rotation speed of the three-duct fan was set to be 1080 $r{\cdot}min^{-1}$, the impeller diameter was 385 mm, the two inlet boundary conditions of the centrifugal fan were set as the pressure inlets with the given pressure of 220 Pa, the three outlet boundaries of the model were set as the pressure outlets with the given pressure of 0 Pa, the convergence residual was set as 0.001 and the iteration steps were set as 500 [22]. The internal airflow velocity vector diagram of the three-duct centrifugal fan is shown in Figure 6b. It can be seen from the figure that the airflow velocity at the upper and lower outlets was high, while the airflow velocity at the middle outlet was slightly lower. The airflow velocity at the upper outlet (regardless of friction resistance) ranged from 7.33 $m{\cdot}s^{-1}$ to 29.3 $m{\cdot}s^{-1}$, and the air volume at the outlet was about 0.55 $kg{\cdot}s^{-1}$; the airflow velocity at the middle outlet ranged from 5.14 $m{\cdot}s^{-1}$ to 24.9 $m{\cdot}s^{-1}$, and the air volume was about 0.42 $kg{\cdot}s^{-1}$; the wind speed of the lower outlet was between 6.6 $m{\cdot}s^{-1}$ and 25.6 $m{\cdot}s^{-1}$, and the air volume was about 0.57 $kg{\cdot}s^{-1}$.

3.3.2. Field Test Verification of Centrifugal Fan Air Flow

The method of spot placement was adopted [23], according to the structural parameters of the three-duct centrifugal fan. The point at the lowest position on the volute area of the centrifugal fan and 50 mm away from the left inlet were taken as the origin points, the radial direction of seed-conveying auger was chosen as the X axis and the vertical direction of the sieve surface was chosen as the Y axis. The air velocity measurement space and the distribution of measuring points are shown in Figure 7.

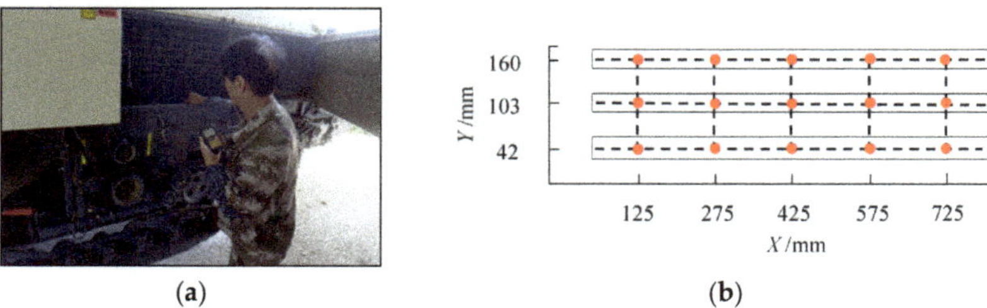

Figure 7. Air velocity measurement site and measurement points distribution. (**a**) Air velocity measurement. (**b**) Measurement points distribution.

The air velocity of each measuring point measured by AR856 digital anemometer is shown in Table 2. The sensitivity and features of the digital anemometer are as follows: airflow speed measurement ranged from 0.3 m·s^{-1} to 45.0 m·s^{-1}, the measurement error was ±3% and the resolution ratio was 0.001 m·s^{-1}. Additionally, 3 planes of 42, 103 and 160 mm in the Y direction corresponded to 3 central measuring planes of air ducts, and each measuring plane was arranged with 5 measuring points at 125, 275, 425, 575 and 725 mm in the X direction. By comparing the results of numerical simulation and experimental study on airflow field, it can be seen that the airflow velocity distribution of each outlet was consistent. The horizontal airflows both at the upper and lower outlets were more uniform, which were conducive to the pre-cleaning of threshed materials and the discharge of impurities in the tail. The middle outlet presented the law of high airflow velocity in the middle and low airflow on both sides, which was conducive to the blowing and stratification of the threshed mixture during the falling process. In addition, the air velocity values were slightly different, the reason for which lied in that the numerical simulation process was completed without considering the gas compression and gas viscosity, and assuming that the whole flow channel was closed. However, in practical work, the cleaning air flow would be attenuated due to the existence of material groups [24].

Table 2. Airflow velocity of measuring points (m·s^{-1}).

Y/mm	X/mm				
	125	275	425	575	725
42	9.7	11.1	10.7	8.8	9.2
103	6.2	5.7	10.1	6.6	6.4
160	9.2	8.6	10.3	7.7	8.1

3.3.3. Layered Impurity-Controlled Cleaning Sieve

There were many stalks in threshed materials of Chinese milk vetch, and the residual materials were mainly glumes, broken stalks and weed seeds. A schematic diagram of the layered impurity-controlled cleaning sieve and the scale sieve's opening is shown in Figure 8, where y is the opening of the scale sieve.

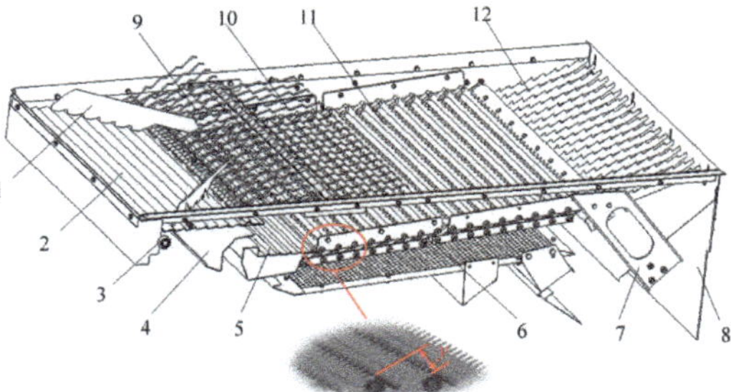

Figure 8. Schematic diagram of the layered impurity-controlled cleaning sieve and the scale sieve's opening: (1) baffle plate; (2) upper shaking plate; (3) bearing; (4) air deflector; (5) lower shaking plate; (6) round-hole sieve; (7) connecting plate; (8) rubber plate; (9) front finger sieve; (10) back finger sieve; (11) scale sieve; (12) saw-teeth sieve.

It was mainly composed of the baffle plate, upper shaking plate, lower shaking plate, air deflector, front finger sieve, back finger sieve, scale sieve, saw-teeth sieve and round-hole sieve. The front and back finger sieves separated the long stalks and residual spikes from the threshed materials under the sieve, the scale sieve was designed as a parallel four-bar linkage structure and the opening of the scale sieve was ranged between 35 mm and 45 mm, which could guide the air flow to blow away glume shell, residual spikes and broken straws, so as to ensure the high-efficiency screening of Chinese milk vetch seeds. The round-hole sieve could further separate the Chinese milk vetch seeds from the mixture of grains and thin stems that were dropped from the scale sieve surface, while the saw-teeth sieve, mounted on the rear of the cleaning sieve, could discharge remaining long straws, residual spikes and broken straws out of the machine backward step by step or multistage. Structural parameters of the layered impurity-controlled cleaning sieve are shown in Table 3.

Table 3. Structural parameters of the layered impurity-controlled cleaning sieve.

Parameters	Design Values
Upper shaking plate length/mm	775
Upper shaking plate width/mm	334
Upper shaking plate (spacing × height)/mm × mm	28 × 6
Front finger sieve length/mm	740
Front finger sieve width/mm	225
Back finger sieve length/mm	680
Back finger sieve width/mm	224
Lower shaking plate length/mm	775
Lower shaking plate width/mm	203
Lower shaking plate (spacing × height)/mm × mm	30 × 4.5
Scale sieve length/mm	754
Scale sieve width/mm	715
Saw-teeth sieve length/mm	640
Saw-teeth sieve width/mm	328
Round-hole sieve length/mm	756
Round-hole sieve width/mm	756
Diameter of round-hole sieve/mm	6

4. Field Experiment and Result Analysis

4.1. Experiment Conditions

The field harvest experiment of Chinese milk vetch was conducted in Yijiang Town, Nanling County, Wuhu City, as shown in Figure 9. The experiment time was from 11–13 May 2020. The Chinese milk vetch variety used for the experiment was Wanzi No. 4, and its yield was 608.95 kg·hm^{-2}. The average height of Chinese milk vetch stems was 330.8 mm, the average height of the bottom pod was 118.5 mm, the average black pod rate was 80.68%, the natural seed loss was 3.31 g·m^{-2} and one thousand seed weight was 3.3~3.5 g. The moisture content of harvested seeds was measured by a PM-8188-A moisture meter, and the moisture meter measurements ranged from 1% to 40%. The sample capacity was 240 mL, the temperature range was 0~40 °C, the measurement accuracy was 0.5% at the basis of the drying method and the water content of harvested seeds was measured three times, with an average value of 10.2%. The Chinese milk vetch plants grew well, and the average length and width of pods picked from 10 different Chinese milk vetch plants were 25.3 mm and 4.2 mm, respectively. The seeds in pods were kidney shaped, and about 3 mm in length.

Figure 9. Combined Chinese milk vetch seed harvester and field test process. (**a**) Three-dimensional diagram of the combined Chinese milk vetch seed harvester. (**b**) Chinese milk vetch field harvest test. (**c**) Materials in the seed bin.

4.2. Experiment Indexes

At present, there is no evaluation standard for the operation of the Chinese milk vetch seed combine harvester. In order to investigate the operation quality of the Chinese milk vetch seed combine harvester, according to leguminosae or cruciferae green manure seed harvest standards such as JB/T 11912-2014, GB/T 5262-2008 and so on, seed loss rate, breakage rate and impurity rate are used as the evaluation indexes for the operation performance of the Chinese milk vetch seed combine harvester. Among them, the machine-harvested seed loss rate is determined by collecting all the seeds and pods, both those fallen in the sampling area and discharged out of the machine with Chinese milk vetch stems, and then removing the natural falling seeds. According to the quantity of harvested Chinese milk vetch seeds and the corresponding harvested area, the yield of Chinese milk vetch seeds per square meter is obtained. Impurities include long and short stalks, grass seeds, pebbles, etc. The broken seeds are incomplete cotyledons (including whole and half seeds), and transverse and broken grains. The specific calculation methods are shown in Equations (2)–(4):

$$Y_1 = \frac{M_{hl}}{M_{ha} + M_{hl}} \times 100\% \quad (2)$$

$$Y_2 = \frac{M_{ei} - M_{ps}}{M_{ei}} \times 100\% \quad (3)$$

$$Y_3 = \frac{M_{ss} - M_{ps}}{M_{ss}} \times 100\% \quad (4)$$

where, Y_1 is the loss rate, %; Y_2 is the breakage rate, %; Y_3 is the impurity rate, %; M_{hl} is the seed loss quantity during harvest of Chinese milk vetch per square meter, g·m^{-2}; M_{ha} is the harvested seed quantity of Chinese milk vetch per square meter, g·m^{-2}; M_{ei} is the sample quality after removal of impurities, g; M_{ps} is the sample quality after removal of impurities and broken seeds, g; M_{ss} is the sample quality, g.

4.3. Experiment Scheme

According to the previous literature search, field experimental research and harvest experience [25,26], taking the loss rate, breakage rate and impurity rate of seeds as the evaluation bases of the operation performance of the Chinese milk vetch seed combine harvester, machine forward speed, rotation speed of the roller, rotation speed of the cleaning fan and the scale sieve's opening were selected as the test variables to carry out the experimental research. Box–Behnken central composite design theory was used to carry out a four-factor three-level quadratic regression response surface test with a total of 29 groups. The test site was about 120 m long and 50 m wide, and the test length of each group was 30 m. Samples were taken for three times in each group after the harvester ran stably.

Experiment factors and codes are shown in Table 4. Among them, X_1~X_4 represented the variables of each factor. According to the design parameters of the seed combine harvester, the relationship between the biological characteristics of Chinese milk vetch during harvest and the threshing and cleaning parameters of the harvester, the value range of each factor was determined, the median value of machine forward speed was set as 4 km·h^{-1} and the low-speed value of 3 km·h^{-1} and high-speed value of 5 km·h^{-1} were selected accordingly. According to the threshing ability characteristics of Chinese milk vetch and referring to the rotation speed of the threshing drum for wheat, soybean and other crops, the middle value of rotation speed of the threshing drum was set as 675 r·min^{-1}, the minimum rotation speed of the threshing drum was 550 r·min^{-1} and the maximum rotation speed was 800 r·min^{-1}. According to the simulation results of the centrifugal fan in the early stage, meanwhile, referring to the test method and standard of seed suspension speed of grain, oil, forage grass, etc., [27], the minimum rotation speed of the cleaning fan was set as 900 r·min^{-1}, the maximum rotation speed was 1260 r·min^{-1} and the intermediate value of 1080 r·min^{-1} was determined. Based on external dimensions of pods and seeds of Chinese milk vetch varieties, the designed opening range of the scale sieve was from 35 mm to 45 mm, the middle value was 40 mm and the corresponding limit values of the opening range were taken as high- and low-level values.

Table 4. Experiment factors and codes.

Factors	Codes		
	−1	0	1
Machine forward speed/(km·h^{-1})	3	4	5
Rotation speed of threshing drum/(r·min^{-1})	550	675	800
Rotation speed of cleaning fan/(r·min^{-1})	900	1080	1260
Scale sieve's opening/mm	35	40	45

The adjustment methods of each parameter in the test were as follows: (1) Machine forward speed adjustment: The driver adjusted the infinitely variable gear to get different forward speeds. The forward speed was accurately detected by the speed sensor installed on the rear wheel and displayed on the instrument panel in the cab. (2) Rotation speed adjustments of the threshing drum and the cleaning fan: the V-belt with continuously variable transmission was adopted, and the moving plate of the belt wheel was adjusted by hydrostatic and manual operation so as to obtain different transmission ratios and different rotation speeds. (3) Scale sieve's opening adjustment: through the position rotation of the adjusting plate installed on the sieve frame, the scale's opening could be adjusted, and the specific scale's opening could be measured by ruler.

4.4. Experiment Results and Analysis

The experimental design, data processing and statistical analysis were conducted by the software Design-Expert V8.0.6.1. The four-factor and three-level responsive section experiment was carried out according to the Box–Behnken design method, which is to select 29 points, including 24 analysis factors and 5 zero estimation errors. The experimental design and response results are shown in Table 5, and the variance analysis of the regression model is conducted in Table 6.

Table 5. Experiment design and response values.

No.	Codes				Response Values		
	X_1	X_2	X_3	X_4	Y_1	Y_2	Y_3
1	−1	0	1	0	10.32	1.35	1.13
2	0	0	−1	1	1.23	1.87	2.93
3	0	1	1	0	12.20	3.01	0.89
4	0	1	−1	0	1.15	3.54	1.58
5	0	0	0	0	4.16	1.72	1.36
6	0	0	1	1	16.47	1.25	2.17
7	1	0	0	1	8.25	1.52	2.45
8	0	−1	1	0	10.69	0.13	1.51
9	1	1	0	0	6.93	2.78	1.32
10	1	0	0	−1	1.95	1.88	1.03
11	0	−1	−1	0	3.22	0.37	1.83
12	0	0	0	0	5.04	2.09	1.39
13	0	−1	0	1	7.36	0.26	2.65
14	0	0	0	0	4.65	1.66	1.28
15	1	0	−1	0	4.26	1.85	2.35
16	−1	0	−1	0	0.96	1.77	1.76
17	1	0	1	0	11.78	1.72	1.38
18	0	1	0	−1	2.13	2.93	0.65
19	0	−1	0	−1	3.46	0.32	0.73
20	0	0	1	−1	8.91	1.5	0.65
21	1	−1	0	0	3.88	0.29	1.57
22	−1	−1	0	0	3.57	0.19	1.38
23	0	0	0	0	5.74	1.74	1.33
24	−1	0	0	1	8.38	1.38	1.97
25	−1	0	0	−1	2.62	1.71	0.43
26	0	0	0	0	5.33	1.51	1.54
27	−1	1	0	0	5.49	2.52	0.93
28	0	0	−1	−1	0.43	1.96	0.78
29	0	1	0	1	9.30	2.37	1.87

Table 6. Variance analysis of regression model.

Indexes	Variance Source	Sum of Squares	F-Value	p-Value	Significance
Y_1	X_1	2.72	2.08	0.1710	
	X_2	2.10	1.61	0.2252	
	X_3	291.26	223.29	<0.0001	***
	X_4	82.64	63.35	<0.0001	***
	Lack of fit	16.78	4.52	0.0796	
	Pure error	1.48			
Y_2	X_1	0.10	2.40	0.1439	
	X_2	20.25	464.38	<0.0001	***
	X_3	0.48	11.01	0.0051	***
	X_4	0.23	5.20	0.0387	**
	Lack of fit	0.43	0.94	0.5761	
	Pure error	0.18			
Y_3	X_1	0.52	27.05	0.0001	***
	X_2	0.49	25.56	0.0002	***
	X_3	1.02	53.03	<0.0001	***
	X_4	7.95	413.18	<0.0001	***
	Lack of fit	0.23	2.39	0.2077	
	Pure error	0.039			

Note: ** indicates that the factors have a significant influence on the test index ($0.01 < p \leq 0.05$), *** indicates that the factors have a very significant influence on the test index ($p \leq 0.01$).

4.4.1. Model Establishment and Significance Verification

According to the sample data in Table 5, the regression models are established in Equations (5)–(7) as follows:

$$\begin{aligned} Y_1 = {} & 4.98 + 0.48X_1 + 0.42X_2 + 4.93X_3 + 2.62X_4 + 0.28X_1X_2 \\ & -0.46X_1X_3 + 0.13X_1X_4 + 0.90X_2X_3 + 0.82X_2X_4 \\ & +1.69X_3X_4 + 0.018X_1^2 + 0.14X_2^2 + 1.67X_3^2 + 0.28X_4^2 \end{aligned} \quad (5)$$

$$\begin{aligned} Y_2 = {} & 1.74 + 0.093X_1 + 1.30X_2 - 0.20X_3 - 0.14X_4 + 0.040X_1X_2 \\ & +0.072X_1X_3 - 0.0075X_1X_4 - 0.073X_2X_3 - 0.13X_2X_4 \\ & -0.04X_3X_4 - 0.1X_1^2 - 0.14X_2^2 + 0.065X_3^2 - 0.11X_4^2 \end{aligned} \quad (6)$$

$$\begin{aligned} Y_3 = {} & 1.38 + 0.21X_1 - 0.20X_2 - 0.29X_3 + 0.81X_4 + 0.05X_1X_2 \\ & -0.085X_1X_3 - 0.03X_1X_4 - 0.093X_2X_3 - 0.17X_2X_4 \\ & -0.16X_3X_4 + 0.025X_1^2 - 0.074X_2^2 + 0.18X_3^2 + 0.10X_4^2 \end{aligned} \quad (7)$$

Through the analysis of Table 6, the regression models of seed loss rate, Y_1, breakage rate, Y_2, and impurity rate, Y_3, show that $p < 0.0001$. Moreover, the significance of each variable influencing the index in the regression equations is judged by the F value. In other words, the smaller the probability value P is, the higher the significance of the corresponding variable is. It can be seen from the F value of each factor that rotation speed of the cleaning fan and the scale sieve's opening have a significant effect on seed loss rate. The descending order of influencing factors is rotation speed of the cleaning fan > scale sieve's opening > machine forward speed > rotation speed of the threshing drum. Meanwhile, rotation speed of the cleaning fan and threshing drum has a significant effect on seed breakage rate, and the descending order is rotation speed of the threshing drum > rotation speed of the cleaning fan > scale sieve's opening > machine forward speed. In addition, all the four factors have a significant effect on seed loss rate, and the descending order is scale sieve's opening > rotation speed of the cleaning fan > machine forward speed > rotation speed of the threshing drum. The determination coefficients, R^2, of Y_1, Y_2 and Y_3 are 0.9580, 0.9723 and 0.975, respectively, showing that the model error is small, and it is reasonable to analyze and predict seed loss rate, breakage rate and impurity rate of the Chinese milk vetch seed combine harvester.

4.4.2. Effect Analysis of Interaction Factors on Harvest Indexes

Based on the established optimization regression model, the influence of machine forward speed, rotation speed of the threshing drum, rotation speed of the cleaning fan and scale sieve's opening on the harvest quality index of the Chinese milk vetch seed combine harvester and the relationship among the factors were analyzed.

The response surfaces of seed loss rate are shown in Figure 10. In Figure 10a, the rotation speed of the threshing drum and scale sieve's opening were both zero, and the increase of machine forward speed had little influence on the seed loss rate when the rotation speed of the cleaning fan was fixed, and when machine forward speed was constant, seed loss rate gradually increased with the increase of the rotation speed of the cleaning fan. In Figure 10b, the machine forward speed was set at a low level, the rotation speed of the cleaning fan was set at a low level, when the value of threshing drum rotation speed was fixed, the seed loss rate increased with the increase of the scale sieve's opening, and when the scale sieve's opening was constant, the rotation speed of the threshing drum had little influence on the seed loss rate. Cause analysis: when the rotation speed of the cleaning fan and the scale sieve's opening increased, the air volume passing through the surface of the scale sieve in unit time would increase, and more seeds on the sieve surface would be blown away by the airflow; meanwhile, seeds that were prepared to pass the cleaning sieve would be blown out of the machine as well, resulting in the increase of seed loss rate.

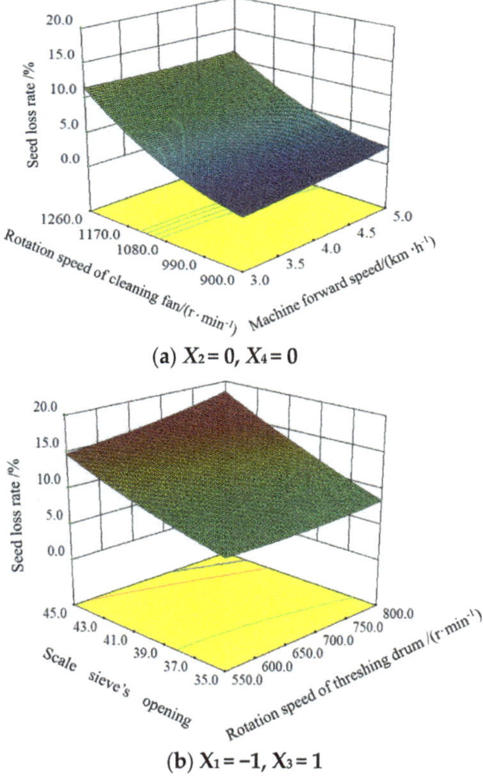

Figure 10. Response surfaces of seed loss rate.

The response surfaces of seed breakage rate are shown in Figure 11. Machine forward speed and the rotation speed of the cleaning fan were both zero, as shown in Figure 11a, when the scale sieve's opening was fixed, and the faster the rotation speed of the threshing drum was, the higher the seed breakage rate was; when the rotation speed of threshing drum was constant, the scale sieve's opening had little influence on the seed loss rate. In Figure 11b, the scale sieve's opening and the rotation speed of the threshing drum were both zero, and when machine forward speed was fixed, the seed breakage rate decreased slightly with the increase of cleaning fan rotation speed; when rotation speed of the cleaning fan was constant, machine forward speed had little influence on the seed breakage rate. Cause analysis: the faster the rotation speed of the threshing drum, the harder the threshing element struck Chinese milk vetch seed, and the higher the possibility of seed breakage. At the beginning, with the increase in the cleaning fan rotation speed, the fan mainly blew away most of the impurities, and the broken seeds flowed into the seed bin through the round-hole screen, and seed breakage rate changed little. However, with further increase of the cleaning fan rotation speed, part of the damaged seeds were blown away, while some complete seeds on the surface of the sieve were also blown away by the fan. According to calculation Formula (3), the breakage rate increased correspondingly.

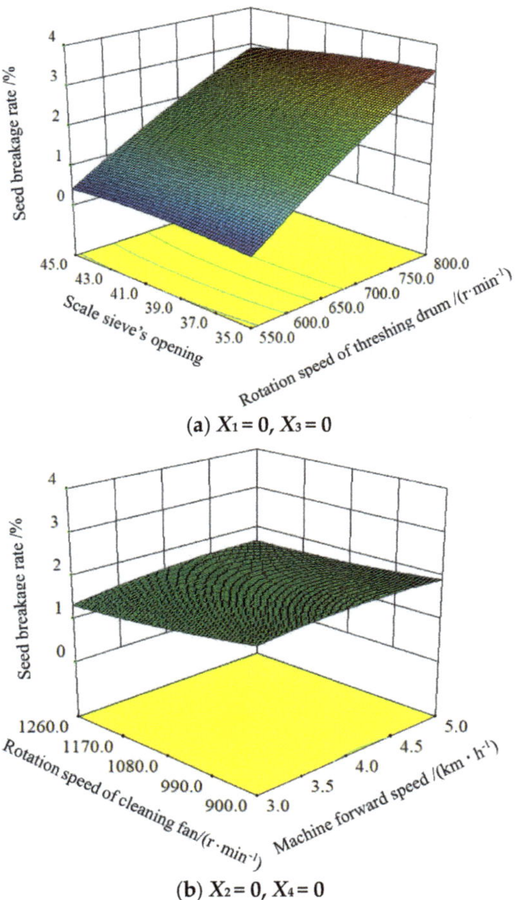

Figure 11. Response surfaces of seed breakage rate.

The response surfaces of seed impurity rate are shown in Figure 12. Rotation speed of the threshing drum and rotation speed of the cleaning fan were both zero, as shown in Figure 12a, when machine forward speed was constant, and the larger the scale sieve's opening was, the higher the seed impurity rate was; when the scale sieve's opening was fixed, with the increase of machine forward speed, the impurity content of seed also increased slightly. In Figure 12b, the machine forward speed was set at a low level, and the scale sieve's opening was set at zero. When the rotation speed of the threshing drum was fixed, seed impurity rate decreased with the increase of cleaning fan rotation speed; when rotation speed of the cleaning fan was constant, the faster the rotation speed of threshing drum was, the smaller the seed impurity rate would be. Cause analysis: when the scale sieve's opening increased, most of the short stalks fell onto the round-hole screen from the gap of the sieve pieces, the number of stalks entering the round-hole sieve surface increased and the screening efficiency decreased, leading to the increase of impurity content. In addition, the lower the rotation speed of the cleaning fan was, the more surplus stalks would fall from the sieve pieces before they were blown away by the fan, resulting in incomplete cleaning separation and higher seed impurity content.

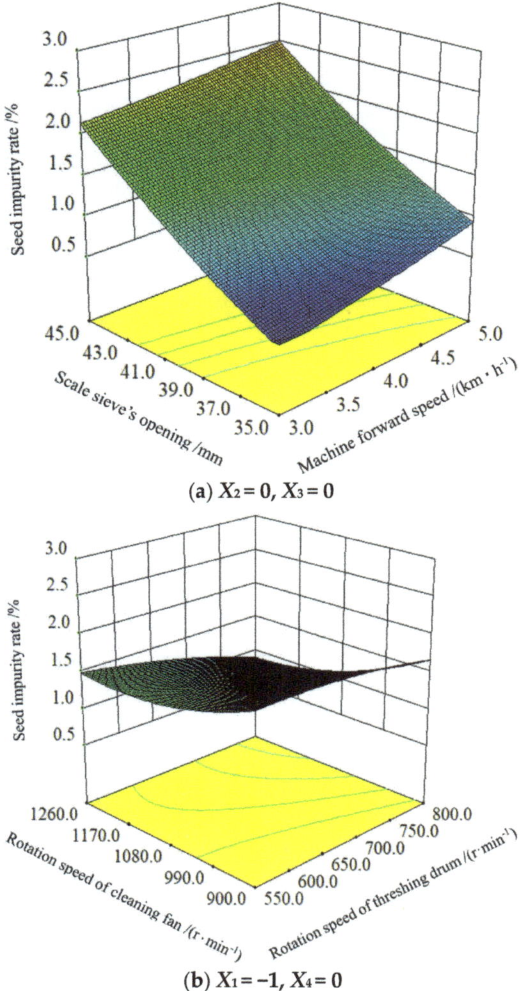

Figure 12. Response surfaces of seed impurity rate.

4.4.3. Parameter Optimization and Verification Test

In this paper, in order to meet requirements of lowest seed loss rate, lowest seed breakage rate and lowest seed impurity rate during the harvest of Chinese milk vetch, parameter optimization of the Chinese milk vetch green manure seed combine harvester was carried out. Design-Expert data analysis software was used to optimize the established total factor quadratic regression model of the three indexes. Constraint conditions were established in Equation (8) as follows:

$$\begin{cases} \min Y_1 \\ \min Y_2 \\ \min Y_3 \\ \text{s.t.} \begin{cases} 3\ \text{km}\cdot\text{h}^{-1} \le X_1 \le 5\ \text{km}\cdot\text{h}^{-1} \\ 550\ \text{r}\cdot\text{min}^{-1} \le X_2 \le 800\ \text{r}\cdot\text{min}^{-1} \\ 900\ \text{r}\cdot\text{min}^{-1} \le X_3 \le 1260\ \text{r}\cdot\text{min}^{-1} \\ 35\ \text{mm} \le X_4 \le 45\ \text{mm} \end{cases} \end{cases} \quad (8)$$

The optimal parameter combination of the Chinese milk vetch seed combine harvester was obtained as follows: the machine forward speed was 3 km·h^{-1}, the rotation speed of the threshing drum was 553.45 r·min^{-1}, the rotation speed of the cleaning fan was 991.77 r·min^{-1} and the scale sieve's opening was 35 mm, then, the seed loss rate, breakage rate and impurity rate predicted by the model were 2.40%, 0.19% and 0.48%, respectively.

In order to verify the accuracy of the above model, the validation test was carried out in Qingyijiang Town, Nanling County, Wuhu City, 17–18 May 2020. Considering the feasibility of the test parameters, the optimized parameters were adjusted so that machine forward speed was 3 km·h^{-1}, the rotation speed of the threshing drum was 550 r·min^{-1}, the rotation speed of the cleaning fan was 990 r·min^{-1} and the scale sieve's opening was 35 mm. Three tests were carried out and the average value was taken as the test verification value. According to the mechanical industry standard of the People's Republic of China (JB/T 11912-2014), the loss rate and damage rate are required to be less than 5%, and the impurity rate must be less than 3%. The seed loss rate, breakage rate and impurity rate measured in the experiment were 2.35%, 0.22% and 0.51%, respectively, which were all lower than the standard.

5. Conclusions

(1) A combined Chinese milk vetch seed harvester was designed, and both its structure composition and working principle were described. Parameter design and simulation analysis were carried out on the key components, such as the flexible anti pod-dropping seedling-lifting header, the longitudinal rod-teeth-type threshing device, the air-sieve-type layered impurity-controlled cleaning device, etc.

(2) The optimization model of harvesting parameters of the Chinese milk vetch green manure seed harvester was established, and the multi-parameter optimization was obtained when seed loss rate, breakage rate and impurity rate were the smallest: machine forward speed was 3 km·h^{-1}, the rotation speed of the threshing drum was 550 r·min^{-1}, the rotation speed of the cleaning fan was 990 r·min^{-1} and the scale sieve's opening was 35 mm. Under these parameter conditions, the field test was carried out, and results showed that the seed loss rate was 2.35%, the breakage rate was 0.22% and the impurity rate was 0.51%, which was better than the loss rate and breakage rate specified in the relevant standards, less than 5%, and the impurity rate, less than 3%.

(3) The production efficiency of the developed Chinese milk vetch seed combine harvester can reach 0.53~0.87 hm^2·h^{-1}, and it can effectively solve the shortage problem of efficient seed harvest equipment in large-scale planting areas of Chinese milk vetch. At present, only one variety of Chinese milk vetch was tested in this study, but the research will help to carry out experiments on different varieties of Chinese milk vetch and other green manure varieties in paddy fields to further verify the operational performance and adaptability of the designed Chinese milk vetch seed combine harvester.

Author Contributions: Conceptualization, Z.Y. and J.Y.; methodology, Z.Y. and H.W. (Hai Wei); software, Z.Y. and X.G.; validation, Z.Y. and H.W. (Huichang Wu); formal analysis, Z.Y. and H.W. (Hai Wei); investigation, Z.Y., T.H. and J.W.; resources, Z.Y. and J.W.; data curation, Z.Y.; writing—original draft preparation, Z.Y. and X.G.; writing—review and editing, Z.Y., X.G. and J.Y.; visualization, J.Y. and X.G.; supervision, H.W. (Huichang Wu); and J.W.; project administration, H.W. (Huichang Wu); funding acquisition, H.W. (Huichang Wu); All authors have read and agreed to the published version of the manuscript.

Funding: This research was funded by China Agriculture Research System—Green Manure, grant number CARS-22; Basic Scientific Research Service Fee of Chinese Academy of Agricultural Sciences, grant number S202012; Innovation Engineering of the Chinese Academy of Agricultural Sciences—Primary Processing Equipment for Major Grain and Economic Crops, grant number 31-NIAM-09.

Institutional Review Board Statement: Not applicable.

Informed Consent Statement: Not applicable.

Data Availability Statement: The data presented in this study are available on-demand from the first author at (youzhaoyan@caas.cn).

Acknowledgments: The authors would like to acknowledge Wuhu Qingyijiang Seed Industry Co., Ltd. for its strong support in the trial production and processing of the Chinese milk vetch seed combine harvester, and for providing the test site for this study.

Conflicts of Interest: The authors declare no conflict of interest.

References

1. Zhou, G.P.; Xie, Z.J.; Cao, W.D.; Xu, C.X.; Bai, J.S.; Zeng, N.H.; Gao, S.J.; Yang, L. Co-incorporation of high rice stubble and Chinese milk vetch improving soil fertility and yield of rice. *Trans. Chin. Soc. Agric. Eng.* **2017**, *33*, 157–163.
2. Chang, H.L.; Ki, D.P.; Ki, Y.J.; Muhammad, A.A.; Dokyoung, L.; Jessie, G.; Pil, J.K. Effect of Chinese milk vetch (*Astragalus sinicus* L.) as a green manure on rice productivity and methane emission in paddy soil. *Agric. Ecosyst. Environ.* **2010**, *138*, 343–347.
3. Liu, Z.Q.; Li, S.J.; Liu, N.; Huang, G.Q.; Zhou, Q. Soil microbial community driven by soil moisture and nitrogen in milk vetch (*Astragalus sinicus* L.)-rapeseed (*Brassica napus* L.) intercropping. *Agriculture* **2022**, *12*, 1538. [CrossRef]
4. Toma, Y.; Sari, N.N.; Akamatsu, K.; Oomori, S.; Nagata, O.; Nishimura, S.; Purwanto, B.H.; Ueno, H. Effects of green manure application and prolonging mid-season drainage on greenhouse gas emission from paddy fields in ehime, southwestern Japan. *Agriculture* **2019**, *9*, 29. [CrossRef]
5. You, Z.Y.; Wu, H.C.; Peng, B.L.; Gao, X.M.; Hu, Z.C.; Zhang, Y.L. Design of double disc ditching and sowing combined working machine of Astragalus smicus for machine-harvested rice stubble field. *Trans. Chin. Soc. Agric. Eng.* **2019**, *35*, 18–28.
6. You, Z.Y.; Zhang, C.; Gao, X.M.; Gao, Y.; Peng, B.L.; Wu, H.C. Design and experiment of incisal-burying and ploughing combined machine for astragalus smicus. *Trans. Chin. Soc. Agric. Mach.* **2020**, *51*, 78–86.
7. Liao, Y.L.; Lu, Y.H.; Nie, J.; Zhao, X.; Xie, J.; Yang, Z.P.; Wu, H.J. Effects of rice combine harvester application to harvesting Chinese milk vetch seed. *Hunan Agric. Sci.* **2014**, *18*, 12–14.
8. You, Z.Y.; Gao, X.M.; Wu, H.C.; Peng, B.L.; Zhang, C.; Chen, Z.K. Calculation of the loss rate of machine-harvested during Chinese milk vetch seed harvesting. *Seed* **2019**, *38*, 140–143.
9. Wang, J.S.; Xiong, Y.S.; Xu, Z.W.; Ma, G.; Wang, Z.M.; Chen, D.J. Improved design and test of key components for longitudianl axial flow combine harvester. *Trans. Chin. Soc. Agric. Eng.* **2017**, *33*, 25–31.
10. Wang, L.J.; Ma, Y.; Feng, X.; Song, L.L.; Chai, J. Design and experiment of segmented vibrating screen in cleaning device of maize grain harvester. *Trans. Chin. Soc. Agric. Mach.* **2020**, *51*, 89–99.
11. Zhang, M.; Jin, C.Q.; Liang, S.N.; Tang, Q.; Wu, C.Y. Parameter optimization and experiment on air-screen cleaning device of rapeseed combine harvester. *Trans. Chin. Soc. Agric. Eng.* **2015**, *31*, 8–15.
12. Jin, C.Q.; Guo, F.Y.; Xu, J.S.; Li, Q.L.; Chen, M.; Li, J.J.; Yin, X. Optimization of working parameters of soybean combine harvester. *Trans. Chin. Soc. Agric. Eng.* **2019**, *35*, 10–22.
13. Shi, R.J.; Dai, F.; Liu, X.L.; Zhao, W.Y.; Qu, J.F.; Zhang, F.W.; Qin, D.G. Design and experiments of crawler-type hilly and mountaineous flax combine harvester. *Trans. CSAE* **2021**, *37*, 59–67.
14. Bruce, D.M.; Hobson, R.N.; Morgan, C.L.; Child, R.D. Threshability of shatter-resistant seed pods in oilseed rape. *J. Agric. Engng. Res.* **2001**, *80*, 343–350. [CrossRef]
15. Mekonnen, G.G.; Josse, D.B.; Martine, B. Effect of a cross-flow opening on the performance of a centrifugal fan in a combine harvester: Computational and experimental study. *Agriculture* **2010**, *105*, 247–256.
16. Shreekant, R.P.; Rameshwar, T.K.; Digvir, S.J.; Noel, D.G. Mechanical damage to soybean seed during processing. *J. Stored. Pro. Res.* **2002**, *38*, 385–394.
17. Chen, N.; Xiong, Y.S.; Chen, D.J.; Xu, J.D.; Zhao, Y. Design and test on the coaxial differential-speed axial-flow threshing rotor of combine harvester. *Trans. Chin. Soc. Agric. Mach.* **2010**, *41*, 67–71.

18. Chinese Academy of Agricultural Mechanization Sciences. *Agricultural Machinery Design Manual (Volume II)*; China Agricultural Science and Technology Press: Beijing, China, 2007.
19. Xu, L.Z.; Yu, L.J.; Li, Y.M.; Ma, Z.; Wang, C.H. Numerical simulation of internal flow field in centrifugal Fan with double outlet and multi-duct. *Trans. Chin. Soc. Agric. Mach.* **2014**, *45*, 78–86.
20. Xu, L.Z.; Li, Y.; Li, Y.M.; Chai, X.Y.; Qiu, J. Research process on cleaning technology and device of grain combine harvester. *Trans. Chin. Soc. Agric. Mach.* **2019**, *50*, 1–16.
21. Shao, W.; Li, Y.M.; Jia, L.H. Numerical simulation of internal flow in centrifugal Fan. *Coal. Mine Mach.* **2006**, *27*, 47–49.
22. Wang, H.H.; Li, Y.M.; Xu, L.Z.; Huang, M.S.; Ma, Z. Simulation and experiment of air flow field in the cleaning device of ratooning rice combine harvesters. *Trans. Chin. Soc. Agric. Eng.* **2020**, *36*, 84–92.
23. Xia, L.L.; Jin, Y.L.; Li, Y.M.; Tang, Z.F. Experimental study of air flow field of air-and-screen cleaning. *J. Agric. Mech. Res.* **2009**, *31*, 188–190.
24. Li, Y.; Xu, L.Z.; Zhou, Y.; Si, Z.Y.; Li, Y.M. Effect of extractions feed-quantity on airflow field in multi-ducts cleaning device. *Trans. CSAE* **2017**, *33*, 48–55.
25. Liu, P.; Jin, C.Q.; Liu, Z.; Zhang, G.Y.; Cai, Z.Y.; Kang, Y.; Yin, X. Optimization of field cleaning parameters of soybean combine harvester. *Trans. Chin. Soc. Agric. Eng.* **2020**, *36*, 35–45.
26. Wu, C.Y.; Ding, W.M.; Zhang, M.; Shi, L.; Lu, Y.; Yu, S.S. Experiment on threshing and cleaning in two-stage harvesting for rapeseed. *Trans. Chin. Soc. Agric. Mach.* **2010**, *41*, 72–76.
27. Hou, H.M.; Cui, Q.L.; Guo, Y.M.; Zhang, Y.Q.; Sun, D.; Lai, S.T.; Liu, J.L. Design and test of air-sweeping suspension velocity testing device for cleaning threshed materials of grain and oil crops. *Trans. Chin. Soc. Agric. Eng.* **2018**, *34*, 43–49.

 agriculture

Article

Research and Experiments of Hazelnut Harvesting Machine Based on CFD-DEM Analysis

Dezhi Ren [1], Haolin Yu [1], Ren Zhang [1], Jiaqi Li [2], Yingbo Zhao [3], Fengbo Liu [2], Jinhui Zhang [4] and Wei Wang [1,*]

[1] College of Engineering, Shenyang Agricultural University, Shenyang 110161, China
[2] Liaoning Agricultural College, Yingkou 115009, China
[3] Liaoning Jinzhou Panghe Economic Development Zone Management Committee, Jinzhou 121400, China
[4] Chaoyang Agricultural Machinery Technology Extension Station, Chaoyang 122000, China
* Correspondence: wangweisyau@126.com; Tel.: +86-188-0404-5818

Abstract: To solve the problem of difficult hazelnut harvesting in mountainous areas of Liaoning, China, a small pneumatic hazelnut harvesting machine was designed, which can realize negative pressure when picking up hazelnut mixtures and positive pressure when cleaning impurities. The key structure and parameters of the harvesting machine were determined by constructing a mechanical model of the whole machine and combining theoretical analysis and operational requirements. To explore the harvesting machine scavenging performance, Liaoning hazelnut No. 3 with a moisture content of 7.47% was used as the experimental object. Firstly, the terminal velocity of hazelnuts and fallen leaves was measured using a material suspension velocity test bench. Secondly, the gas–solid two-phase flow theory was applied comprehensively, and the motion state, particle distribution, and air flow field distribution of hazelnuts from the inlet to the outlet of the pneumatic conveying device were simulated and analyzed using the coupling of computational flow fluid dynamics method (CFD) and discrete element method (DEM) to evaluate the cleaning performance from the perspective of the net fruit rate of hazelnuts in the cleaning box. Finally, a Box–Behnken design experiment was conducted with the sieve plate angle, the distance of the sieve plate, and the air flow velocity as factors and the net fruit rate of hazelnuts as indicators to explore the influence of the three factors on the net fruit rate of hazelnuts. The parameter optimization module of Design-Expert software was used to obtain the optimal combination of parameters for the factors. The experimental results show that the test factors affecting the test index are the following: the air flow rate, the angle of the screen plate, and the distance of the screen plate. The best combination of parameters was an air flow velocity of 14.1 m·s^{-1}, a sieve plate angle of 55.7°, and a distance of the sieve plate of 33.2 mm. The net fruit rate of hazelnuts was 95.12%. The clearing performance was stable and can guarantee the requirements of hazelnut harvester operation, which provides a certain theoretical basis for the design of a nut harvester.

Keywords: pneumatic; CFD-EDM; simulation analysis; harvester

Citation: Ren, D.; Yu, H.; Zhang, R.; Li, J.; Zhao, Y.; Liu, F.; Zhang, J.; Wang, W. Research and Experiments of Hazelnut Harvesting Machine Based on CFD-DEM Analysis. *Agriculture* **2022**, *12*, 2115. https://doi.org/10.3390/agriculture12122115

Academic Editor: Francesco Marinello

Received: 19 November 2022
Accepted: 7 December 2022
Published: 9 December 2022

1. Introduction

According to statistics, China has produced 130,000 tons of hazelnuts in recent years, making it the second largest hazelnut producer in the world, with wild mountain hazelnut production accounting for 23% of the total. The picking and sorting of mountain forest hazelnuts is an important part of hazelnut mechanization, and the design of its cleaning device directly affects the working performance of the harvester. Most of the hazelnut harvesters in other countries are adapted from vibratory forest fruit harvesters for flat orchard operations [1–3]. In China, hazelnuts are grown in mountainous areas, so small hazelnut harvesters are suitable for hazelnut harvesting in mountainous area. At present, the hazelnut harvesting operation is divided into two parts, picking and impurity sorting, and the manual operation faces problems such as high labor intensity and low operational efficiency. Therefore, the mechanization of hazelnut harvesting can solve the above problems.

The agricultural material cleaning process is typically a combined effect of the gas–solid two-phase flow field [4,5]. It is of great significance to study the air-and-screen cleaning device to analyze the air flow distribution of the cleaning shoe and to explore the motion law of agricultural materials on the screen surface, which not only provide the theoretical basis for designing and optimizing the existing typical cleaning unit, but also give theoretical inspiration to look for new cleaning methods. The use of computational fluid dynamics (CFD) for the computation of turbo machinery flows has significantly increased in recent years [6,7]. Furthermore, combined with measurements, CFD provides a complementary tool for simulating, designing, optimizing, and analyzing the flow field inside a turbo machine [8]. The coupling of DEM and CFD provides a means of momentum and energy exchange between solids and fluids, which, in principle, removes the need for some of the semi-empirical approximations employed in CFD solid–fluid models, and is attracting increasing interest from industries. This enables the investigation of fluidized beds, pneumatic conveying, filtration, solid–liquid mixing, and many other systems. Effective modeling of the solid–fluid flow requires methods for adequately characterizing the discrete nature of the solid phase and representing the interaction between solids and fluids. DEM-CFD models reported in the literature have largely been applied to the simulation of fluidized beds and, more recently, to the pneumatic transport of particles [9–16]. Many industrial processes involve complex geometries, often with moving parts, and complex fluid dynamics. For example, the design of pneumatic seed collecting and discharging devices and horizontal seed supply tubes [17–21] coupled with DEM-CFD techniques have also been used as tools to optimize the design of machine structures [22–27].

The air-absorbing hazelnut harvester designed in this paper includes two parts: a conveying pipeline pick-up box and a scavenging box. Because the gas–solid two-phase flow formed during the pneumatic conveying of hazelnuts is more complicated, and at the same time the structure of the scavenging box has a greater impact on the operating effect of this machine, this paper first measured the suspension velocity of typical hazelnuts and fallen leaves by using a suspension test bench to provide data support for the design of the scavenging device. Then, the discrete element method (DEM), computational fluid dynamics method (CFD), and gas–solid two-phase flow theory were used to simulate the flow–solid coupling analysis of the harvester device from the air flow distribution of the cleaning box and the movement law of the material on the screen, in order to obtain the relationship between the movement characteristics of hazelnuts in the conveying process, the flow field distribution characteristics, and the structural parameters of the cleaning box. By analyzing the motion state of the mixture material during the conveying process and the air flow field distribution law of the hazelnut harvester, the effects of the sieve plate angle, the distance of the sieve plate, and the air flow velocity on the net fruit rate of hazelnut were specifically studied, and the potential mechanism was analyzed. The optimal operating parameters of the harvester were determined, and the performance of the harvester was evaluated by prototype tests.

2. Materials and Methods

2.1. Test Materials and Their Basic Physical Characteristics

Hazelnuts are widely grown in the mountainous area of Liaoning, and in this study, Liaoning hazelnut No. 3 grown in Huanren, Liaoning, was used as the sample. These samples were harvested in September 2022 with a cumulative sample size of 20 kg. The instrument used for the moisture content determination test of hazelnuts and their mixes was the JH-H5 moisture dryer, and the experimental results showed that the water content of hazelnut was 7.47% and the water content of the leaves was 0.53%. The mixture was classified into two states, hazelnut and leaf, using an electronic balance (BS200S-MEI) and a statistical classification.

The mass and size of the hazelnuts were measured, and according to the measurements it can be concluded that the mass of hazelnuts is between 2 g and 3 g in 70% of fruit. The volumes of hazelnuts were measured, and they can be considered as round; the diameter distribution was between 23 mm and 30 mm in 83% of fruit. Hazelnut harvesters also have leaves mixed in during the harvesting process, as shown in Figure 1.

Figure 1. Hazelnut, mixed sample chart: (**a**) hazelnut; (**b**) leaves.

2.2. Design of the Main Components of the Hazelnut Harvester

The hazelnut harvester achieves separation of the picking device and the wind selection device by using the unloading device, so that the wind of the two devices does not interfere with each other. At the same time, the mixture from the picking device can be transported to the wind sorting device for secondary wind sorting. When the pick-up device picks up, the mixture collides with the screen plate and breaks up the material. The dust-like impurities are sucked out by the negative-pressure air flow of the centrifugal fan, and the rest of the hazelnut mixture will fall into the discharge device and further fall into the wind selection box device for secondary wind selection. The hazelnuts slide down to the hazelnut collection box by gravity, and the remaining leaves and other debris are wind-selected to the miscellaneous collection box by the sieve leaf plate under positive-pressure wind. The angle and position of the sieve plate at each level inside this hazelnut harvester can be freely adjusted to harvest different kinds of nuts. The working principle is shown in Figure 2.

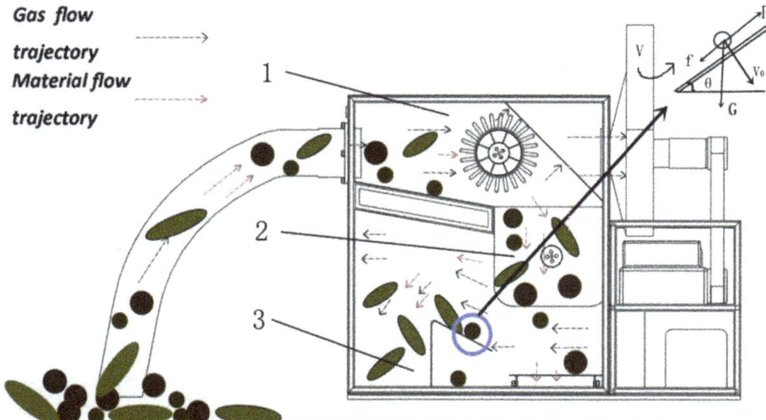

Figure 2. Hazelnut harvester gas–material flow trajectory diagram: 1. material picking device, 2. wind sorting device, 3. discharge device.

The designed multi-purpose picking head device includes a harvesting device, transmission device, support device, collection device, and other structural components. The pick-up device includes eccentric vibration ring, brush head, and other structural components. The collection device includes a fan, collection box, and collection funnel structural components. The transmission device includes a 57 series three-phase hybrid stepper motor, transmission shaft, and other structural components. The support device includes frame, bearing seat, front and rear part of the grip handle, and other structural components. When the hazelnut harvester is working, the handheld multifunctional picking head can pick hazelnuts on the tree, and the brush head part can rotate under eccentric vibration, relying on inertia to remove the hazelnut fruits from the tree. In addition, a small fan on the multi-purpose picking head is designed to provide cooling for the working motor and to help the brush head get rid of twisted branches. This fan can also blow away impurities at any time during the working process and blow down the already combed hazelnut fruit into the collection pipe for the next sorting step of the hazelnut harvester.

The main components of said hazelnut harvester also include structural components such as a picking device, wind selection device, power device, and support device. The picking device includes structural components such as a picking duct, deflector, air regulating valve, fish scale sieve plate (primary sieve plate), and cylindrical roller brush. The wind sorting device includes structural components such as the unloading device, the sieve leaf plate (secondary sieve plate), the hazelnut collection box, and the impurity collection box. The power and support device includes a centrifugal fan, clutch, diesel engine, battery, frame, universal wheel, and other structural components. The hazelnut harvesting machine structure display is shown in Figure 3.

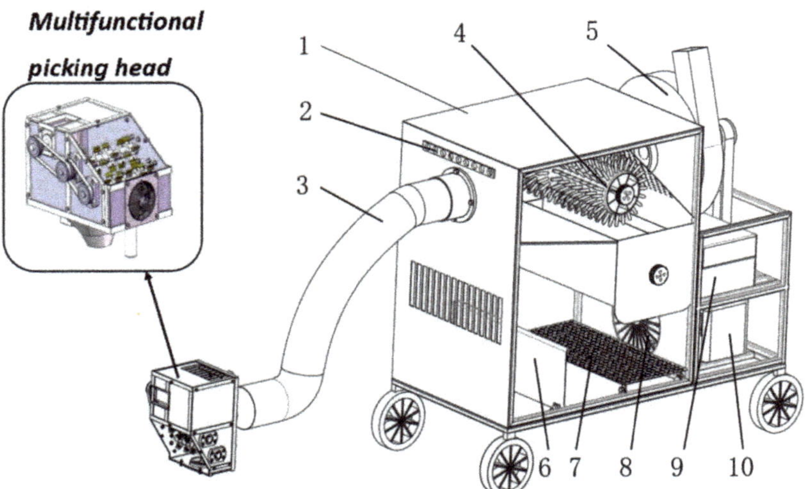

Figure 3. Hazelnut harvesting machine structure diagram: 1. machine outer casing, 2. air regulating valve, 3. conveying pipeline, 4. cylinder roller brush, 5. centrifugal fan, 6. sieve leaf plate, 7. vibrating screens, 8. positive-pressure fan, 9. gasoline engine, 10. storage battery.

The discharging device mainly consists of a cavity shell, a fixed blade, and an adjusting blade, as shown in Figure 4. Six cavities of equal volume are formed inside the discharge device, and the regulating blades are made of rubber to prevent damage to the hazelnuts. When working, the particle mixture falls into the discharge device, and when the cavity is opposite to the lower outlet, hazelnuts and fallen leaves are discharged from the discharge device.

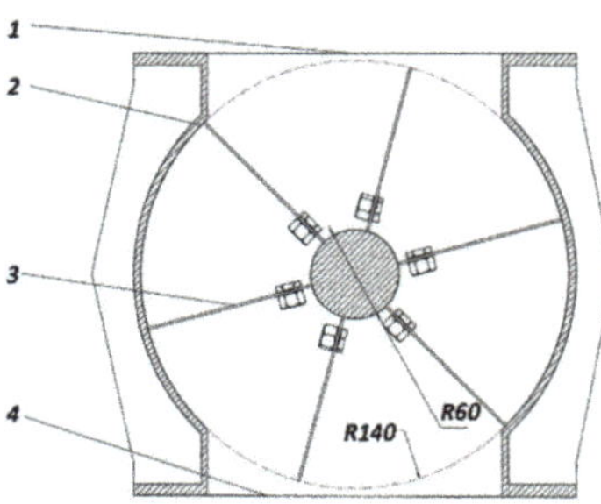

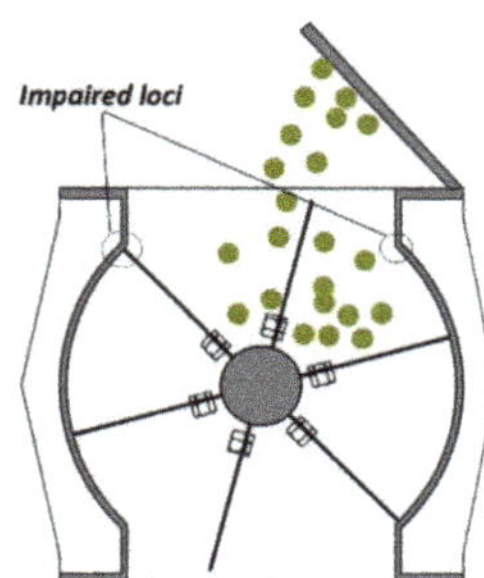

Figure 4. Discharge device structure and hazelnut collision schematic: 1. material feed inlet, 2. chamber housing, 3. adjusting blades, 4. material outlet.

The discharging device mainly plays the role of conveying the particle mixture and locking the gas, and the discharging device has an important influence on the working performance of the hazelnut harvesting machine. If discharging is too slow, it can cause hazelnuts and leaves to accumulate in the picking device and reduce conveying efficiency. If discharging is too fast, it is easy to cause the particle mixture to be too late for sorting, and also cause the pressure loss at the connection between the discharge device and the harvester to increase. During the design process, it was found that most of the particle mixture fell on the right side of the cavity shell after hitting the screen plate. When the blade rotates clockwise, the blade, the right damage point, and hazelnuts touch more at the same time. When the blade rotates counterclockwise, the hazelnuts, the left damage point, and the blade touch less at the same time, so the discharging device uses counterclockwise rotation to discharge. The discharging device structure and hazelnut collision are shown in Figure 4.

The discharge volume of the discharge device should meet the requirements of pneumatic conveying volume, which can be calculated by using Formula (1).

$$Gs = 0.06n\Psi\, i\gamma s \tag{1}$$

In the formula, Gs is the discharge volume, kg/h; n is the impeller speed, 25 r/min; and Ψ is the impeller filling factor, 0.6~0.8.

The value of 0.8 for is granular materials; i is the effective discharge volume of the impeller, m^3; and γs is the capacity of the conveyed material, 850 kg/m^3.

$$i = (R - r)[\pi(R + r) - \xi z\,] \tag{2}$$

In the formula, R is the radius of the outer edge of the impeller, 0.14 m; r is the radius of the root of the impeller, 0.06 m; ξ is the thickness of the blade, 0.01 m; z is the number of blades, 6; and L is the blade length, 0.23 m.

Gs = 640 kg/h is obtained by Equations (1) and (2). Considering that impurities are picked up together with the picking process, the unloading volume should be 1.2~1.5 times the pneumatic conveying volume, and the obtained Gs value meets the design requirements.

The wind sorting device is one of the main structures of the hazelnut harvester. In order to improve the net fruit rate of the wind sorting device, the air flow sorting device with negative-pressure picking and positive-pressure cleaning was designed according to the fluid flow principle. The picking device mainly consists of a fish scale sieve plate and air regulating valve. The fish scale sieve plate is at a 30° angle, and according to the size of the picking chamber design the rectangular fish scale sieve plate length is 600 mm and the width is 250 mm. The aperture of the fish scale sieve is a 15 mm, 15 × 10 arrangement. According to the characteristics of the material through mechanical analysis, it is concluded that the screen leaf plate is at an angle of 55° with the bottom. The aperture of the vibrating screen plate is a 17 mm, 15 × 30 arrangement, mainly to distinguish the size of the hazelnuts.

In order to explore the mechanical mechanism affecting the hazelnut picking and sorting process, the critical velocity of the sucking up and sorting of hazelnuts and fallen leaves is calculated according to the principle of gas–solid two-phase flow, as follows.

$$F_x = \frac{1}{8}\Pi\mu\rho y^2 v_q^2 \tag{3}$$

$$F_z = \Pi g\rho x y^2 \tag{4}$$

$$G = m_{\max} g \tag{5}$$

$$G = F_x + F_z \tag{6}$$

Simplification of Equations (3) and (6) of critical conditions for hazelnut suspension velocity:

$$v_q = \sqrt{\frac{8m_{max}g - 2\Pi g\rho x y^2}{\Pi\mu\rho y^2}} \tag{7}$$

$$v = kv_q \tag{8}$$

In the formula, F_x is the attractive force (N); F_z is the resistance (N); μ is the resistance constant; ρ is the air density ($kg{\cdot}m^{-3}$); x is the short axis diameter of the material (mm); y is the long axis diameter of the material (mm); v_q is the flow velocity of the theoretical air flow ($m{\cdot}s^{-1}$); v is the flow velocity of the actual air flow ($m{\cdot}s^{-1}$); m is the mass of the material (g); m_{max} is the maximum mass of the material (g); g is the acceleration of gravity ($m{\cdot}s^{-2}$); and k is the reliability coefficient.

According to the test that measured the average mass of hazelnuts, m is 3.61 g, the average mass of fallen leaves is 2.3 g, the drag coefficient C is 0.6, the projected area of hazelnut in the direction of motion S is 3.1 mm^2 and fallen leaves is 12.56 mm^2, the critical condition of hazelnut suspension speed that can be calculated from the above formula is 15.6 $m{\cdot}s^{-1}$, and the critical value of fallen leaves suspension speed is 4.92 $m{\cdot}s^{-1}$.

To verify the theoretical calculation value, the suspension speed experiment was conducted at the Agricultural Machinery Laboratory of Shenyang Agricultural University. The instruments used were the material suspension speed test bench of model PS-20 and an electronic balance (BS200S-MEI). In order to ensure the accuracy of the test, each group of materials was tested five times repeatedly, three measurement points were selected and averaged for each group of tests, and the final suspension speed of each material was similar to the theoretical calculated value, which can be used as a basis for design.

The hazelnut gravitational component force $gcos\theta$ can be considered to be balanced with the Magnus effect force F_M; in the direction of hazelnut motion by the joint action of the differential pressure force generated by the air flow and $gsin\theta$, the differential equation for the motion of the hazelnut is obtained according to D'Alembert's principle:

$$\frac{dv_{sx}}{dt} = \frac{F_M}{m} + gcos\theta \tag{9}$$

$$\frac{dv_{sy}}{dt} = \frac{k\rho S(v - v_s)^2}{m} - gsin\theta \tag{10}$$

In the equation, F_M is the Magnus effect (N) and v_s is the hazelnut velocity ($m{\cdot}s^{-1}$).

In the sorting process, hazelnuts will be in contact with the sieve plate, in addition to the above forces, but are also affected by the sieve plate support force F_N and friction resistance F_f, according to D'Alembert's principle to obtain the differential equation of hazelnut motion, as follows:

$$F_f = \mu F_N \tag{11}$$

$$\frac{dv_{sx}}{dt} = \frac{F_M + F_N}{m} + gcos\theta \tag{12}$$

$$\frac{dv_{sy}}{dt} = \frac{k\rho S(v - v_s)^2}{m} - gsin\theta + \frac{\mu F_N}{m} \tag{13}$$

According to Equation (13), the velocity of hazelnut movement in the sorting device is influenced by the average velocity of gas flow v, the initial average velocity of hazelnut v_s, and the angle θ between the sieve plate and the horizontal direction. Therefore, the factors affecting hazelnut sorting mainly include gas flow velocity and sieve plate angle, which are controlled by fan speed and sieve plate angle in the experiment. The theoretical analysis of fluid dynamics provides the theoretical basis for the simulation and experimental design.

2.3. *Theoretical Analysis of CFD-DEM Coupling*

Fluent is one of the most functional and applicable CFD software in recent years, and it has good applications mainly in industries related to fluids, heat transfer, and chemical reactions. The discrete element simulation software Rocky is a general CAE software based on the discrete element method, which can be used to simulate and analyze the mechanical behavior of granular materials and their effects on material handling equipment. It has been widely used in agricultural machinery, mining equipment, chemical and food grade pharmaceuticals, and other fields Many processes in various industries involve the simultaneous flow of fluids and particles. In these cases, it is important to consider the fluid flow in order to first obtain the correct particle behavior. This is then determined by the particle-level interactions between the fluid, the particles, and the boundaries. Therefore, it is obvious that a modeling approach is needed to deal with particle fluid systems, and there are two methods commonly used to solve them: Eulerian and Lagrangian methods.

In the Eulerian approach, both the fluids and solid phases are treated as interpenetrating continua in a computational cell that is much larger than the individual particles, but still small compared to the size of the process scale. Therefore, continuum equations are solved for both phases with an appropriate interaction term to model them. This in turn means that constitutive equations for inter- and intraphase interactions are needed. Since the volume of a phase cannot be occupied by the other phases, the concept of phasic volume fraction is introduced. Location-based mapping techniques are applied, and local mean variables are used in order to obtain conservation equations for each phase. The advantage of this approach is its reasonable computational cost for practical application problems, making it the most used granular-fluid modeling technique in use today.

In the Lagrangian approach, the fluid is still treated as a continuum by solving the Navier–Stokes equations, while the dispersed phase is solved by tracking a large number of particles through the flow field. Each particle (or group of particles) is individually tracked along the fluid phase by the result of forces acting on them by numerically integrating Newton's equations that govern the translation and rotation of the particles. This approach is made considerably simpler when particle–particle interactions can be ignored. This requires that the dispersed second phase occupies a low volume fraction, which is not the reality in the majority of the industrial applications. Due to the fact that no particle interaction is resolved, the model is inappropriate for modeling applications where the volume fraction of the second phase cannot be ignored, such as fluidized beds. For applications such as these, particle–particle interactions need to be taken into account when solving the dispersed phase. Now, numerous authors have published their work using the Euler–Lagrange type of model to study granular flow [28–31].

The coupled CFD-DEM approach is an effective alternative for modeling particulate fluid systems because it captures the discrete nature of the particle phase while maintaining computational tractability. This is achieved by solving the fluid flow at the cell level rather than at the detailed particle level. By reducing the required fluid calculations, this technique expands the range of devices and processes that can be studied with numerical simulations. In the coupled CFD-DEM approach, the fluid flow is obtained by the traditional continuous medium approach, providing information to calculate the fluid forces acting on individual particles, while the particle motion is obtained by using the discrete particle approach. The gas phase is numerically simulated by CFD, the particles are solved by the DEM method, and the exchange of energy is coupled through the gas–solid phase interaction. Currently, Rocky has two methods to perform the coupling between particles and fluid: the multiphase coupling approach relies on the Eulerian Multiphase Model of Fluent, where the material particles are represented by another dedicated phase, and the multiphase approach supports an arbitrary number of fluid phases; the single coupling approach is achieved by setting the fluid domain in Fluent as a porous medium, which enables the material particles to influence.

When the Multiphase Model is set to Eulerian in the Fluent case, the averaged mass conservation equation is given by

$$\frac{\partial}{\partial t}\left(\alpha_f\rho_f\right)+\nabla\cdot\left(\alpha_f\rho_f u\right)=0 \tag{14}$$

whereas the averaged momentum conservation equation is written as

$$\frac{\partial}{\partial t}\left(\alpha_f\rho_f u\right)+\left(\alpha_f\rho_f uu\right)=-\alpha_f\nabla p+\nabla\cdot\left(\alpha_f\Pi_f\right)+\alpha_f\rho_f g+F_{p\to f} \tag{15}$$

where α_f stands for the fluid volume fraction, p is the shared pressure, ρ_f is the fluid density, u is the fluid phase velocity vector, and T_f is the stress tensor of the fluid phase, defined as

$$\Pi_f=\mu_f\left(\nabla u+\nabla u^T\right)+\left(\lambda_f-\frac{2}{3}\mu_f\right)\nabla\cdot u\mathrm{II} \tag{16}$$

In Equation (15), $F_{p\to f}$ represents the source term of momentum from an interaction with the particulate phase, calculated according to the expression

$$F_{p\to f}=-\frac{\sum_{p=1}^{N}F_{f\to p}}{V_c} \tag{17}$$

where V_c is the computational cell volume, N is the number of particles inside the computational cell volume, and $F_{f\to p}$ accounts for the forces generated by the fluid on the particles.

When the Multiphase Model is turned off in the Fluent case, Rocky adapts the Fluent setup to treat the DEM particles as a porous media and to assign to the fluid phase momentum and energy source terms (that account for fluid–particle interactions) calculated by Rocky during coupled simulations. The porosity distribution of the domain is a function of the concentration of the solid phase as the simulation progresses.

Considering a single-phase flow through a porous medium and assuming that there is no mass transfer between phases, the averaged mass conservation equation of the fluid phase is given by

$$\frac{\partial}{\partial t}\left(\gamma\rho_f\right)+\nabla\cdot\left(\gamma\rho_f u\right)=0 \tag{18}$$

where γ is the porosity of the medium. Likewise, the averaged momentum conservation equation is

$$\frac{\partial}{\partial t}\left(\gamma\rho_f u\right)+\nabla\cdot\left(\gamma\rho_f uu\right)=-\gamma\nabla p+\nabla\cdot\left(\gamma\Pi_f\right)+\gamma\rho_f g+F_{p\to f} \tag{19}$$

and the averaged energy conservation equation is

$$\frac{\partial}{\partial t}\left(\gamma\rho_f h_f\right)+\nabla\cdot\left(\gamma\rho_f u h_f\right)=\gamma\frac{\partial p}{\partial t}+\gamma\Pi_f:\nabla u-\nabla\cdot\gamma q_f+Q_{p\to f} \tag{20}$$

The porosity γ is defined as the relative volume occupied by the void spaces of the porous region. As a single-phase coupled simulation runs, Rocky estimates the porosity of each cell as

$$\gamma=1-\alpha_s \tag{21}$$

where α_s is the local volume fraction of the solid phase at the current time step.

2.4. Hazelnut Harvester Simulation Parameters' Setting

The establishment of a discrete element model is mainly divided into two kinds, which are geometric model establishment and particle model establishment. According to the design parameters, the 3D modeling software SolidWorks was used to build the 3D model of the hazelnut harvester and to simplify its model parts. The simplified 3D model was imported into Rocky, and a fluid domain model was created. The fluid domain model was meshed with a structured meshing method, with a total of 2.57 million mesh cells and 460,000 nodes. The fluid domain model is shown in Figure 5.

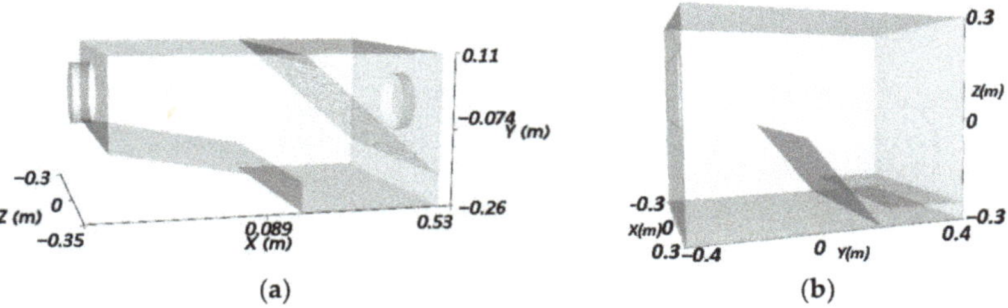

Figure 5. Harvester fluid domain model: (**a**) pickup device fluid domain model; (**b**) fluid domain model of wind separator.

Based on Rocky software, a three-factor, three-level Box–Behnken simulation experiment was conducted with the sieve plate angle, distance of the sieve plate, and air flow rate as factors and hazelnut net fruit rate as the index. The simulation experimental results provide guidance and comparative verification for subsequent field tests. The establishment of the material particle model in Rocky is the basis of the simulation analysis, and the realistic degree of its model directly affects the simulation results. In this paper, we take Liaoning hazelnut No. 3 as the research sample, through the dimensional analysis of hazelnut and leaf drop, where the average diameter of hazelnuts is 22 mm as a round ball, and the average diameter of leaf drop is 36 mm as a disc shape with a thickness of 0.5 mm. In Rocky, the stem model of particles can be created directly, and the shape of leaf drop is sphero-polygon, with a vertical aspect ratio of 1.8, a horizontal aspect ratio of 0.1, and an angle number of 80; the shape of the hazelnuts is spherical. The particle models are shown in Figure 6. In order to distinguish the size of hazelnuts, their diameter was set at 80% of hazelnuts above 18 mm and the rest at 20%.

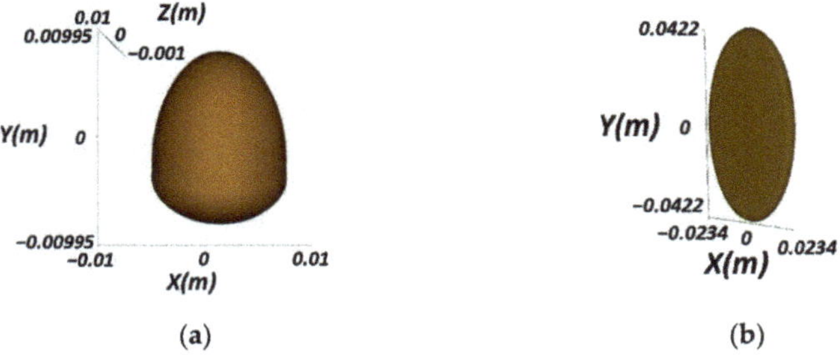

Figure 6. Particle mixture model: (**a**) hazelnut model; (**b**) leaf model.

The material inlet was used to set the simulation particle entry position, direction, and time. A particle inlet in the Geometries module was created, and the entry port was set as a rectangular inlet according to the actual geometry of the feeding device. The length was set as 0.15 m, width as 0.25 m, alignment angle as 90°, and incline angle as 270°. The inlet at the feeding port of the picking device and the feeding port of the sorting device was set. According to the simulation needs, the feeding time was set to 0~2.5 s. The number of particles was estimated according to the actual machine output, the hazelnuts and leaves in the mixture picked up by the machine were sieved, and the feeding volume was set to 1.6 kg/s for hazelnuts and 0.15 kg/s for leaves. Particle entry simulation parameters are shown in Table 1.

Table 1. Particle inlet settings.

Project Name	Project Setting
Entry Point	Inlet
Stop Time	2.5 s
Mass Flow Rate (Hazelnut)	1.6 kg/s
Mass Flow Rate (Leaves)	0.15 kg/s

In order to ensure the authenticity of the simulation test and reduce the simulation time, hazelnuts and leaves were selected as the study objects in this simulation. The simulation test only considers the interaction between hazelnut, leaf, and hazelnut harvester, ignoring the influence of other impurities on the simulation. Each part of the hazelnut harvester is endowed with material characteristics, and the entire device is made of structural steel. The values of material density, Poisson's ratio, Young's modulus, and dynamic and static friction factors of the particles directly affect the simulation results. We imported the data from the hazelnut stacking angle experiment and the static friction experiment into the EDEM database for analysis and obtained these simulation parameters. The parameters of material mechanical properties and interparticle contact parameters of hazelnuts and leaves are shown in Table 2.

Table 2. Simulation parameters.

Type	Density/(g·cm^{-3})	Poisson's Ratio	Young's Modulus/Pa
Hazelnut	5.5	0.41	3.1×10^9
Leaf	0.7	0.38	2.71×10^5
Type	**Dynamic Friction Coefficient**		**Static Friction coefficient**
Hazel–Hazel	0.06		0.45
Hazel–Leaf	0.06		0.4
Leaf–Leaf	0.05		0.42

2.5. Box–Behnken Experimental Design

The experiment was conducted in October 2022 in a hazelnut orchard in the Heishan region of Liaoning Province, China, as shown in Figure 7.

The experiment was designed according to GB/T5667-2008 "Agricultural Machinery Production Test Methods" and with reference to the GB/T5262-2008 "General Provisions for the Determination of Test Conditions of Agricultural Machinery" standard. Each group of tests collected 1 m^2 of fallen hazelnut surface, and each group was repeated three times, setting the picking mouth 20 mm above the ground, taking the average value as the test results recorded and analyzed, with the net fruit rate of hazelnut as the evaluation index. After the first harvest the overall weight in the hazelnut collection box was weighed as M_{total}, the weight of hazelnuts was manually selected as M_{Hazel}. Then, the total mass of the hazelnut collection box and the miscellaneous collection box in the hazelnut picking and sorting machine was weighed as M_{pick}, the surface of the plot was harvested manually again using manual harvesting, and the mass of the manual harvesting was M_{miss}. The hazelnut net fruit rate Y was calculated by the following equation.

$$Y = M_{Hazel}/M_{total} \times 100\%$$

The factors influencing the hazelnut harvester were selected and combined with the previous CFD-DEM simulation analysis of the sieve plate angle, the distance of the sieve plate, and the air flow velocity factors. The sieve plate angle was set at 51–59°, the distance of the sieve plate was 17–46 mm, and the air flow velocity was 9–18 m·s^{-1}, and a three-factor, three-level Box–Behnken test was conducted. The experimental factor coding table is shown in Table 3.

Figure 7. Machine field trials: (**a**) hazelnut picking and sorting test; (**b**) harvester operating site.

Table 3. Experimental factor and level table.

Level	Factor		
	Sieve Plate Angle *A*/(°)	Distance between Sieve Plates *B*/mm	Air Flow Velocity *C*/(m·s^{-1})
−1	51	17	9
0	55	31.5	13.5
1	59	46	18

3. Results and Discussion

3.1. Basic Physical Characteristics of the Experimental Samples

The basic physical parameters of the particles were derived from the statistics and measurements of the harvested samples, as shown in Table 4.

Table 4. Measurement of physical parameters of hazelnut harvesting material.

Type	Quantity/Each	Average Length/cm	Average Diameter/cm	Average Mass/g	Density/(g·cm^3)
Simple fruit	1894	1.83	1.73	2.8	1.3
Leaf	521	4.5	3.6	1.07	0.7

3.2. Analysis of Simulation Results

3.2.1. Terminal Speed Analysis of the Pick-Up Device

In Fluent, the SIMPLE algorithm is used, the mesh model is a hexahedral mesh, and the simulation model is a standard k-epsilon model. The inlet is set as the velocity inlet, the velocity is consistent with the initial velocity of the Rocky particle inlet, the velocity is 20 m·s^{-1}, and the fan port is set as the pressure outlet condition. In Rocky, the particle inlet initial velocity is set to be the same as the inlet wind velocity in Fluent, in the direction normal to the inlet plane. Air flow velocity clouds were obtained for the hazelnut harvester at three air flow speeds, as shown in Figure 8. In the simulation experiment, there was no adhesion between the hazelnut and the leaf, and a contact model without sliding was used between the particles and the wall, with the gravitational acceleration direction along the negative direction of the Z-axis. The feeding time was 2.5 s, total feeding 1.7 kg, total simulation time 3 s, and interval time 0.05 s.

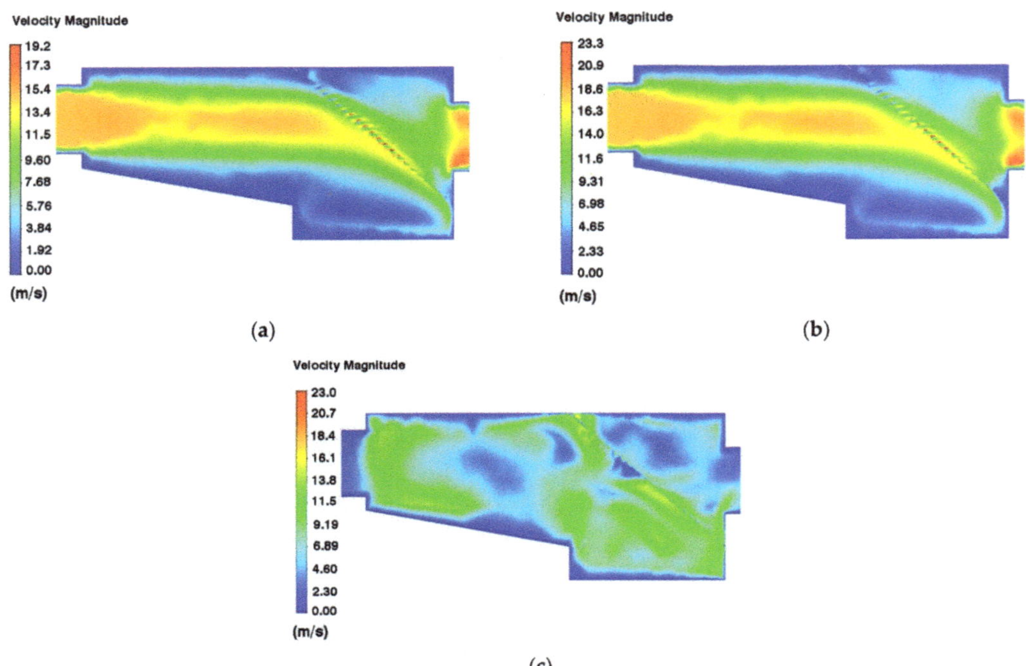

Figure 8. Velocity contour map of the pick-up device: (**a**) air velocity 15; (**b**) air velocity 18; (**c**) air velocity 21.

3.2.2. Analysis of the Working Process of the Sorting Device

The particle flow trajectory inside the hazelnut harvester corresponding to 0~3 s time was obtained by coupling simulation with Fluent fluid dynamics software and Rocky discrete element software. The working process is shown in Figure 9. At t = 0.55 s, the particles began to fall by the discharge device, and the particle mixture was blown to the sieve plate by the positive-pressure wind. After the leaves and hazelnuts hit the sieve plate, the lighter leaves were blown to the miscellaneous collection box in front, the gravity of the hazelnuts themselves was greater than the blowing force of the air flow, and the hazelnuts fell by gravity to the hazelnut collection box. At t = 1.65 s, the particle mixture was further separated through the sieve plate. The lighter leaves were partly blown to the debris collection box in front and partly blown out through the air outlet, and more hazelnuts fell by gravity to the vibrating screen waiting for further sorting. At t = 2.55 s, the number of falling particles reached 1.7 kg. The hazelnut collection box had a small number of leaves, and the vibrating sieve further separated the hazelnuts by size. The device in line with the expected design effect.

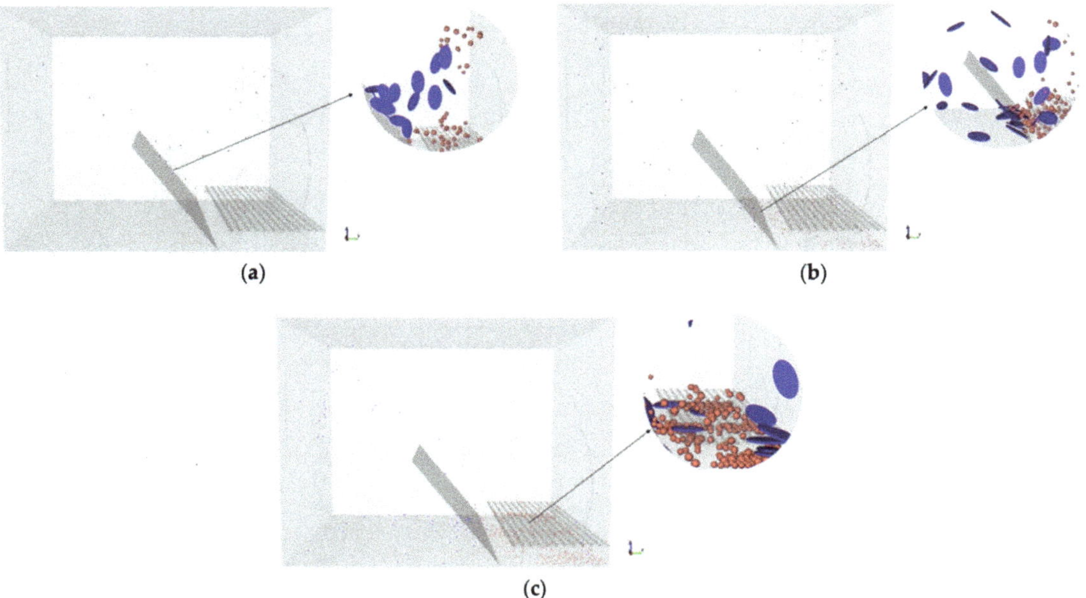

Figure 9. The working process of wind sorting device: (**a**) start sorting at t = 0.55 s; (**b**) during sorting at t = 1.65 s; (**c**) end sorting at t = 2.55 s.

Simulation of the whole process was completed for a total of 3 s with a sieve leaf plate angle of 50~60 °, air flow velocity of 10~20 m·s^{-1}, and distance of the sieve plate of 15~50 mm for seventeen groups of simulation. The results show that the particles with different densities can be sorted by the screening of the baffles. The comparison curves of the total mass of material with time between the simulation experiment of the sorting device and the field experiment are shown in Figure 10. The trend of the total mass of the particle mixture in the pick-up device with time is shown in Figure 11. The results show that the sorting effect is most obvious, and the net fruit rate is high at the sieve plate angle of 55°, air flow velocity of 15 m·s^{-1}, and distance of the sieve plate of 30 mm.

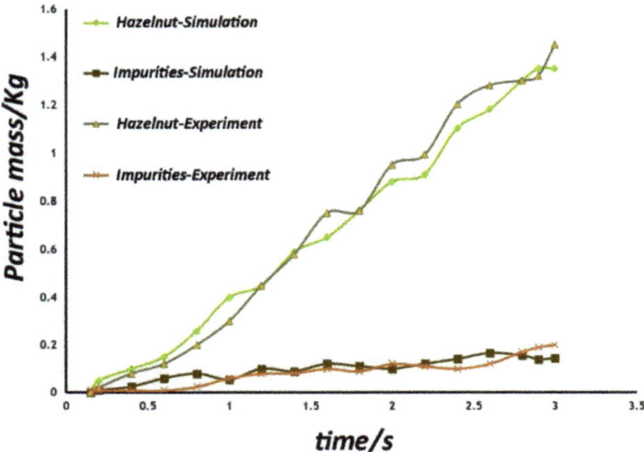

Figure 10. Variation in the mass profile of each material in the sorting device.

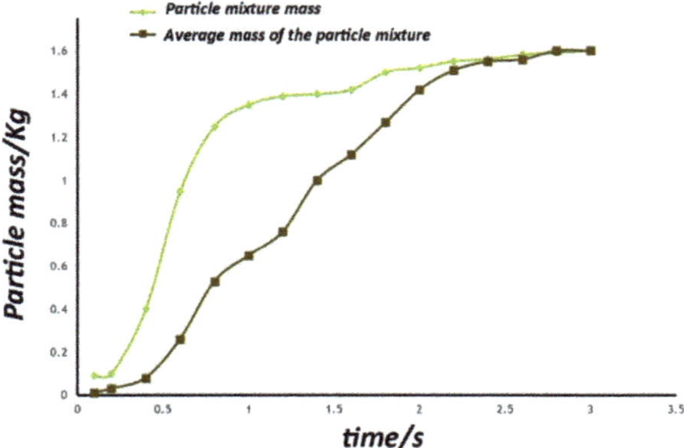

Figure 11. Variation in the total material mass profile in the pick-up device.

The coupled CFD-DEM simulation analysis shows that after the granular mixture material goes down from the discharge device, the mixture material collides with the screen plate, the leaves are blown to the impurity collection box with the air flow, and the hazelnuts flow to the hazelnut collection box under the action of gravity. The air flow in the designed sorting device model follows the expected trajectory and the particle mixture is effectively separated. The CFD-DEM simulation experiment obtained the distribution and movement of particles in the separation device, and the overall process of simulation is reasonable and more consistent with the actual working conditions, which can be used to simulate the harvesting and sorting process of hazelnuts.

3.3. Analysis of Experimental Design Results

According to the data samples in Table 5, the quadratic polynomial regression model of net fruit rate with three factors was obtained using the data processing software Design-Expert 8.0, and the regression equation was the following:

$$Y = 94.76 + 1.64A + 1.11B + 2.35C - 1.13AB - 1.16AC + 1.55BC - 3.69A^2 - 4.79B^2 - 8.87C^2$$

Table 5. Experiment design and result.

Test Number	Factor			Test Result
	A	B	C	Net Fruit Rate Y
1	51	31.5	9	76.9%
2	59	31.5	18	84.3%
3	55	46	18	86.2%
4	55	31.5	13.5	93.3%
5	55	31.5	13.5	95.3%
6	55	17	18	81.6%
7	55	17	9	79.1%
8	55	31.5	13.5	95.5%
9	51	31.5	18	83.9%
10	51	46	13.5	87.4%
11	59	46	13.5	88.1%
12	55	46	9	77.5%
13	59	17	13.5	87.4%
14	59	31.5	9	83.7%
15	51	17	13.5	82.2%
16	55	31.5	13.5	95.3%
17	55	31.5	13.5	94.4%

In the formula, Y is the net fruit rate of hazelnuts; A is the sieve plate angle; B is the distance of the sieve plate; and C is the air flow velocity.

The regression model ANOVA and significance test results are shown in Table 6. From Table 6, it is clear that the fit of the net fruit rate model is highly significant ($p < 0.01$). The regression equation misfit was not significant, and it was a good fit with the actual situation. The *p*-values of the sieve plate angle, distance of the sieve plate, and air flow velocity could determine the effects of the three test factors on the net fruit rate of hazelnuts. The *p*-value of regression term B was less than 0.05 and the effect was significant, the *p*-values of regression terms A and C were less than 0.01 and the effects were highly significant, and the effects of the other terms were significant or highly significant. The effects of the test factors on the net fruit rate were air flow velocity, sieve plate angle, and distance between sieve plates in descending order.

Table 6. Analysis of variance.

Source	Net Fruit Rate			
	SS	**df**	**F**	***p*-Value ***
Model	632.23	9	78.37	<0.0001
A	21.45	1	23.93	0.0018
B	9.90	1	11.05	0.0127
C	44.18	1	49.29	0.0002
AB	5.06	1	5.65	0.0491
AC	10.24	1	11.42	0.0118
BC	9.61	1	10.72	0.0136
A^2	57.41	1	64.05	<0.0001
B^2	96.71	1	107.89	<0.0001
C^2	331.08	1	369.37	<0.0001
Residual	6.27	7		
Lack of fit	2.89	3	1.13	0.4361
Error	3.39	4		
Sum	638.51	16		

* $p < 0.01$ means extremely significant, $0.01 < p < 0.05$ means significant, $p > 0.05$ means not significant, SS means Sum of Squares, df means Degree of Freedom, F means F-value

Using Design-Expert 8.0 software to process the data and analyze the relationship between the test index and factors, the effect of sieve plate angle, air flow velocity, and distance of the sieve plate on the net fruit rate can be obtained, and the response surface is shown in Figure 12, fixing the factors of sieve plate angle, air flow velocity, and distance of the sieve plate as the 0 level, respectively, and analyzing the interaction between the remaining two factors on the net fruit rate according to the response surface plot. The effect of the interaction between the remaining two factors on the net fruit rate was analyzed according to the response surface.

The response surface of the interaction between air flow velocity and distance of the sieve plate on the net fruit rate is shown in Figure 12a. When the air flow rate is certain, the net fruit rises first and then falls with the increase in distance of the sieve plate; when the distance of the sieve plate is certain, the net fruit rises first and then falls with the increase in air flow velocity. The response surface of the interaction effect of sieve plate angle and distance of the sieve plate on the net fruit rate is shown in Figure 12b. When the sieve plate angle is certain, with the increase in distance of the sieve plate, the net fruit rises first and then falls; when the distance of the sieve plate is certain, with the increase in sieve plate angle, the net fruit rises first and then falls. The response surface of the interaction between air flow velocity and sieve plate angle on the net fruit rate is shown in Figure 12c. When the air flow velocity is certain, with the increase in sieve plate angle, the net fruit rises first and then decreases. When the sieve plate angle is certain, the net fruit rate also rises first and then decreases with the increase in air flow velocity. Based on the interaction effect analysis, it can be seen that when any of the factors of air flow velocity, sieve plate angle, and distance of the sieve plate are fixed, the interaction of the remaining two factors is first increased to a certain value so that the net fruit effect in the impurity sorting process is significant, and then the net fruit rate decreases as the two factors continue to increase.

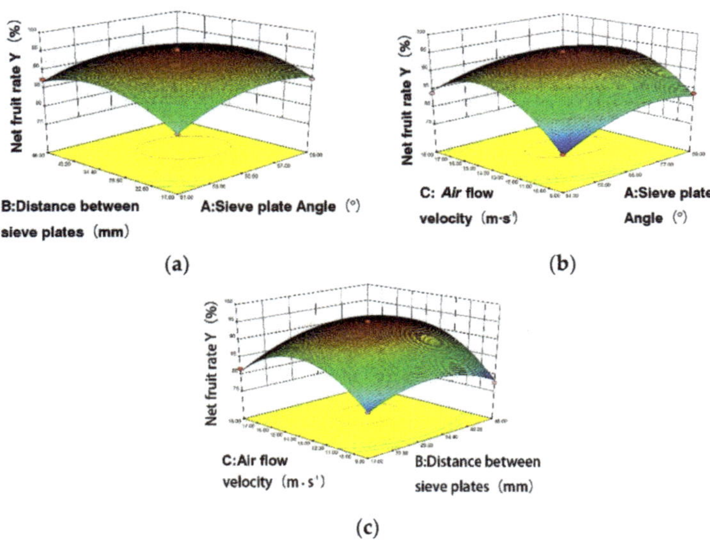

Figure 12. Response surface curve of interaction factors for test indexes: (**a**) the effect of AB interaction on Y; (**b**) the effect of AC interaction on Y; (**c**) the effect of BC interaction on Y.

3.4. Parameter Optimization and Comparison Experiments

In order to obtain the optimal parameters for each factor, the established three-factor orthogonal test was optimized by using Design-Expert 8.0 software. In order to obtain the optimal parameters for each factor, a three-factor orthogonal test was established using Design-Expert numerical analysis software for parameter optimization. Based on the analysis of the test results and the optimization of the influencing factors, the optimal combination of factors was obtained: the angle of the sieve plate angle was 55.7°, the air flow velocity was 14.1 m·s^{-1}, and the distance of the sieve plate was 33.2 mm, which resulted in a prediction of 95.12% of the net fruit yield of hazelnuts.

Based on the optimal parameters obtained, five replicate tests were performed, and the test result was 94.25%. The error rate was 3.4%. It can be seen that under the optimal parameters, the field prototype test results are similar to the simulation test results and can meet the requirements of hazelnut picking operation.

4. Discussion

Hazelnuts and leaves were selected from the particle mixture for the study through statistics and analysis of the mixture. The critical value of particle suspension velocity was obtained by the experiment, and it is an important parameter to measure to know whether the material could be separated. A coupled CFD-DEM method was used to simulate the effect of a pneumatic hazelnut harvester on the net fruit rate of hazelnuts under different combinations of operating parameters. The coupled CFD-DEM simulation was verified to describe the motion of the particle mixture in the sorting device well through a field experiment. Through analyzing the fluid dynamics inside the hazelnut harvester, the study proved that the optimization of air flow velocity, sieve plate angle, and distance of the sieve plate could improve the working performance of the hazelnut harvester. The working principle of the hazelnut harvester was analyzed from the point of the movement of the sorting mixture. The results of the comparison experiment show that the hazelnut harvester with the optimized combination of operating parameters can effectively sort the particle mixture and reduce the impurity rate and energy consumption. Working performance experiments show that the optimized hazelnut harvester could pick and sort the particle mixture well under real field conditions.

In the follow-up research, the combination of various devices should be considered in order to further improve the net fruit rate. For example, HASATSAN researched and designed a high horsepower tractor-driven hazelnut harvester and used a precision-improving interchangeable vibrating screen set to work simultaneously with three harvesting hoses. The harvester was constructed to harvest a particle mixture in any geographic condition with a vacuuming system; TURBO-VAC has designed an integrated hazelnut harvester that used a diesel engine to drive a fan that rotates at high speed to pick and sort particles in a single pass. By refining the functions of each component and the components combining and cooperating with each other, we will be able to further improve the performance of the hazelnut harvester. In this paper only one hazelnut variety was used as the experimental material, and the actual suspension speed of different varieties of hazelnuts varied greatly; hence, further research is needed in terms of applicability. In addition, the vibration of the whole machine and the environment of the actual workplace will have some influence on the cleaning situation; these are influences which need to be considered in future experimental research.

5. Conclusions

This paper designs a pneumatic hazelnut harvester for mountains areas. In the experiments, the suspension velocity of hazelnuts was obtained as 15.6 $m \cdot s^{-1}$ and the critical value of the suspension velocity of leaves was 4.92 $m \cdot s^{-1}$. A coupled CFD-DEM method was used to simulate the effect of a pneumatic hazelnut harvester on the net fruit rate of hazelnuts under different combinations of operating parameters. In addition, a Box–Behnken design experiment was conducted with the sieve plate angle, the distance of the sieve plate, and the air flow velocity as factors and the net fruit rate of hazelnuts as indicators to explore the influence. The order of the factors affecting the index was air flow velocity > the sieve plate angle > distance of the sieve plate. The parameter optimization module of Design-Expert software was used to obtain the optimal combination of parameters for the factors: the sieve plate angle was 55.7°, the air flow velocity was 14.1 $m \cdot s^{-1}$, and the distance of the sieve plate was 33.2 mm. The research results of this paper could provide a new design idea for optimizing the picking structure and sorting mechanism of the hazelnut harvester.

Author Contributions: Conceptualization: D.R. and H.Y.; software: H.Y. and R.Z.; validation: Y.Z. and H.Y.; supervisory role: W.W.; writing—review and editing: D.R.; investigation: J.Z.; data curation: F.L. and J.L.; funding acquisition and supervision: W.W. All authors have read and agreed to the published version of the manuscript.

Funding: This research was mainly supported by the Liaoning Xingliao Talent Program for Science and Technology Innovation Leaders (XLYC2002009).

Institutional Review Board Statement: Not applicable.

Informed Consent Statement: Not applicable.

Data Availability Statement: The datasets generated during and/or analyzed during the current study are available from the corresponding author upon reasonable request.

Conflicts of Interest: The authors declare no conflict of interest.

References

1. Loghavi, M.; Mohseni, S.H. The effects of shaking frequency and amplitude on detachment of lime fruits. *Iran Agric. Res.* **2006**, *24*, 27–38.
2. Polat, R.; Gezer, I.; Guner, M. Mechanical harvesting of pistachionuts. *J. Food Eng.* **2007**, *79*, 1131–1135. [CrossRef]
3. Peterson, D.L.; Whiting, M.D.; Wolford, S.D. Fresh-market quality tree fruit harvester. *Appl. Eng. Agric.* **2003**, *19*, 539–543.

4. Zhou, Z.Y.; Kuang, S.B.; Chu, K.W.; Yu, A.B. Assessments of CFD–DEM models in particle–fluid flow modelling. *J. Fluid Mech.* **2010**, *661*, 482–510. [CrossRef]
5. Yu, A.B.; Xu, B.H. Particle-scale modelling of gas–solid flow in fluidisation. *J. Chem. Technol. Biotechnol.* **2003**, *78*, 111–121. [CrossRef]
6. Muggli, F.A.; Holbein, P.; Dupont, P. CFD calculation of a mixed flow pump characteristic from shutoff to maximum flow. *ASME J. Fluids Eng.* **2002**, *124*, 798–802. [CrossRef]
7. Gebrehiwot, M.G.; de Baerdemaeker, J.; Baelmans, M. Effect of a cross-flow opening on the performance of a centrifugal fan in a combine harvester. *Biosyst. Eng.* **2010**, *105*, 247–256. [CrossRef]
8. Karim, A.A.; Nolan, P.F. Modelling reacting localized air pollution using Computational Fluid Dynamics (CFD). *Atmos. Environ.* **2011**, *45*, 889–895. [CrossRef]
9. Kawaguchi, T.; Tanaka, T.; Tsuji, Y. Numerical simulation of two-dimensional fluidized beds using the discrete element method (comparison between the two- and three-dimensional models). *Powder Technol.* **1998**, *96*, 129–138. [CrossRef]
10. Rong, D.G.; Horio, M. Behavior of particles and bubbles around immersed tubes fluidized bed at high temperature and pressure: A DEM simulation. *Int. J. Multiph. Flow* **2001**, *27*, 89–105. [CrossRef]
11. Ibsen, C.H.; Helland, E.; Hjertager, B.H.; Solberg, T.; Tadrist, L.; Occelli, R. Comparison of multi-fluid and discrete particle modelling in numerical predictions of gas particle flow in circulating fluidised beds. *Powder Technol.* **2004**, *149*, 29. [CrossRef]
12. Chu, K.W.; Yu, A.B. Numerical simulation of the gas–solid flow in three-dimensional pneumatic conveying bends. *Ind. Eng. Chem. Res.* **2008**, *47*, 7058–7071. [CrossRef]
13. Chu, K.W.; Wang, B.; Vince, A.; Yu, A.B.; Barnett, G.D.; Barnett, P.J. CFD–DEM study of the effect of particle density distribution on the multiphase flow and performance of dense medium cyclone. *Miner. Eng.* **2009**, *22*, 893–909. [CrossRef]
14. Chu, K.W.; Wang, B.; Yu, A.B.; Vince, A. CFD–DEM modelling of multiphase flow in dense medium cyclones. *Powder Technol.* **2009**, *193*, 235–247. [CrossRef]
15. Gao, X.J. Study on Precision Seeding Technology and Device Based on High-Speed Centrifugal Filling-Clearing. Ph.D. Thesis, China Agricultural University, Beijing, China, 2021. (In Chinese with English Abstract).
16. Brosh, T.; Kalman, H.; Levy, A.; Peyron, I.; Ricard, F. DEM-CFD simulation of particle comminution in jet-mill. *Powder Technol.* **2014**, *257*, 104–112. [CrossRef]
17. Ren, B.; Zhong, W.Q.; Chen, Y.; Chen, X.; Jin, B.S.; Yuan, Z.L.; Lu, Y. CFD-DEM simulation of spouting of corn-shaped particles. *Particuology* **2012**, *10*, 562–572. [CrossRef]
18. Ding, L.; Yang, L.; Wu, D.H.; Li, D.Y.; Zhang, D.X.; Liu, S.R. Simulation and experiment of corn air suction seed metering device based on DEM-CFD coupling method. *Trans. Chin. Soc. Agric. Mach.* **2018**, *49*, 48–57.
19. Guzman, L.; Chen, Y.; Landry, H. Coupled CFD-DEM simulation of seed flow in an air seeder distributor tube. *Processes* **2020**, *8*, 1597. [CrossRef]
20. Arzu, Y.; Vedat, D.; Adnan, D. Comparison of computational fluid dynamics-based simulations and visualized seed trajectories in different seed tubes. *Turk. J. Agric. For.* **2020**, *44*, 599–611.
21. Pasha, M.; Hassanpour, A.; Ahmadian, H.; Tan, H.S.; Bayly, A.; Ghadiri, M. A comparative analysis of particle tracking in a mixer by discrete element method and positron emission particle tracking. *Powder Technol.* **2015**, *270*, 569–574. [CrossRef]
22. Landry, H.; Thirion, F.; Lague, C.; Roberge, M. Numerical modeling of the flow of organic fertilizers in land application equipment. *Comput. Electron. Agric.* **2006**, *51*, 35–53. [CrossRef]
23. Mori, Y.; Sakai, M. Development of a robust Eulerian-Lagrangian model for the simulation of an industrial solid-fluid system. *Chem. Eng. J.* **2021**, *406*, 126841. [CrossRef]
24. Pei, C.L.; Wu, C.Y.; England, D.E.; Byard, S.; Berchtold, H.; Adams, M. DEM-CFD modeling of particle systems with long-range electrostatic interactions. *AIChE J.* **2015**, *61*, 1792–1803. [CrossRef]
25. Jiang, Z.H.; Rai, K.T.; Tsuji, T.; Washino, K.; Tanaka, T.; Oshitani, J. Upscaled DEM-CFD model for vibrated fluidized bed based on particle-scale similarities. *Adv. Powder Technol.* **2020**, *31*, 4598–4618. [CrossRef]
26. Takabatake, K.; Sakai, M. Flexible discretization technique for DEM-CFD simulations including thin walls. *Adv. Powder Technol.* **2020**, *31*, 1825–1837. [CrossRef]
27. Liu, S.J.; Li, Y.W.; Hu, X.Z. Effect of particle volume fraction on the performance of deep-sea mining electric lifting pump based on DEM-CFD. *J. Mech. Eng.* **2020**, *56*, 257–264, (In Chinese with English Abstract).
28. Hoomans, B.; Kuipers, J.; Van Swaaij, W. Granular dynamics simulation of segregation phenomena in bubbling gas-fluidised beds. *Powder Technol.* **2000**, *109*, 41–48. [CrossRef]
29. Xu, B.; Yu, A.; Chew, S.; Zulli, P. Numerical simulation of the gas-solid flow in a bed with lateral gas blasting. *Powder Technol.* **2000**, *109*, 13–26. [CrossRef]
30. Ye, M.; Van der Hoef, M.; Kuipers, J. A numerical study of fluidization behavior of geldart a particles using a discrete particle model. *Powder Technol.* **2004**, *139*, 129–139. [CrossRef]
31. Ye, M.; Van der Hoef, M.; Kuipers, J. From discrete particle model to a continuous model of geldart a particles. *Chem. Eng. Res. Des.* **2005**, *83*, 833–843. [CrossRef]

Article

Parameter Optimization of Reciprocating Cutter for Chinese Little Greens Based on Finite Element Simulation and Experiment

Wei Wang [1], Xiaolan Lv [2,3] and Zhongyi Yi [2,3,*]

1 Key Laboratory of Modern Agricultural Equipment and Technology, Ministry of Education, Jiangsu University, Zhenjiang 212013, China
2 Institute of Agricultural Facilities and Equipment, Jiangsu Academy of Agricultural Science, Nanjing 210014, China
3 Key Laboratory for Protected Agricultural Engineering in the Middle and Lower Reaches of Yangtze River of Ministry of Agricultural and Rural Affairs, Nanjing 210014, China
* Correspondence: 20140006@jaas.ac.cn

Abstract: Optimizing the working performance of the cutting device for harvesting Chinese little greens is crucial to reducing energy consumption in cutting and improving cutting quality. To explore the mechanical characteristics of leafy vegetables in cutting, the dynamic process of cutting Chinese little greens with a cutter was simulated numerically by using the finite element method based on theoretical analysis. In the numerical simulation, the response-surface methodology (RSM) and central composite rotatable design (CCD) were used to describe the influence rule of sliding–cutting angle (X_1), oblique angle (X_2), and the average cutting speed (X_3) on cutting stress. Then, the stress distribution pattern produced by the cutting blade and the stalks were evaluated by using different working parameters. Subsequently, taking the minimum cutting stress as the target value, the best combination of cutter structure and working parameters were obtained: the sliding–cutting angle was 29°, the oblique angle was 38°, and the average cutting speed was 500 mm/s. At the condition of optimal parameter combinations, the ultimate cutting stress of the upper cutting blade was 0.95 Mpa and that of the bottom cutting blade was 0.77 Mpa. A cutting test was carried out by using a bench test of the cutting performance, and the mechanical properties of cutting at different cutting speeds were studied. Test results showed that at the optimal cutting speed of 500 mm/s, the cutting stress on the cutter was relatively small and the cutting effect reached the best value. The finite element simulation of cutting the little greens reduced the test cost and provided a reference for the development of a cutting device with low power consumption.

Keywords: cutting system; numerical simulation; response-surface methodology; parameter optimization; mechanical properties

Citation: Wang, W.; Lv, X.; Yi, Z. Parameter Optimization of Reciprocating Cutter for Chinese Little Greens Based on Finite Element Simulation and Experiment. *Agriculture* **2022**, *12*, 2131. https://doi.org/10.3390/agriculture12122131

Academic Editor: Wen-Hao Su

Received: 18 October 2022
Accepted: 6 December 2022
Published: 12 December 2022

Publisher's Note: MDPI stays neutral with regard to jurisdictional claims in published maps and institutional affiliations.

1. Introduction

Chinese little greens, whose scientific name is non-heading cabbage, are rich in nutrients and are one of the important vegetable varieties in China. With a large planting density and a short growth cycle, Chinese little greens can be harvested 18–25 days after sowing [1]. However, at present, the harvest of Chinese little greens is still completed manually with low harvesting efficiency and high work intensity. Mechanized harvesting operation is the key to solving this problem.

The stress–strain process for cutting failure of the plant stalk fiber layer can be divided into three stages: elastic deformation, plastic deformation, and shear failure. First, the cutter presses the stalk to bend and deform elastically. Then, the cutter continues to bend the stalk to produce plastic deformation and the fiber tensile stress continues to increase. Then, the fibers near the blade break off and eventually fail. Eventually, the entire fiber layer slips and breaks, causing shear damage to the entire stalk [2]. Studies have found that the mechanical properties during cutting of crop stalks are closely related to the working parameters of the cutting device [3,4].

Relevant scholars studied cutting characteristics using the quasi-static experimental method [5]. Cui et al. [6] designed a shear fixture on a universal materials tester, and selected the blade distance, sliding–cutting angle, skew cutting angle, and shearing angle as test factors to study the shearing characteristics of lettuce stems, and obtained the best cutting combination parameters. Zhang et al. [7] also reported similar results in a shear stress test of a single rice stem, in which the peak cutting force decreased with the sliding–cutting angle. Esgici et al. [8] used a material testing machine to study the influence of grapevine diameter and age on the cutting force, and found that the cutting force was proportional to the vine diameter and age. Du et al. [9] established a mechanical model of the cabbage cutting process, and studied the influence of sliding–cutting angle and cutting speed on fracture, and verified and analyzed it through experimental data. In a study of cutting red bean stalks, the shear stress generated by the 28° bevel angle was smaller than that generated by the 0° bevel [10]. Clementson and Hansen [11] showed that the cutting force of the machete is smaller than that of the chopper in sugarcane harvesting. However, the quasi-static experimental method cannot fully reflect the cutting characteristics of crop stalks, and the influence factors such as the motion parameters of the cutting blade are ignored.

To simulate the actual operation mode of the cutting process, scholars have analyzed the effect of cutting and feeding speed on the cutting characteristics by developing a stalk-cutting test bench [12–14]. Johnson et al. [15] conducted an in-depth study on the cutting energy of the cutting tool under different working parameters and found that the cutting energy of miscanthus stalk was proportional to the cutting speed. Mathanker et al. [16] used the same equipment to evaluate the effect of cutting speed and blade bevel on cutting energy, and found that the specific cutting energy increased with the increasing cutting speed. The research of Allameh and Alizadeh found that the cutting speed of the shear parts had a significant impact on cutting energy. When the cutting speed increased from 1.5 to 2.5 m/s, the cutting energy grew by 77%. At the same time, the interaction between the cutting angle and the bevel angle had a significant impact on the cutting energy [17]. Zhao et al. [18] selected the blade angle, blade shape, cutting speed, and cutting angle as the test factors and studied the effects of working parameters on power consumption by means of mathematical statistics on a self-designed testing system, and obtained the optimal parameter combination for cutting performance of maize stalks under no-support cutting status.

In general, the cutting process of plant stalks belongs to the category of typical high-speed collision and penetration. The contact process between the cutting blade and the stalk forms a randomly nonlinear interaction relationship; hence, it is difficult for traditional physical methods to analyze the cutting mechanism between them. At present, numerical simulation technology has become one of the important tools for solving engineering practice problems [19]. The dynamic simulation analysis can efficiently simulate the non-linear dynamic contact conditions in stalk cutting, and visualize the force and deformation conditions during the cutting process. Meng et al. [20] established a simulation model and studied the influence of working conditions of the circular cutter to determine the best cutting parameters. Yang et al. [21] established an analytical three-dimensional model of the cutting system by using the ANSYS/LS-DYNA software f, and provided a reference for the optimization of the cutter's parameters and the cutting method. Huang et al. [22] studied the dynamic characteristics of the reciprocating cutting system, and evaluated the dynamic response of the cutting system by factors such as cutting speed and inclination angle by numerical simulation and cutting experiments. Qiu et al. [23] found that numerical simulation technology could accurately simulate the cutting process of sugarcane stalks, and analyzed the influence of external and internal factors in the cutting process on the cutting quality, and finally determined the best working parameters based on the simulation results and orthogonal experiments. In summary, although most of the existing studies on numerical simulation of the cutting process of plant stems are focused on the cutting method of a circular saw blade, it shows that using computer technology can effectively solve stem-cutting problems.

Research on crop stalk-cutting devices has matured, but related research is mainly concentrated on sugarcane, hemp, miscanthus, and other crops. Therefore, it is necessary to study the structural and working parameters of the cutting devices that specially match with the mechanized harvesting of Chinese little greens.

2. Materials and Methods

2.1. Experiment Influencing Factors and Evaluation Indexes

During the cutting process, the cutter continues to be subjected to nonlinear force from the plant stalk until the stalk is cut off. The force for cutting plant stalks is the cutting force, which is closely related to cutting energy consumption. Large cutting resistance will inevitably cause greater power consumption in cutting. Therefore, the cutting stress on the cutter is selected as the evaluation index for cutting power consumption and cutting quality [24,25].

The reciprocating cutting system of the cutting machine includes a drive motor, an underdriving gear set, an eccentric wheel mechanism, a blade holder, and a cutting blade [14]. The reciprocating cutting blade is driven by an eccentric cam mechanism to create a reciprocating linear movement with fast cutting frequency and excellent dynamic balance ability while it is working. Its structure is shown in Figure 1.

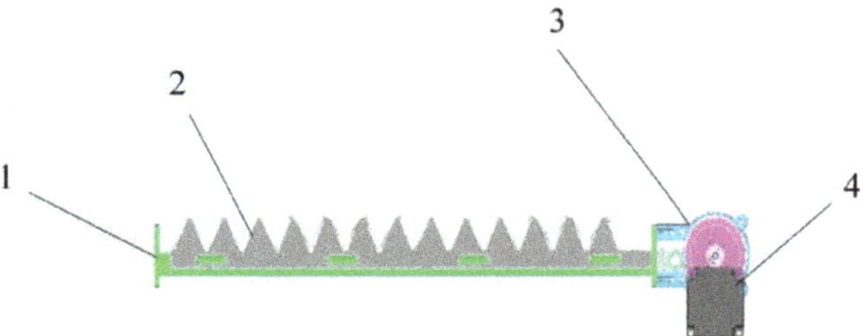

Figure 1. The structure drawing of the reciprocating cutter system: 1—cutter holder; 2—cutter; 3—transmission mechanism; 4—motor.

In order to facilitate the theoretical analysis of the cutting process, it is assumed that the stalk is a homogeneous body, and the cutting blade is always in the same plane during the cutting process, regardless of the vibration during the cutting process. When the cutter first contacts the stalk, the stalk deforms elastically at the contact point. Since the absolute speed change is large at this time, the cutting resistance rises rapidly. There is a positive correlation between cutting feed and cutting resistance before reaching the allowable limit. When this limit is exceeded, the stalk undergoes plastic deformation. The length of the line contact between the blade and the stalk continues to change, and the cutting force presents a fluctuating process. At this time, the force of the stalk on the cutter is mainly horizontal cutting force F_a, vertical cutting force F_p, and inertial force F_n. In addition, *V1* and *V2* are the cutting speeds of the upper and lower blades, respectively, and β is the angle of the blade.

In the *XY* plane, the moving direction of the unit cutting force P is the same as the contact point of the cutting blade and the stalk. The movement path of the contact point is a composite of the lateral movement of the cutting blade and the forward movement of the harvesting machine. The cutting resistance on the cutting blade is the integral of the unit cutting force in the contact length between the cutter and the stalk (Figure 2b).

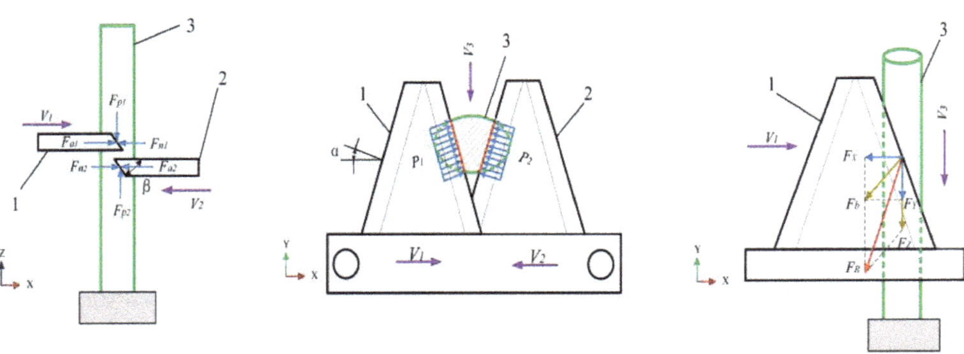

Figure 2. Reciprocating cutter-system cutting force diagram: 1—upper cutting blade; 2—bottom cutting blade; 3—stalk.

Therefore, the sliding friction force at the contact point can be decomposed into the component forces in the *X* and *Y* directions, and the direction is opposite to the relative movement direction [23]. For convenience in the analysis, the cutting resistance force F_R at the contact point is decomposed into three mutually perpendicular component forces, which are the horizontal force F_x along the cutting direction, the horizontal force F_y along the feed direction, and the force F_z perpendicular to the *XOY* plane, as shown in Figure 2c. Then, the cutting force F_R is expressed as follows:

$$F_R = \sqrt{F_x^2 + F_y^2 + F_z^2} \tag{1}$$

In this study, the smaller vertical component F_z was ignored, and only the horizontal direction forces F_x and F_y were considered. To avoid the effect of the cutting force that is changed with the stalk diameter of Chinese little greens, according to previous studies, the maximum cutting stress was taken as the evaluation index of the cutting property, and the cutting stress σ was calculated by Equation (2) [6]:

$$\sigma = \frac{F_{max}}{A} \tag{2}$$

where the F_{max} is the maximum cutting force in horizontal direction, N; and A is the cross-sectional area of the stalk at the cutting position, mm^2.

The factors that affect the shear stress are the structural and movement parameters of the cutter [26–28]. In this study, the three factors, sliding–cutting angle, oblique angle, and average cutting speed, were taken as the factors that affect the cutting mechanical property was analyzed.

2.2. Finite Element Modeling and Analysis

2.2.1. Geometric Model

The three-dimensional geometric model of the reciprocating cutting system unit was established by applying the modeling software SolidWorks. In order to improve the accuracy of the solution and shorten simulation operation time, the cutting system model was appropriately optimized, and a cutting unit was extracted for calculating the simulation when constructing the physical model of the cutting system [23,29]. In Figure 3, the simulation model includes three parts: the upper cutting blade, the bottom cutting blade, and the stalk of Chinese little greens.

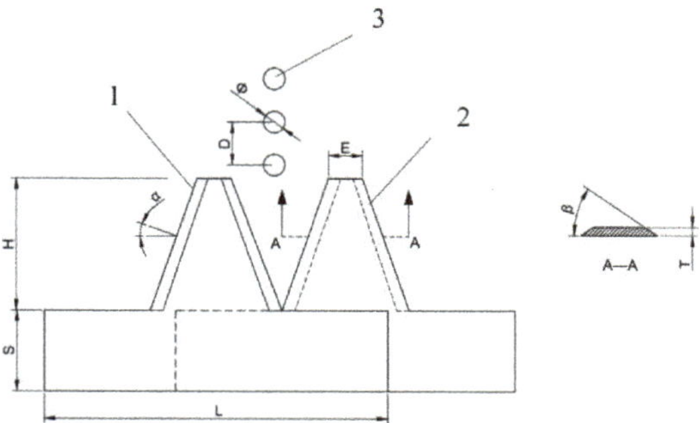

Figure 3. The structural diagram of the geometric model: 1—upper cutting blade; 2—bottom cutting blade; 3—stalk.

The cutting blade was regarded as the same integral rigid body, whose features such as the drive motor and transmission components were omitted, but the oblique angle was retained. In the modeling of the greens, it grew with no inclination angle to the ground, and no branches on the stems. The difference between cortex and xylem was ignored. At the same time, each stem to be cut was assumed to be relatively independent, and there was no implicated force between the stems. In addition, the gap between the cutting blade and stem was narrowed as much as possible to reduce the computing time and the amount of calculation. In Table 1, the primary geometric parameters are listed. The three-dimensional geometric model of the reciprocating cutting system unit is shown in Figure 3. The model was saved in Parasolid (*.x_t) format.

Table 1. Primary parameters of the geometric model.

Parameters	Values	Parameters	Values
Tool holder width S/mm	19	Cutter top width E/mm	8
Cutter unit length L/mm	80	Cutter thickness T/mm	2
Cutter edge height H/mm	31	Oblique angle $\beta/°$	35, 40, 45
Sliding–cutting angle $\alpha/°$	20, 25, 30	Stalk diameter Φ/mm	5

2.2.2. Material Property Parameters

The ANSYS Workbench has a rich database of materials, and structural steel material was selected as the material for the cutter. The cutting process of the stalk is essentially a state of penetration destruction, and large deformation and destruction inevitably occur as the material fails. The model of the stem adopts linear elastic anisotropic material properties; the constitutive parameters of the materials are shown in Table 2 [30].

Table 2. Main material parameters of the geometric model.

Material	Stalk	Blade
Density ρ/ kg·m^{-3}	800	7850
Young's modulus E_X/Mpa	20.5	2.0×10^5
Young's modulus E_Y/Mpa	20.5	2.0×10^5
Young's modulus E_Z/Mpa	3.5	2.0×10^5
Shear modulus G_{XY}/Mpa	7.88	7.7×10^4
Shear modulus G_{XZ}/Mpa	1.35	7.7×10^4
Shear modulus G_{YZ}/Mpa	1.35	7.7×10^4
Poisson's ratio μ_{XY}	0.35	0.3
Poisson's ratio μ_{XZ}	0.3	0.3
Poisson's ratio μ_{YZ}	0.3	0.3

2.2.3. Meshing

The ANSYS Workbench contains multiple built-in meshing methods. In this simulation, a hexahedral meshing method was adopted. In order to ensure the accuracy of the simulation, proper mesh densification was performed on the part where the cutter contacts during meshing [31,32]. The model element size of the cutter and the stalk were both 0.8 mm, and the element size of the dense part of the stalk was 0.6 mm. The total number of elements in the cutting model was 8280 together with 11,842 nodes. The finite element model meshing is shown in Figure 4.

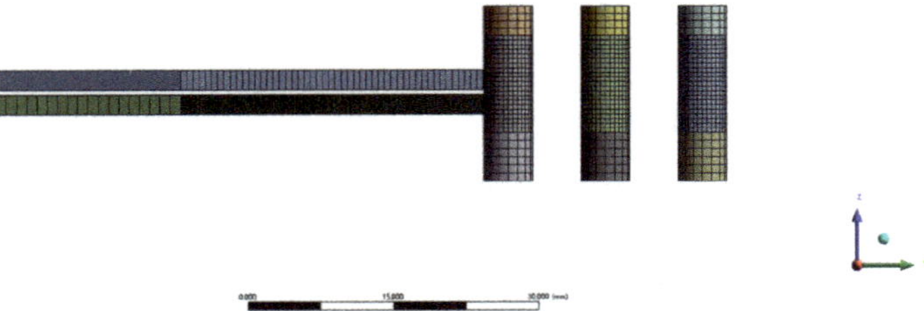

Figure 4. The finite element model.

2.2.4. Loads and Constraints

Initial conditions and boundary constraints were determined according to actual working conditions. It was assumed that there was hard soil. The constraint of the ground on the stalks was assumed to be a cantilever beam constraint. The stalks were constrained in the X and Z directions at its bottom to limit the displacement in both the X and Z directions. The speed in the Y direction was given, and then the same speed of the harvesting machine was simulated. The influence of the mechanical vibration of the cutter during the cutting process was ignored, and it was assumed that the cutting blade always moves in the same plane.

The constraint of the cutting blade was the displacement constraint in the fixed Y and Z directions; that is, the cutting blade only moves in the opposite direction of the X direction, and the speeds of the upper cutter and the bottom cutter were equal in reverse. During the cutting process, the interaction between the cutting blade and the stalk belongs to the category of stab. Hence, the contact type was defined as surface-to-surface erosion contact. The dynamic friction coefficient between the cutter and the stalk was set to 0.38, and the static friction coefficient was set to 0.4 [20].

2.3. Orthogonal Test Design

The central–composite test design was adopted [33]. By taking the maximum cutting equivalent stress σ1 of the upper cutting blade and the maximum cutting equivalent stress σ2 of the bottom cutting blade as the target value, then the cutting blade sliding–cutting angle, oblique angle, and the average cutting speed were the three factors used to design an orthogonal test with three factors and five levels. The sliding–cutting angle is an important parameter that affects the shape of the blade, which is the angle between the absolute motion direction and the normal direction of the cutting edge [7]. The larger the sliding–cutting angle is, the smaller the cutting force, but a sliding–cutting angle that is too large is not conducive to stable clamping, so its value ranges from 20° to 30°. The oblique angle is the edge angle of the cutter. If the oblique angle is too small, the service life of the cutter is reduced, and if it is too large, the cutting resistance is increased. In this paper, the oblique angle of the cutter ranges from 35° to 45° [6]. The average cutting speed is the ratio of a single cutting displacement to the reciprocating movement time of the cutter. It is an important indicator to measure the cutting performance. Some studies have found that when the cutting speed is too fast, the cutting resistance increases, obviously, and when the cutting speed is too small, the cutting motion becomes difficult [20]. Therefore, the average cutting speed was set at 300 to 500 mm/s. The test was divided into 20 groups, and the coding table of test factor levels is shown in Table 3.

Table 3. Levels and codes of experimental variables.

Level	Test Factors		
	Sliding–Cutting Angle X_1 (°)	**Oblique Angle X_2 (°)**	**Average Cutting Speed X_3 (mm/s)**
−1.682	16.59	31.59	231.82
−1	20	35	300
0	25	40	400
1	30	45	500
1.682	33.41	48.41	568.18

2.4. Experiment and Methods

In order to verify the accuracy of the numerical simulation results and detect the cutting effect under the best parameters of the reciprocating cutting system, the cutting performance experiment was carried out.

The Chinese little greens variety used were Nanjing Yongxin, with a growth cycle of 35 days. The average height of the selected Chinese little greens selected in the experiments was 180 ± 30 mm, the length of the cutting position was 10 ± 3 mm, and the diameter of the stalk was 7 ± 2 mm. Before the cutting test, five Chinese little greens roots were randomly weighed and recorded. They were then dried continuously in an oven for 24 h (ASABE standard, 2012). The equation for calculating moisture content follows:

$$M = \frac{M_L}{M_W} \times 100\% \tag{3}$$

where M is the moisture content, %; M_L is the weight lost, g; and M_W is the sample weight, g.

The leaves of the root were removed, and the whole plant was transplanted into a seedling tray and fixed. The test sample is shown in Figure 5a. A reciprocating cutting stress measurement system for the Chinese little greens was built [34]. The test system was mainly composed of a stalk-feeding device, a cutting device, and a cutting stress measurement system. The specific working parameters of the test bench are shown in Table 4.

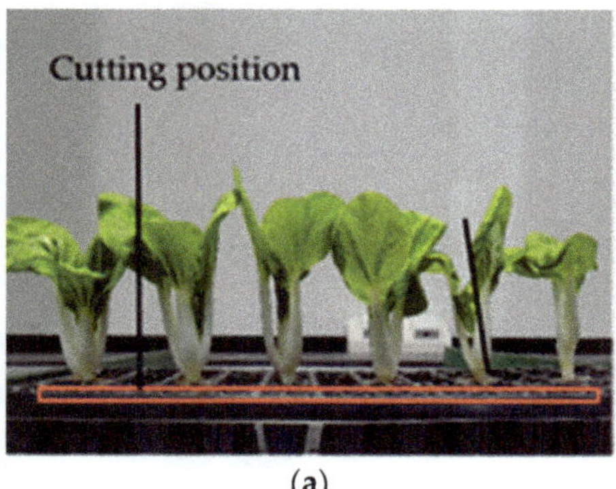

(**a**)

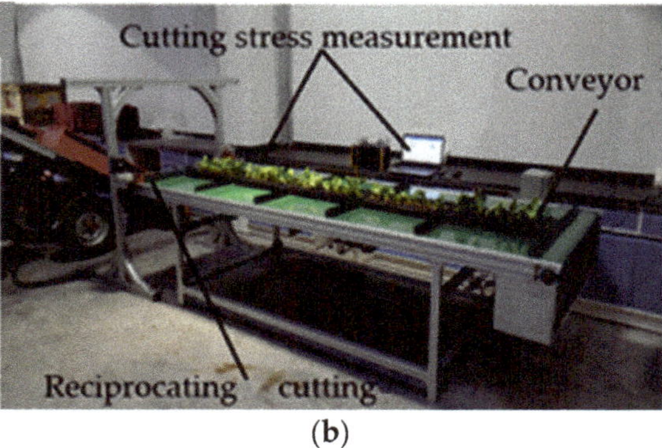

(**b**)

Figure 5. Schematic diagram of measurement system of cutting stress: (**a**) test sample; (**b**) cutting stress measurement system.

Table 4. Parameters of the reciprocating cutting test system.

Parameters	Values
Average cutting speed/mm·s^{-1}	0~1500
Reciprocating cutting stroke/mm	30
Average feeding speed/mm·s^{-1}	0~1000
Resistance strain measurement range/Mpa	0~100
Resistance strain gauge sensitivity factor	2.17 ± 1%
Data collection frequency/Hz	500
Data acquisition channel	1~4

The stalk-feeding device includes a frequency converter, an AC motor, a conveyor belt, and a stalk-fixing seedling tray. The stalk-fixing tray of greens was placed on the center of the conveyor belt, which was powered by an AC motor to drive the seedling tray forward. The frequency converter was used to adjust the feeding speed of the stems, and the specific structure is shown in Figure 6a. The cutting device was fixed on the frame by fastening bolts, including a controller, a DC motor, a cutter, a transmission device, and a frame. In operation, the double eccentric wheel mechanism was driven by the stepping motor. The phase difference between the two eccentric wheels was transmitted to the eccentric shaft through the reduction gear, thereby driving the upper and bottom blades to make a reciprocating linear motion. The average cutting speed can be adjusted by the DC motor, and its structure is shown in Figure 6b. The test system mainly includes a resistance strain gauge and a DH5902N solid data acquisition system. A set of adjacent blades for each of the upper and bottom blades was selected, and the surface of the blades was polished and cleaned. A set of strain gauges at the center of the blades was installed in rectangular distribution, so that the principal stresses of the blades in the X and Y directions were collected. The X direction was in line with the cutting direction, and the Y direction was in line with the stalk-feeding direction. The strain gauge was connected to the data acquisition system through wiring. It sends the mechanical signals of the cutter to the dynamic signal acquisition and analysis system for real-time data recording of stress data. The specific structure is shown in Figure 6c.

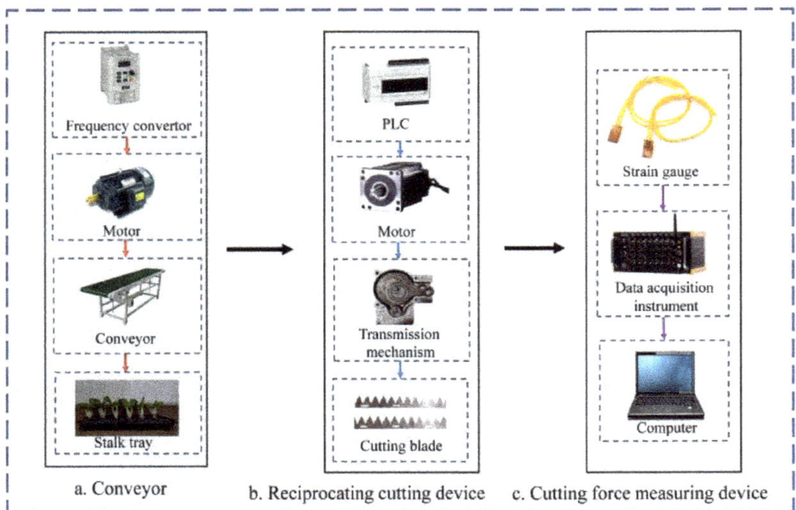

Figure 6. Structural diagram of the cutting force measurement system.

Before the test, stalks with a diameter of 5 mm were selected, and the leaf crown on the top of the parsley was constructed and fixed on the conveying device. By adjusting the distance between the cutting blade and the conveying device by adjusting the fastening bolt between the blade holder and the frame, the appropriate cutting position was determined. In the test, start the test system and motor, and adjust the frequency converter to adjust the control frequency, and keep the stalk-feeding speed at a constant value of 200 mm/s. The structural parameters of the cutting blade were kept unchanged, but the frequency operation of the cutting motor was adjusted by controller to obtain different average cutting speeds. After the various systems of the test bench enter stable operation, the motor of the conveying device was started [34]. The installation and wiring diagram of the resistance strain gauge sensor is shown in Figure 7. The sensor collected four sets of data for the cutting normal stress of the upper and bottom blades in real time, and then the data were transmitted to the PC terminal after being processed by the dynamic data acquisition system. Considering that the maximum cutting stress was an influencing factor that affects cutting power consumption and effect, the maximum cutting stress was taken as the test result. Each test was repeated three times, and the average value of the ultimate cutting stress was taken as the reference value and compared with the results of numerical simulation [24].

(a)

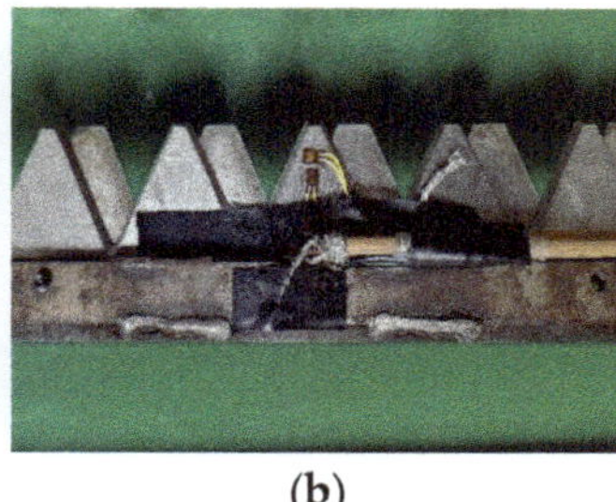
(b)

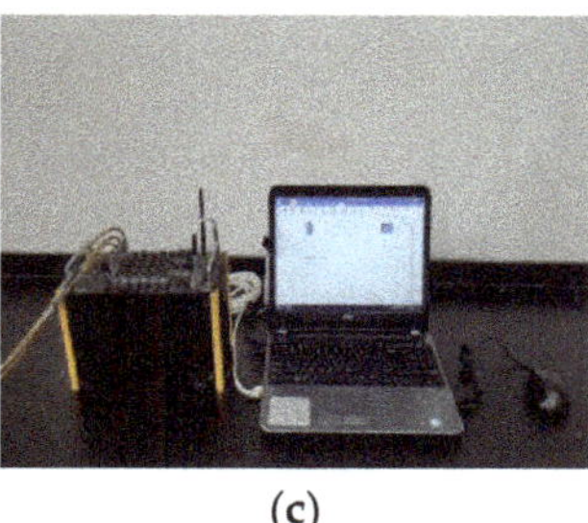
(c)

Figure 7. Schematic diagram of arrangement and wiring of strain gauge: (**a**) installation locations of the strain gauges on the upper blade; (**b**) installation locations of strain gauges on the bottom blade; (**c**) wire connection between the data acquisition instrument and computer.

3. Results

3.1. Post-Processing Results and Analysis of Numerical Simulation

After calculating the numerical simulation, the calculation results underwent post-processing [23]. The equivalent stress distribution cloud diagram was obtained in the cutting process when the sliding–cutting angle was 20°, the oblique angle was 35°, and the average cutting speed was 300 mm/s. Figure 8 reflects the dynamic change in equivalent stress during the cutting process. It can be seen from Figure 8a that when t is 0 ms, the cutting blade and the stalk were out of touch, and the equivalent stress between the cutting blade and the greens was 0 Mpa. Then the cutting blade moved toward the stalk at 300 mm/s, and contacted the stalk at 25.9 ms. Figure 8b (t = 26.8 ms) shows the cloud map of the equivalent stress distribution at the initial cutting stage. The cutting blade compressed the stalk locally to produce significant buckling and plastic deformation, and the fiber tensile stress continued to increase, and the shear strain exceeded the tensile strength of the fibers. Hence, the unit was damaged and failed, the fibers of the stalk broke at the blade edge, and then the cutter gradually cut into the stalk. At this time, both the upper and bottom cutting blades had stress concentration, and the maximum equivalent stresses appeared at the contact point on the stalk, which were 0.31 and 0.68 Mpa, respectively. The sheared part of the stalk showed the maximum equivalent stress, which was 3.93 Mpa, which was consistent with the actual working conditions. As shown in Figure 8c (t = 30.3 ms), at the stage of stalk rupture, when the blade cut into about one-half the diameter of the stalk, and the stem tissues at the tip of the blade were further bent and deformed, which eventually caused the entire fiber layer to slip and break, resulting in shear damage to the entire stem. At this stage, stress concentration occurred throughout the cutting edge. The reason might be that the stalk exerted greater squeezing and friction on the cutter at a deeper position. At this time, the cutting cross section had a tendency to crack, and the surface of the stubble of the cracked part was uneven with poor quality. Therefore, the relevant cutting parameters should be optimized to reduce this phenomenon. Figure 8d reflects the process of separating stalks from the stubbles after the cutting was completed. During this process, the stalks were cut and separated. The stress concentration of the cutting blade and the stalks gradually disappeared, then reaching the minimum. The effect force was significantly reduced, but the cutter still had residual stress.

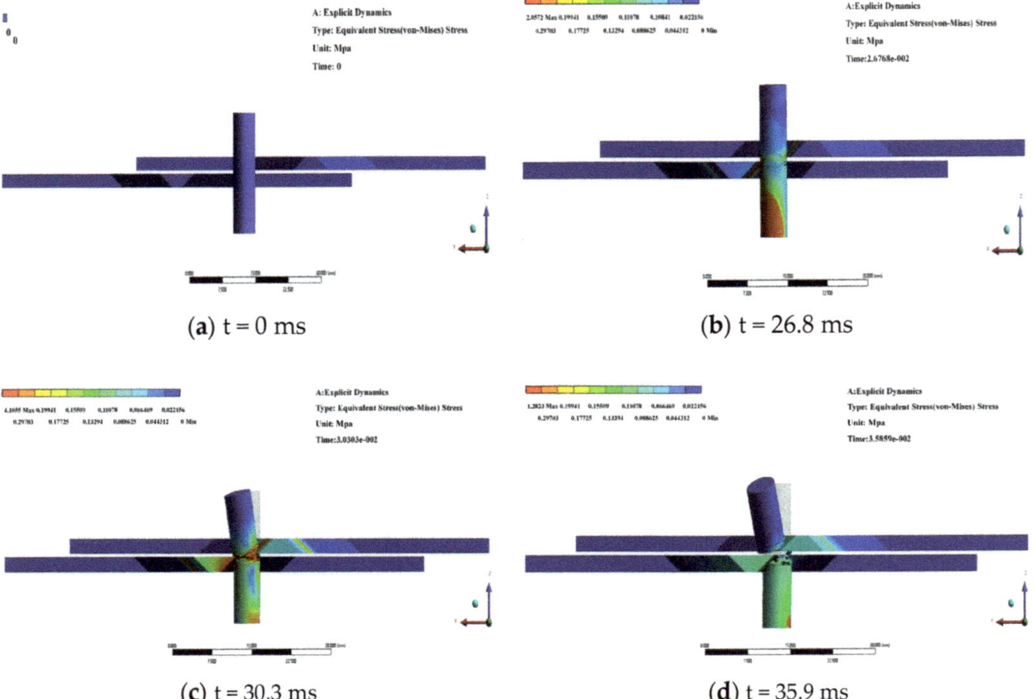

(**a**) t = 0 ms (**b**) t = 26.8 ms

(**c**) t = 30.3 ms (**d**) t = 35.9 ms

Figure 8. The effective stress cloud diagram of cutting unit.

The equivalent stress curves are shown in Figure 9. It can be seen that the cutting blade and the stalk contacted each other at about 26 ms, and the cutting operation was completed at about t = 36 ms. The entire cutting process lasted about 10 ms, during which the cutting equivalent stress of the process continued to change dynamically, which was consistent with the actual working conditions of the shearing process. There was no interaction between the cutter and the stalk after separation; however, since the cutter was still affected by the residual stress, the stress of the cutter at this stage was not zero. When t = 28.1 ms, the maximum equivalent stress of the upper cutting blade was 1.05 Mpa. The maximum equivalent stress of the lower cutting blade occurred at t = 36.6 ms, which was 0.90 Mpa. The maximum equivalent stress of the cutting blade is much lower than 355 Mpa, which is the tool material yield limit. It means that the cutter would not undergo significant plastic deformation.

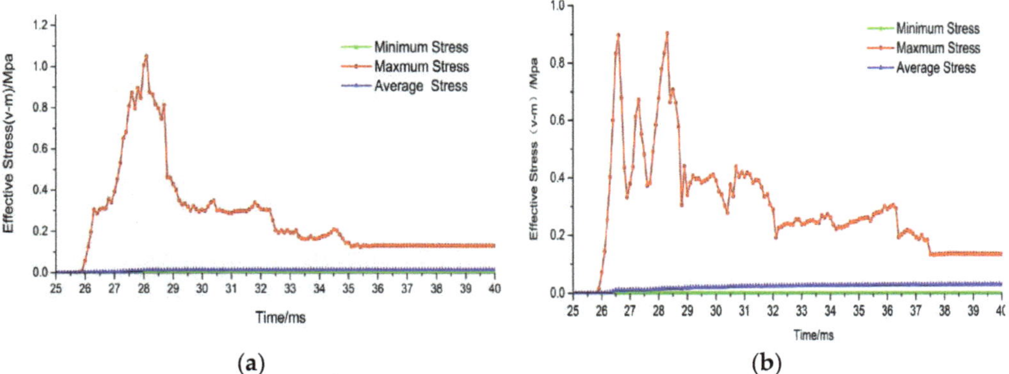

Figure 9. Equivalent stress curves of cutting blade: (**a**) upper cutting blade; (**b**) bottom cutting blade.

Figure 10 shows cloud diagrams for the maximum equivalent stress distribution of the cutting blade during the cutting process. Taking the description of Figure 10a as an example, it can be seen that area A, where the upper cutting blade contacts the stalk, received the greatest reaction stress, and here was the peak value of local stress. In addition, the stress mainly occurred in the edge area of the cutting blade, which showed that the method and position of the strain gauges used in this study were reasonable.

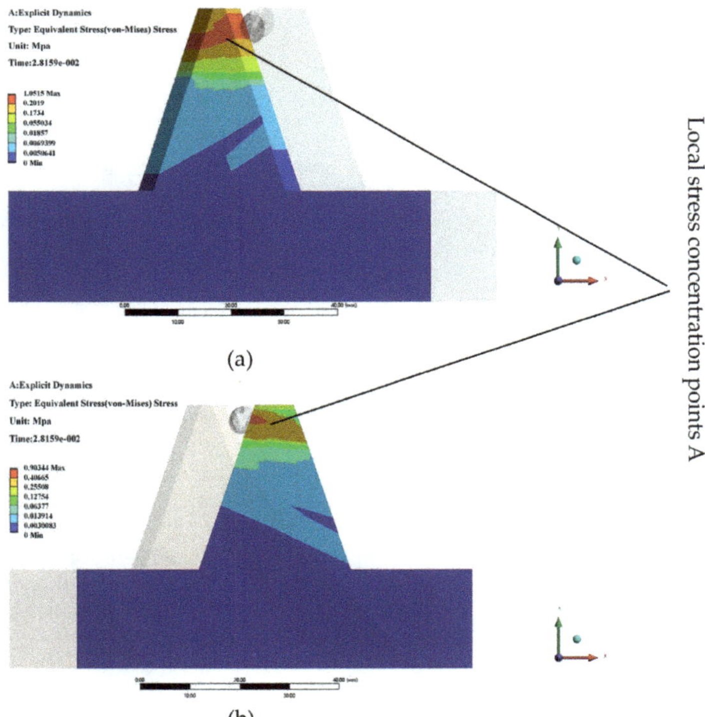

Figure 10. Maximum equivalent stress distribution cloud diagram: (**a**) upper cutting blade; (**b**) bottom cutting blade.

3.2. Orthogonal Test Results and Significance Analysis

According to the test method described in Section 2.2, the numerical simulation of the cutting orthogonal test was carried out using Design-Expert 10.0.7 software. A total of 20 sets of simulation tests were performed; the test results are shown in Table 5.

Table 5. Design and results of the orthogonal test in numerical simulation.

Test Number	Test Factors			Test Indicators	
	X_1 (α) °	X_2 (β) °	X_3 (V) mm/s	Y_1 (σ_1) Mpa	Y_2 (σ_2) Mpa
1	0.000	0.000	0.000	1.06	0.83
2	−1.000	−1.000	1.000	1.11	1.02
3	−1.000	1.000	−1.000	1.58	1.65
4	0.000	0.000	0.000	1.07	0.92
5	1.000	1.000	1.000	1.19	1.10
6	1.000	−1.000	1.000	0.90	0.81
7	0.000	0.000	0.000	1.26	0.72
8	−1.000	1.000	1.000	1.48	1.60
9	1.000	−1.000	−1.000	1.17	1.17
10	0.000	1.682	0.000	1.62	1.74
11	0.000	−1.682	0.000	1.25	1.17
12	−1.682	0.000	0.000	1.34	1.31
13	0.000	0.000	−1.682	1.20	1.10
14	0.000	0.000	0.000	1.10	0.96
15	−1.000	−1.000	−1.000	1.05	0.90
16	1.000	1.000	−1.000	1.24	1.17
17	0.000	0.000	0.000	1.08	0.95
18	0.000	0.000	0.000	1.07	0.92
19	1.682	0.000	0.000	1.12	1.10
20	0.000	0.000	1.682	0.99	0.82

The test results were analyzed by multiple regression fitting using Design-Expert 10.0.7 software, and the regression equation between maximum cutting equivalent stresses and each influencing factor were established. The insignificant terms were then removed. Regression equations are shown in Equations (4) and (5):

$$Y_1 = 1.11 - 0.08X_1 + 0.14X_2 - 0.052X_3 - 0.067X_1X_2 + 0.11X_{22} \quad (4)$$

$$Y_2 = 0.88 - 0.093X_1 + 0.19X_2 - 0.061X_3 - 0.13X_1X_2 + 0.10X_{12} + 0.19X_{22} \quad (5)$$

The variance analysis of the regression equation is shown in Tables 6 and 7. According to the variance analysis of the maximum cutting equivalent stresses Y_1 and Y_2 of the cutter, it can be seen that the significance level p values of the two models were both less than 0.01, indicating that the models were extremely significant. The values for the lack-of-fit items of the regression model were all greater than 0.05, indicating that the regression model had a high fitting accuracy. The coefficients of determination R^2 for the two models were 90.4% and 93.9%, respectively, indicating 90.4% and 93.9% of the total variation in the limit cutting force can be explained by this model, so that both models had high reliability [35–37]. Therefore, the structural and working parameters of the reciprocating cutting system can be optimized and analyzed by models.

It can be seen from the results of the variance analysis that various factors had different effects on the indicators. The p value reflected the degree of influence for each parameter of the regression equation. The smaller p value had more significant effect. The oblique angle had the greatest influence, followed by the sliding–cutting angle and average cutting speed, and these three factors were significant items ($p < 0.05$).

Table 6. Variance analysis results for Y_1.

Source	Sum of Squares	Freedom	Mean Square	*F*-Value	*p*-Value
Model	0.61	9	0.068	10.48	0.0005
X_1	0.087	1	0.087	13.43	0.0044
X_2	0.26	1	0.26	40.04	<0.0001
X_3	0.037	1	0.037	5.75	0.0375
X_1X_2	0.036	1	0.036	5.63	0.0392
X_1X_3	0.0098	1	0.0098	1.51	0.2469
X_2X_3	0.00045	1	0.00045	0.069	0.7975
X_1^2	0.016	1	0.016	2.52	0.1437
X_2^2	0.16	1	0.16	25.05	0.0005
X_3^2	0.0028	1	0.0028	0.44	0.5214
Residual	0.065	10	0.0064		
Lack of Fit	0.036	5	0.0071	1.22	0.4149
Pure Error	0.029	5	0.0058		
Cor. Total	0.68	19			

Table 7. Variance analysis results for Y_2.

Source	Sum of Squares	Freedom	Mean Square	*F*-Value	*p*-Value
Model	1.46	9	0.16	17.18	<0.0001
X_1	0.12	1	0.12	12.56	0.0053
X_2	0.49	1	0.49	51.53	<0.0001
X_3	0.051	1	0.051	5.35	0.0433
X_1X_2	0.14	1	0.14	14.31	0.0036
X_1X_3	0.031	1	0.031	3.31	0.0990
X_2X_3	0.0018	1	0.0018	0.19	0.6718
X_1^2	0.15	1	0.15	16.36	0.0023
X_2^2	0.53	1	0.53	56.20	<0.0001
X_3^2	0.0041	1	0.0041	0.44	0.5227
Residual	0.094	10	0.00093		
Lack of Fit	0.052	5	0.010	1.21	0.4207
Pure Error	0.043	5	0.0086		
Cor. Total	1.56	19			

According to Tables 6 and 7, it can be seen that the interaction terms of factor X_1 and X_2 had a significant impact on the two indicators ($p < 0.05$), and the three-dimensional response surface of the two-factor interaction effect are shown in Figures 11 and 12 [38,39]. The change rate of the test index for the upper cutter and bottom cutter along the factor X_2 direction was faster than that along the factor X_1 direction. At the same time, as the sliding–cutting angle increased, the cutting resistance on the cutter gradually decreased. The reason is that the cutting has a sliding progression during the cutting process. The larger sliding–cutting angle of the cutter can make more tangential slip. When the oblique angle of the cutter was 35~38°, the stress on the cutter gradually decreased, and then when the oblique angle continued to increase, the stress on the cutting knife gradually increased. It can be seen from the contour map that the rate of change in the test index along the factor X_2 direction is faster than for the factor X_1, which means that the oblique angle has a more significant influence on the cutting stress than the sliding angle.

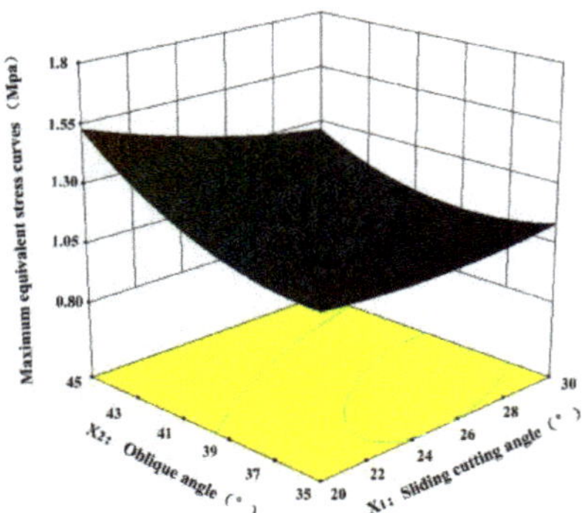

Figure 11. Effects of the sliding–cutting angle and the oblique angle on maximum equivalent stress of the upper cutting blade.

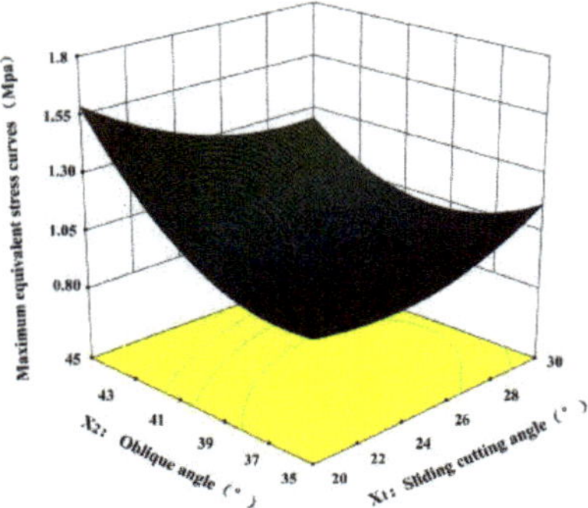

Figure 12. Effects of the sliding–cutting angle and the oblique angle on maximum equivalent stress of the bottom cutting blade.

3.3. Regression Model Optimization

In order to optimize the working parameters of the reciprocating cutting system, the maximum cutting stress of the upper cutter and the bottom cutter were set as the optimal object, and the optimization model was established as follows:

$$\begin{cases} f = minY_1 \\ f = minY_2 \\ \begin{cases} 20^\circ \leq X_1 \leq 30^\circ \\ 35^\circ \leq X_2 \leq 45^\circ \\ 300\ \text{mm/s} \leq X_3 \leq 500\ \text{mm/s} \end{cases} \end{cases} \quad (6)$$

The model was optimized by using Design-Expert 10.0.7 software's data optimization module, Optimization and the optimal cutting parameter combination for the reciprocating cutting system of greens was obtained: sliding–cutting angle was 29°, oblique angle was 38°, and average cutting speed was 500 mm/s. The maximum equivalent stress in cutting of the upper cutter and the bottom cutter is the minimum value, and their predicted values are 0.95 and 0.77 Mpa, respectively.

3.4. Analysis of Cutting Performance Test Results

The root moisture calculated by Equation (3) was 88.7% ± 2.1%, and the difference in moisture content between samples was small, indicating that the root moisture content had little effect on the test results. In order to verify the accuracy of the numerical simulation results and the actual cutting effect, the cutting performance test was carried out according to the equipment and methods described in Section 2.4.

The resistance strain gauge collected normal stress of both blades and obtained four sets of data. Figure 13 shows the maximum normal cutting stress results when the sliding–cutting angle is 29°, the oblique angle is 38°, and the average cutting speed is 300, 400, and 500 mm/s. In three sets of physical tests, the maximum cutting stress in the X direction of each group is less than the maximum cutting stress in the Y direction, which is consistent with the numerical simulation results. As the average cutting speed decreases, the maximum cutting stress shows an increasing trend, and the test index is the minimum when the average cutting speed is 500 mm/s. The physical test verifies the reliability of the test platform. The parameters of the Chinese little greens cutting device were optimized.

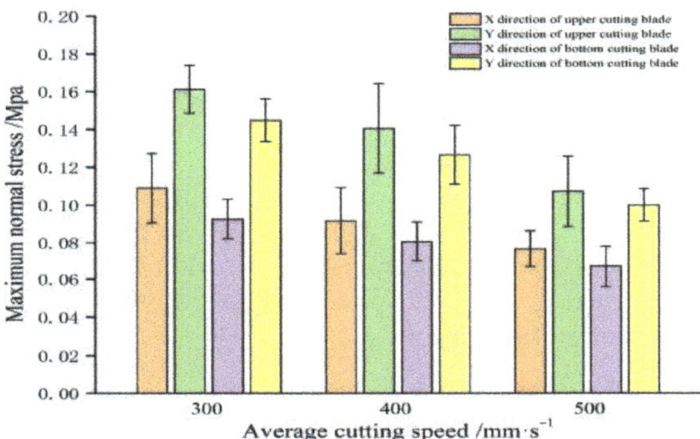

Figure 13. Maximum normal cutting stress at different average cutting speeds used in the physical tests.

4. Discussion

The cutting mechanical properties of the stalks of Chinese little greens were investigated by numerical simulation technology, and then the optimal cutting combination parameters were obtained. Finally, the simulation results were verified to be sensitive and reliable by physical experiments. The results of this study showed that the sliding–cutting angle, oblique angle, and the average cutting speed all had a significant effect on the maximum cutting stress, and the optimal cutting combination parameters were 29°, 38°, and 500 m/s, respectively. Results indicated that dynamic simulation techniques are simple and effective compared to traditional physical testing methods.

We found that the oblique edge angle had the most significant effect on the maximum cutting stress, reaching the optimum value at 38°. When the oblique angle was 35~38°, the cutting stress was proportional to the oblique angle. This is because the squeezing and rubbing effect of the stalk fibers on the cutter become stronger when the blades are gradually cut into the stalk, and the resistance of the cutting blade to destroy the stalk fibers increases, so stress concentration is prone to occur. Furthermore, the cutting stress was inversely proportional to sliding angle. This result is comparable to the results of the study by Cui et al. [6] and Zhang et al. [7], which proved that the stalk of Chinese little greens has similar cutting mechanical properties to that of lettuce and rice. However, a sliding angle that is too large will cause problems such as unstable clamping and cutting failure. The cutting stress is smaller for high-speed cutting progress, and it was in agreement with millet stalk [40]. In addition, the cutting power consumption was proportional to the cutting speed in some studies [15,18]. This result may be due to the inconsistent physical properties of apple tree branches and greens. It also may be because the cutting-stress speed curve is a U-shaped curve. When cutting progresses at low speed, the cutting blades generate impact kinetic energy, and the impact force at the moment of contact with the stalk will cause the fiber structure to rupture and reduce the cutting resistance. When a certain threshold is reached, however, the cutting stress increases with the increase in cutting speed.

However, it is worth noting that the explicit dynamic analysis technology still has certain limitations, such as an inability to realize the simulation of continuous and reciprocating cutting progress, which has a certain error with the actual operating conditions. In addition, mechanical vibration during cutting also affects simulation accuracy. Considering the short harvest period for vegetables and the cultivation mode of facility agriculture, the effect of constitutive parameters among plant materials, such as differences in moisture content on cutting resistance, was also not considered in this study. Further studies would help to optimize cutting parameters according to material constitutive parameters.

5. Conclusions

A simulation model of the cutting system for Chinese little greens was established, and the influence law of the working parameters of the cutting device on cutting characteristics was studied. The optimal cutting parameters were obtained by using the method of combining numerical simulation and orthogonal experiment. The result showed that significant influence on the maximum cutting stress was related, in decreasing order, to the oblique angle, the sliding–cutting angle, and the average cutting speed. The ultimate cutting equivalent stress of the upper cutting blade was 0.95 Mpa, and that of the bottom cutter was 0.77 Mpa, at the condition of optimal parameter combinations. Physical tests of stalk cutting showed that the test index was the smallest under the optimal parameter combination. The optimal parameter combination of the cutter in tests significantly reduces the cutting resistance, which provides a reference for the optimization of the cutting device for the Chinese little greens.

Author Contributions: Conceptualization, Z.Y. and W.W.; methodology, X.L. and W.W.; software, W.W.; formal analysis, W.W.; investigation, Z.Y.; writing—original draft preparation, X.L.; writing—review and editing, W.W.; supervision, W.W.; funding acquisition, X.L. All authors have read and agreed to the published version of the manuscript.

Funding: This research was supported by the Jiangsu Agricultural Science and Technology Innovation Fund (No. CX (19)2025) and the Jiangsu Modern Agricultural Equipment and Technology Demonstration Promotion Fund (No. NJ2020–23).

Institutional Review Board Statement: Not applicable.

Informed Consent Statement: Not applicable.

Data Availability Statement: Not applicable.

Conflicts of Interest: The authors declare no conflict of interest.

References

1. Liu, D.; Xiao, H.R.; Yang, G.; Jin, Y.; Yang, G.; Cao, G.Q.; Zhang, J.F.; Liu, M. Design and experiment of the orderly harvester of Chinese little greens. *Int. Agric. Eng. J.* **2018**, *27*, 295–306.
2. Shah, D.U.; Reynolds, T.P.S.; Ramage, M.H. The strength of plants: Theory and experimental methods to measure the mechanical properties of stems. *J. Exp. Bot.* **2017**, *68*, 4497–4516. [CrossRef] [PubMed]
3. Zhou, D.; Jing, C.; She, J.; She, J.K.; Tong, J.; Chen, Y.X. Temporal dynamics of shearing force of rice stem. *Biomass Bioenergy* **2012**, *47*, 109–114. [CrossRef]
4. Wang, Y.; Yang, Y.; Zhao, H.M.; Liu, B.; Ma, J.T.; He, Y.; Zhang, Y.T.; Xu, H.B. Effects of cutting parameters on cutting of citrus fruit stems. *Biosyst. Eng.* **2020**, *193*, 1–11. [CrossRef]
5. Gan, H.; Mathanker, S.; Momin, M.A.; Kuhns, B.; Stoffel, N.; Hansen, A.; Grift, T. Effects of three cutting blade designs on energy consumption during mowing-conditioning of Miscanthus Giganteus. *Biomass Bioenergy* **2018**, *109*, 166–171. [CrossRef]
6. Cui, Y.J.; Wang, W.Q.; Wang, M.H.; Ma, Y.D.; Fu, L.S. Effects of cutter parameters on shearing stress for lettuce harvesting using a specially developed fixture. *Int. J. Agric. Biol. Eng.* **2021**, *14*, 152–158. [CrossRef]
7. Zhang, C.L.; Chen, L.Q.; Xia, J.F.; Zhang, J.M. Effects of blade sliding cutting angle and stem level on cutting energy of rice stems. *Int. J. Agric. Biol. Eng.* **2019**, *12*, 75–81. [CrossRef]
8. Esgici, R.; Pekitkan, F.G.; Ozdemir, G. Cutting parameters of some grape varieties subject to the diameter and age of canes. *Fresenius Environ. Bull.* **2019**, *28*, 167–170.
9. Du, D.D.; Wang, J.; Qiu, S.S. Analysis and test of splitting failure in the cutting process of cabbage root. *Int. J. Agric. Biol. Eng.* **2015**, *8*, 27–34.

10. Boyda, M.G.; Omakli, M.; Sayinci, B.; Kara, M. Effects of moisture content, internode region, and oblique angle on the mechanical properties of sainfoin stem. *Turk. J. Agric. For.* **2019**, *43*, 254–263. [CrossRef]
11. Clementson, C.L.; Hansen, A.C. Pilot study of manual sugarcane harvesting using biomechanical analysis. *J. Agric. Saf. Health* **2008**, *14*, 309–320. [CrossRef] [PubMed]
12. Vu, V.D.; Ngo, Q.H.; Nguyen, T.T.; Nguyen, H.C.; Nguyen, Q.T.; Nguyen, V.D. Multi-objective optimization of cutting force and cutting power in chopping agricultural residues. *Biosyst. Eng.* **2020**, *191*, 107–115. [CrossRef]
13. Ma, P.B.; Li, L.Q.; Wen, B.Q.; Xue, Y.H.; Kan, Z.; Li, J.B. Design and parameter optimization of spiral-dragon type straw chopping test rig. *Int. J. Agric. Biol. Eng.* **2020**, *13*, 47–56. [CrossRef]
14. Cheng, S.; Zhang, B.; Li, X.W.; Yin, G.D.; Chen, Q.M.; Xia, C.H. Bench cutting tests and analysis for harvesting hemp stalk. *Int. J. Agric. Biol. Eng.* **2017**, *10*, 56–67. [CrossRef]
15. Johnson, P.C.; Clementson, C.L.; Mathanker, S.K.; Grift, T.E.; Hansen, A.C. Cutting energy characteristics of Miscanthus x giganteus stems with varying oblique angle and cutting speed. *Biosyst. Eng.* **2012**, *112*, 42–48. [CrossRef]
16. Mathanker, S.K.; Grift, T.E.; Hansen, A.C. Effect of blade oblique angle and cutting speed on cutting energy for energy cane stems. *Biosyst. Eng.* **2015**, *133*, 64–70. [CrossRef]
17. Allameh, A.; Alirzadeh, M.R. Specific cutting energy variations under different rice stem cultivars and blade parameters. *Idesia* **2016**, *34*, 11–17. [CrossRef]
18. Zhao, J.L.; Huang, D.Y.; Jia, H.L.; Zhuang, J.; Guo, M.Z. Analysis and experiment on cutting performances of high-stubble maize stalks. *Int. J. Agric. Biol. Eng.* **2017**, *10*, 40–52.
19. Stopa, R.; Komarnicki, P.; Kuta, ł.; Szyjewicza, D.; Słupska, M. Modeling with the finite element method the influence of shaped elements of loading components on the surface pressure distribution of carrot roots. *Comput. Electron. Agric.* **2019**, *167*, 105046. [CrossRef]
20. Meng, Y.; Wei, J.D.; Wei, J.; Chen, H.; Cui, Y.S. An ANSYS/LS-DYNA simulation and experimental study of circular saw blade cutting system of mulberry cutting machine. *Comput. Electron. Agric.* **2019**, *157*, 38–48. [CrossRef]
21. Yang, W.; Yang, J.; Liu, Z.H.; Liang, Z.X.; Mo, J.L. Dynamic simulation experiment of one-blade cutting sugarcane process. *Trans. Chin. Soc. Agric. Mach.* **2011**, *27*, 150–156.
22. Huang, H.D.; Wang, Y.X.; Tang, Y.Q.; Zhao, F.; Kong, X.F. Finite element simulation of sugarcane cutting. *Trans. Chin. Soc. Agric. Eng.* **2011**, *27*, 161–166.
23. Qiu, M.M.; Meng, Y.M.; Li, Y.Z.; Shen, X.B. Sugarcane stem cut quality investigated by finite element simulation and experiment. *Biosyst. Eng.* **2021**, *206*, 135–149. [CrossRef]
24. Xie, L.X.; Wang, J.; Cheng, S.M.; Zeng, B.S.; Yang, Z.Z. Optimisation and finite element simulation of the chopping process for chopper sugarcane harvesting. *Biosyst. Eng.* **2018**, *175*, 16–26. [CrossRef]
25. Luo, Y.Q.; Ren, Y.H.; Zhou, Z.X.; Huang, X.M.; Song, T.J. Prediction of single-tooth sawing force based on tooth profile parameters. *Int. J. Adv. Manuf. Technol.* **2016**, *86*, 641–650. [CrossRef]
26. Liu, Q.; Mathanker, S.K.; Zhang, Q.; Hansen, A.C. Biomechanical properties of miscanthus stems. *Trans. Asabe* **2012**, *55*, 1125–1131. [CrossRef]
27. Maughan, J.D.; Mathanker, S.K.; Grift, T.; Hansen, A.C. Impact of Blade Angle on Miscanthus Harvesting Energy Requirement. In Proceedings of the 2013 Kansas City, Kansas City, MI, USA, 21–24 July 2013.
28. Moiceanu, G.; Voicu, P.; Paraschiv, G.; Voicu, G. Behaviour of miscanthus at cutting shear with straight knives with different edge angles. *Environ. Eng. Manag. J.* **2017**, *16*, 1203–1209. [CrossRef]
29. Yang, W.; Zhao, W.J.; Liu, Y.D.; Chen, Y.Q.; Yang, J. Simulation of forces acting on the cutter blade surfaces and root system of sugarcane using FEM and SPH coupled method. *Comput. Electron. Agric.* **2020**, *180*, 105893. [CrossRef]
30. Liu, D.; Xiao, H.R.; Jin, Y.; Cao, G.Q.; Shen, C.; Yang, G.; Zhang, J.F.; Liu, M. Mechanical properties test and analysis on the stalks of Chinese little greens. *Int. Agric. Eng. J.* **2018**, *27*, 38–43.
31. Salarikia, A.; Ashtiani, S.M.; Golzarian, M.R.; Mohammadinezhad, H. Finite element analysis of the dynamic behavior of pear under impact loading. *Inf. Process. Agric.* **2017**, *4*, 64–77. [CrossRef]
32. Ahmadi, E.; Barikloo, H.; Kashfi, M. Viscoelastic finite element analysis of the dynamic behavior of apple under impact loading with regard to its different layers. *Comput. Electron. Agric.* **2016**, *121*, 1–11. [CrossRef]
33. Hasanzadeh, R.; Mojaver, M.; Azdast, T.; Park, C.B. Polyethylene waste gasification syngas analysis and multi-objective optimization using central composite design for simultaneous minimization of required heat and maximization of exergy efficiency. *Energy Convers. Manag.* **2021**, *247*, 114713. [CrossRef]
34. Zhao, J.; Guo, M.Z.; Lu, Y.; Huang, D.Y.; Zhuang, J. Design of bionic locust mouthparts stubble cutting device. *Int. J. Agric. Biol. Eng.* **2020**, *13*, 20–28. [CrossRef]
35. Xie, L.X.; Wang, J.; Cheng, S.M.; Zeng, B.S.; Yang, Z.Z. Optimisation and dynamic simulation of a conveying and top breaking system for whole-stalk sugarcane harvesters. *Biosyst. Eng.* **2020**, *197*, 156–169. [CrossRef]
36. Hu, H.N.; Li, H.W.; Wang, Q.J.; He, J.; Lu, C.Y.; Wang, Y.B.; Wang, C.L. Performance of waterjet on cutting maize stalks: A preliminary investigation. *Int. J. Agric. Biol. Eng.* **2019**, *12*, 64–70. [CrossRef]
37. Wang, J.; Yang, W.; Sun, X.; Li, X.; Tang, H. Design and experiment of spray and rotary tillage combined disinfection machine for soil. *Int. J. Agric. Biol. Eng.* **2019**, *12*, 52–58. [CrossRef]

38. Hu, H.N.; Li, H.W.; Wang, Q.J.; He, J.; Lu, C.Y.; Wang, Y.B.; Liu, P. Anti-blocking performance of ultrahigh-pressure waterjet assisted furrow opener for no-till seeder. *Int. J. Agric. Biol. Eng.* **2020**, *13*, 64–70. [CrossRef]
39. Huang, J.; Tian, K.; Shen, C.; Zhang, B.; Liu, H.L.; Chen, Q.M.; Li, X.W.; Ji, A. Design and parameters optimization for cutting-conveying mechanism of ramie combine harvester. *Int. J. Agric. Biol. Eng.* **2020**, *13*, 94–103. [CrossRef]
40. Zhang, Y.Q.; Cui, Q.L.; Guo, Y.M.; Li, H.B. Experiment and analysis of cutting mechanical properties of millet stem. *Trans. Chin. Soc. Agric. Mach.* **2019**, *50*, 146–155.

Article

Research on Flexible End-Effectors with Humanoid Grasp Function for Small Spherical Fruit Picking

Fu Zhang [1,2,3,*], Zijun Chen [1], Yafei Wang [3], Ruofei Bao [1], Xingguang Chen [1], Sanling Fu [4], Mimi Tian [1] and Yakun Zhang [1]

1 College of Agricultural Equipment Engineering, Henan University of Science and Technology, Luoyang 471003, China
2 Collaborative Innovation Center of Machinery Equipment Advanced Manufacturing of Henan Province, Luoyang 471003, China
3 Key Laboratory of Modern Agricultural Equipment and Technology, Ministry of Education, Jiangsu University, Zhenjiang 212013, China
4 College of Physical Engineering, Henan University of Science and Technology, Luoyang 471023, China
* Correspondence: zhangfu@haust.edu.cn

Abstract: The rapid, stable, and undamaged picking of small-sized spherical fruits are one of the key technologies to improve the level of intelligent picking robots and reduce grading operations. Cherry tomatoes were selected as the research object in this work. Picking strategies of two-stage "Holding-Rotating" and finger-end grasping were determined. The end-effector was designed to separate the fruit from the stalk based on the linear motion of the constraint part and the rotating gripper. This work first studied the human hand-grasping of cherry tomatoes and designed the fingers with sinusoidal characteristics. The mathematical model of a single finger of the gripper was established. The structural parameters of the gripper were determined to meet the requirements of the grabbing range from 0 to 61.6 mm. Based on the simulation model, the constraint part was set to 6 speeds, and the fruit sizes were set to 20 mm, 30 mm, and 40 mm, respectively. When the speed was 0.08m/s, the results showed that the grabbing time was 0.5381 s, 0.387 s, and 0.2761 s, respectively, and the maximum grabbing force was 0.9717 N, 3.5077 N, and 4.0003 N now of clamping, respectively. It met the picking requirements of high speed and low loss. The criterions of two-index stability and undamaged were proposed, including the grasping index of the fixed value and the slip detection of variance to mean ratio. Therefore, the control strategy and algorithm based on two-stage and two-index for rapid, stable, and non-destructive harvesting of small fruit were proposed. The results of the picking experiment for seventy-two cherry tomatoes showed that the picking success rate was 95.82%, the average picking time was 4.86 s, the picking damage rate was 2.90%, the browning rate was 2.90% in 72 h, and the wrinkling rate was 1.49% in 72 h, which can meet the actual small spherical fruit picking requirements. The research will provide an idea for the flexible end-effectors with humanoid grasp function and provides a theoretical reference for small spherical fruit picking.

Keywords: small spherical fruits; end-effector; design; mathematical model; slip detection

Citation: Zhang, F.; Chen, Z.; Wang, Y.; Bao, R.; Chen, X.; Fu, S.; Tian, M.; Zhang, Y. Research on Flexible End-Effectors with Humanoid Grasp Function for Small Spherical Fruit Picking. *Agriculture* **2023**, *13*, 123. https://doi.org/10.3390/agriculture13010123

Academic Editors: Cheng Shen, Zhong Tang and Maohua Xiao

Received: 29 November 2022
Revised: 19 December 2022
Accepted: 20 December 2022
Published: 2 January 2023

1. Introduction

Due to the rich nutrients and unique flavor, the demand for cherry tomatoes is increasing [1–3]. However, harvesting is the most time-consuming and laborious part of its production. Therefore, realizing mechanization and automation is a significant development direction of fruit harvesting robots. However, the growing environment of cherry tomatoes is complex, and the skin is fragile. The quality of harvest directly affects the economic benefits. Its delicate and fragile biological characteristics make mechanized undamaged harvesting face great challenges [4–7].

In recent years, scholars had developed different fruit-picking end-effectors [8,9]. Xiong et al. [10] designed a strawberry end-effector with cutting function, and the success rate of picking reached 97.1%. Tested and compared three traditional methods of collecting mushrooms by hand, Huang et al. [11] developed a vacuum end-effector, and the results showed that the picking rate of the target mushroom reached 100% by the bending picking method. To reduce the scrapes of picking, Miao et al. [12] developed an apple end-effector with a compliant mechanism whose success rate is about 95.3%. It ensures the constant output force required for low destructive fruit clamping. Roshanianfard et al. [13] created a five-fingered pumpkin manipulator with a grabbing radius of 76.2-265 mm, which was electrically driven and had an internal shock-grabbing mode. In addition, the end-effectors of litchi [14], kiwi [15], chrysanthemum [16], blueberry [17], green pepper [18], cucumber [19], and other fruits and vegetables were designed and studied.

However, fruits and vegetables have different biological characteristics, sizes, and growth conditions. The design of specific end-effectors has gradually become one of the essential research directions based on their factors. According to the principle of bionics, Wang et al. [20] made the citrus picking end actuator with occlusion mode by imitating the snake head. At 46° as the optimum harvest posture, the harvest rate reached 74%. By studying the grasping behavior of human hands during tomato picking, Wang et al. [21] developed a new 16-channel data acquisition electronic glove, which showed the maximum pressure contribution from the distal regions of the thumb, middle finger, and index finger by picking each fruit. Gao et al. [22] developed a pneumatic-controlled end-effector. It could pick cherry tomatoes continuously and steadily. The average time of picking single cherry tomato was 6.4 s, and the highest success rate was 84%. In summary, the design of an end-effector according to the characteristics of fruits and vegetables can improve the success rate of picking and reduce damage [23–26].

In this work, fingers with sinusoidal contours were designed, and the picking strategy of the two-stage "Holding-Rotating" picking strategy of finger-end grasping was determined. The end-effector was designed to separate the fruit from the stalk based on the linear motion of the constraint part and the rotating gripper. The clamping regulating system based on clamping pressure feedback was built to realize undamaged picking. The double threshold picking index was proposed, which included the grasping index of the fixed value and the slip detection of variance to mean ratio. The performance of the end effector was verified by cherry tomatoes picking test.

2. Design of End-Effector System

2.1. Design of the Gripper Inspired by Human Hand Grasping

This work compared the grasping experiments of different size spherical fruits by hand. When picking larger fruit, such as apples and oranges, each finger joint rotated to make the palm close to the fruit, realizing the enveloping grasping. In contrast, when picking small fruit, such as cherry tomatoes and plums, only the ends of thumb, index finger, and middle finger contact with the fruit. The rotation angle of the metacarpophalangeal joint was the largest. The distal and proximal phalangeal joints turned slightly. Fingers formed a more fixed arc shape in the grasping process. The comparison of fruit grasping in different volumes shows in Figure 1.

Figure 1. Comparison of fruit grasping in different sizes: (**a**) grasping of larger fruit; (**b**) grasping of small fruit.

Therefore, inspired by the dynamic harvesting of small fruits by human hands, the gripper was designed, which has three features, including sinusoidal contour, single-joint rotation, and finger-end grasping. The equation of the sinusoidal curve refers to the outer shape, and its shape is shown in Figure 2a. The equation of the sinusoidal curve of the single finger is Equation (1):

$$y = 12\sin\left[\frac{1}{30}(x - 20\pi)\right] + 6\sqrt{3},\ x \in [0, 45\pi] \quad (1)$$

Figure 2. The design of gripper: (**a**) opening and closing of gripper; (**b**) the constraint part.

The design of gripper is shown in Figure 2. Three fingers were in each of the three holes of the constraint part. The fingers were fixed to the chassis by the "L-shaped" steel parts. The steel parts can produce elastic deformation. The fingers close when the constraint part advances. When the constraint part retreats, the fingers gradually recover to their initial positions. Due to the sine curve is smooth, continuous, and differentiable, the constraint part moves smoothly in contact with the fingers. On the one hand, it is intended to transform the finger opening and closing into the constraint behavior, which realizes the stable movement and precise control of the fingers. On the other hand, the problems of grasping caused by rotational speed, torque, and precision of joint motor are avoided, to a certain extent.

2.2. Design and Working Principle of End-Effector

The end-effector designed in this paper was used to harvest small spherical fruits. The picking strategy of "Holding-Rotation" was used to realize picking. It was mounted on a picking robot based on binocular stereo vision. The picking robot guided the end-effector to aim at the fruit employing binocular vision. The gripper closed, grabbed the fruit, then rotated with the fruit to separate the fruit from the stalk (Figure 3).

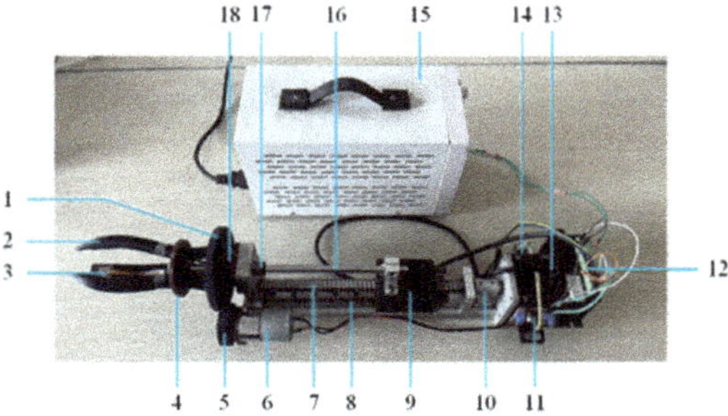

Figure 3. The design of end-effector for small spherical fruit picking. Note: 1, Chassis; 2, Finger; 3, Pressure sensors; 4, Constraint Part; 5, Gear; 6, DC Motor; 7, Screw; 8, Slide track; 9, Slider; 10, Coupling; 11, DC motor drivers; 12, Stepper motor driver; 13, Stepper motor; 14, Arduino control panel; 15, Power; 16, Pole; 17, Linear bearings; 18, Change the channel.

During the holding stage, the gripper opens and closes to achieve fruit gripping. Its essence is to restrain the movement of fingers by linear movement of constraint part. Therefore, the end-effector rotates the screw through a stepper motor, driving the pole and the slider in reciprocating linear motion along the Slide track. Rotating stage. The DC motor drives the gear to turn after grasping the fruit. Then, the gripper and the fruit are rotated synchronously, separating the fruit from the stalk. The pressure sensors are FSR-402 sensors, which can obtain the gripping force in real time, and the measuring range is 0.1~100 N. The stepper motor is two-phase four-wire 42-stepper motor (42BYGH24). The torque is 0.13 N, and the step angle is 1.8°. The stepper motor drive model is DM542C. The DC motor is the planetary DC decelerating motor with the working voltage of 24 V and a rotating speed of 300 r/min.

The picking time was taken as a technical index to realize the quick picking in this work. It was set to 1.5 s, which means from the beginning of the clamping to the separation of the stalks, including 1 s for the clamping phase and 0.5 s for the rotation phase.

3. Simulation and Analysis of End-Effector Motion

3.1. Establishment of the Single-Finger Kinematics Model and Setting of Key Component Parameters

According to the size statistics of cherry tomatoes planted in the standardized greenhouse, their diameters are about 20~40 mm. The kinematics model of the gripper was established first to meet the size requirements, then the parameters of relative parts were determined. Due to the gripper consists of three symmetrical fingers, the single finger was modeled and analyzed. Figure 4 shows the schematic representation of the structural parameters of gripper's the single finger. The rectangular coordinate systems *xoy* and *XOY* were created. The center of the chassis was set to the origin of *XOY*. The connection between the fingers and the chassis were set to the origin of *XOY*. The axes were established in the direction of pole motion and perpendicular to it, respectively.

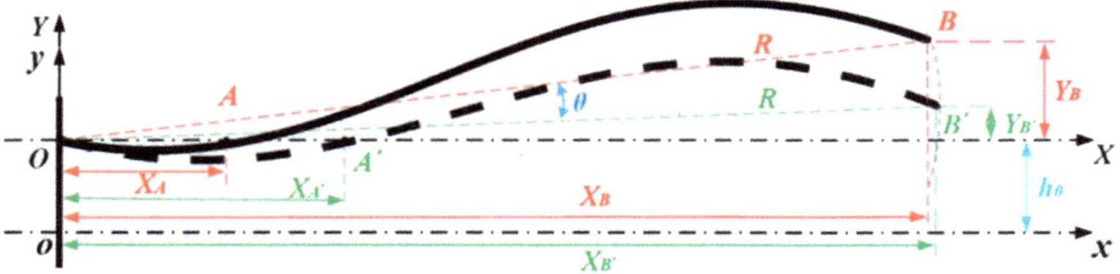

Figure 4. Schematic diagram of the motion structure parameters of gripper's the single finger. Note: Curve *OB* is the initial position of the finger; *A* and *A′* represent the contact point between the constraint part and the finger before and after polymerization, respectively; *B* and *B′* represent the position of the finger end before and after polymerization, respectively; Arc *BB′* is the motion trajectory of the finger end; θ is the rotation angle of the finger; h_0 is the distance from the finger bottom to the center of the chassis; *R* is the distance from the finger bottom to end.

According to the grasping characteristics, the opening and closing of the fingers could be regarded as the rotation around *O*. *BB′* was the circular arc with *O* as the center and *R* as the radius. *Q* and *Q′* were set to random points on the *OB* and *OB′*, respectively. The homogeneous coordinates of *Q* in the *xoy* coordinate system is:

$$Q_{xoy} = \begin{bmatrix} x \\ y \\ 1 \end{bmatrix} = T \times R \times Q_{XOY} \tag{2}$$

where $Q_{XOY} = \begin{bmatrix} X \\ Y \\ 1 \end{bmatrix}$ is the homogeneous coordinate of Q in XOY; $R = \begin{bmatrix} \cos\theta & \sin\theta & 0 \\ -\sin\theta & \cos\theta & 0 \\ 0 & 0 & 1 \end{bmatrix}$ is the rotation matrix; $T = \begin{bmatrix} 1 & 0 & 0 \\ 0 & 1 & h_0 \\ 0 & 0 & 1 \end{bmatrix}$ is the translation matrix of the position of the XOY relative to XOY. After rotating θ clockwise, the equation for OB' under XOY is:

$$X\sin\theta + Y\cos\theta = 12\sin\left[\frac{1}{30}(X\cos\theta - Y\sin\theta - 20\pi)\right] + 6\sqrt{3},\ \theta \in [0, 45\pi\cos\theta] \quad (3)$$

The coordinate of the Q' on the OB' in XOY is:

$$Q'_{xoy} = \begin{bmatrix} x_{Q'} \\ y_{Q'} \end{bmatrix} = \begin{bmatrix} X\cos\theta + Y\sin\theta \\ h - X\sin\theta + Y\cos\theta \end{bmatrix} \quad (4)$$

When the movement of the constraint part caused the finger to rotate θ, X_A was always on the x-axis and $y = 0$. The movement distance of constraint part X_A was set to L. From Equation (3), the relationship between θ and L can be expressed as a complex nonlinear equation, which in MATLAB can be described by the solve function as:

$$f(\theta, m) = L\sin\theta - 12\sin\left[\frac{1}{30}(L\cos\theta - 20\pi)\right] - 6\sqrt{3},\ \theta \in [0,\ 45\pi\cos\theta] \quad (5)$$

The n, the open range of the finger end, can be expressed as:

$$\begin{cases} n = 2y_{B'} \\ y_{B'} = h - X_B\sin\theta + Y_B\cos\theta \end{cases} \quad (6)$$

Figure 5 was inferred from Equations (5) and (6). With the increase in L, θ and n changed synchronously, and showed a negative correlation. As the constrain part advanced, the finger's movement could be divided into three stages. Stage one was the constraint part did not touch the finger, and θ and n did not change. Stage two was a gripper closing stage. As the constraint part advanced to the L_C, it touched the finger and caused it to start closing. Then, L increased, θ increased, and n decreased. When the constraint reaches L_D, θ and n reached the maximum and minimum, respectively. Stage three was gripper opening stage. As the constraint part pass through L_D, θ decreased and n increased.

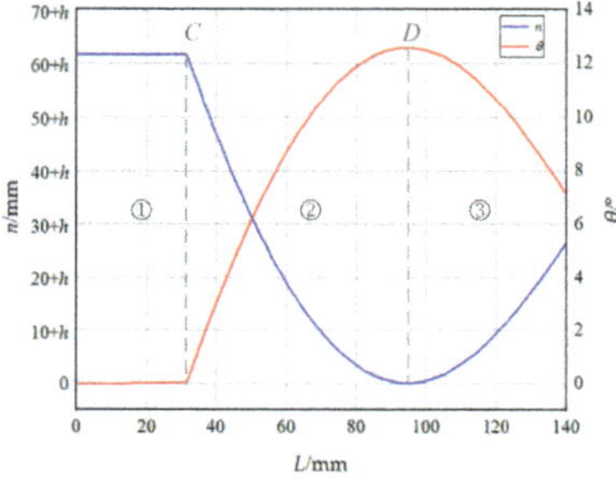

Figure 5. The relationship between L and θ and n. Note: C and D are the two points through which the constraint part moves from chassis to finger ends. ①, ② and ③ are three stages divided by the changes in n and θ.

Therefore, stage one can be used as the stopping region of constraint the part in the natural state. The constraint part does not encounter fingers, so that the parts can be well protected. In stage two, the gripper end had the maximum and minimum gripping range.

This stage can be used as the movement region of the constraint part in the grasping process. Stage three cannot be used as the region of movement of the constraint part.

Therefore, stage two was discussed to adapt the cherry tomato sizes. From Equation (1), when $y = 0$, $L_C = 31.4$ mm, $\theta_{\min}= 0°$, and $n_{max}= 2h+20.78$ mm. Equation (5) was calculated by the chain rule as:

$$z = \frac{\mathrm{d}\theta}{\mathrm{d}L} = -\frac{F'_L}{F'_\theta} \tag{7}$$

Let $z = 0$, and L can be calculated as:

$$L = \frac{30}{\cos\theta}\left(\frac{2\pi}{3} + \arccos\frac{5\sin\theta}{2\cos\theta}\right) \tag{8}$$

When the above formula was associated with $f(\theta, m)= 0$, $L_D = 94.64$ mm, $\theta_{max} = 12.45°$, and $n_{\min}= 2h-40.64$ mm. Thus, the range of n was $2h-40.64$ mm to $2h+20.78$ mm. The open range of the finger end was only related to h. Therefore, when h was determined to be 20.4 mm, the open range of the gripper end was about 0~61.6 mm. It can meet the sizes requirement of cherry tomato picking.

According to the above analysis, the length of a single finger was 140 mm. As the fingers were printed in 3D, the thicknesses were set to 5 mm, and the widths were set to 10mm for grasping reliability. The length of the pole should be greater than 94.64 mm because the active regions of the constraint part were stage one and two. Therefore, the length of the pole was set to 120 mm because it was to be fixed to the slider. The length of screw was set to 150 mm for security during system operation. The pitch was 10 mm. The interior radius of the constraint part was set to 21 mm to ensure that the finger reaches the above grabbing range, which is slightly larger than h. Outer radius was set to 25 mm. The Radius of the Chassis was set to 50 mm to mesh with the gears of the DC motor. The width of steel shrapnel is the same as the width of the finger. The parameters of key parts of the end-effector were determined according to h. (Table 1).

Table 1. The table of the key parts parameters.

Parts	Parameters	Value/mm
Pole	Length	120
Fingers	Length	140
	Thickness	5
	Width	10
Chassis	Radius	50
Constraint part	Interior radius	21
	Outer radius	25
Screw	Length	150
	Pitch	10
Steel shrapnel	Width	10

3.2. *Design and Analysis of Finger End*

The stability of fruit grasping is one of the key techniques for fast and undamaged fruit picking. Additionally, the finger end is the only part in contact with the fruit. It is a crucial factor in achieving a stable grasp.

3.2.1. Design of the Finger End

The human finger pulp has the biological characteristics of a thicker cuticle and subcutaneous fat pad. In grasping small fruits, the last finger pulps gradually form the concave geometric surface that are fitted to the fruit surface. Hand-picking realizes the grasping characteristics of large contact areas, strong envelopment, and flexible contact. Therefore, the finger ends were designed as the geometric surface close to the spherical surface of the cherry tomato. The parameters of the geometric surface are shown in

Equation (7) (unit: mm). The model is shown in Figure 6. Its surface was covered with the 3 mm soft material called Dragon Skin 10 for flexible contact.

$$\begin{cases} x^2+y^2+z^2=40 \\ -5 \le x \le 5 \\ y>0 \\ -10 \le z \le 10 \end{cases} \tag{9}$$

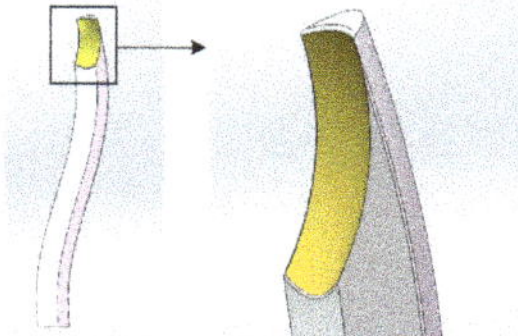

Figure 6. The design of finger end surface. Note: the yellow area is the surface.

3.2.2. Feasibility Analysis of Grasping

The gripper is rotated synchronously with the fruit to separate the fruits from the stalk. During this period, the fingers contact the target fruit, and the two objects do not allow to slide against each other to achieve stable grasping. Thus, the whole system [27,28] is in equilibrium. The following relationships of mechanical and dynamic need to be satisfied:

$$\begin{cases} Gf = \omega \\ J^T f = \tau \end{cases} \tag{10}$$

$$\begin{cases} G^T u = \dot{x} \\ J\dot{q} = \dot{x} \end{cases} \tag{11}$$

where ω is the external force vector; τ is the joint torque vector; f is the total contact vector acting on the fruit; u is the velocity vector; $\dot{q}$ is the joint velocity vector; $\dot{x}$ is the contact velocity vector; J is the gripper Jacobian matrix and determinant; and G is the gripping matrix of the gripper.

Therefore, if the object is required to be constrained entirely, then the relative acceleration vector in the contact force space must be zero. The finger ends and the target fruit are relatively static. For a given τ, the above-mentioned feasible velocity must be zero in all directions of binding force action. For this reason, for a given system state and external force, the amplitude λ of normal contact force should be greater than 0 ($\lambda_C > 0$) to satisfy the linear complementary condition of contact constraint. λ_C includes normal contact force F_N and friction force F_f. It must satisfy the constraint condition $F_f \le \mu F_N$. (Figure 7).

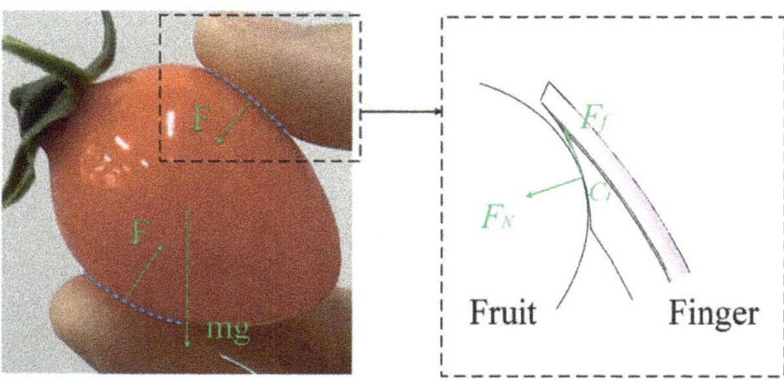

Figure 7. The feasibility analysis of gripper gripping. Note: arrows and dashed lines say finger designs are inspired by hand grips.

So, at the contact point C_i, the contact force $F_{C_i} = \left[F_{C_i,n} \quad F_{C_i,t}\right]^T$ satisfies the constraint condition:

$$F_{C_i,t} = \sqrt{\lambda_{i,t_1}^2 + \lambda_{i,t_2}^2} \le \mu F_{C_i,n} \tag{12}$$

where λ_{i,t_1} and λ_{i,t_2} are special solution vectors of λ_C. They correspond to the magnitude and curl of the friction at C_i. λ_i is the amplitude component of the normal contact force at C_i. Let $f_i \times \lambda_i = F_{C_i,n}$, and the constant factor $f_i \ge \sqrt{\lambda_{i,t_1}^2 + \lambda_{i,t_2}^2}/\mu\lambda_i$. For $\lambda = \lambda_C + a\lambda_0$, all normal contact force amplitudes $\lambda_i > 0$ were guaranteed. The dynamic model of grasping constraint can be satisfied by the feasible solution of contact force. a is a scalar quantity.

3.3. Simulation and Analysis of the Gripper

ADMAS 2018 software was used to carry out an analysis of motion simulation, exploring the movement characteristics of the gripper. It would provide the theoretical basis for realizing fast and undamaged control algorithms. The end-effector clamping principle is that constrain part controls the opening and closing of three fingers. In the simulation, only the relevant parts of the gripper were studied. Cherry tomatoes are non-standard spheres. So, the ball was used instead of actual fruit in the simulation model. The physical properties of cherry tomatoes were taken as the parameters of the ball. The fruit-holding process was idealized in simulation (Figure 8).

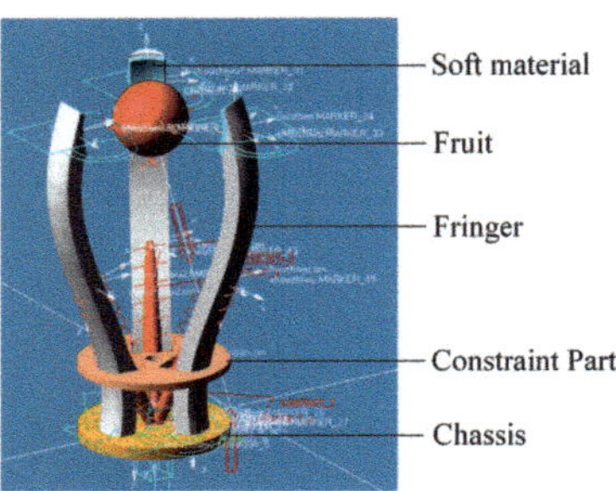

Figure 8. The Adams simulation model of gripper.

First, the ball was deactivated to verify the rationality of the gripper design. The constraint part moved from the chassis to the end of the mechanical finger. Figure 9 shows how the open range of the finger ends varies with the distance of the constrain part. The open range of the finger ends were from 72.8 mm to 9.8 mm, respectively, which met the actual grasping demand of cherry tomato sizes.

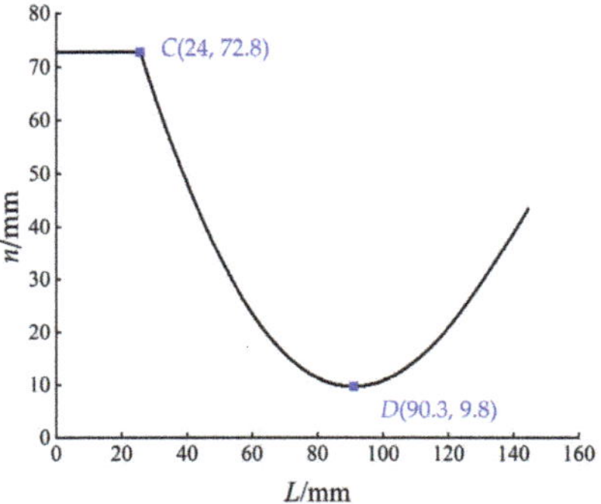

Figure 9. The simulation result of distance of finger end. Note: C is where the constraint part touches the finger. D is the position at which the finger ends converge to the minimum.

The ball diameters were set to three sizes, 20 mm, 30 mm, and 40 mm, respectively, to explore the motion parameters of the constraint part. The balls were activated in the simulation. The distance was 24 mm from the constraint part to the chassis, and it was set to the beginning position. The constraint part and fingers did not touch each other under this parameter. The simulation results are shown in Table 2. When the finger ends connected the ball, the distances between the constraint part and the chassis were 66.9 mm, 54.9 mm, and 46.2 mm, respectively. The moving distances of constraint part were 22.2 mm, 30.9 mm, and 42.9 mm, respectively. It is shown that the movement distance of the constraint part was related to the diameter of the ball when gripping. The larger the fruit sizes, the smaller the movement distance. There was a negative correlation between fruit sizes and the movement distances of the constraint part.

Table 2. Parameters of constraint part movement on holding cherry tomatoes.

r/mm	L_1/mm	L_2/mm	ΔL/mm
20		66.9	42.9
30	24	54.9	30.9
40		46.2	22.2

Note: r is the fruit size; L_1 is the beginning distance of the constraint part from the chassis; L_2 is the distance from the constraint part to the chassis when the finger ends touch the fruit; ΔL is the moving distance.

The target required a holding time of 1s. In this paper, the maximum moving distance ΔL was set to 50 mm, and 50 mm/s was the minimum velocity of the constraint part. Based on this, for the convenience of calculation and electrical control, this section discussed the holding force of the finger ends on the three fruit sizes at six speeds. The speeds were 60 mm/s, 70 mm/s, 80 mm/s, 90 mm/s, and 100 mm/s, respectively.

From the mechanical structure of the gripper, it is easy to know that the gripping force of the finger ends on the fruit increases continuously from the touching to the gradual clamping process. Therefore, this work only needs to explore the maximum gripping force amplitude of contact moment between the fingers and the fruit. The simulation step of contact moment was adjusted to 10,000. The definition of the maximum gripping force amplitude of contact moment is:

$$F_s = max[F_{\Delta t}] \tag{13}$$

where Δt is the time of contact moment, Δt = 0.001 s; $F_{\Delta t}$ is the gripping force of contact moment, N; $[F_{\Delta t}]$ is the gripping force set of contact moment; $max[F_{\Delta t}]$ is the maximum gripping force amplitude of contact moment. The simulation results are shown in Figure 10 and Table 3.

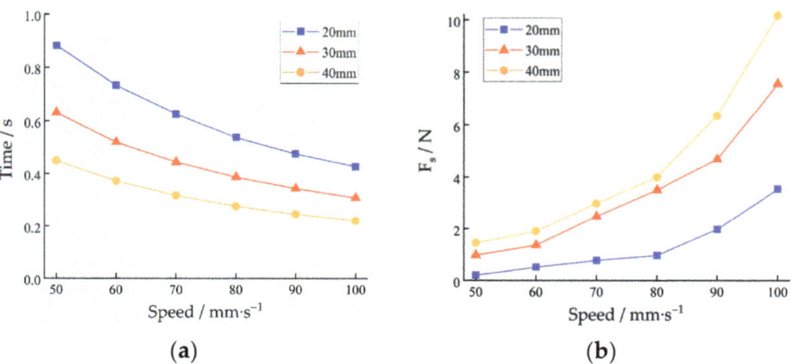

Figure 10. The simulation results: (**a**) the relationship between speed and time; (**b**) the relationship between speed and F_s

Table 3. Results of motivation simulation.

Speed /mm·s^{-1}	Time/s			F_s/n		
	20 mm	30 mm	40 mm	20 mm	30 mm	40 mm
50	0.8826	0.6314	0.4490	0.2074	0.9792	1.4509
60	0.7320	0.5190	0.3724	0.5230	1.3712	1.9064
70	0.6256	0.4435	0.3169	0.7816	2.4866	2.9877
80	0.5381	0.3872	0.2761	0.9717	3.5077	4.0003
90	0.4755	0.3432	0.2449	1.9838	4.6872	6.3619
100	0.4271	0.3079	0.2200	3.5447	7.5650	10.1627

At the same speed, the holding time increased with the decrease in fruit size. Due to gripping small-size fruit, the moving distance of constraint part was long. Under the same fruit size, the larger the speed of constraint part was, the shorter the holding time was. The holding time of three kinds of cherry tomatoes was less than 1 s, which reached the expected target. At the same speed, F_s increased with fruit sizes. The minimum values of F_s were 0.2074 N, 0.9792 N, and 1.4509 N at the speed of 50 mm/s, respectively. At the same fruit size, F_s increased with the speed of the constraint part. When the constraint part speed was 0.1 m/s, F_s reached 10.1627 N [22,29] with 40 mm, which produced a risk of damaging the cherry tomatoes.

The indicators of time and gripping force should be considered to ensure a fast and undamaged grab. When the speed of the restrained part is 0.08 m/s, the comprehensive simulation results show that the clamping time were 0.5381 s, 0.3872 s, and 0.2761 s, respectively, and the maximum gripping force were 0.9717 N, 3.5077 N, and 4.0003 N, respectively, for three fruit sizes. This speed meets the requirement of fast and undamaged when picking. Therefore, 0.08 m/s can be used to the actual speed of the constrain part, then the program and algorithm can be designed.

4. Control Method of Undamaged Picking

Due to the complex growing environment, size, and different postures, the values of piking force cannot be determined in the actual picking process. The movement distance and speed of the constrain part cannot be set as the only condition for picking control. If the clamping force is too small, the finger and the fruit will slide [30–34], and the picking will fail. Too much clamping force will damage the fruit. Thus, the feedback signals should be added. It is the best choice to make the manipulator and fruit synchronous rotation with the smallest clamping force. It is a critical state where no slip just happens. This strategy can be used as the control target of undamaged picking.

Therefore, the experiment designed the two-stage "Holding-Rotating" picking strategy. The clamping stage is the key to realizing undamaged picking, and the clamping force directly affects the degree of fruit damage. The rotation stage is another key to achieving the separation of the fruit handle, and it is the stage where the fruit slides relative to the manipulator. It is an effective way to design the undamaged picking algorithm to explore the minimum clamping force and slip criterion through the clamping experiment.

4.1. Experiment and Analysis of Clamping Fruit

The pressure on the sensor's surfaces changes as the gripper grasps and rotates the fruits. Its output signals will also vary according to working principle. Therefore, this experiment mainly collected the corresponding time signal changes of sensor outputs to explore the undamaged grab strategy in the picking process. The precise mathematical relationship between the grasping force and the output of the sensor's electrical signal had not been established.

Figure 11 is the experiment of grasping and slipping. Fixing cherry tomato, constrain part at 0.08m/s advanced to close fingers and clamped fruit. Then, the gripper was rotated through the DC motor, producing a slide between it and the fruit. Three pressure sensors were calibrated first, and their output values were collected in real-time. The sampling frequency of sensors was set to 0.2 kHz. The aim was to explore the law of grasping and find the basis for relative slip. The Dragon Skin 10 with 3 mm was covered between the finger ends and the sensors.

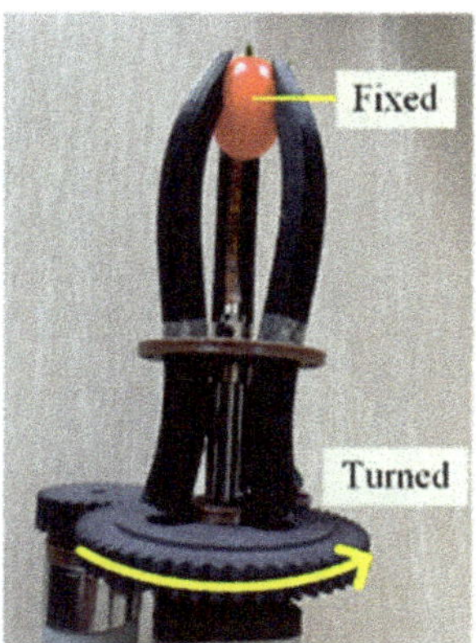

Figure 11. Experiment of grabbing fruit.

Figure 12 shows the output values of sensors during the experiment. The changes in sensor outputs had the following five stages according to the results. (I) The fingers closed with the constrain part motivation. However, the finger ends did not touch the fruit. The output values were zero. (II) The constraint part moved, and the ends grasped the fruit. The output values increased with the increase in the clamping force. (III) When the constraint part stopped moving, the output values were maintained at a relatively stable value. (IV) The slip happened when the gripper was rotated. The output values fluctuated greatly. (V) When the gripper stopped, the slip disappeared. The output values were maintained at a relatively stable value.

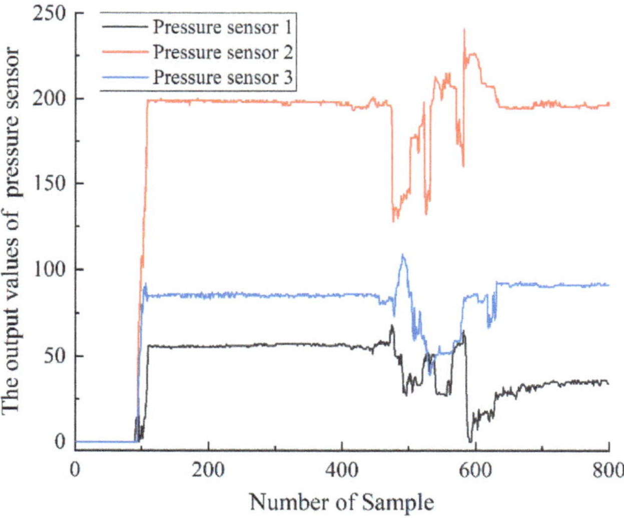

Figure 12. The output values of three pressure sensors.

The output values of the three sensors were inconsistent according to the above results. The reason is the asymmetry on the cherry tomato surface. In addition, fingers clamped the fruit in different positions. However, the trends of output values were the same during the picking process. Obviously, in combination with the above picking strategy, it was a key that minimum gripping force was set during III, and it was another key that slip was judged during IV. The other three stages need not be judged. Therefore, the above two keys were studied to be critical for achieving undamaged and stable picking based on the sensor output values.

The greater the pressure on the sensor surface, the smaller its resistance, and the greater its output values. Therefore, the output values were set during III. In addition, the output values of the three sensors were different under stable clamping. Therefore, when the output of any sensor is greater than the minimum clamping force, it is judged to have reached the minimum clamping force condition. During IV, the values of the three sensors did not have the specific change rule, which cannot judge slip directly. Still, their fluctuation was obvious, and the numerical discreteness degree became higher. Therefore, this work introduced five statistical statistics to find the judgment basis of slip. They were adjacent difference (*AD*), average (*A*), average deviation (*AD*), variance (*DX*), and standard deviation (*SD*), respectively. They are shown in Equations (14)–(18).

$$AD = x(i+1) - x(i) \tag{14}$$

$$A = \frac{\sum_{i=1}^{n} x(i)}{i} \tag{15}$$

$$AD = \frac{\sum_{i=1}^{n} |x(i) - A|}{i} \tag{16}$$

$$DX = \frac{\sum_{i=1}^{n} (x(i) - A)^2}{i} \tag{17}$$

$$SD = \sqrt{DX} \tag{18}$$

where *i* represents the sampling frequency; *x(i)* represents the output value; *n* represents the total amount of data collected since III. The sensor output values were analyzed according to the above formulas in stages III and IV. The results are shown in Figure 13.

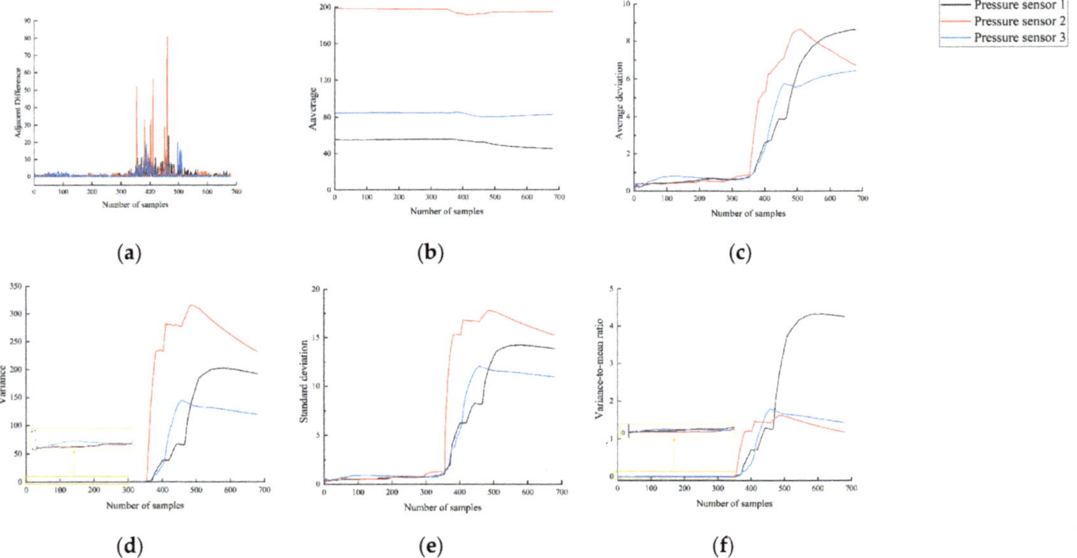

Figure 13. Statistical analysis of pressure sensor output: (**a**) Adjacent difference; (**b**) Average; (**c**) Average deviation; (**d**) Variance; (**e**) Standard deviation; (**f**) Variance-to-mean ratio.

It can be seen from Figure 13a that the difference in adjacent data fluctuates greatly. When a little external interference is encountered in the picking, the adjacent data difference will increase, resulting in wrong judgments. Compared with the smooth trend of the average in Figure 13b, the average deviation, the variance, and the standard deviation had upward inflection points, as shown in Figure 13c–e. However, these three statistics occurred in small fluctuations before the 30th sampling. The reason is that the original output values of the sensor fluctuated instantly during the III to IV. The trend of these three statistics was first up, then down in the 30th–210th sampling of sensor three. The reason is that the pressure sensor three output values appeared to float in the early stage. These three statistics happened small range rise at the 290th–310th sampling of pressure sensor one, corresponding to the frequency of the original value that appeared a small hill. The reasons might be due to the irregular shape of cherry tomatoes or the small gripping force. The inevitable slight fluctuation affected the mean deviation, variance, and standard deviation, and it also affected the correct judgment of slip.

To reduce the influence of unavoidable slight fluctuations of data and realize more reliable slip detection, this study proposed the "Variance-to-Mean Ratio (*VMR*)" as the slip criterion, and it is shown in Formula (19). Figure 13f is the result of the variance-to-mean ratio.

$$VMR = \frac{SD}{A} \tag{19}$$

The variance-to-mean ratio remained near 0 and did not fluctuate significantly during phase III. It eliminates the judgment that values fluctuate due to fluctuations in the raw data. In addition, it also had significant inflection points to judge slide during V. The reason is that averages simply flattened out small data fluctuations but did not change the trend of the standard deviation. Therefore, the variance-to-mean ratio can be used as the sliding criterion. However, the corresponding samplings of the three sensors were different when the rising inflection points occurred. For example, sensor one changed first, followed by the other two. It shows that the three fingers slide for the cherry tomato at different times during rotation. The first appeared inflection point should be used as the judgment of the slip criterion.

4.2. Design of Algorithm for Undamaged Harvesting

The two-stage "Holding-Rotating" picking strategy was used to achieve fast and undamaged picking. During the holding stage, when the gripper clamps the fruit with the minimum gripping force, the constrain part stops. α was set to the value of the sensor output corresponding to the minimum holding force. During the rotation stage, the gripper is rotated to separate the fruit from the stalk. In case of a slip, the constraint part advances, the fingers are closed, and the clamping force is increased to prevent sliding. The rotation of the gripper is realized by the DC motor. The running time of the DC motor was set to t_M. The rising increment was set to γ, which was used as the direct judgment indicator of slip. It is shown in Equation (20). ρ was set to the number of incremental successive occurrences.

$$\gamma = VMR(i+1) - VMR(i) \tag{20}$$

When γ occurs ρ times, it is judged to have produced slip. Let the forward distance of the constraint part be l when slip occurs. Based on multi-group clamping test, when $\alpha = 30$, $\gamma = 0.1$, $\rho = 3$, $t_M = 0.5$ s, and $l = 0.5$ mm, it had a better judging and gripping effect. Therefore, according to the two-index judgment, the grabbing index α and the slide index γ, the control algorithm of the two-stage and two-index was designed for fast, stable, and undamaged harvesting of small fruits. The flow chart of control algorithm is shown in Figure 14.

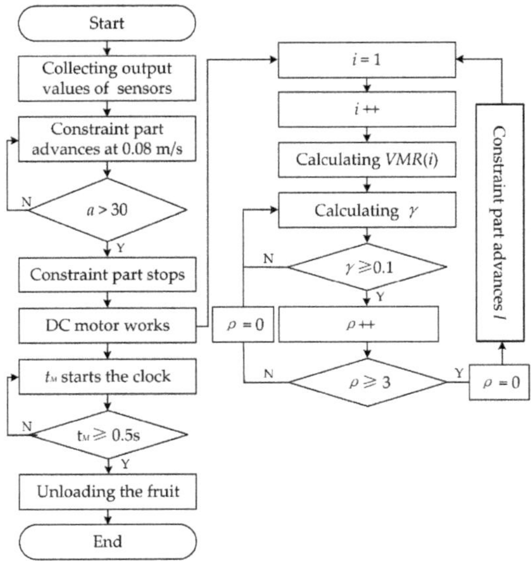

Figure 14. Flow chart of control algorithm.

5. Experiment and Analysis of Picking

5.1. Material and Method

The picking experiment was conducted on 6 July 2022. The end-effector was mounted at the end of the arm based on the cartesian coordinate system. The maximum axial velocity at the manipulator's end is set to 0.8 m/s. Guided by ZED 2 (STEREOLABS, England) binocular stereo camera, the end-effector was sent to the position where it could pick the target fruit. Then, the end-effector worked and picked the fruit. In the harvesting experiment, the Arduino Due controller was used to realize the two-stage and two-index harvesting strategy.

A total of 72 cherry tomatoes were tested to verify the harvesting performance of the end-effector. There are two damage types to cherry tomatoes. (1) The skin of the cherry tomato is damaged, and the internal tissues are exposed. Then, the fruit develops mold or decay. (2) The cherry tomatoes skin is not damaged, but its cells are damaged. In this case, an enzymatic reaction occurs. The color of the pressed area darkens. When the tomato cells breathe faster, the tomatoes' water will be lost, leading to atrophy. Thus, picking success rate, the average picking time, the picking damage rate, the browning rate, and the wrinkling rate were used as measures of rapid and undamaged picking. The definitions are as follows:

$$T = \frac{\sum_1^h T_0 - T_1}{h} \tag{21}$$

$$S = \frac{r}{h} \times 100\% \tag{22}$$

$$D = \frac{f}{h} \times 100\% \tag{23}$$

$$B_t = \frac{p}{r-f} \times 100\% \tag{24}$$

$$C_t = \frac{q}{r-f} \times 100\% \tag{25}$$

where T is the average time of picking, s; T_0 is the beginning time of picking, s; T is the time of unloading, s; S is the success rate,%; r is the number of fruits picked successfully; h is the total number of fruits; D is the damage rate, %, the damaged fruit was clipped by the end-effector; f is the damaged number; B_d is the browning rate, %; p is the browning number; C_d was the wrinkling rate, %; q was the wrinkling number; and t is the hours after picking, t = 24 h, 48 h, 72 h, respectively. The picked tomatoes were stored at an average temperature of 28 °C. Browning and wrinkling tomatoes were recorded. The picking experiment is shown in Figure 15.

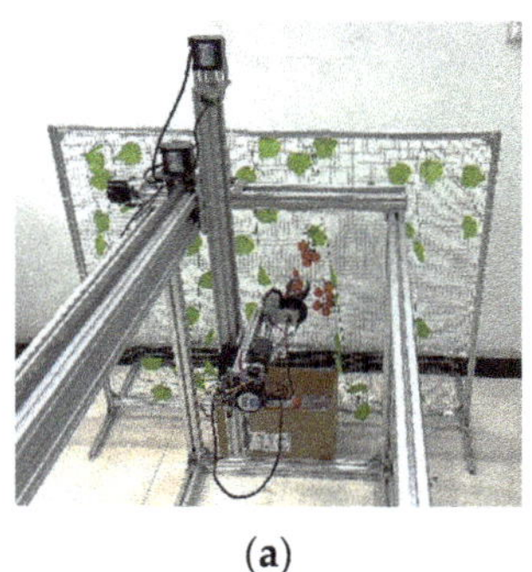

(**a**)

(**b**)

Figure 15. Experiment of cherry tomatoes picking: (**a**) The system of picking robot; (**b**) Gripper picking.

5.2. Analysis of Picking Results

The statistical results of the picking test are shown in Table 4. The results of the picking experiment for 72 cherry tomatoes showed that the average picking time was 4.86 s, the picking success rate was 95.82%, and the picking damage rate was 2.90%. The damage was mainly caused by two factors. (1) The edges of the fringe ends were in contact with the tomato surface. The pressure feedback signals were inaccurate or missing. Judgment was an error. (2) Under the influence of the stalk, the branches, and the leaves, the ends clipped them. The finger did not fully attach to the fruit surface. The gripping force is too large during the rotation.

Table 4. Results of picking experiment.

Item		Number	Result/%
Success		69	95.83
Damaged		2	2.90
Browning	24 h	0	0
	48 h	1	1.49
	72 h	2	2.90
Wrinkling	24 h	0	0
	48 h	0	0
	72 h	1	1.49

The browning rates at 24 h, 48 h, and 72 h were 0%, 1.49% and 2.90%, respectively. The wrinkling rates at 24 h, 48 h, and 72 h were 0%, 0%, and 1.49%, respectively. The risk of browning and wrinkling increased with time after picking. Of course, browning and wrinkling were not only related to the gripping force, but also the ripeness of tomato fruit. In this work, the mature tomatoes were tested, and the effects of different maturity were not studied. Different samples of damaged cherry tomatoes are shown in Figure 16.

Therefore, based on the success rate, picking time, damage rate, browning rate, and wrinkle rate, the end-effector developed in this paper had a good fast and undamaged clamping effect, and excellent picking performance.

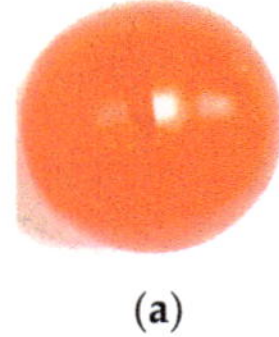
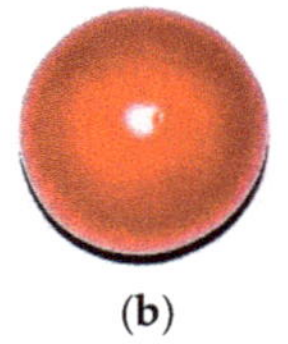

(**a**) (**b**) (**c**)

Figure 16. Examples of picking tomatoes: (**a**) Damaged fruit; (**b**) Browning fruit; (**c**) Wrinkling fruit.

6. Conclusions

(1) The two-stage "Holding-Rotating" picking strategy of the finger ends grasping was determined. The end-effector was designed to separate the fruit from the stalk based on the linear motion of the constraint part and the rotating gripper.

(2) The mathematical model of the gripper single finger was established, and the gripper structural parameters were determined to meet the requirements of the grabbing range from 0~61.6 mm. 80 mm/s was set by the constraint part through the simulation test for achieving undamaged and fast requirements.

(3) The statistical principle was used to analyze the sensor's output to study the rule of stable grasping and slip through the two-stage picking test. The criterion of two-index stability and non-loss is proposed, which includes the grasping index of a fixed value and the slip detection of dynamic variance-to-mean ratio. Therefore, the control strategy and algorithm based on two-stage and two-index for rapid, stable, and non-destructive harvesting of small fruit were proposed. Seventy-two cherry tomatoes were picked. The results of the picking experiment for 72 cherry tomatoes showed that the picking success rate was 95.82%, the average picking time was 4.86 s, the picking damage rate was 2.90%, the browning rate was 2.90% in 72 h, and the wrinkling rate was 1.49% in 72 h.

Author Contributions: Conceptualization, F.Z. and Z.C.; methodology, Z.C., M.T. and R.B.; software, Z.C. and R.B.; validation, F.Z. and Z.C.; formal analysis, F.Z. and Z.C.; investigation, Z.C., X.C. and M.T.; resources, Z.C. and M.T; data curation, F.Z., Z.C. and R.B.; writing—original draft preparation, F.Z. and Z.C.; writing—review and editing, F.Z., Y.Z., Y.W. and Z.C.; visualization, F.Z. and Y.Z.; supervision, F.Z. and S.F.; project administration, F.Z. and S.F.; funding acquisition, F.Z. and S.F. All authors have read and agreed to the published version of the manuscript.

Funding: This work was supported by the National Natural Science Foundation of China (grant no. 52075149 and 51905155), and the Scientific and Technological Project of Henan Province (grant no. 212102110029), Key Laboratory of Modern Agricultural Equipment and Technology (Ministry of Education), High-tech Key Laboratory of Agricultural Equipment and Intelligence of Jiangsu Province (Grant No. JNZ201901), and the Colleges and Universities of Henan Province Youth Backbone Teacher Training Program (grant no. 2017GGJS062).

Institutional Review Board Statement: Not applicable.

Informed Consent Statement: Not applicable.

Data Availability Statement: Not applicable.

Conflicts of Interest: The authors declare no conflict of interest.

References

1. Simiele, M.; Argentino, O.; Baronti, S.; Scippa, G.S.; Chiatante, D.; Terzaghi, M.; Montagnoli, A. Biochar Enhances Plant Growth, Fruit Yield, and Antioxidant Content of Cherry Tomato (*Solanum lycopersicum* L.) in a Soilless Substrate. *Agriculture* **2022**, *12*, 1135. [CrossRef]
2. Lipan, L.; Issa-Issa, H.; Moriana, A.; Zurita, N.M.; Galindo, A.; Martín-Palomo, M.J.; Andreu, L.; Carbonell-Barrachina, Á.A.; Hernández, F.; Corell, M. Scheduling Regulated Deficit Irrigation with Leaf Water Potential of Cherry Tomato in Greenhouse and Its Effect on Fruit Quality. *Agriculture* **2021**, *11*, 669. [CrossRef]

3. Distefano, M.; Steingass, C.B.; Leonardi, C.; Giuffrida, F.; Schweiggert, R.; Mauro, R.P. Effects of a Plant-Derived Biostimulant Application on Quality and Functional Traits of Greenhouse Cherry Tomato Cultivars. *Food Res. Int.* **2022**, *157*, 111218. [CrossRef] [PubMed]
4. Xiao, X.; Wang, Y.N.; Jiang, Y.M. End-Effectors Developed for Citrus and Other Spherical Crops. *Appl. Sci.* **2022**, *12*, 7945. [CrossRef]
5. Shi, G.K.; Li, J.B.; Kan, Z.; Ding, L.P.; Ding, H.Z.; Zhou, L.; Wang, L.H. Design and Parameters Optimization of a Provoke-Suction Type Harvester for Ground Jujube Fruit. *Agriculture* **2022**, *12*, 409. [CrossRef]
6. Guo, T.Z.; Zheng, Y.F.; Bo, W.X.; Liu, J.; Pi, J.; Chen, W.; Deng, J.Z. Research on the Bionic Flexible End-Effector Based on Tomato Harvesting. *J. Sens.* **2022**, *2022*, 2564952. [CrossRef]
7. Kim, J.Y.; Pyo, H.R.; Jang, I.; Kang, J.; Ju, B.K.; Ko, K.E. Tomato Harvesting Robotic System Based on Deep-Tomatos: Deep Learning Network Using Transformation Loss for 6D Pose Estimation of Maturity Classified Tomatoes with Side-Stem. *Comput. Electron. Agric.* **2022**, *201*, 107300. [CrossRef]
8. Vrochidou, E.; Tsakalidou, V.N.; Kalathas, I.; Gkrimpizis, T.; Pachidis, T.; Kaburlasos, V.G. An Overview of End Effectors in Agricultural Robotic Harvesting Systems. *Agriculture* **2022**, *12*, 1240. [CrossRef]
9. Ji, W.; Tang, C.C.; Xu, B.; He, G.Z. Contact Force Modeling and Variable Damping Impedance Control of Apple Harvesting Robot. *Comput. Electron. Agric.* **2022**, *198*, 107026. [CrossRef]
10. Xiong, Y.; Ge, Y.Y.; Grimstad, L.; From, P.J. An Autonomous Strawberry-Harvesting Robot: Design, Development, Integration, and Field Evaluation. *J. Fileld Robot.* **2019**, *37*, 202–224. [CrossRef]
11. Huang, M.S.; He, L.; Choi, D.; Pecchia, J.; Li, Y.M. Picking Dynamic Analysis for Robotic Harvesting of Agaricus Bisporus Mushrooms. *Comput. Electron. Agric.* **2021**, *185*, 106145. [CrossRef]
12. Miao, Y.B.; Zheng, J.F. Optimization Design of Compliant Constant-Force Mechanism for Apple Picking Actuator. *Comput. Electron. Agric.* **2020**, *170*, 105232. [CrossRef]
13. Roshanianfard, A.; Noguchi, N. Pumpkin Harvesting Robotic End-Effector. *Comput. Electron. Agric.* **2020**, *174*, 105503. [CrossRef]
14. Cao, X.M.; Zou, X.J.; Jia, C.Y.; Chen, M.Y.; Zeng, Z.Q. RRT-Based Path Planning for an Intelligent Litchi-Picking Manipulator. *Comput. Electron. Agric.* **2019**, *156*, 105–118. [CrossRef]
15. Williams, H.A.; Jones, M.H.; Nejati, M.; Seabright, M.J.; Bell, J.; Penhall, N.D.; Barnett, J.J.; Duke, M.D.; Scarfe, A.J.; Ahn, H.S.; et al. Robotic Kiwifruit Harvesting Using Machine Vision, Convolutional Neural Networks, and Robotic Arms. *Biosyst. Eng.* **2019**, *181*, 140–156. [CrossRef]
16. Wang, R.; Zheng, Z.; Lu, X.; Gao, L.; Jiang, D.; Zhang, Z. Design, Simulation and Test of Roller Comb Type Chrysanthemum (Dendranthema Morifolium Ramat) Picking Machine. *Comput. Electron. Agric.* **2021**, *187*, 106295. [CrossRef]
17. Brondino, L.; Borra, D.; Giuggioli, N.R.; Massaglia, S. Mechanized Blueberry Harvesting: Preliminary Results in the Italian Context. *Agriculture* **2021**, *11*, 1197. [CrossRef]
18. Li, Y.X.; Li, B.J.; Jiang, Y.Y.; Xu, C.R.; Zhou, B.D.; Niu, Q.; Li, C.S. Study on the Dynamic Cutting Mechanism of Green Pepper (*Zanthoxylum armatum*) Branches under Optimal Tool Parameters. *Agriculture* **2022**, *12*, 1165. [CrossRef]
19. Van Henten, E.J.; Schenk, E.J.; Van Willigenburg, L.G.; Meuleman, J.; Barreiro, P. Collision-Free Inverse Kinematics of the Redundant Seven-Link Manipulator Used in a Cucumber Picking Robot. *Biosyst. Eng.* **2010**, *106*, 112–124. [CrossRef]
20. Wang, Y.; Yang, Y.; Yang, C.H.; Zhao, H.M.; Chen, G.B.; Zhang, Z.; Fu, S.; Zhang, M.; Xu, H.B. End-Effector with a Bite Mode for Harvesting Citrus Fruit in Random Stalk Orientation Environment. *Comput. Electron. Agric.* **2019**, *157*, 454–470. [CrossRef]
21. Wang, J.N.; Li, B.X.; Li, Z.G.; Zubrycki, I.; Granosik, G. Grasping Behavior of the Human Hand during Tomato Picking. *Comput. Electron. Agric.* **2021**, *180*, 105901. [CrossRef]
22. Gao, J.; Zhang, F.; Zhang, J.X.; Yuan, T.; Yin, J.L.; Guo, H.; Yang, C. Development and Evaluation of a Pneumatic Finger-like End-Effector for Cherry Tomato Harvesting Robot in Greenhouse. *Comput. Electron. Agric.* **2022**, *197*, 106879. [CrossRef]
23. Xiang, C.Q.; Yang, H.; Sun, Z.Y.; Xue, B.; Hao, L.N.; Rahoman, M.D.; Davis, S. The Design, Hysteresis Modeling and Control of a Novel SMA-Fishing-Line Actuator. *Smart Mater. Struct.* **2017**, *26*, 037004. [CrossRef]
24. Ahmed, S.; Ounaies, Z. A Study of Metalized Electrode Self-Clearing in Electroactive Polymer (EAP) Based Actuators. In *Electroactive Polymer Actuators and Devices*; SPIE: Bellingham, WA, USA, 2016. [CrossRef]
25. Cai, S.B.; Tang, C.E.; Pan, L.F.; Bao, G.J.; Bai, W.Y.; Yang, Q.H. Pneumatic Webbed Soft Gripper for Unstructured Grasping. *Int. J. Agric. Biol. Eng.* **2021**, *14*, 145–151. [CrossRef]
26. Ji, W.; Zhang, J.W.; Xu, B.; Tang, C.C.; Zhao, D.A. Grasping mode analysis and adaptive impedance control for apple harvesting robotic grippers. *Comput. Electron. Agric.* **2021**, *186*, 106210. [CrossRef]
27. Roderick, W.R.; Da Cutkosky, M.R.; Lentink, D. Bird-Inspired Dynamic Grasping and Perching in Arboreal Environments. *Sci. Robot.* **2021**, *6*, abj7562. [CrossRef]
28. Mablekos-Alexiou, A.; Cruz, L.; Bergeles, C. Friction-Inclusive Modeling of Sliding Contact Transmission Systems in Robotics. *IEEE Trans. Robot.* **2021**, *37*, 1252–1267. [CrossRef]
29. Xie, H.B.; Kong, D.Y.; Wang, Q. Optimization and Experimental Study of Bionic Compliant End-Effector for Robotic Cherry Tomato Harvesting. *J. Bionic Eng.* **2022**, *19*, 1314–1333. [CrossRef]
30. James, J.W.; Lepora, N.F. Slip Detection for Grasp Stabilization with a Multifingered Tactile Robot Hand. *IEEE Trans. Robot.* **2021**, *37*, 506–519. [CrossRef]

31. Li, Q.; Kroemer, O.; Su, Z.; Veiga, F.F.; Kaboli, M.; Ritter, H.J. A Review of Tactile Information: Perception and Action through Touch. *IEEE Trans. Robot.* **2020**, *36*, 1619–1634. [CrossRef]
32. Zou, L.L.; Yuan, J.; Liu, X.M.; Li, J.G.; Zhang, P.; Niu, Z.R. Burgers Viscoelastic Model-Based Variable Stiffness Design of Compliant Clamping Mechanism for Leafy Greens Harvesting. *Biosyst. Eng.* **2021**, *208*, 1–15. [CrossRef]
33. Costanzo, M.; De Maria, G.; Natale, C. Two-Fingered in-Hand Object Handling Based on Force/Tactile Feedback. *IEEE Trans. Robot.* **2020**, *36*, 157–173. [CrossRef]
34. Jiang, C.P.; Zhang, Z.; Pan, J.; Wang, Y.C.; Zhang, L.; Tong, L.M. Finger-Skin-Inspired Flexible Optical Sensor for Force Sensing and Slip Detection in Robotic Grasping. *Adv. Mater. Technol.* **2021**, *6*, 2100285. [CrossRef]

Article

Simulation Analysis and Experiments for Blade-Soil-Straw Interaction under Deep Ploughing Based on the Discrete Element Method

Jin Zhang [1,2], Min Xia [1], Wei Chen [1], Dong Yuan [1], Chongyou Wu [1] and Jiping Zhu [1,*]

[1] Nanjing Institute of Agricultural Mechanization, Ministry of Agriculture and Rural Affairs, Nanjing 210014, China
[2] Jiangsu Yanjiang Institute of Agricultural Sciences, Nantong 226012, China
* Correspondence: zhujiping@caas.cn; Tel.: +025-84346259

Abstract: The desirable sowing period for winter wheat is very short in the rice-wheat rotation areas. There are also lots of straw left in harvested land. Deep rotary tillage can cover rice straw under the surface to increase soil organic matter. Clarifying the effect of the rotary tillage blade on the soil and straw, as well as analyzing the movement patterns and forces on the straw and soil, are essential to investigate the deep rotary tillage process in order to solve the problems of energy consumption and poor straw burial effect of deep tillage and deep burial machinery. In this study, we built the interaction model of rotary blade-soil-straw through the discrete element method to conduct simulation and identified the factors that affect the power consumption and operation quality of the rotary blade. The simulation process reflects the law of rotary blade-soil-straw interaction, and the accuracy of the simulation model has been verified by field trials. The simulation test results show that the optimized structural parameters of the rotary tillage blade were 210 mm, 45 mm, 37° and 115° (R, H, α and β) designed based on this theoretical model can cultivate to a depth of 200 mm. The operating parameters were 8π rad/s for rotational speed and 0.56 m/s for forward speed, respectively; the simulated and field comparison tests were conducted under the optimal combination of parameters, and the power, soil breaking rate, and straw burial rate were 1.73 kW, 71.34%, and 18.89%, respectively; the numerical error rates of simulated and field test values were 6.36%, 5.42%, and 8.89%, respectively. The accuracy of the secondary model was verified. The simulation model had good accuracy at all factor levels. The model constructed in this study can provide a theoretical basis and technical reference for the interaction mechanism between rotary tillage and soil straw, the optimization of machine geometry, and the selection of operating parameters.

Keywords: rotary blade soils; straw burying; discrete element method; interaction

Citation: Zhang, J.; Xia, M.; Chen, W.; Yuan, D.; Wu, C.; Zhu, J. Simulation Analysis and Experiments for Blade-Soil-Straw Interaction under Deep Ploughing Based on the Discrete Element Method. *Agriculture* **2023**, *13*, 136. https://doi.org/10.3390/agriculture13010136

Academic Editor: Tao Cui

Received: 28 November 2022
Revised: 22 December 2022
Accepted: 27 December 2022
Published: 5 January 2023

1. Introduction

Rotary tillers are widely used in the seed bed preparation segment of agricultural production [1]. The study demonstrated that a rotation tillage depth of 200 mm could fully mix the straw into the soil and significantly improve the sowing quality and wheat yield [2]. However, there are a significant number of challenges, such as insufficient tillage depth (less than 150 mm), unsatisfactory straw burying effect (low straw burying rate, insufficient straw burying depth), high power consumption, etc., in rice-wheat rotation areas. Moreover, long-term shallow rotational tillage results in a thinning and low-fertility soil tillage layer, which in turn leads to lower crop yields [3]. Simultaneously, inadequate straw mulching resulted in the distribution of straw in the upper part of the tillage layer, which allowed gaps between the top and bottom of the soil tillage layer, thus leading to segregation of the soil. It is easy to cause the next crop to fall dry and die, reducing the seeding rate in the cultivation area of two crops a year [4]. The concentration of a huge number of agricultural machines in the busy season, coupled with the increase in power consumption, is prone to air contamination [5,6], which is not in accordance with the concept of green development of agricultural machinery and decelerates the achievement of the carbon peaking and carbon neutral development goals.

To meet the above challenges and realize deep plowing, four solutions are applied in production practice as follows. (1) Increase the radius of the rotary blade. The test proved that the machine could achieve a tillage depth of 300 mm, but there are problems with the huge mechanism, heavy tillage load, and low reliability of the machine [7]. (2) Double-axis front shallow and behind deep rotary tillage. The machine sets two rows of knife rollers at the front and behind, the former for the first operation, the latter again, and finally, two operations to achieve deep tillage. In this solution, the soil accumulates in the gap between the front and rear blade axes of the machine, which interferes with the soil and prevents the rear blade axes from sinking into the ground. Moreover, the structure is complicated and heavy and consumes much energy [8,9]. (3) Submerged soil reversal rotary tillage. This solution is a tillage method in which the cutter shaft is reversed and submerged below the ground surface. This form of mechanism resulted in the transmission box and frame being difficult to enter the soil and congestion problems [10–17]. (4) Diagonal submerged soil reversal rotary tillage. This solution effectively solves the problem that the transmission box cannot enter the soil in the above two solutions. The blade roller is angled in the horizontal plane during the operation of the machine, and tillage is carried out obliquely, but there is a significant lateral force, which is not favorable for the operator to control the direction [18,19].

The structure of key components is unreasonable, and the equipment design lacks academic guidance. The above methods restrict the research progress of deep rotary tillage. The systematic study of rotary blade-soil-straw interaction in deep rotary tillage operation is insufficient, and the design of rotary tiller parameters and whole machine design lack theoretical support. This study synthesizes the research results in related fields of agricultural engineering and conducts research and discussion on the mechanism of rotary blade-soil-straw interaction in deep rotary tillage operation based on the discrete element method. The Discrete Element Method (DEM) is a numerical simulation method for basic research on solving discontinuous media problems [20]. It is different from the Finite Element Method (FEM) for solving continuous media problems. The basic principle of this method is to separate the research object into a set of rigid units. In order to make each unit conform to Newton's second law, the motion equation of each unit is calculated by means of central difference, and the whole motion state and parameters of the research object are obtained. This method is favored by most researchers in the field of agricultural engineering. In this paper, the discrete element method is used to deal with the fact that the soil model is a discontinuous medium [21,22].

We have developed a discrete element model of rotary blade-soil-straw for issues such as soil breakage rate, straw displacement, and power consumption during observation of rotary tillage operation. After that, the prototype was tested according to the 3D model of the machine designed on the basis of this study, and field verification tests were done. Finally, the validity of the model was determined by comparing the results of field tests and simulation analysis. The structural arrangement of the remainder of this article is as follows: Part 2 describes the analysis of the relevant phenomena during the simulation test. Part 3 shows the relevant results. Part 4 presents the discussion of the test. Part 5 concludes and summarizes the whole article. In this study, a blade-soil-straw interaction model was established for rice-wheat rotation. The model can provide a theoretical basis and technical reference for the interaction mechanism between rotary tillage and soil straw, the optimization of machine geometry, and the selection of operating parameters.

2. Materials and Methods

2.1. Rotational Tillage Mechanism

2.1.1. Structure of Rotary Blades

The tractor is connected to the rotary tiller by a towing device and transmits power through a universal coupling. The schematic diagram of the rotary tiller is shown in Figure 1a. The rotary tiller moves forward with a tractor. The rotary tiller shaft rotates through the gearbox, following the tractor power output shaft to drive the rotary tiller. The scraper impacts and levels the scattered soil to achieve a flat cultivated land. In order to study the interaction rule of rotary tillage blade-soil-straw more intuitively, this paper first studies the operation mechanism of rotary tillage single blade-soil-straw and then optimizes the form and structure parameters of rotary tillage single blade.

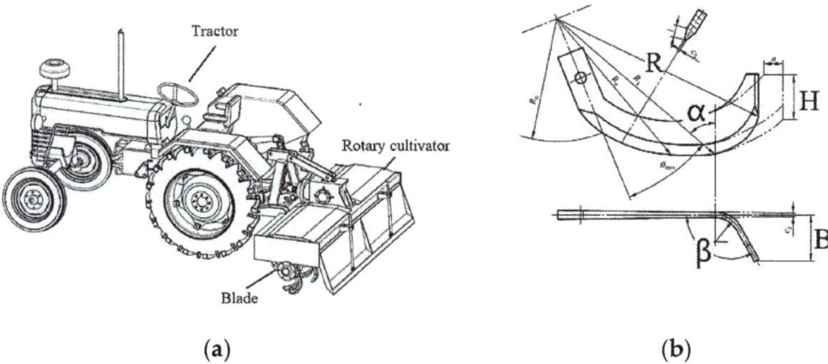

Figure 1. Standard rotary tiller structure.

The blade is the key working part of the rotary tiller. The shape and structural parameters of the blade badly affected the quality and power consumption of the rotary tiller. The blade in this paper is designed and researched based on a standard rotary tillage cutter according to the research objectives, as shown in Figure 1b. The remaining are implemented in accordance with GBT 5669-2017 Rotary Tillage Machine Blades and Holders. The main structural parameters are shown in Table 1.

Table 1. The main structural parameters.

Symbol	Name	Numerical Value
R	Radius of gyration	150~245 mm
B	Working breadth	35~55 mm
H	End cutter height	35~55 mm
α	Included angle	35~55°
β	Bending angle	110~130°

2.1.2. Mechanisms of Rotary Blades

The values of forward velocity and linear velocity of the blade endpoint determine the normal progress of soil cutting operation. The blade endpoint presents a composite motion; therefore, arrangements of velocity parameters will have diverse trajectories.

Let V_d—the tangential velocity of the blade endpoint, m/s;

$$V_d = R\omega \tag{1}$$

$$\lambda = \frac{V_d}{V_m} \tag{2}$$

The blade velocity ratio λ and the motion trajectory are shown in Figure 2.

(1) When $\lambda < 1$, that is, the forward velocity V_m is greater than the blade linear velocity V_d, its motion trajectory is a short cycloid, the direction of the horizontal component of the linear velocity at any point on the cycloid is consistent with the forward direction, and it is hardly difficult to realize the operation of post throwing soil blocks.

(2) When $\lambda = 1$, that is, the two values are equal, its motion trajectory is a cycloid curve, The horizontal velocity at any point is zero, and the same as above cannot realize soil throwing.

(3) When $\lambda > 1$, that is, the forward velocity V_m is less than the blade's linear velocity V_d, its motion trajectory is a complementary cycloid. The horizontal partial velocity direction of any point below the maximum cross chord is opposite to the forward direction, exact converse above, the soil block can be reprojected effectively at present.

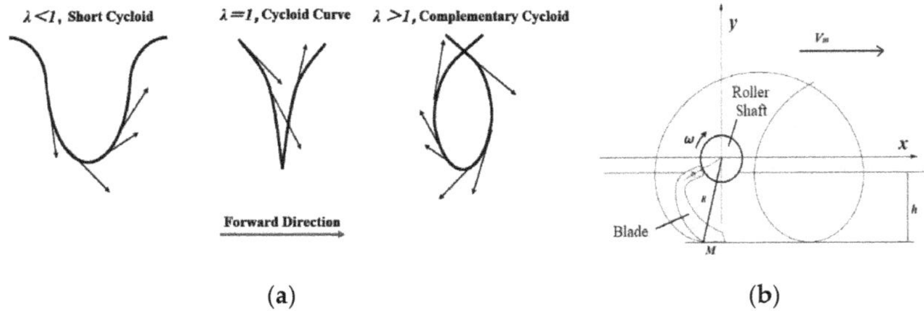

Figure 2. Blade end point motion track.

Set point M is set as the blade cutting point; it will meet the normal working conditions of the rotary cultivator from the beginning of soil penetration to the end of soil throwing and leaving the ground; there are:

$$x = R\cos\alpha t + V_m t \tag{3}$$

$$y = R\sin\cot t = R - h \tag{4}$$

where R is the radius of gyration of a blade, with the unit of m, V_m is forward velocity, with the unit of m/s, ω is angular velocity, with the unit of rad/s, t is time, with the unit of s, h is tillage depth, with the unit of m.

To meet the condition of throwing soil backward, the horizontal partial velocity V_x of the absolute velocity at any point on the absolute motion track of the blade is less than 0, according to the above equation, let:

$$V_x = \frac{d_x}{d_t} = v_m - 2\omega\sin\omega t < 0 \tag{5}$$

$$\sin\cot = \frac{R}{R-h}\sin\omega t = \frac{R}{R-h} \tag{6}$$

$$V_m < (R-h)\omega \tag{7}$$

The movement locus of the blade endpoint is shown in Figure 2b. As can be seen from the above equation: R increases, H increases, but the increased torque leads to increased power consumption, which leads to a decrease in speed and, therefore, productivity. Therefore, generally, R = 215, 245, 260 mm, and in this study, a range of 150 to 240 mm was chosen in order to explore whether a small radius could be used for rototilling under ground level.

2.2. The Discrete Element Model of Rotary Blade-Soil–Straw

2.2.1. Soil and Straw Model Parameters

The discrete element parameters consist of the intrinsic parameters (material density, Poisson's ratio, shear modulus), material contact parameters (coefficient of restitution, coefficient of static friction, dynamic friction coefficient), and contact model parameters. The parameters employed in this article are taken from the reference [21]. The relevant parameters are shown in Table 2.

The increase in the number of particles in the software will lead to a decline in the geometric level of the simulation calculation speed [22]. In addition, too small particles will lead to an uncertain impact on the model under the micro force. Therefore, the size of the soil model is generally larger than the real soil particles in the simulation calculation. Without affecting the simulation test, the soil particle radius selected in this paper is 5 mm, 10 mm, 15 mm, and 30 mm, as shown in Figure 3.

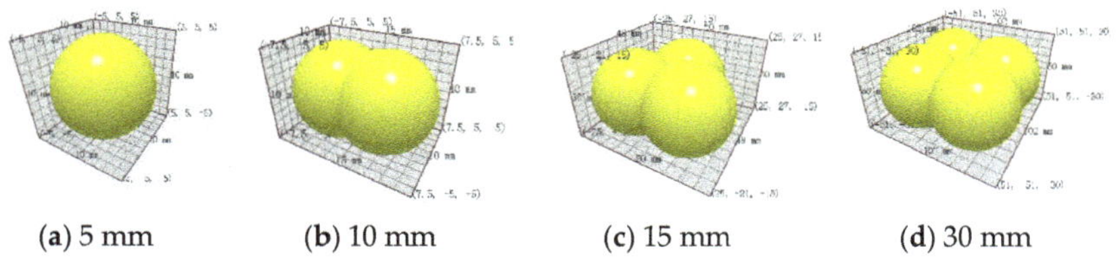

(**a**) 5 mm (**b**) 10 mm (**c**) 15 mm (**d**) 30 mm

Figure 3. Soil particles with different sizes.

In this paper, the rotary tillage blade mainly covers the straws scattered on the ground during operation and will not crush and cut the straws. For the purpose of improving the simulation efficiency, the straws established in this paper are also simplified to a certain extent. It is mainly composed of 10 particles with a radius of 5 mm. The center distance between adjacent particles is 5 mm. The total length of the straw model is 55 mm. The simulation model of straw is shown in Figure 4.

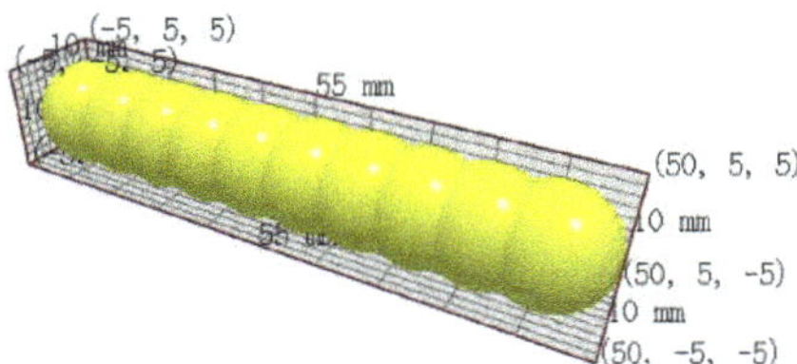

Figure 4. Straw model.

Table 2. Material physical properties parameters.

Material	Density (g/cm^3)	Poisson's Ratio	Shear Modulus (Pa)
Soil	1.668	0.38	1.0×10^6
Straw	0.227	0.40	1.0×10^6
65Mn	7.800	0.30	7.9×10^{10}

2.2.2. Contact Model Settings

Edem software 2018 integrates many options to apply to different material types of contact. We selected Hertz-Mindlin with Bonding along with the characteristics that the soil particles are difficult to reconnect after fracture in this paper. At the same time, the most commonly used contact theoretical model Hertz-Mindlin (no slip), is selected for ordinary contact calculation.

(a) Hertz-Mindlin (no slip)

In this model, the normal force component model is based on the Hertzian contact theory. The tangential force component model is based on the work of Mindlin-Deresiewicz. Both normal and tangential forces have damping components, where the damping coefficient is related to the coefficient of restitution. The tangential friction force follows the Coulomb friction law, and the rolling friction adopts the research results of Sakaguchi. The relevant calculation principles are described in the literature [23–28].

The normal force F_n is expressed as follows:

$$F_n = \frac{4}{3}E * \sqrt{R^*}\delta_n{}^{\frac{3}{2}} \tag{8}$$

$$\frac{1}{E^*} = \frac{(I - v_i^2)}{E_i} + \frac{(1 - v_j^2)}{E_j} \tag{9}$$

$$\frac{1}{R^*} = \frac{1}{R_i} + \frac{1}{R_j} \tag{10}$$

where F_n is the normal force, with the unit of N, E^* is equivalent to Young's modulus, with the unit of Pa, R^* is the equivalent radius, with the unit of m, δ_n is the amount of normal overlap, with the unit of m, $v_{i,j}$ is Poisson's ratio, $E_{i,j}$ is Particle Young's modulus, with the unit of Pa, $R_{i,j}$ is Particle radius, with the unit of m.

The normal damping force $F_n{}^d$ is expressed as follows:

$$F_n^d = -2\sqrt{\frac{5}{6}}\beta\sqrt{S_d m^*}v_n{}^{\overrightarrow{rel}} \tag{11}$$

$$\beta = \frac{-\ln e}{\sqrt{\ln^2 e + \pi_n{}^2}} \tag{12}$$

$$S_n = 2E^*\sqrt{R^*\delta_n} \tag{13}$$

where $F_n{}^d$ is Normal damping force, with the unit of N, β is viscous damping coefficient, e is coefficient of restitution, Sn _is normal stiffness, with the unit of N/m, m^* is equivalent mass, with the unit of g, $m_{i,j}$ is Particle mass, with the unit of g, $v_n^{\overrightarrow{rel}}$ is Normal relative velocity, with the unit of m/s.

The expression for the tangential force F_t is as follows:

$$F_t = -S_t\delta_t \tag{14}$$

$$S_t = 8G^*\sqrt{R^*\delta} \tag{15}$$

where F_t is the normal force, N,S_t is tangential stiffness, with the unit of N/m, δ_t is tangential overlap, with the unit of m, G^* is equivalent shear modulus, with the unit of Pa.

The tangential damping force $F_t{}^d$ is expressed as follows:

$$F_t^d = -2\sqrt{\frac{5}{6}}\beta\sqrt{S_t m^*}v_t{}^{\overrightarrow{rel}} \tag{16}$$

where $F_t{}^d$ is angential damping force, with the unit of N, S_t is tangential stiffness, with the unit of N/m, $v_t^{\overrightarrow{rel}}$ is Tangential relative velocity, with the unit of m/s.

The total tangential force is limited by the Coulomb friction force. When the tangential force reaches the Coulomb friction force, that is, $F_t > \mu_s F_n$, relative sliding occurs between the particles, where $_s$ is the static friction coefficient. To simplify the contact model, The model does not take into account the sliding friction of the contact model. The rolling friction is expressed in terms of the contact torque of the particle model, and the expression is as follows:

$$\tau_i = -\mu_r F_n R_i \omega_i \tag{17}$$

where τ_i is torque, with the unit of N·m, μ_r is coefficient of rolling friction, R_i is the distance from the contact point to the centroid, with the unit of m, ω_i—angular velocity of particles at the contact point, with the unit of r/min.

(b) Hertz-Mindlin with Bonding

Hertz-Mindlin with Bonding contact model bonds particles together. Since this model includes Hertz-Mindlin (no slip), the default Hertz-Mindlin (no slip) contact model needs to be deleted from the software Creator Tree list when generating the bond.

Hertz-Mindlin with Bonding contact model uses finite-sized "bonding" bonds to bond particles. The bond can withstand the normal and tangential motion between the contacting particles to generate normal and tangential shear stress until the bond breaks at the critical maximum stress. The particles interact as rigid spheres, and no new bonds are generated, which is in line with the soil stress mechanical behavior characteristics of remaining loose after crushing [29–31].

In this model, the first generated particles interact with each other according to the Hertz-Mindlin contact model, and the connection starts after the set bond generation time point. After the particles are connected, the force and torque between the particles are both set to 0, according to the following Stepwise adjustment at each time step:

$$\delta F_n = -v_n S_n A \delta_t \tag{18}$$

$$\delta F_t = -v_t S_t A \delta_t \tag{19}$$

$$\delta M_n = -\omega_n S_t J \delta_t \tag{20}$$

$$\delta M_t = -\omega_t S_n \frac{J}{2} \delta_t \tag{21}$$

$$A = \pi R_B^2 \tag{22}$$

$$J = \frac{1}{2} \pi R_B^4 \tag{23}$$

where $F_{n,t}$ is normal and tangential force, with the unit of N, $v_{n,t}$ is normal and tangential velocity, with the unit of m/s, $S_{n,t}$ is normal and tangential stiffness, with the unit of N/m, A is the cross-sectional area of bonding bond, with the unit of m^2, δ_t is time step, with the unit of s, $M_{n,t}$ is normal and tangential moments, with the unit of N·m, $\omega_{n,t}$ is normal and tangential angular velocities, with the unit of r/min, J is the moment of inertia, with the unit of m^4, R_B is the bond-bond radius.

When the normal and tangential stresses exceed preset critical values, the bonded bonds will break, the particles will interact as rigid spheres, and no new bonds will be formed. Spawn bonds in this model can create cohesive forces when particles are generated, even if the particles are not in contact, so the contact radius should be set larger than the radius of the particle body.

Table 3 lists the parameters adopted in this study. In this article, we use Edem's dynamic generation method to generate soil particle beds layer by layer, and the model is shown in Figure 5.

The particle contact model can quantify the interaction between discrete elements within a given contact range. Adding a contact model to the particle model is a key step in the test. Appropriate contact model to accelerate simulation test.

Table 3. Contact parameters.

Contacts	Coefficient of Restitution	Coefficient of Static Friction	Dynamic Friction Coefficient
Soil-soil	0.6	0.6	0.40
Soil-straw	0.5	0.5	0.05
Soil-65Mn	0.6	0.6	0.05
Straw-straw	0.3	0.3	0.01
Straw-65Mn	0.3	0.3	0.01

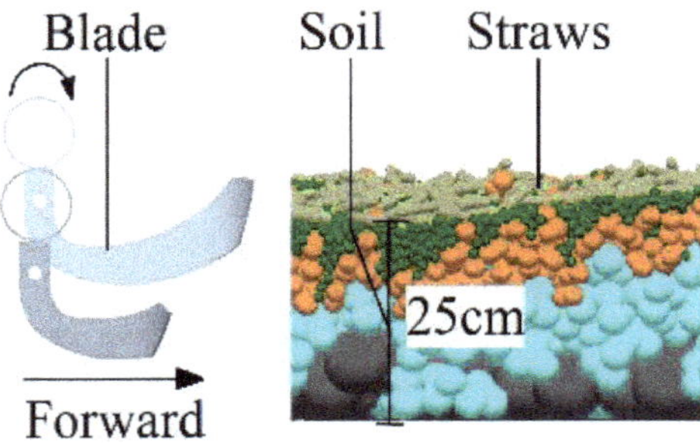

Figure 5. The discrete element model.

2.2.3. Initial Conditions for Simulation

In this simulation test, in order to ensure that the tillage depth was equal to 200 mm, the rotation center was set to the corresponding position into the soil according to the radius of rotation of each blade, as shown in Figure 6.

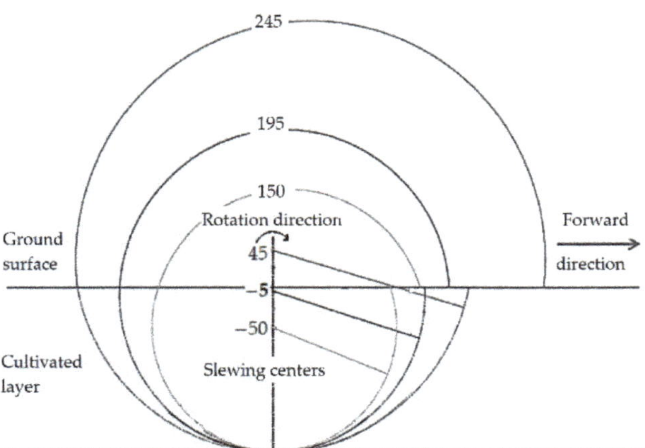

Figure 6. Initial position of the rotary blade.

According to the research objectives of this article, combined with the existing research experience and actual working conditions, the main indicators were determined as the power (P), the rate of crushed soil (RC), and the rate of straw burial (RB). Figure 1b shows that the value of B is fixed when R and β are certain, so the factor B is removed. The initial conditions of the experiment are shown in Table 4, and the factor level design is shown in Table 5.

Table 4. Initial conditions.

Radius (mm)	Center of Rotation (mm)	Revs (rad/s)	Velocity (m/s)
150	−50	8π	0.56
195	−5	8π	0.56
245	45	8π	0.56

Table 5. Factor levels.

Levels	R (mm)	H (mm)	α (°)	β (°)
−1	150	35	35	110
0	195	45	45	120
1	245	55	55	130

In this study, for convenience of calculation, the ratio of the number of bond-bond breaks per unit volume to the number of bond bonds at the beginning is used as the index of soil breakage. The sampling position is set 500 mm from the center of rotation after rotary tiller operation, and 1000 mm is set by grid-grid function after EDEM treatment. 1000 mm × 500 mm × 250 mm grid solver calculates the bonding bond between particles as a problem to be solved.

The value of power (P) is selected from the data within 2 s of the operation stabilization stage. Each blade is tested three times, and the average value is obtained, which is directly exported after EDEM software post-processing.

The crushed soil rate (RC) is measured by the ratio of bond breakage between particles in the tillage area. In this study, for the convenience of calculation, the ratio of the number of soil bond breakage per unit volume to the number of bonded bonds at the initial time was used as the indicator of crushed soil rate. The sampling position was set at 500 mm from the center of rotation after the rotary tillage knife operation, and the grid solver of 1000 mm × 500 mm × 250 mm was set using the grid function of EDEM post-processing, and the particle-to-particle bond was calculated as the problem to be solved.

Straw burial rate (RB) is the percentage of straw residual mass on the ground post-tillage to the pre-tillage. In the post-processing of the EDEM software, L0 is the straw layer. The grid solver was divided into four layers (L1, L2, L3, L4) in the vertical direction, and the percentage of the total straw mass from L1 to L4 to the initial L0 mass was calculated as the straw burial rate.

2.3. *Simulation*

To better observe the movement of soil and straw in the simulation, we selected the velocity clouds from the simulation with different angles of blade rotation and placed them in Figure 7.

The velocity of the model is represented by the color-changing legend. The velocity ranges from green to red is 0–1 m/s, which means the velocity is the same as the forward direction. The velocity ranges from green to blue is 0–−1 m/s, which means the velocity is opposite to the forward direction. L refers to the whole tillage layer, L0 represents the straw, and L1–L4 represents the soil from 0–50 m, 50–100 mm, 100–150 mm, and 150–200 mm below the horizontal surface, respectively.

It can be seen from L: At 0° the rotary blade touches the soil, the soil and straw on the side edge are first disturbed by the force of the rotary blade, L1 soil, and L0 straw splash. The disturbed area has a definite initial velocity and appears yellow, representing a low value; At 90°, the rotary blade has fully entered the soil. The color of the soil and straw starts to change, which indicates that the number of particles gaining velocity starts to increase. At L, L0, and L1, it is clearly observed that the soil and straw near the rototiller knife are red and dark blue in color, with the majority of blue particles, which indicates that the velocity starts to increase and most of the soil acquires the initial velocity backwards.

Concurrently, the color of the soil and straw on the other side was light blue, indicating that the latter was significantly more disturbed than the former, and the velocity was negative. At 180°, the rotary blade was about to break out of the soil. The soil near the front section of the rotary tiller is dark blue and blue, which means that the rotary tiller throws the soil in the lower part of the plow layer backwards and upwards. The color of the soil near the side edge of the rototiller is red, illustrating that the soil in this position moves forward with the rototiller. At this time, the color of the soil with greater disturbance on the left side changes to yellow, indicating that this part of the soil starts to move forward and gradually returns to its original state. The color of the thrown soil and straw particles changed after the rotary tillage knife came out of the soil, indicating that the soil and straw continued to move in the direction of the speed obtained when thrown and finally fell into the ground. The color of the disturbed area is dark green, demonstrating that the velocity returns to 0.

L0 and L1 received a significant disturbance. The soil close to the center is taken to L2 and L3 when the blade starts to cut the soil, and the soil far from the center is thrown upward and backward and finally falls into L2 and L4. L2 and L3 are disturbed similarly to L1 and L0. The soil far from the center is thrown out and finally falls into L1 and L3; For L4, the soil disturbance is minimal, but a small part of the soil far from the center is still affected by the force and acquires a certain initial velocity and flow.

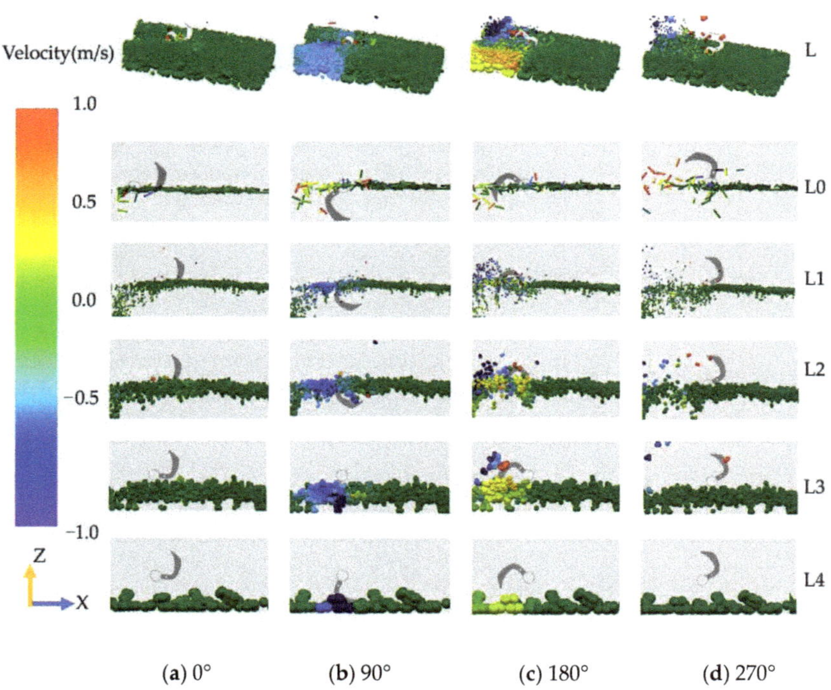

Figure 7. Velocity of soil and straw in different tillage layers during single blade operation.

The migration of the soil between L1 and L4 gives a side view of the mixing state of the tillage layer after rotary tillage. The more the layers migrate, the further they are from their original position and the better the mixing of the tillage layer. When the rototiller knife enters the soil again, it continues to disturb the soil after tillage, causing the straw to rebound and thus causing it to continue to migrate in the direction of the initial velocity. The soil is significantly displaced by the action of the rotary cutter, and the movement of the soil particles causes more significant displacement of the straw as the mechanical properties of the straw material are lower than those of the soil particles. The movement of the soil particles causes the straw to move more significantly due to the lower mechanical properties of the straw material than the soil particles. When the force is greater than the ultimate strength of the bond, the soil particles are separated and fractured, which means that the soil particles that were originally connected by the bond are separated and move to a different layer; as the tillage continues to advance, some of the particles will be separated by secondary cutting. During the tillage process, the soil is disturbed in different areas of the rototiller, with a clear range of disturbance away from the center. The surface of the tilled soil shows some undulations, mainly due to the fact that the tillage process has changed the surface form of the original soil, and there is a certain friction effect between the soil particles, which makes it difficult to recover the initial form.

2.4. Field Trials

In a comparison of the power consumption and operational effects of EDEM simulation and actual soil entry conditions, in field trials, the length of the experiment area is 200 m, and the width is 15 m. The experiment area belongs to the plain area with two crops a year, with an average annual precipitation of about 600 mm, and the soil texture is loam. The first crop of the experiment field is wheat, the vegetation coverage before tillage is 302.4 g/m^2, the soil firmness of 0–250 mm thick is 2098.0 kPa, the absolute soil moisture content is 21.63%, the average volume mass is 1.67 g/cm^3, and the temperature during the experiment is 28 °C. We connected the torque sensors and the farm machinery test box to the rotary tiller and the tractor and debugged them to ensure that the test system could complete the work normally, as shown in Figure 8.

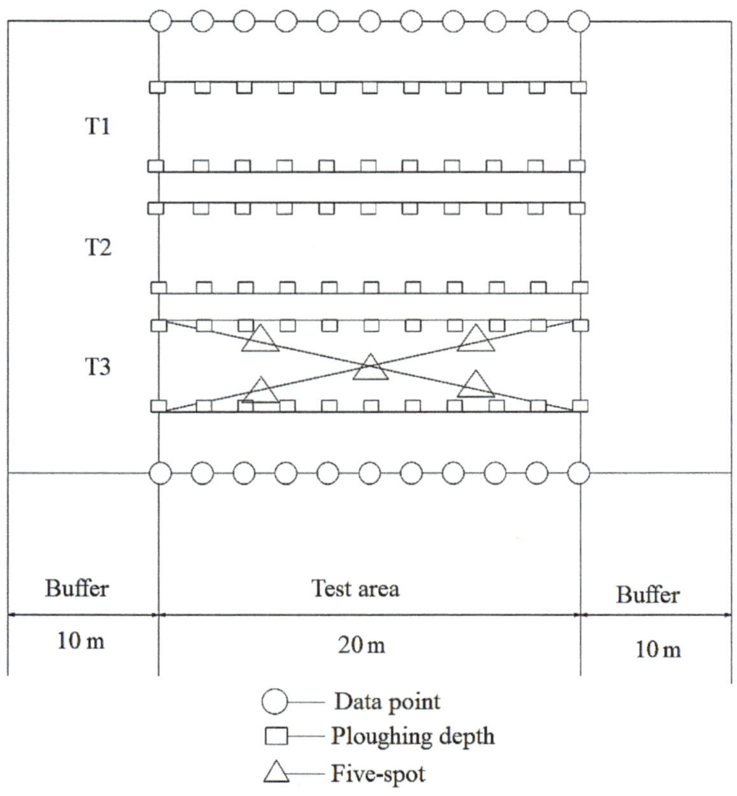

Figure 8. Data acquisition hint.

The power (P) is measured by a torque sensor mounted between the tractor and the coupling.

The rate of soil crushed (RC) is determined by measuring a 200 mm × 200 mm × 200 mm full tilled layer of soil on ploughed land. The size of the sampled soil is divided into three classes according to their side lengths: 0–4 cm, 4–8 cm, and 8 cm or more, and the percentage of soil mass from 0–4 cm to the total mass is the rate of soil fragmentation, measured one point per stroke.

The straw burial rate (RB)was determined by separating and weighing the straw from the soil at 0–5 cm, 5–10 cm, 10–15 cm, 15–20 cm, and 20–25 cm. During the sampling process, samples were taken strictly in accordance with the sampling frame and the boundaries of the tillage layer. Straw beyond the boundaries was cut off with scissors, keeping only the straw within the boundaries of the sampling area, and finally, the weight of each layer was used for statistical analysis.

The rotary tiller was connected to the tractor through a three-point suspension, and the torque sensor was installed at the end of the power take-off shaft. The rotary tiller was connected to the tractor through a three-point suspension, and the torque sensor was installed at the end of the power take-off shaft. The rototiller is mounted in the middle of the blade shaft, and to facilitate the same experimental conditions as the simulation, the soil is treated as a monopoly with a height of 250 mm and a width of 200 mm to facilitate the entry of a 150 mm radius rototiller stick into the soil. The structure diagram is shown in Figure 9.

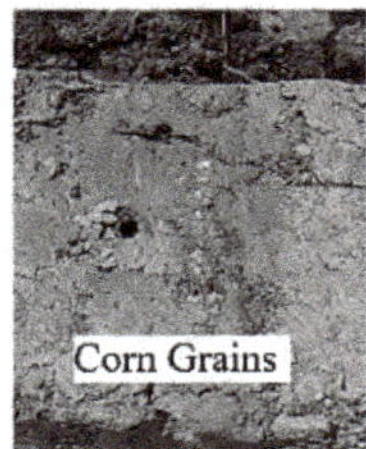

Figure 9. Schematic diagram of field test component links.

3. Results

3.1. Simulation Results

3.1.1. Results of Box-Behnken Design

Simulation tests were conducted according to the Box-Behnken design, with the power (P) and the rate of crushed soil (RC), and the rate of straw burial (RB) as the evaluation indexes, and the test results are shown in Table 6.

Table 6. Design and results of Box-Behnken Design.

Std	Run	Factors				Responses		
		A:R(mm)	B:H(mm)	C:α(°)	D:β(°)	P(kW)	RC(%)	RB(%)
1	29	150	35	45	120	1.47	49.50	09.60
2	22	240	35	45	120	2.86	60.28	14.09
3	10	150	55	45	120	1.52	58.20	11.50
4	20	240	55	45	120	2.92	61.08	14.56
5	6	195	45	35	110	1.27	69.50	17.60
6	18	195	45	55	110	1.35	64.29	18.09
7	16	195	45	35	130	1.29	64.29	15.34
8	14	195	45	55	130	1.17	65.09	13.98
9	15	150	45	45	110	1.59	56.40	07.90
10	5	240	45	45	110	2.97	69.09	12.45
11	3	150	45	45	130	1.51	52.40	04.50
12	13	240	45	45	130	2.78	62.30	15.36
13	8	195	35	35	120	1.18	45.32	13.89
14	7	195	55	35	120	1.50	67.09	17.90
15	4	195	35	55	120	1.16	45.80	17.60
16	9	195	55	55	120	1.52	64.90	13.98
17	24	150	45	35	120	1.56	52.10	09.30
18	25	240	45	35	120	2.92	60.09	14.09
19	21	150	45	55	120	1.46	54.30	08.90
20	23	240	45	55	120	2.91	67.90	13.35
21	19	195	35	45	110	1.24	53.09	17.39
22	26	195	55	45	110	1.56	64.20	15.39
23	27	195	35	45	130	1.13	59.20	13.50
24	1	195	55	45	130	1.51	69.40	17.36
25	12	195	45	45	120	1.22	69.23	18.40
26	17	195	45	45	120	1.31	67.50	19.19
27	2	195	45	45	120	1.18	65.30	15.75
28	28	195	45	45	120	1.2	68.20	17.90
29	11	195	45	45	120	1.21	66.30	17.75

3.1.2. ANOVA for Quadratic Model

According to the data samples in Table 6, the quadratic model of power (P), rate of crushed soil (RC), and straw burial rate (RB) was calculated by Design-Expert 12.0 software:

- Power (P)

According to the ANOVA results in Table 7, the Model F-value of 135.01 implies the model is significant. There is only a 0.01% chance that an F-value this large could occur due to noise. p-values less than 0.0500 indicate model terms are significant. In this case, A, B, D, A^2, B^2 are significant model terms. Values greater than 0.1000 indicate the model terms are not significant. Removing the insignificant term, the regression equation of the power consumption model in this study.

$$P = 1.22 + 0.6875A + 0.1242B - 0.0492D + 0.9322A^2 + 0.0697B^2 \quad (24)$$

The ANOVA results showed that the interaction of the factors in this model was not significant, and only the radius (R), the end cutter height (H), and the bending angle (β) were significant among the factors. It can be found from Figure 10a that when H = 45, α = 45, β = 120, the power consumption value is the lowest when R is 180, which is 1.12 kW, and as the radius R gradually increases, the power consumption increases significantly, and the analysis of the reason is due to the increase of torque, which leads to the increase of power consumption; when the radius R is lower than 180 and starts to decrease, the reason for the increase of power consumption is that the center of the blade sinks too deep into the tillage layer, which causes the blade to repeatedly Cutting soil, unable to throw out the soil fully, causing an increase in power consumption. Figure 8b,c show that the power consumption decreases with the increase of H and increases with the decrease of β, respectively.

Table 7. ANOVA for Quadratic model of Power.

Source	Sum of Squares	df	Mean Square	F-Value	p-Value	
Model	11.79	14	0.8422	135.01	<0.0001	
A-R	5.67	1	5.67	909.20	<0.0001	**
B-H	0.1850	1	0.1850	29.66	<0.0001	**
C-α	0.0019	1	0.0019	0.3006	0.5922	
D-β	0.0290	1	0.0290	4.65	0.0489	*
AB	0.0000	1	0.0000	0.0040	0.9504	
AC	0.0020	1	0.0020	0.3246	0.5779	
AD	0.0030	1	0.0030	0.4849	0.4976	
BC	0.0004	1	0.0004	0.0641	0.8038	
BD	0.0009	1	0.0009	0.1443	0.7098	
CD	0.0100	1	0.0100	1.60	0.2261	
A^2	5.64	1	5.64	903.50	<0.0001	**
B^2	0.0315	1	0.0315	5.05	0.0413	*
C^2	0.0078	1	0.0078	1.25	0.2824	
D^2	0.0129	1	0.0129	2.07	0.1718	
Residual	0.0873	14	0.0062			
Lack of Fit	0.0772	10	0.0077	3.05	0.1468	
Pure Error	0.0101	4	0.0025			
Cor Total	11.88	28				

** indicates highly significant ($p < 0.01$), * indicates significant ($p < 0.05$), and p-values greater than 0.1000 indicate the model terms are not significant.

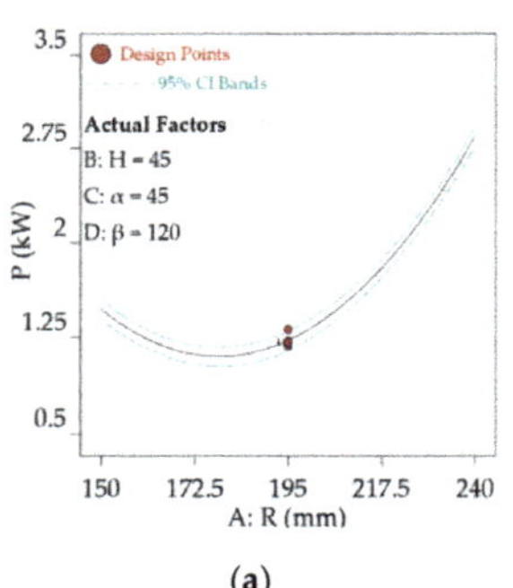

(a)

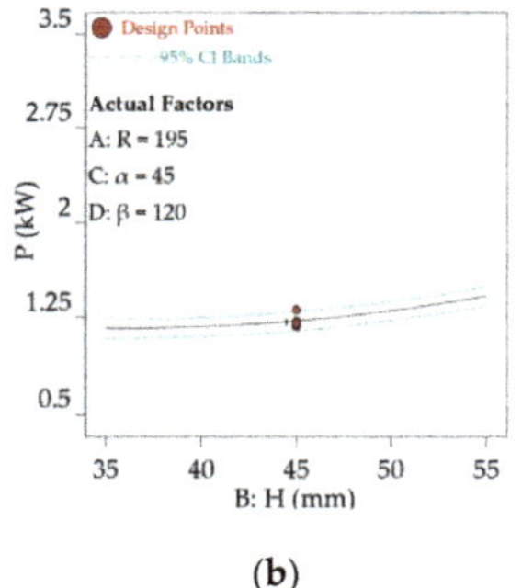

(b)

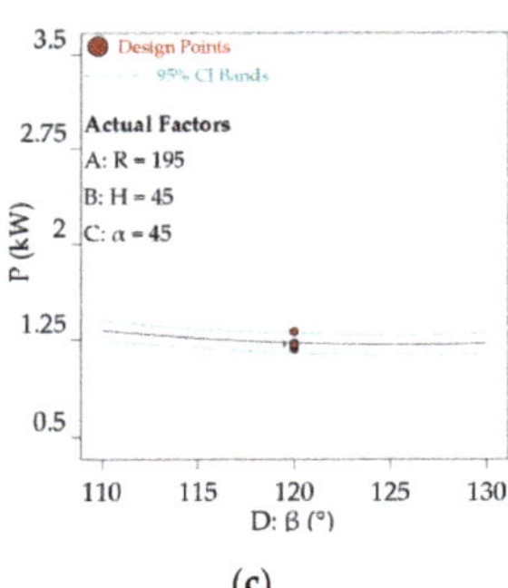

(c)

Figure 10. Power consumption value at different factors.

- Rate of crushed soil (RC)

According to the ANOVA results in Table 8, the Model F-value of 4.00 implies the model is significant. There is only a 0.70% chance that an F-value this large could occur due to noise. p-values less than 0.0500 indicate model terms are significant. In this case, *A*, *B*, A^2, B^2 are significant model terms. Values greater than 0.1000 indicate the model terms are not significant. Removing the insignificant term, the regression equation of the rate of crushed soil model in this study.

$$RC = 67.31 + 4.82A + 5.97B - 5.52A^2 - 6.22B^2 \quad (25)$$

Table 8. ANOVA for Quadratic model of Rate of crushed soil.

Source	Sum of Squares	df	Mean Square	F-Value	p-Value	
Model	1175.90	14	83.99	4.00	0.0070	
A-R	278.79	1	278.79	13.28	0.0027	**
B-H	428.17	1	428.17	20.39	0.0005	**
C-α	1.26	1	1.26	0.0601	0.8100	
D-β	1.26	1	1.26	0.0601	0.8100	
AB	15.60	1	15.60	0.7431	0.4032	
AC	7.87	1	7.87	0.3747	0.5503	
AD	1.95	1	1.95	0.0927	0.7653	
BC	1.78	1	1.78	0.0849	0.7751	
BD	0.2070	1	0.2070	0.0099	0.9223	
CD	9.03	1	9.03	0.4301	0.5226	
A^2	197.89	1	197.89	9.42	0.0083	**
B^2	251.03	1	251.03	11.96	0.0038	**
C^2	74.75	1	74.75	3.56	0.0801	**
D^2	0.2051	1	0.2051	0.0098	0.9227	
Residual	293.96	14	21.00			
Lack of Fit	284.39	10	28.44	11.88	0.0146	
Pure Error	9.57	4	2.39			
Cor Total	1469.86	28				

** indicates highly significant ($p < 0.01$), and p-values greater than 0.1000 indicate the model terms are not significant.

The ANOVA results showed that the interaction of the factors in this model was not significant. The radius (R) and the end cutter height (H) were significant among the factors. It can be found from Figure 11 that when H = 45, α = 45, β = 120, RC gradually increases with the increase of R and H. The curve stops growing at R = 195 and H = 45, respectively. The value of RC hovers around 68%. It indicates that the space for disturbing the soil at the same tillage depth under the action of a single blade is limited, and the scanning trajectory formed by its rotation is approximated.

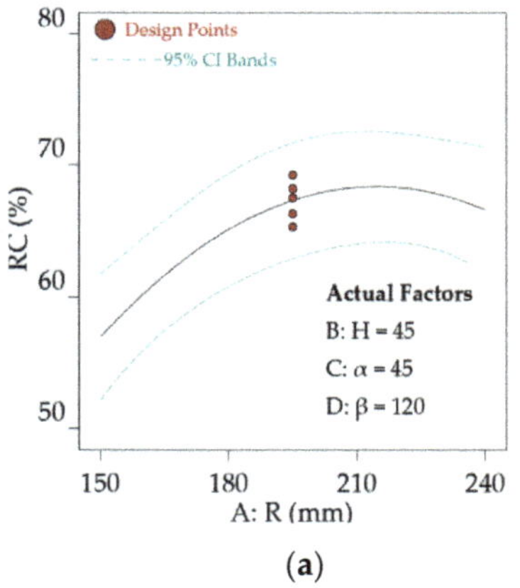

(a)

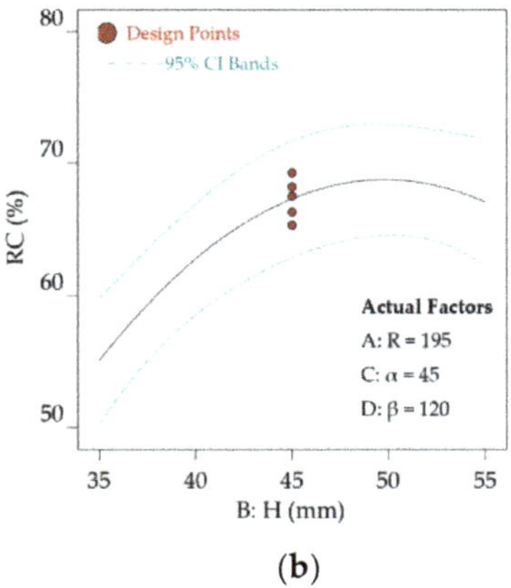

(b)

Figure 11. Rate of crushed soil value at different factors.

- Rate of straw burial (RB)

According to the ANOVA results in Table 9, the Model F-value of 11.95 implies the model is significant. There is only a 0.01% chance that an F-value this large could occur due to noise. p-values less than 0.0500 indicate model terms are significant. In this case, A, AD, BC, A^2, D^2 are significant model terms. Values greater than 0.1000 indicate the model terms are not significant. Removing the insignificant term, the regression equation of the rate of crushed soil model in this study is:

$$RB = 17.80 + 2.68A + 1.58AD - 1.91BC - 5.60A^2 - 1.44D^2 \quad (26)$$

Table 9. ANOVA for Quadratic model of Rate of straw burial.

Source	Sum of Squares	df	Mean Square	F-Value	p-Value	
Model	337.33	14	24.10	11.95	<0.0001	
A-R	86.40	1	86.40	42.86	<0.0001	**
B-H	1.78	1	1.78	0.8823	0.3635	
C-α	0.4107	1	0.4107	0.2037	0.6586	
D-β	6.42	1	6.42	3.19	0.0959	
AB	0.5112	1	0.5112	0.2536	0.6224	
AC	0.0289	1	0.0289	0.0143	0.9064	
AD	9.95	1	9.95	4.94	0.0433	*
BC	14.55	1	14.55	7.22	0.0177	*
BD	8.58	1	8.58	4.26	0.0581	
CD	0.8556	1	0.8556	0.4244	0.5253	
A^2	203.40	1	203.40	100.90	<0.0001	**
B^2	1.34	1	1.34	0.6656	0.4282	
C^2	4.12	1	4.12	2.05	0.1746	
D^2	13.49	1	13.49	6.69	0.0215	*
Residual	28.22	14	2.02			
Lack of Fit	21.72	10	2.17	1.33	0.4194	
Pure Error	6.51	4	1.63			
Cor Total	365.55	28				

** indicates highly significant ($p < 0.01$), * indicates significant ($p < 0.05$), and *p*-values greater than 0.1000 indicate the model terms are not significant.

The ANOVA results showed that the interaction of the factors in this model was significant. The R, Rβ, and Hα were significant among the factors. The trend of RB with R is shown in Figure 12a, which shows that the RB value increases and then decreases, with a maximum at R = 200 and a model prediction of 17.78%. This is due to the upper limit of the maximum straw burial capacity of a single blade role at the same tillage depth. Too small a blade radius causes the straw not to be fully tilled into the ground, and conversely, too large a radius causes the straw to be thrown farther by the blade, also preventing effective mulching. Figure 12b presents the interaction between R and β. It is clear that the effect of changing R on RB is similar to that shown in Figure 12a. Changing the value of β, when β is 110, the maximum value of RB is slightly less than that at 120. When β increases to 130, the extreme value of the variation curve decreases further, and the value of R is taken to 217.5. Figure 12c presents the interaction curve of H with α. It can be seen that the magnitude of α affects the trend of RB with H. This is because α affects the position of the bending line and, thus, the direction of the end line. α is 35, and RB decreases with increasing H. It can be seen that the larger H is at this angle, the more straw can be buried by a single blade. On the contrary, for α of 55, RB behaves in the opposite way to the above. The intersection points of the two are near H = 45.

3.2. Parameter Optimization and Comparison Test

The optimization function in Design-Expert 12.0 software was applied for the quadratic model in this study. The optimal parameters of the blade were obtained by solving the regression model with the conditions that the broken soil rate was taken as the minimum value of 65%, the minimum straw burial rate was 15%, and the power consumption was the lowest, as shown in Figure 13. The theoretical optimal parameters were obtained as the combination of radius (R) 209.761 mm, end face height (H) 45.8743 mm, pinch angle (α) 37.607°, and bending angle (β) 113.209°. In accordance with the actual fabrication process and practice, the values were taken as R = 210 mm, H = 45 mm, α = 37°, and β = 115°, etc.

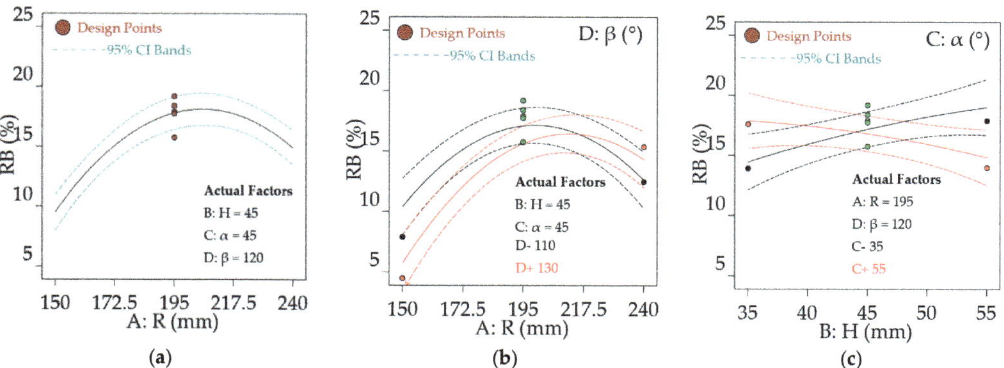

Figure 12. Rate of straw burial value at different factors.

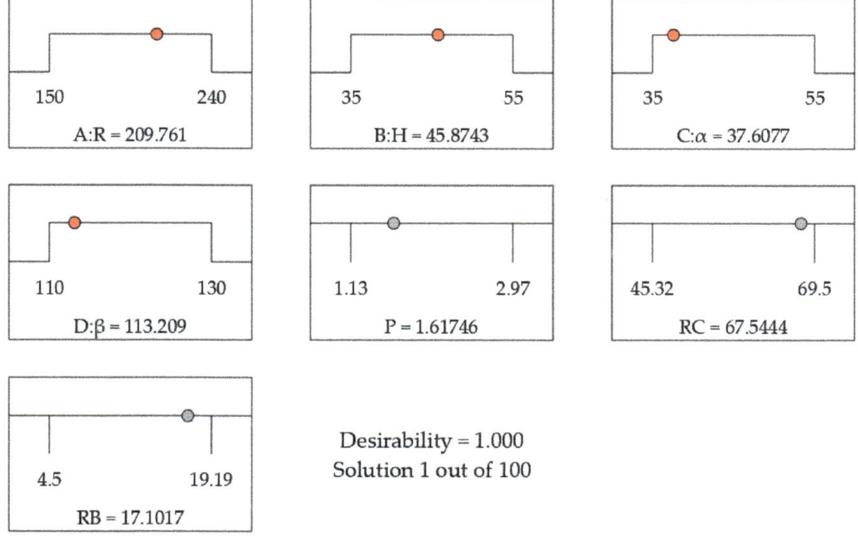

Figure 13. Parameter optimization of theoretical values.

A comparison test between the EDEM simulation and field was conducted under the optimal parameters to verify the reliability of the optimized parameters. Table 10 shows the results of this comparative test. The simulation test results showed that the power, soil breakage rate, and straw burial rate were 1.62 kW, 67.47%, and 17.21%, respectively; the field results showed that the power, soil breakage rate, and straw burial rate were 1.73 kW, 71.34% and 18.89%, respectively. The average error rates of the simulated and field test values were 6.36%, 5.42%, and 8.89%, respectively, which indicated that the model was a decent fit.

Table 10. Comparison of simulation and test results under optimal parameters.

Responses	Field Test	Simulation	Error (%)
P (kW)	1.73	1.62	6.36
RC (%)	71.34	67.47	5.42
RB (%)	18.89	17.21	8.89

4. Discussion

(a) Rotary blade radius R has the most obvious influence on the power consumption of single blade operation, while other indexes have no significant influence. If deep tillage is to be realized with small radius blades, the whole blade roll must be trapped in the inner part of the tillage layer in structural design, causing congestion and increasing power consumption; Conversely, using a blade with a large radius results in an increase in torque and a sharp increase in power consumption due to blade wear. The rotary tillage radius used in this paper is 210 mm, which can meet the purpose of 200 mm tillage depth and reduce power consumption at the same time. The reason for this is that a good mechanical structure and excellent operating performance make it stand out. Half of the rotary tiller rollers of this length fall into the tillage layer, which not only causes a sudden increase in power consumption caused by congestion but also achieves a good soil-cutting effect and reduces blade life loss;

(b) The H had the most obvious effect on the straw cover rate. The reason is that they are close to the end of the blade and come into contact with the straw earlier, which has the most obvious impact on spraying and burying of straw. One of the objectives of this paper is to increase the straw cover rate and depth in order to better meet the agronomic requirements, reduce the straw residue on the surface of cultivated land, and bury it as far as possible to a deeper position in the tillage layer, so as to prevent the phenomenon of shelving and airing, and provide a good seeding bed for seeding. Considering comprehensively, 45 mm can meet the agronomic demand:

(c) The α-influenced trend of each operation is different, so we can know by analyzing it. In the range of 35–55 with the increase of value, power consumption P first decreases and then increases and reaches the minimum value at 36. The reason for these changes is α. The size has different effects on the shape of the end of the blade. Too small a value will cause the forward cutting edge to turn outwards and the angle of penetration to become larger, resulting in flat cutting of the blade, which cannot effectively exert the soil cutting effect of the positive cutting edge of the blade. On the contrary, excessive value results in the form of a hook down on the positive cutting edge, which is easy to cause excavation of the positive cutting edge when cutting soil, and the soil carrying effect is improved, which is not conducive to the crushing of soil blocks;

(d) Bending angle β affected trend of each index is basically the same; only the RB is greatly affected. Based on the research results, 115 degrees is selected as the optimal solution. The change of its value results in the change of bending degree of the forward cutting edge. The larger the value, the closer the blade approaches the straight blade, the higher the cutting performance, and the lower the throwing effect. The smaller the value, the opposite.

It is the first time, to our knowledge, that the effect of rototill blades on soil and straw after deep rototilling has been analyzed using tillage soil stratification to describe the effect of rototill blades on soil and straw in a rice-wheat rotation. This study reveals the interaction between the rototill blade and the soil and straw, a finding that will assist in the development of technologies for deep rototill straw burial operations. In addition, the design of deep rototill blades with improved blade-breaking energy and increased straw mulching rates could lead to the provision of good seedbeds.

5. Conclusions

In this study, a discrete meta-simulation model of rotary tillage knife-soil-straw interaction was developed. The interaction law of soil-touching components in deep tillage operation was explored, and the field experiment results also verified its usability. The optimized structural parameters of the rotary tillage blade were 210 mm-45 mm-37°-115° (R-H-α-β). The simulated and field comparison tests were conducted under the optimal combination of parameters, and the power, soil breaking rate, and straw burial rate were 1.73 kW, 71.34%, and 18.89%, respectively; the numerical error rates of simulated and field test values were 6.36%, 5.42%, and 8.89%, respectively. The accuracy of the secondary model was verified.

Note that the aim of the study was to investigate the interaction of the rotary single blade with the soil straw in deep rototilling operations, applicable to optimizing the structural parameters of the rotary blade, and our experiments found that the soil movement was more consistent with the simulation results, which confirmed the validity of the simulation model. Papers exploring the arrangement of the rototill knives and the operating parameters have been published in the literature and will be the focus of our future work. Deep rototill mulching of straw will also be a hot research issue in the tillage chain.

Author Contributions: Conceptualization, J.Z. (Jin Zhang) and C.W.; methodology, J.Z. (Jiping Zhu); software, J.Z. (Jin Zhang); validation, W.C., M.X. and D.Y.; formal analysis, J.Z. (Jin Zhang); investigation, M.X.; resources, W.C.; data curation, J.Z. (Jin Zhang); writing—original draft preparation, J.Z. (Jin Zhang); writing—review and editing, J.Z. (Jiping Zhu); visualization, J.Z. (Jiping Zhu); supervision, W.C.; project administration, M.X.; funding acquisition, J.Z. (Jiping Zhu). All authors have read and agreed to the published version of the manuscript.

Funding: The author thanks Jiping Zhu of Nanjing Institute of Agricultural Mechanization, Ministry of Agriculture and Rural Affairs for funding this work through the National Key Research and Development Program of China (2020YFD1000802), Jin Zhang of Jiangsu Yanjiang Institute of Agricultural Sciences for Science Fund for Young Scholars of Jiangsu Yanjiang Institute of Agricultural Sciences (YJ2022006), Liu Jian of Jiangsu Yanjiang Institute of Agricultural Sciences Jiangsu Province Science and Technology Project (BE2020388), Jiangsu Province Modern Agricultural Machinery Equipment and Technology Demonstration and Extension Project (NJ2020-04).

Institutional Review Board Statement: Not applicable.

Data Availability Statement: The data presented in this study are available within the article.

Acknowledgments: Thanks are due to Zhenggang Liu for assistance with the experiments and to Yurong Zhang for valuable discussion.

Conflicts of Interest: The authors declare no conflict of interest.

References

1. Du, Z.H.; Chen, Y.Y.; Zhang, J.; Han, X.M.; Geng, A.J.; Zhang, Z.L. Development Status and Prospects of Rotary Farm Machinery in Domestic and Abroad. *J. Chin. Agric. Mech.* **2019**, *40*, 43–47.
2. Zhou, J.H.; Wang, J.Y.; Meng, F.Y.; Tong, G.X.; Mei, L.; Liu, G.M.; Wang, Y.; Luo, J.; Xie, C.Y. Effects of Tillage Methods on Sowing Quality Yield and Benefit of Wheat. *Crops* **2022**, *4*, 199–204.
3. Li, M.L.; Wang, W.D.; Han, K.; Wang, S.; Huang, Y.C.; Wang, Y.; Jia, B.Y. Influences of Rotary Tillage Depth, nitrogen Application Rate and Planting Density on Yield and Nitrogen Distribution and Utilization of Rice. *Jiangsu Agric. Sci.* **2021**, *49*, 55–61.
4. Wang, L. Several Problems Needing Attention in Returning Corn Straw to the Field. *Sci. Technol. Inf.* **2011**, *28*, 28–356.
5. Su, R.J. Environmental Pollution and Prevention of Agricultural Machinery. *Hebei Agric. Mach.* **2010**, *06*, 28–29.
6. Yan, B.J. Brief Analysis on Environmental Pollution of Agricultural Mechanization and Countermeasure. *Farm Mach.* **2008**, *14*, 51–52.
7. Xia, X.D.; Wu, C.Y.; Zhang, R.L.; Qiao, F.S. Development Prediction of Large and Efficient Rotary Tillage Machinery. *Rural. Mech.* **1999**, *1*, 75–78.
8. Liang, X.Z.; Zhu, S.; He, Y.P. On the Design of Rotary Cultivation Parts for Biaxial Layered Soil Cultivators. *Agric. Equip. Technol.* **2020**, *46*, 42–44.
9. Yang, Z.; Ge, X.Y.; Yang, D.H.; Lu, R.C.; Zhang, R.H. Research on Drive Improvement Design of Double-axle Rotary Tillage Duplex Operator. *Agric. Equip. Technol.* **2014**, *40*, 13–15.

10. Liu, Y.Q.; Sang, Z.Z. Mathematical Model of Submerged Reverse Rotary Tiller and Parameters Optimization. *Trans. Chin. Soc. Agric. Eng.* **2000**, *16*, 88–91.
11. Li, B.Q.; Chen, C.Y.; Liang, J. Proof of Theoretic Model for Latent Soil of Up-cut Rotary Cultivation and Optimization of Working Parameters. *J. Jiangsu Univ. (Natl. Sci. Ed.)* **2005**, *26*, 203–205.
12. Li, B.; Chen, C. An Analysis on Collision of Throwing Soil with Cover of Latent Soil of Up-cut Rotary Cultivation. *Trans. Chin. Soc. Agric. Mach.* **1999**, *06*, 41–45.
13. Chen, C.Y.; Gao, J.M. 3-d Scan and Data Treatment of Submerged Reverse-rotary Tiller. *J. Agric. Mech. Res.* **2007**, *29–31*, 56.
14. Yan, J.C.; Li, H.C.; Hu, J.P. Study on Calculation of the Up-cut Rotary's Throwing Soil Volume and Distribution. *J. Agric. Mech. Res.* **2014**, *36*, 29–33.
15. Liu, Y.Q.; Sang, Z.Z. Effect of the Parameters of Submerged Reverse-rotary Blade on Energy Consumption. *J. Jiangsu Univ. Sci. Technol.* **2001**, *04*, 15–18.
16. Jin, Y.; Gao, J.M. The Dynamic Balance of Cutter Roller Simulation Analysis of Oblique Rotary—Based on Pro/e and Adams. *J. Agric. Mech. Res.* **2013**, *35*, 160–162, 167.
17. Kong, L.D.; Sang, Z.Z.; Wang, G.L. The Development of Three Dimensional Convexity of Oblique Rotary Tillage. *Trans. Chin. Soc. Agric. Mach.* **2000**, *31*, 28–31.
18. Kong, L.D.; Sang, Z.Z.; Wang, G.L. Experimental Study on Oblique Rotary Tillage. *Trans. Chin. Soc. Agric. Mach.* **2000**, *2*, 31–34.
19. Kong, L.D.; Sang, Z.Z. The Research of Oblique Rotary Tillage. *J. Taiyuan Heavy Mach. Inst.* **1999**, *31*, 310–313, 343.
20. Zeng, Z.W.; Ma, X.; Cao, X.L.; Li, Z.H.; Wang, X.C. Critical Review of Applications of Discrete Element Method in Agricultural Engineering. *Agric. Eng.* **2021**, *11*, 29–34.
21. Fang, H.M.; Ji, C.Y.; Farman, A.C.; Guo, J.; Zhang, Q.Y.; Chaudhry, A. Analysis of Soil Dynamic Behavior During Rotary Tillage Based on Distinct Element Method. *Trans. Chin. Soc. Agric. Mach.* **2016**, *47*, 22–28.
22. Mustafa, U.; John, M.F.; Chris, S. 3D DEM tillage simulation: Validation of a hysteretic spring (plastic) contact model for a sweep tool operating in a cohesionless soil. *Soil Tillage Res.* **2014**, *144*, 220–227.
23. Fang, H.M.; Ji, C.Y.; Zhang, Q.Y.; Guo, J. Force Analysis of Rotary Blade Based on Distinct Element Method. *Trans. Chin. Soc. Agric. Eng.* **2016**, *32*, 54–59.
24. Sakaguchi, H. Plugging of the Flow of Granular Materials During the Discharge from a Silo. *Int. J. Mod. Phys. B* **1993**, *7*, 1949–1963. [CrossRef]
25. Cundall, P.; Strack, O.A. Discrete Numerical Model for Granular Assemblies. *Geotechnique* **1979**, *29*, 47–65. [CrossRef]
26. Hertz, H. On the Contact of Elastic Solids. *J. Fur Die Reine Und Angew. Math.* **1882**, *92*, 156–171. [CrossRef]
27. Mindlin, R.D.; Deresiewicz, H. Elastic Spheres in Contact Under Varying Oblique Forces. *J. Appl. Mech.* **1953**, *16*, 327–344. [CrossRef]
28. Mindlin, R.D. Compliance of Elastic Bodies in Contact. *J. Appl. Mech.* **1949**, *16*, 259–268. [CrossRef]
29. Tsuji, Y. Lagrangian Numerical Simulation of Plug Flow of Cohesionless Particles in a Horizontal Pipe. *Powder Technol.* **1992**, *71*, 239–250. [CrossRef]
30. Potyondy, D.; Cundall, P.A. Bonded-particle Model for Rock. *Int. J. Rock Mech. Min. Sci.* **2004**, *41*, 1329–1364. [CrossRef]
31. Zhang, J.; Chen, W.; Tashi, N.; Zhu, J.P.; Ding, Y.; Yuan, D.; Yao, K.H.; Xia, M. Design and Analysis of Energy-saving Forward Deep rotary Tillage Machine. *Agric. Mech. Asia* **2021**, *52*, 4935–4949.

 *agriculture* 

Article

Intelligent Algorithm Optimization of Liquid Manure Spreading Control

Pengjun Wang, Yongsheng Chen, Binxing Xu, Aibing Wu, Jingjing Fu, Mingjiang Chen * and Biao Ma

Nanjing Institute of Agricultural Mechanization, Ministry of Agriculture and Rural Affairs, Nanjing 210014, China
* Correspondence: chenmingjiang@caas.cn

Abstract: The growth of field crops needs appropriate soil nutrients. As a basic fertilizer, liquid manure provides biological nutrients for crop growth and increases the content of organic matter in crops. However, improper spraying not only reduces soil fertility but also destroys soil structure. Therefore, the precise control of the amount of liquid manure is of great significance for agricultural production and weight loss. In this study, we first built the model of spraying control, then optimized the BP neural network algorithm through a genetic algorithm. The stability and efficiency of the optimized controller were compared with PID, fuzzy PID and BPNN-PID control. The simulation results show that the optimized algorithm has the shortest response time and lowest relative error. Finally, platform experiments were designed to verify the four control algorithms at four different vehicle speeds. The results show that, compared with other control algorithms, the control algorithm described here has good stability, short response time, small overshoot, and can achieve an accurate fertilizer application effect, providing an optimization scheme for research on the precise application of liquid manure.

Keywords: precision agriculture; fertile field machinery; flow control; application of liquid manure; GA-BPNN-PID control

Citation: Wang, P.; Chen, Y.; Xu, B.; Wu, A.; Fu, J.; Chen, M.; Ma, B. Intelligent Algorithm Optimization of Liquid Manure Spreading Control. *Agriculture* **2023**, *13*, 278. https://doi.org/10.3390/agriculture13020278

Academic Editors: Cheng Shen, Zhong Tang and Maohua Xiao

Received: 5 January 2023
Revised: 18 January 2023
Accepted: 19 January 2023
Published: 23 January 2023

1. Introduction

The spraying of liquid manure is an important direction in precision agriculture research. This technology can address the issues of high labor intensity, low efficiency of throwing liquid organic manure, and uneven artificial fertilization [1–3]. Liquid manure contains a large amount of soluble nutrients and a variety of bioactive substances, such as amino acids and trace elements. The reasonable application of liquid manure can promote plant growth, increase yield, reduce fertilizer cost, raise soil organic matter content, and add soil chemical and physical latent capacity. Studies have shown that precision spraying technology can increase crop yield by 8–19% on average, reduce fertilizer application by approximately 30%, and improve soil quality [4–6]. At present, although the amount of liquid manure used is increasing annually worldwide, crop yields have not increased correspondingly. An analysis of liquid manure application indicates that the main reason for this is the large difference in the amount of liquid manure application per mu of land. Especially at the edge of the plot, the excessive application of liquid manure on some plots and the insufficient application of liquid manure on others due to manual control of fertilizer application or changes in vehicle speed directly affects the normal growth of crops in subsequent periods. In addition, excessive fertilization damages the soil structure and affects the growth environment and survival rate of crops. Hence, to increase the utilization coefficient of liquid manure and ensure the normal use of land resources, a more accurate and reasonable liquid manure spreading control technology should be adopted.

The main component of liquid manure is biogas slurry, which can be used to spread the base fertilizer in the field. Generally, the amount of basic fertilizer per hectare is 55 L. When biogas slurry is used as a basic fertilizer in the field, its dosage should be strictly controlled to prevent ammonia poisoning. Therefore, the precise application of liquid manure is fundamental and represents an important field of research within precision agriculture for the rapid and accurate adjustment of the amount of liquid manure fertilizer applied, as well as to maintain the set requirements. Research on precision fertilization mainly involves the optimization of algorithms and the establishment of a fertilization system model. At present, research on the control modeling of the fertilization process mainly includes nonlinear and linear artificial intelligence models. Within nonlinear control optimization, feedback linearization transformation control, adaptive regulation, and other methods are commonly used. Since over the traditional PID (proportion, integral, derivative) control method it is difficult to deal with complex nonlinear and hysteretic spraying control in the fertilizer transportation process, there is a need to develop a more intelligent optimization algorithm in order to precisely adjust and control the spraying process [7].

In the process of liquid manure spraying, most spraying controllers internally need an experience PID control strategy. However, in the actual spraying course, the driving speed and intelligent algorithm have a significant impact on the control system, leading to problems, including time variation and the hysteresis of the controlled object of the system, that cannot be solved. In this context, the traditional PID control algorithm is usually ineffective for variable spraying. A previous study [8] used the fuzzy neural network PID control of particle swarm optimization to adjust the fertilization system and found that the control algorithm had a small overshoot, excellent constancy, and a short rise time, which can bring about the aim of a good fertilization system. In another study [9], the authors proposed a fuzzy PID control method according to a genetic algorithm to address the problem of a long response time in the control process of an electric proportional valve. Their experimental verification identified the control method as superior to the traditional PID operating system together with the fuzzy PID control system. Another study [10] used traditional PID control to adjust the deviation and deviation rate of the PH value in fertilizer, and found that the cognoscenti PID control exhibited fine control behavior. Chang, C. et al. [11] aimed at the problem of pressure fluctuation caused by high-frequency opening and closing of valves during liquid fertilizer point application; they designed a high-frequency intermittent fertilizer supply system to optimize the system, and the parameters of the PID algorithm were adjusted using the critical proportion method. The system had certain stability. Li, T. et al. [12] put forward a way to precisely fertilize maize in view of the wavelet BP neural network algorithm to compute the non-linear matter of fertilization, which improved the accuracy of the optimal fertilization amount. In another study [13], the nonlinear model of the elastic BP neural network along with mixture gray wolf optimization was used for predictive control. The corresponding simulation dispatch demonstrated that the controller was capable of effectively diminishing the jam due to nonlinearity and exhibited a good performance. Xiuyun et al. [14] designed a variable-rate deep application system for liquid fertilizer based on ZigBee. ZigBee was used for networking and realizing short-distance wireless communication between upper and lower computers and achieved accurate fertilization by controlling the frequency of the variable-frequency pump. The control system was optimized using the incremental integral derivative algorithm, and the control effect was good. A previous study [15] proposed a simplified and linearized deep construction model based on theoretical considerations using frequency domain identification technology estimation. According to the dynamic changes of the system in the process of fertilization, the worst case of stability of the depth control system was found, and the parameters of the fertilization model were determined to achieve shallow and deep fertilization. Guangkun [16] designed a variable-rate fertilizer injection operating system for fluidity fertilizer that optimized the operating system by using a PID algorithm. With the wheeled-point fertilizer injector as the carrier, the authors achieved a good variable-rate fertilizer application effect. However, owing to the simple control algorithm, the control of the flow and pressure continued to have a large optimization space. Accordingly, in studies on the variable-rate spraying process control system of liquid manure, there are many control methods for valves and variable-frequency pumps, with most research methods optimizing the control of valves with PWM or using a PID algorithm. The fuzzy control and neural network control algorithms were studied based on PID control. In a traction-type liquid manure fertilizer applicator [10] with slow current velocity monitoring retroaction, the calculation of liquid fertilizer concentration, together with the reaction time of the solenoid valve to modify the opening of valve in line with the need, were discovered as significant elements to consider in the alternating-quantity spread manure control system [12,17,18].

With this kind of situation of automatically spread manure, the aim of this study is improve the accuracy of the amount of liquid manure in the process of spraying, as follows: (i) optimize the parameters of the spraying process; (ii) apply the BPNN–PID algorithm for optimizing dominant weight parameters; (iii) simulate and check the conventional PID dominants and BPNN–PID dominants, as well as BPNN–PID dominants optimized using a genetic algorithm on the MATLAB/Simulink plate; (iv) evidence the practicability as well as stability of the recommended algorithm. The outcome identified that the optimized dominate algorithm had fewer errors and shorter system response time than before.

The remainder of this paper is organized as follows. In the first part of this study, the control process of liquid manure spraying is analyzed and a transfer function model is established. The second part constructs the classical PID pilot together with the BPNN–PID pilot and optimizes the BPNN–PID using the genetic algorithm. First, the control signal is received and collected. After that, the BPNN system is built, as well as a genetic algorithm is used to multilayer the neural network. The third part contrasts the pilot impression of three pilot methods by way of software simulation along with test checking, as well as, lastly, achieving the experimental fruit of the liquid manure spraying control system based on the present investigation.

2. Materials and Methods

2.1. Working Principle of Liquid Manure Spreader

The liquid manure spreader used in this study was towed using a tractor, and the liquid manure was spread to the field ground through 12 spray pipes. The operating breadth of the fluidity manure spout pipe was 6 m and the driving velocity of the tractor used for towing varied from 1 to 2.5 m/s. An important part of the system for controlling the change in liquid manure flow was the electric proportional valve, which was installed at the front end of the fertilizer distributor. The distributor used in this study was a mass-production product with a uniform distribution. We did not need to consider its distribution uniformity and blockage but instead only needed to consider the open–shut sizes of the electric proportional gate controlled via computer to achieve accurate control of the liquid manure distribution. Based on the requirements for the accurate application of liquid manure, the principle of system flow control was based on the demand for liquid manure in the field. Turning on the solenoid valve was transformed by vehicle speed. The application amount was monitored using a fluid meter. Then, the data were coupled back to the regulator. The control formed a closed–cycle coupled-back uniform machine by analyzing and comparing the actual time runoff, the magnitude of fertilizer called for by the current plot, and the current vehicle speed, achieving an accurate control of the current plot spraying amount. The response accuracy and stability of the closed–cycle coupled-back uniformity machinery were the main influences of the exact fluidity manure spray control system in this study.

2.2. Equipment System Components

The formation of the manure spreading command equipment is exhibited in Figure 1. During fertilization, the chief command pattern, with the host computer as the main control unit, was used to control the open–shut the solenoid valve (Proportional solenoid valve, Shanghai Kaiweixi Co., Shanghai, China). An angular speed sensor (Analog output, Beijing Tianyuhengchuang Co., Beijing, China) was used to collect the wheel speed and convert it into the vehicle driving speed. The instantaneous flow rate of liquid manure was collected through the flowmeter as input data. The controller output the current signal to the solenoid valve after algorithm conversion. The solenoid valve was used to control valve opening, and the output of the system was the liquid manure flow. Then, the real–time data collected with the fluid meter were coupled back to the regulator, through which the closed–loop negative feedback control was realized.

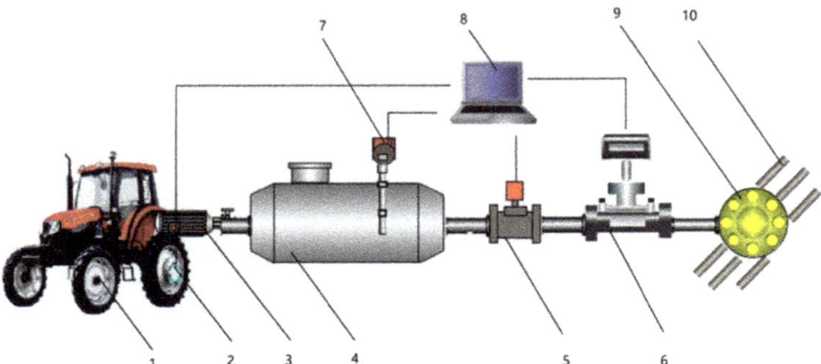

Figure 1. Structure diagram of fertilizer throwing control device: 1: tractor; 2: angular speed sensor; 3: pump; 4: liquid manure storage tank; 5: electric proportional valve; 6: flow meter; 7: liquid level meter; 8: monitoring and control by computer; 9: distributor; 10: sprinkler pipe.

2.3. Control System Model Construction

The operation of the main liquid manure spraying system decreased the response time of the operating system. This established the liquid manure spraying operating system transfer function. The liquid manure spreading input operating channel model was collected in real-time using an angular velocity sensor. The controller output the current signal to the solenoid valve after conversion. A solenoid valve was used to control valve opening. The final output of the system was the instantaneous current speed of liquid manure. Current speed was coupled back to the control system through the flowmeter in the flow chart of the operate channel, as exhibited in Figure 2. Finally, control via the inverse back-coupling closed cycle was conducted through the control system.

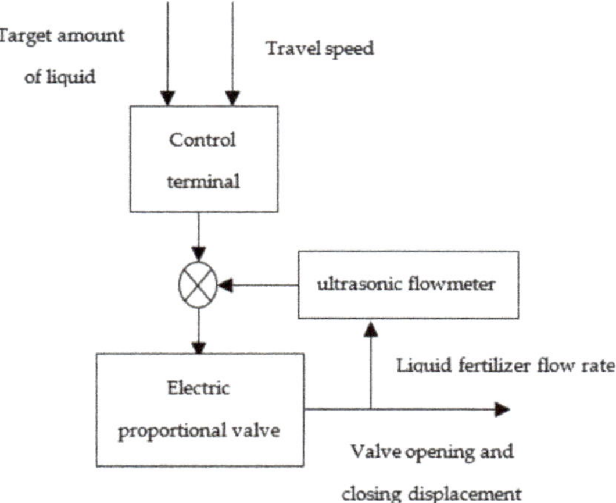

Figure 2. Control system diagram of liquid manure spreading.

In the light of input/output relationships in the operating system flow chart, the system input/output relationships are presented (Equation (1)):

$$V = f(q, v) = q \cdot 10^{-4} \cdot w \cdot v \tag{1}$$

where V (L/min) is liquid manure volume exportation from the precision spray operating channel, F (q, v) is the transversion function, vehicle running speed is v(m/s), Q is the target fertilization amount input into the system according to the amount of liquid fertilizer used in corn fields in Tai'an, Shandong Province (q was set as 55 L/hm^2), and W (m) is the width of fertilization. According to the vehicle conditions in this study, w = 6 m.

According to the liquid manure spraying operating channel in Figure 2, the system's input feedback data were from the instantaneous discharge registration through the flowmeter. The current signal was the control system signal exported through the retroaction passageway. The control system switched signals and compared them with the target fertilizer amount and vehicle speed input to implement the inverse back-coupling control.

Thus, the control model feedback link use indicates Equation (2):

$$H(s) = \frac{v(s)}{V(s)} = \frac{700}{s \cdot 5 \cdot w} \cdot e^{-\tau s} \tag{2}$$

where t (s) is the delay time of preference transmission in the feedback link. The transfer function aleatory variable after the Laplace transform is S, and the transfer function inverse back-coupling chain is H.

The discharge monitoring of the operating channel via flowmeter in this study was real-time online. The retroaction latency of link time could be left out in the light of the actual hardware situation.

In the light of actual demand for a liquid manure precise spraying operating channel, the solenoid valve was our major control aim. We chose a Shanghai Kaiweixi Co. VB7200 solenoid valve. Figure 3 shows the signal control chart.

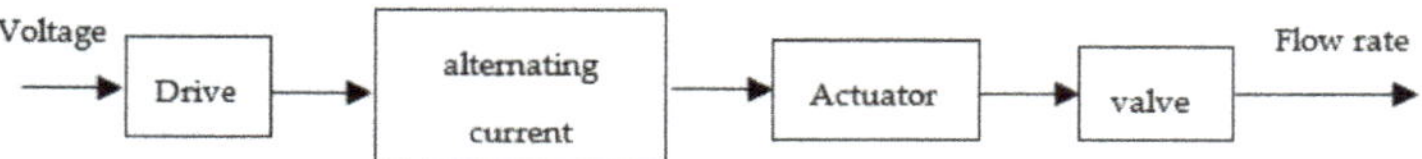

Figure 3. Signal control chart of solenoid valve.

In Figure 3, propel module input/output signals were electricity semaphores. Transfer function was a ratio and delay segment. The relationship was as follows (Equation (3)):

$$G_1(s) = \frac{I_{out}}{I_{in}} = k_s \cdot e^{\tau s} \tag{3}$$

where k_s is the magnify quotient of the translator. I_{in} as well as I_{out} are current signal input/output with the propel module, respectively, and $G_1(s)$ is the propel module's transfer function.

Drive module current signal transmission delay (τ) was set to $\tau < 0.02$ s. Therefore, τ was left out in the system, such that the propel module's transfer function indicated a proportional sector [19].

The solenoid valve had a DC motor. The electricity signal was the dominant input signal. The angle of the electrical machinery's axle was the output. The DC motor's electrocircuit of semaphore control included an armature loop balance and induction of the rotor, as well as the balance of the axle couples of electrical system. Equation (4) demonstrates the balance equation:

$$\begin{cases} i_a(t) - E = R_a \cdot u(t) + \frac{du(t)}{dt} L_a \\ E = \omega_n \cdot k_{ep} = \dot{\theta}_n \cdot k_{ep} \\ u(t) \cdot k_T - M_1 = \dot{\omega}_n \cdot J_n = \ddot{\theta}_n J_n \end{cases} \tag{4}$$

Time (s) is expressed as t. Entering electricity in the DC motor is $i_a(t)$(mA). The motor's electromotance is expressed by E (V). The armature voltage is expressed by u(t) (V). Total armature resistance is expressed by R_a (Ω). The armature's inductance is expressed by L_a(H). k_{ep} is the back electromotive force coefficient. The angle of electrical machinery's axle is expressed by θ_n (°). The motor torque coefficient is displayed as K_T. Motor load torque is displayed as M_1 (N···m, $M_1 = f \cdot \ddot{\theta}_n$, where f is the friction coefficient). The rotational inertia of the rotor's moment of inertia is expressed by J_n (kg···m^2). Finally, angular speed of the motor rotor is displayed as ω_n (rad/s).

The DC motor transfer function in the solenoid valve was obtained by transforming Equation (4) using a Laplace transformation, as shown in Equation (5):

$$\begin{aligned} G_2(s) &= \frac{\theta_n(s)}{I_\alpha(s)} = \frac{k_T}{k_{ep}k_T s + R_\alpha(J_n+f)s^2 + L_\alpha(J_n+f)s^3} \\ &= \frac{0.51}{2.2\times10^{-3}\cdot s + 5.8\times10^{-5}\cdot s^2 + 8.7\times10^{-5}\cdot s^3} \end{aligned} \tag{5}$$

where $\theta_n(s)$ is the angle of the electrical machinery's axle with the Laplace transform function. $I_\alpha(s)$ is the Laplace transform function of the motor input current, and $G_2(s)$ is the motor's transfer function. We can see Table 1 for simulation parameters [9].

Table 1. Simulation parameters.

K_T	k_{ep}	R_a	J_n	f	L_α
0.51	4.32×10^{-3}	1.02×10^{-4}	0.27	0.3	1.527×10^{-4}

The gear set was the main component of the reducer. The displacement output of the valve element was the axis speed of DC motor after deceleration. The valve element displacement, 0–18 mm, is expressed as X. The valve's opening was translocation of the valve element.

The transmission relevance of decelerator input/output is expressed as the reduction ratio, which adopts the proportional dominant mode. The transfer function is expressed as follows (Equation (6)):

$$G_3(s) = \frac{X(s)}{\theta_n(s)} = \frac{Y}{2\pi p} = 8.19 \times 10^{-6} \tag{6}$$

where the decelerator gear ratio is displayed as p. The lead of the drive rod is expressed with Y (mm). The spool traversed by the Laplace function is expressed by $X(s)$. The reducer transfer function is demonstrated with $G_3(s)$.

During this research, the flow and opening of the solenoid valve were linear under the same-pressure operating mode. Therefore, Equation (7) shows the relevance within flow and opening:

$$G_4(s) = \frac{V(s)}{X(s)} \tag{7}$$

The flow as well as the valve opening's transfer function is $G_4(s)$.

In Figure 3, the plant of the front path is guided by the solenoid valve. Equation (8) shows its transfer function:

$$G(s) = \frac{V(s)}{I_\alpha(s)} = G_1(s)\cdot G_2(s)\cdot G_3(s)\cdot G_4(s) \tag{8}$$

where the transfer function of solenoid valve is defined by $G(s)$.

As can be seen from the control model as well as the functions of each control activity, in this research, Equation (9) is the closed–cycle retroaction control transfer function of the precision broadcast application control system:

$$G_z(s) = \frac{G(s)}{H(s)G(s)+1} = \frac{4.17}{64.3s + 16.9s^2 + 25.4s^3 + 2.92} \tag{9}$$

The liquid manure spreading system's dominant transfer function is $G_z(s)$.

2.4. PID Control Method

The PID control systematic is a classic control systematic. Figure 4 shows its operating system picture. The difference between the actual output signal of the controlled object and the given signal and tracking is the primary PID control method. The control rate (u) of the proportional (P), integral (I), and differential (D) systems was used as the main index for the analysis of the performance of liquid manure spraying. When the flowmeter feeds back the actual flow to the PID controller, the PID controller first reflects the difference between the actual flow and the demand flow in proportion. The output u (t) is directly proportional to the input deviation e (t), which can quickly reflect the deviation and thus reduce the deviation. Then, the function of the differential link can reflect the change trend (change rate) of the deviation signal, and can introduce an effective early correction signal into the system before the value of the deviation signal becomes too large, in order to speed up the action speed of the system and reduce the adjustment time. Finally, the integration link is used to eliminate the static error and improve the error-free degree of the system. The setting parameters of PID are provided by experience, through which the opening of the electromagnetic proportional regulating valve can be adjusted quickly. Through the feedback value of the flowmeter, the opening of the electromagnetic proportional valve is continuously adjusted to realize the PID control of liquid manure spraying. However, owing to the influence of nonlinearity and time delay in the process of liquid manure spreading, the precision of utilizing the classical PID dominant method to adjust liquid manure spreading is poor. Tracking variations in the application amount cannot be performed well with liquid manure spreading [20].

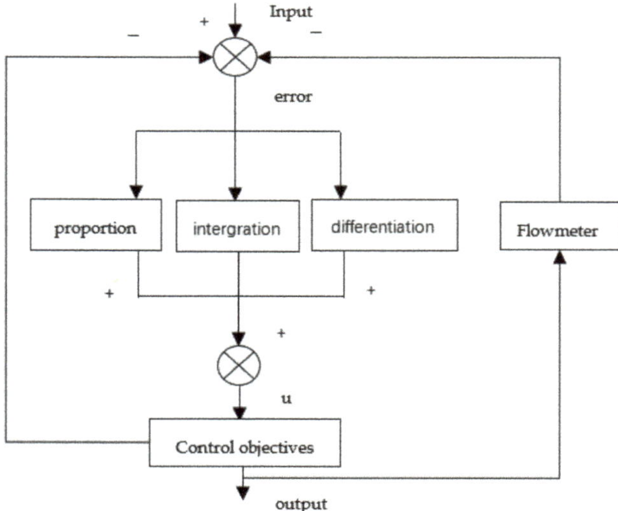

Figure 4. PID dominant method block diagram.

2.5. Devise of BPNN–PID Controller

In this study, a BPNN–PID controller was designed according to the control requirements of an alternating–quantity scatter fertility control system for liquid manure. Figure 5 exhibits a flow chart of BPNN–PID control system.

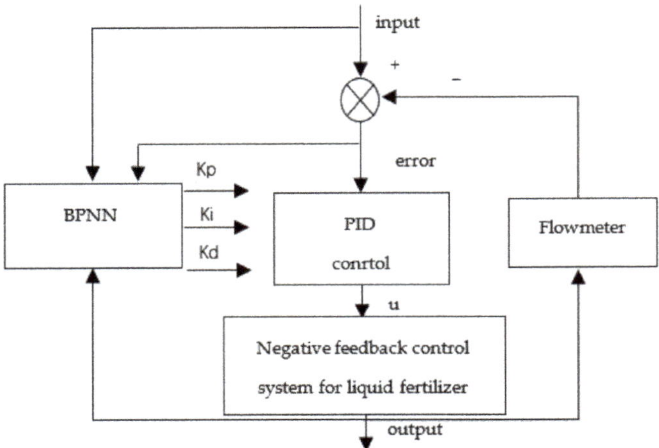

Figure 5. BPNN–PID dominant system flow chart.

The BPNN is a three–layer front-feed network. It includes three layers of network structure: the input layer, hidden layer, and output layer. The quantities of panel points in the import and export layers are decided by the dimensions of the input and output vectors, respectively. The concealed layer has a significant role in the ability as well as the composition of the system [21].

The BPNN–PID regulator was applied with three layers to define neurons with proportional, integral, and differential functions to integrate the regulation of PID control into the network. Through a teacher-free learning mode and online self-learning, adjustment could be carried out according to the control effect to ensure that the system performed well.

The first layer was the input layer, wherein the input was the preset fertilizer demand. The hidden layer was the second layer, wherein the entry was as follows (Equation (10)):

$$\begin{cases} net_i^2(k) = \sum\limits_{j=1}^{3} N_j^{(1)}(k) \cdot \theta_{ij}^{(2)} \\ N_j^{(2)}(k) = f\left[net_i^{(2)}(k)\right] \\ (i = 1, 2, \ldots, 9) \end{cases} \tag{10}$$

where θ_{ij} is the hidden layer weight divisor. $f[\cdot]$ is the activation function in the hidden layer. The output layer was selected as third layer, with the following input and output: $f(x) = tanr(x)$.

The output layer was the last layer, and the input and output can be seen from Equation (11):

$$\begin{cases} net_{cj}^{(3)}(k) = \sum\limits_{i=1}^{9} \theta_{ci}^{(3)} N_i^{(2)}(k) \\ N_c^{(3)}(k) = h\left[net_c^{(3)}(k)\right] \\ (c = 1, 2, 3) \\ N_1^{(3)}(k) = k_p \\ N_2^{(3)}(k) = k_i \\ N_3^{(3)}(k) = k_d \end{cases} \tag{11}$$

Among $\theta_{ci}^{(3)}$, Li is the weight factor of the export layer. $F[\cdot]$ is the activation function of the output layer, wherein $h(x) = i(x)/(i - x + ix)$ was selected.

After the BPNN–PID control outputs, KP, KI, and KD were determined; these were substituted for the bulking PID expression and calculated. Substitute PID control was used to memorize and maintain the system condition during above moments to minimize the impact of possible errors. Therefore, the optimal control signal was obtained by using incremental PID, as follows (Equation (12)):

$$\begin{cases} Mu(k) & = K_I e(k) + K_P[e(k) - e(k-1)] \\ & +K_D[e(k) - 2e(k-1) + e(k-2)] \\ & u(k) = u(k-1) + M_u \end{cases} \tag{12}$$

where $M_u(k)$ is the pilot signal spike, time is K, the scale factor is K_p, the integration factor is K_I, and the differentiation factor is K_D.

2.6. Neural Network PID Control Algorithm Optimized by Genetic Algorithm

The study procedure of the BPNN mainly involves constantly adjusting the weight arguments of every network layer, followed by use of the weight ratio to count the optimum dominant arguments. Hence, the BPNN must study and constantly renew the weighting ratio array of every network layer.

2.6.1. Updating BPNN Weight Factor

In light of the BPNN study method, we delimit the target cost function L:

$$L = \frac{1}{2}e^2(k) = \frac{1}{2}[d(k) - O(k)]^2 \tag{13}$$

When the target cost function L was determined, a search was performed in the direction of the negative gradient. The objective was to minimize the target function L and add an inertial term to accelerate seek convergence to the global minima, as below:

$$\Delta\theta_{cj}^{(3)}(k+1) = \beta\Delta\theta_{cj}^{(3)}(k) - \eta\frac{\partial u(k)}{\partial\theta_{cj}^{(3)}} \tag{14}$$

where θ is the study rate and β is the ratio of inertia.

Thus, it was further concluded that the modified formula for the weight factor of the export layer of BPNN is:

$$\begin{cases} \Delta\theta_{cj}^{(3)}(k+1) = \beta\Delta\theta_{cj}^{(3)}(k) + \eta\gamma_c^{(3)}D_i^{(2)}(k) \\ \gamma_c^{(3)} = \frac{\partial u(k)}{\partial D_i^{(3)}(k)}\cdot f'\left[net_i^{(3)}(k)\cdot e(k+1)sgn\left(\frac{\partial o(k+1)}{\partial u}\right)\right] \\ (c = 1,2,3) \end{cases} \tag{15}$$

According to over–calculation, the right formula for the hidden layer weight ratio was obtained as below:

$$\begin{cases} \Delta\theta_{ij}^{(2)}(k+1) = \beta\Delta\theta_{ij}^{(2)}(k) + \eta\gamma_i^{(2)}D_j^{(1)}(k) \\ \gamma_i^{(2)} = \sum_{c=1}^{3}\gamma_c^{(3)}\theta_{ci}^{(3)}(k)\cdot g'\left[net_i^{(2)}(k)\right] \\ (i = 1,2,3,4,\ldots,9) \end{cases} \tag{16}$$

where we set:

$$f_0(x) = f(x)[1 - f(x)]; f_0(x) = 1 - f_2(x)/2 \tag{17}$$

Therefore, the BPNN–PID control may be summarized as below: setting initialization of the three–layer system of BPNN arguments, together with inertia factor and study rate; using a BPNN control to process obtained error value; inputting the determined liquid fertilizer flow into the algorithm; and, after that, calculating the input/output of every layer via BPNN. Calculation of the three–layer arguments k_p, $k_{i,}$ as well as k_d of the PID control was performed using Expression (10). The export control signal is denoted by q (k).

The systematic sustained renewal of q (k) along with the weight ratio of network on basis of the renew error value. A schematic of the control strategy is displayed in Figure 6.

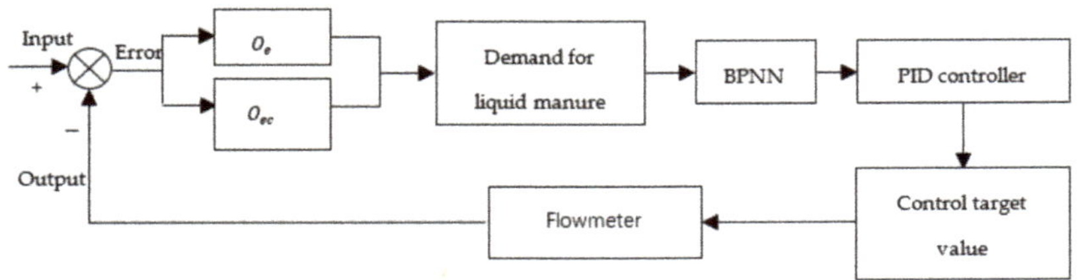

Figure 6. Control schematic diagram of BPNN–PID regulator.

2.6.2. BPNN–PID Algorithm Is Optimized using Genetic Algorithm

Recommending the Genetic Algorithm

A genetic algorithm screens individuals based on the chosen adaptation function combined with choice, intersection, and variation within the genetic process, retains those with good fitness value and eliminates those with inadaptable ones. The recent generation inherits information from the former generation and is better than the former generation. This loop was duplicated until the situation was met.

Optimizing BPNN Algorithm by Improving Genetic Algorithm

GA–BPNN–PID uses the genetic algorithm to optimize the weight and threshold of BPNN so that the optimized BPNN can better realize the function output. Although the GA–BPNN–PID control method is larger than the traditional method, the optimized BPNN weight and threshold can provide better control effect for the liquid manure spraying control algorithm. Both simulation and platform tests have verified that the algorithm has small overshoot, good stability, short rise time, and can better achieve the effect of precise spraying.

In this study, optimizing the BPNN has three parts: determination of the BPNN composition, optimization via genetic algorithm, then prediction by BPNN. The corresponding processes are illustrated in Figure 7. The structure of BPNN was determined according to the input and output parameters. In this study, the input was vehicle speed and the outputs were Kp, Ki, Kd. According to the above input and output data, we set the BPNN structure as 1-6-3, that is, the input layer had 1 node, the hidden layer had 6 nodes, and the output layer had 3 nodes, with a total of 19 weights and 9 thresholds. Therefore, the individual coding length of the genetic algorithm was 28. The BPNN prediction used a genetic algorithm to obtain the initial weight and threshold assignment of the optimal individual to the network. Finally, the prediction function was output after training. The GA–BPNN–PID structure diagram is shown in Figure 8.

The essential factors of BPNN optimized using genetic algorithm t included the initial population, fitness function, select action, crossover operator, and revisited mutations.

The genetic algorithm majorization main theory of the BPNN–PID dominant algorithm involved optimizing the initial value as well as the threshold of BPNN using a genetic algorithm, such that BPNN optimization could better forecast function export. Simultaneously, genetic algorithm optimization of the BPNN input state was used to reduce the network output and was insensitive to the system import, which was likely lag. The corresponding stages were as below:

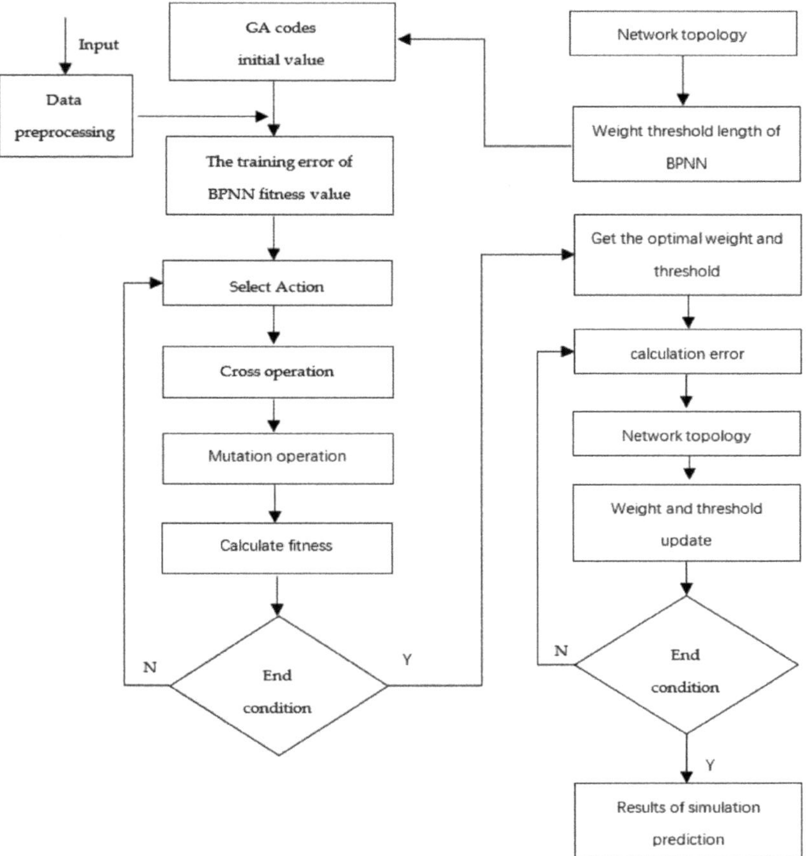

Figure 7. Flow chart of GA–BPNN.

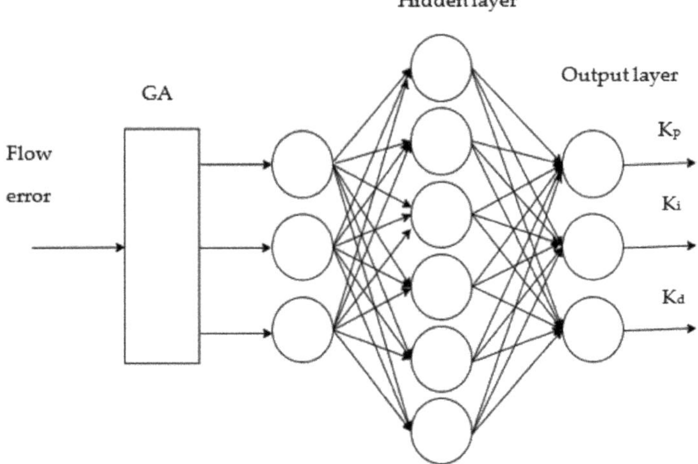

Figure 8. GA–BPNN–PID structure diagram.

(1) Population initialization performed personality coding according to the actual amount or number coding. Every personality was the actual amount or number bunch. It was composed of four sections: link weight of import layer as well as hidden layer, the hidden layer threshold value, the link weight of the hidden layer and the export layer, and export layer threshold. When solving continuous optimization problems, the commonly used binary coding methods have a long code string, and the encoding and decoding operations take up more time. Therefore, this study did not use binary encoding and decoding. During this optimization, we used real numbers for coding. The real number used was the vehicle speed. The optimized algorithm could overcome the uncertainty of the continuous optimization process and improve the response time and accuracy of the optimized algorithm.

(2) An individual obtained an initial weight value and threshold value in the BPNN. The output of the forecast machinery was predicted after training the BPNN on the training data. The absolute value of the error among the forecast export and expected output, as well as E, was compared with the personality fitness price Y. The count formula is as follows (Equation (18)):

$$Y = k\left[\sum_{i=1}^{n} abs(O_i - h_i)\right] \tag{18}$$

where n is the quantity of network export panel points, o_i is the BPNN's desired export of the ith panel point, h_i is the forecast export of the ith panel point, and k is a ratio.

(3) The choice manipulation of a genetic algorithm is the choice policy according to the fitness proportion. The choice percentage p_i of every personality i was such that (Equation (19)):

$$g_i = \frac{k}{G_i}; p_i = \frac{g_i}{\sum_{j=1}^{N} G_i} \tag{19}$$

where G_i is the fitness price of personality i. Before individual selection, the reciprocal of the fitness proportion was calculated. The factor is k, and N is the amount of the personality in the population.

(4) In the light of optimizing the population initialization process, real number coding was adopted according to the individual; therefore, a real number crossover valve was adopted for the crossover operation. The overlapping operation of the k-th chromosome b_k as well as c-th chromosome b_c at position j is as below (Equation (20)):

$$b_{kj} = b_{ij}d + b_{kj}(1-b); b_{cj} = b_{kj}d + b_{cj}(1-d) \tag{20}$$

where d is a stochastic digit among (0–1).

(5) In the mutation operation, the j-th gene b_{ij} of i-th personality was chosen with mutation. The mutation operation method was as below (Equation (21)):

$$b_{ij} = \begin{cases} (b_{ij} - b_{max})\cdot G(f) + b_{ij} & r \geq 0.5 \\ (b_{min} - b_{ij})\cdot G(f) + b_{ij} & r < 0.5 \end{cases} \tag{21}$$

where b_{max} is the top margin of gene b_{ij}, b_{min} is the bottom margin of gene b_{ij}; $g(f) = r_2(1 - f/f_{max})$, r_2 is the stochastic digit, and f is the present iteration digit. F_{max} are the maximum evolution times, and r is stochastic digit among [0–1]. Its control principle is exhibited in Figure 9.

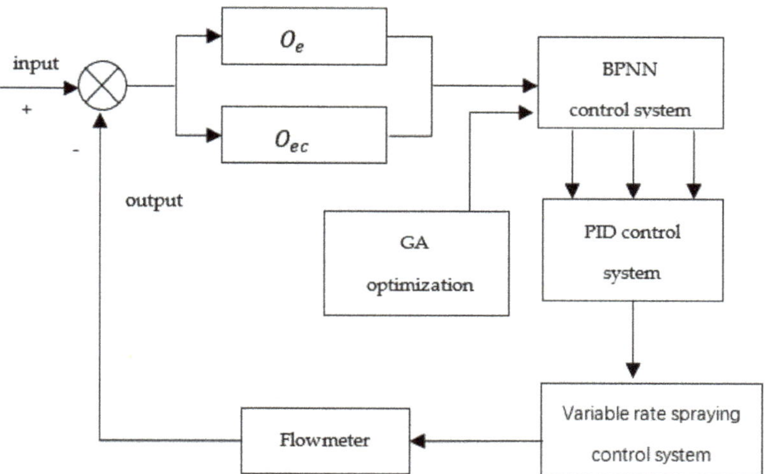

Figure 9. GA–BPNN–PID control schematic diagram.

2.7. *Simulation Analysis of Spraying Control System*

2.7.1. Modeling and Simulation of PID Control System

To achieve a control system model for liquid manure precise spraying, the classic PID control simulation variable spraying control system model of liquid manure was built using the Simulink simulation module in MATLAB software. The amplitude of import step signal was disabled by two. PID controller arguments were adjusted. The waveform of the system output was analyzed.

In the PID control simulation model, when t = 0, the import amplitude step signal was 2. Simulation time was disabled at 20 s. The K_P, K_I, and K_D of PID control were set. Finally, the oscilloscope output a waveform. The simulated waveforms are displayed in Figure 10.

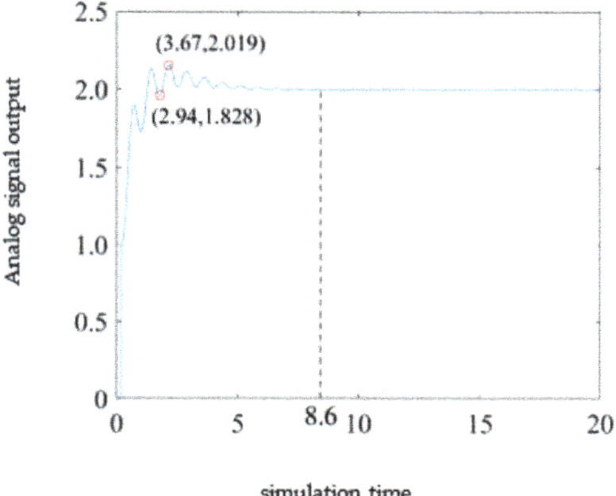

Figure 10. Simulation waveform of PID control model.

As shown in Figure 9, the model response time was 8.6 s, the overshoot was 0.019, and there was some oscillation before the system operation reached stability. Finally, $K_P = 42$, $K_I = 0.51$, and $K_D = 0.01$ were selected according to the empirical trial method.

2.7.2. Modeling and Simulation of PID Control System

The simulation model of the fuzzy PID control system was established in the simulation module. The input signal was a saving signal of 2. The simulation process was as follows: input a step signal with amplitude of 2 at the time of t = 0, set the simulation time as 60 s, input variables of fuzzy controller as fuzzed error e (k) and error change rate ec (k). The fuzzy controller output the PID parameter compensation value after fuzzing and optimized the initial parameters through the compensation value. Finally, the control system simulation waveform was obtained [9]. The simulation waveform is shown in Figure 11. Figure 11 shows that the response time of fuzzy PID control was 7.21 s, and the overshoot was 0.035.

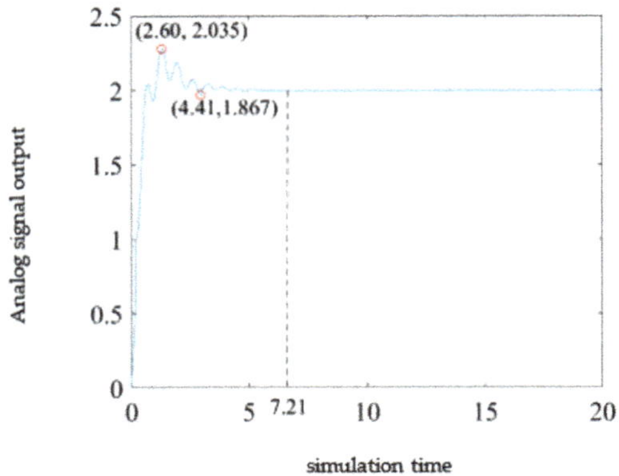

Figure 11. Simulation waveform of fuzzy PID control model.

2.7.3. Modeling and Simulation of BPNN–PID Control System

The simulation model of the BPNN–PID dominant system was established in the Simulink simulation module, and the input signal was also a step signal with an amplitude of 2. In the simulation process, the import amplitude step signal with an amplitude of 2 was time t = 0. Simulation time was disabled at 20 s. The input variables of the fuzzy regulator were the error e (k) and error rate of change ec (k) processed by the neural network, and the BPNN-PID regulator output compensation values of PID parameters after calculating K_P, K_I, and K_D through BPNN. Initial parameters were optimized through compensation values, and the control system simulation waveform was obtained.

The built BPNN–PID control system model is displayed in Figure 12, while its simulation waveform is shown in Figure 13.

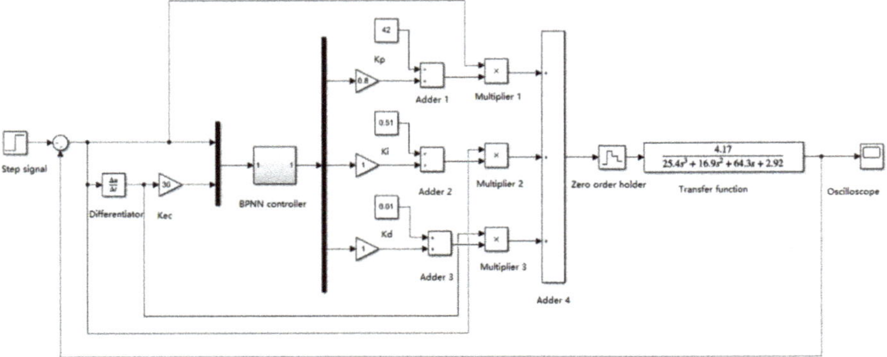

Figure 12. BPNN–PID control system model.

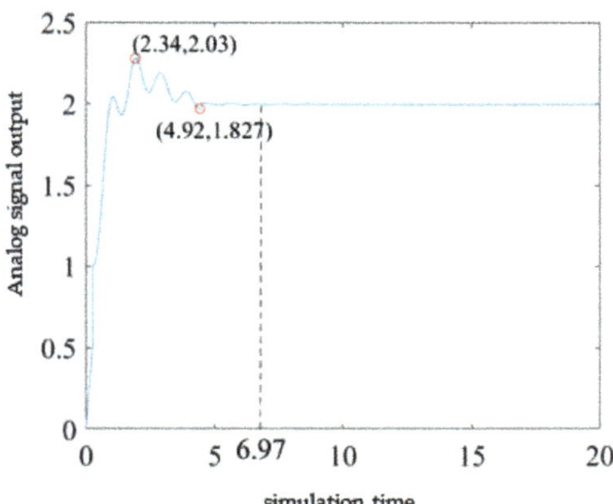

Figure 13. Simulation waveform of BPNN–PID control model.

As shown in Figure 13, the reaction time of the BPNN–PID control was 6.97 s, the overshoot was 0.03, and there was some oscillation before the system ran stably. Compared with the classical PID control, the overshoot increased by 0.011, while the system response time decreased by 1.63 s. Compared with the fuzzy PID control, the overshoot decreased by 0.005, while the system response time decreased by 0.24 s.

2.7.4. Simulation Analysis of BPNN–PID Control System Optimized by Genetic Algorithm

A model-based alternating–quantity spraying dominant system of liquid manure genetic algorithm was programmed using MATLAB to optimize BPNN–PID control.

The absolute error integral criterion was used to judge the performance index of each generation of individuals optimized using the genetic algorithm. The optimization process was ended when the population iteration reached the required performance index. If the required index was not reached, the optimal individual in the last generation of the population was considered as the result of the control model simulation analysis.

The control system input a step signal with amplitude of 2, and then output the compensation values ΔK_P, ΔK_I, and ΔK_D of the BPNN regulator corresponding to individual chromosomes before randomly generating the initial population. The individual chromosome in the population was optimized using the genetic algorithm operator, and the population was iterated to the maximum genetic algebra.

The iterative optimization process of the best individual of genetic algorithm is exhibited in Figure 14.

The optimized BPNN algorithm was imported into the PID controller, and the control model was simulated. The simulation results are exhibited in Figure 15.

As shown in Figure 15, for BPNN–PID control in view of genetic algorithm optimization, the response time of the system was 4.68 s and the overshoot was 0.027. After the system ran stably, a small disturbance was observed. Compared with the classical PID control, the overshoot increased by 0.008, while the response time was reduced by 3.92 s. Compared with the fuzzy PID control, the overshoot increased by 0.008, while the response time was reduced by 2.53 s. Compared with the BPNN–PID control, the overshoot was reduced to 0.003 and the response time decreased by 2.29 s. To summarize, the genetic–algorithm–optimized BPNN-PID control system had faster response, a smaller overshoot, and a better overall control effect.

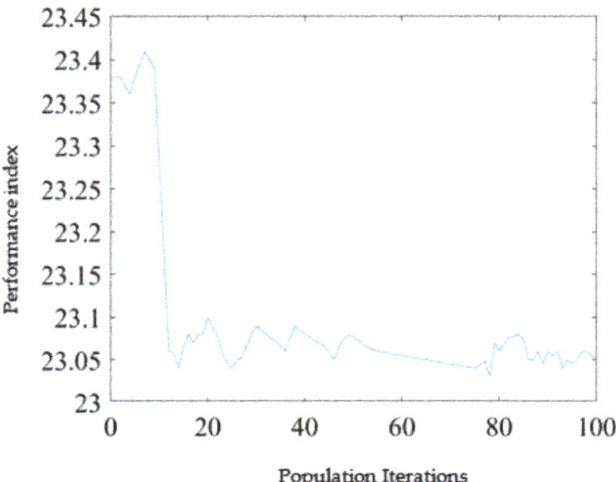

Figure 14. Iteration results of optimal population of genetic algorithm.

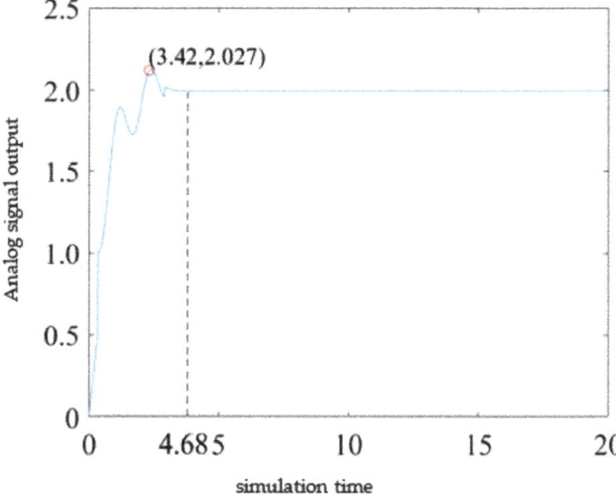

Figure 15. GA–BPNN–PID control system simulation waveform.

3. Results

3.1. Test Materials and Platforms

The variable application test of liquid manure was conducted in November 2021 and September 2022 at the test site of Tai'an Yimeite Co. Ltd. The test platform included a self-priming jet pump, monitoring and control via computer, electric proportional valve, flowmeter, distributor, and sprinkler pipe (Figure 16). The max lift of the self-priming jet pump was 15 m, the maximum absorption was 10 m, and the maximum flow was 81 L/min. The controller used the Siemens PLCS7–200 series, and the control software used the force control configuration software. The maximum diameter of the solenoid valve core was 18 mm. The maximum flow under 0.15 MPa pressure was 100 L/min.

Figure 16. Test platform. 1: pump; 2: monitoring and control via computer; 3: electric proportional valve; 4: flow meter; 5: distributor; 6: sprinkler pipe.

The control object of the test was a proportional solenoid valve. The experiment material was clean water without suspended solids. The measurement and verification of the control accuracy of the flow rate of liquid manure distribution were carried out for classical PID control, fuzzy PID control, BPNN–PID control, and GA–BPNN–PID control.

The solenoid valve and control system of the experiment platform were powered by an onboard DC. The controller and actuator converted 48 V DC to 24 V DC by switching the power supply before sending them to the dominant system. The experiment platform was 1.8 m tall and 1.2 m broad.

3.2. Control Accuracy Analysis of Control System

In this study, the spreading accuracy of the control system was reflected by the flow error, flow control stability, and valve group response time under the condition of vehicle speed change. The open–shut processes of the solenoid valve were the main reasons for the fertilizer flow error. A degree of error was observed in the liquid supply of the self-priming jet pump, as well as the measurement of the flowmeter. Flowmeter error [22–24] was the main error. The potential main causes for the flow error are as follows: (i) the influence of sediment; (ii) the fluid contained many bubbles; (iii) the wave height being uneven during the flow of the liquid to be measured. In addition, the ultrasonic flowmeter was also possibly subject to external interference, such as the electromagnetic environment, which affects system accuracy.

In this control system, the flow error of the liquid manure mainly originated from the fluid. The absolute error of the fluid flow represents the difference between the reading value of the flowmeter and the actual flow, and the flow proportional error is the ratio of absolute error to actual flow.

Therefore, the flow error was calculated as follows (Equation (22)):

$$\begin{cases} \Delta_0 = \left| V_i - V_j \right| \\ \Delta_1 = \frac{\Delta_0}{V_i} \times 100\% \end{cases} \tag{22}$$

where Δ_0 (L/min) is the absolute error of the system flow, Δ_1 (%) is the relative error of the system flow, V_i (L/min) is the actual flow, and V_j (L/min) is the flow rate read by the flowmeter.

3.3. Experiment Results and Analysis

3.3.1. Control System Stability Experiment

The data measured by the ultrasonic flowmeter were transmitted to the upper computer through 4–20 mA current, while the actual liquid fertilizer flow of spraying and fertilization was measured using the test. We used the actual driving speed [25] of the traction liquid fertilizer spreader in the field. In the test, the vehicle speed was set to 1, 1.5, 2.0, and 2.5 m/s in the control system, and the flow relevant to every vehicle speed group was determined. Each flow group was measured 10 times, and the average value was taken after removing the highest and lowest values. Using a spraying amount of 300 L/hm^2, the acknowledge fertilization quantities in the light of the above four different velocities were determined to be 10.8, 16.2, 21.6, and 27 L/min, according to Equation (1).

The system output flow was measured at four vehicle speeds. Six measurements were taken for each vehicle speed state. The duration of each measurement was 2 min. The mean value of the six measurements was considered with measurement flow of the velocity. Simultaneously, the absolute and proportional errors of every control system were noted. The experiment outcomes are listed in Table 2.

Table 2. Flow error of control system.

Speed (m/s)	Calculated Flow Rate (L/min)	PID			Fuzzy PID			BPNN-PID			GA-BPNN-PID		
		Actual Flow (L/min)	Absolute Error (l/min)	Relative Error (%)	Actual Flow (L/min)	Absolute Error (L/min)	Relative Error (%)	Actual Flow (L/min)	Absolute Error (L/min)	Relative Error (%)	Actual Flow (L/min)	Absolute Error (L/min)	Relative Error (%)
1	10.8	11.38	0.58	5.37	11.26	0.46	3.7	11.14	0.34	3.17	10.87	0.07	0.67
1.5	16.2	17.09	0.89	5.49	15.57	0.63	3.9	16.78	0.58	3.56	16.35	0.15	0.92
2	21.6	22.88	1.28	5.93	22.42	0.82	3.7	22.39	0.79	3.65	21.86	0.26	1.21
2.5	27	28.39	1.39	5.2	27.91	0.91	3.4	27.83	0.83	3.09	27.29	0.29	1.06

As shown in Table 2, the flow error of liquid manure controlled by classical PID was higher than that of BPNN—PID control and GA-BPNN–PID. The PID flow control average relative error was 5.50%. The maximum absolute error was 0.91 L/min. The average relative error of the flow controlled by fuzzy PID was 3.67%. The maximum absolute error was 0.83 L/min. The average relative error of the flow controlled by BPNN-PID was 3.37%. The maximum absolute error was 0.83 L/min. The average relative error of GA–BPNN-PID was 0.97%. The maximum absolute error was 0.29 L/min. The experiment outcomes demonstrated that the relative error of the GA-BPNN-PID on the system flow control was the smallest. In comparison with PID control, the relative error was reduced by 4.53 percentage points, and compared with BPNN–PID control, the relative error was reduced by 2.4 percentage points. The control system exhibited the highest stability.

3.3.2. Variable Control Test

In the variable control test, a data acquisition card was used to collect the angular velocity sensor data. The square wave signal generated by the angular velocity sensor was used to calculate the vehicle speed. The gathered square wave signal was installed and stored through the programmable signal generator. In this test, a programmable signal generator was used to simulate the electrical signal when speed changed, and the classical PID, fuzzy PID control, BPNN–PID control, and GA–BPNN–PID control were used for variable control of liquid manure flow.

In this test, the discharge of liquid manure was determined using a flowmeter. Test data were noted after system stability. Test results for the liquid manure spreading control under the condition of vehicle speed change are shown in Figures 17–20. During the spreading control system, the flow regulation's average response time with the classical PID control was 5.02 s, the fuzzy PID control was 4.19 s, the BPNN-PID control was 3.97 s, and the GA–BPNN–PID control was 2.85 s. These results indicate that the actual response time of GA–BPNN–PID control was 2.17 s less than that of PID control, 1.34 s less than that of fuzzy PID control, and 1.12 s less than that of BPNN–PID control. The GA–BPNN–PID had the fastest response speed and better flow control stability.

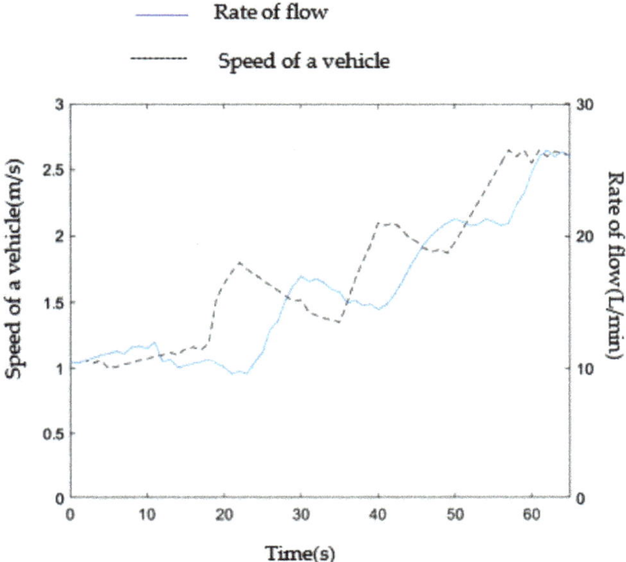

Figure 17. Test results of liquid manure flow by PID control.

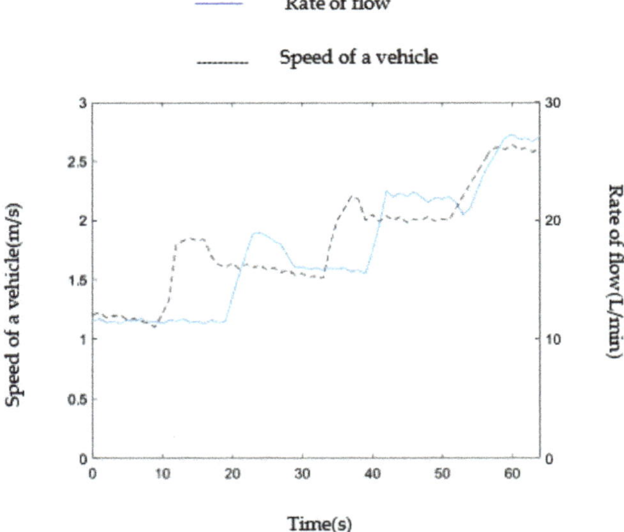

Figure 18. Test results of liquid manure flow by fuzzy PID control.

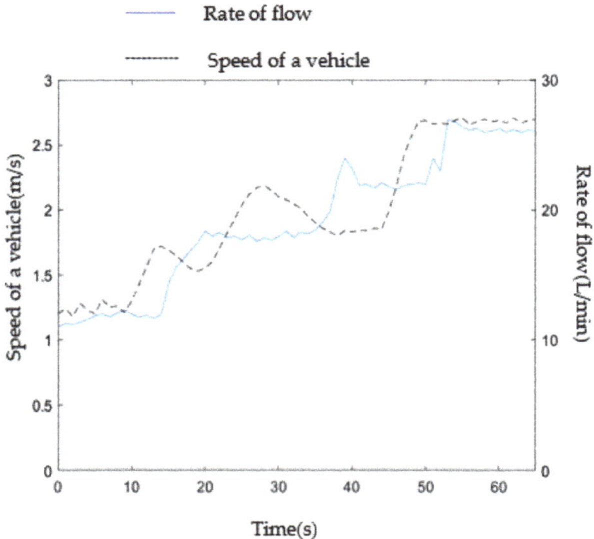

Figure 19. Test results of liquid manure flow by BPNN–PID control.

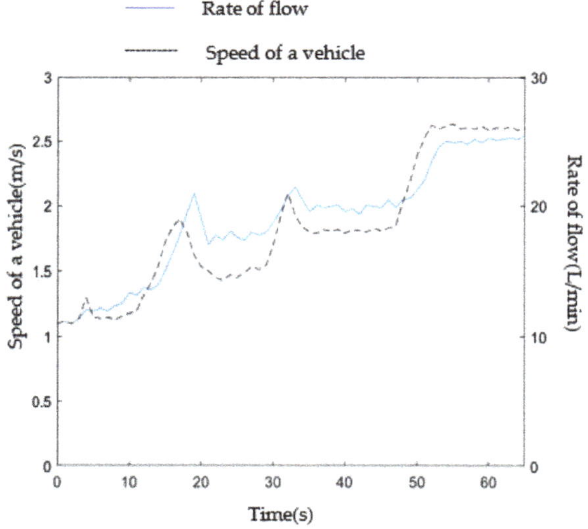

Figure 20. Test results of liquid manure flow by GA–BPNN–PID control.

4. Conclusions

In this study, spraying precision control technology was evaluated for use in a traction liquid manure variable spraying control system, and a control system model was built. Simulation analysis and tests were carried out for liquid manure flow control under the classical PID, fuzzy PID, BPNN–PID, and GA–BPNN–PID control modes. The following conclusions were drawn:

(1) The GA–BPNN–PID optimized the dominant fertilizer system. This family has a variable-rate fertilizer-spraying control system model. GA–BPNN was used to optimize PID parameters, which enhanced the stability of the control system.

(2) The negative feedback adjusting procedure of variable-rate liquid manure spraying control system was studied. The control system was modeled and simulated using MATLAB. The outcome shows that the response time for classical PID control to achieve stability was 8.6 s. The response time for fuzzy PID control to achieve stability was 7.21 s. The response time for BPNN–PID control to achieve stability was 6.97 s. The response time of GA–BPNN–PID control to achieve stability was 4.68 s. Thus, GA–BPNN–PID had the shortest response time and best stability.
(3) Our experimental results show that for the variable distribution control system of liquid manure, the flow average relative error of liquid manure controlled by classical PID was 5.02%, the flow average relative error of liquid manure controlled by fuzzy PID was 3.37%, the flow average relative error of liquid manure controlled by BPNN–PID was 3.10%, and the flow average relative error of liquid manure controlled by GA–BPNN–PID was 1.07%. The practice response time of classical PID control was 5.02 s, the practice response time of BPNN-PID control was 3.97 s, 1.34 s less than that of fuzzy PID control, and the practice response time of GA–BPNN–PID control was 2.85 s. Compared with the classical PID control, the practice response time of GA–BPNN–PID control was reduced by 2.17 s, and the relative error was reduced by 3.95 percentage points. Compared with BPNN-PID control, the practice response time was reduced by 1.12 s, and the relative error was reduced by 2.03 percentage points. Therefore, the practice control effect of GA-BPNN-PID control was the best.

Taken together, the results presented in this study provide a reference for the fertilization control of maize in the field.

Author Contributions: This study was conceptualized by P.W. and M.C. Test setup was completed by P.W., B.X., A.W. and B.M. The software was designed by P.W. and A.W. B.X. and J.F. provided resources and P.W. curated the data. The original draft of the manuscript was prepared by P.W. and Y.C., B.M. reviewed and edited the manuscript. All authors have read and agreed to the published version of the manuscript.

Funding: This research was funded by the basic scientific research business expenses of CAAS, No. S202106-03, the Jiangsu Modern Agricultural Machinery Equipment and Technology Demonstration and Promotion Project (NJ2021-23), and the Jiangsu Modern Agricultural Machinery Equipment and Technology Demonstration and Promotion Project (NJ2022-26).

Institutional Review Board Statement: Not applicable.

Data Availability Statement: All relevant data presented in the article are stored according to institutional requirements and, as such, are not available on-line. However, all data used in this manuscript can be made available upon request to the authors.

Acknowledgments: We are grateful to Wei Zhao for his help in project management. We also thank Taian Yimeite Company for providing the experimental conditions for us to successfully complete this experiment.

Conflicts of Interest: The authors declare no conflict of interest.

Abbreviations

PID	Proportion, integral, derivative
BPNN	BP neural network
BPNN-PID	BP neural network PID control system
GA-BPNN-PID	BP neural network PID control system optimized by genetic algorithm

References

1. Han, Y.; Jia, R.; Tang, H. Research status and development suggestions of precision variable-rate fertilization machine of the article. *Agric. Eng.* **2019**, *9*, 1–6.
2. Ma, B.; Ming, J.C.; Ai, B.W.; Jing, J.F.; Zhi, C.H.; Bin, X.X. Working Mechanism and Parameter Optimization of a Crushing and Impurity Removal Device for Liquid Manure. *Agriculture* **2022**, *12*, 1228. [CrossRef]

3. Jin, C.; Bin, Z.; Shu, J.Y. Research on present situation and the development countermeasures of variable rate fertilization technology in China. *J. Agric. Mech. Res.* **2017**, *39*, 1–6.
4. Zhang, L.; He, Y.; Yang, H.; Tang, Z.; Zheng, X.; Meng, X. Analysis of the relationship between the development of liquid fertilizer machinery and modern agriculture. *J. Chin. Agric. Mech.* **2021**, *42*, 34–40.
5. Meng, H.Q.; Xu, M.G.; Lü, J.L.; He, X.H.; Li, J.W.; Shi, X.J.; Chang, P.; Wang, B.R.; Zhang, H.M. Soil pH Dynamics and Nitrogen Transformations under Long-Term Chemical Fertilization in Four Typical Chinese Croplands. *J. Integr. Agric.* **2013**, *12*, 2092–2102. [CrossRef]
6. Yang, X.J.; Yan, H.R.; Zhang, T.; Dong, X.; Sun, X. Development status and trend of dual-purpose self-propelled boom sprayer in paddy field and dry field in China. *Agric. Eng.* **2020**, *10*, 1–5.
7. Zhu, D.L.; Ruan, H.C.; Wu, P.T.; Li, J.H.; Lu, L.Q. Remote fuzzy PID control strategy for fertilizer conductivity of water-fertilizer machine. *J. Agric. Mach.* **2022**, *53*, 186–191.
8. Bai, Y.Q.; Xiong, K.H.; Zhi, C.W. Development of variable-rate spraying system for high clearance wide boom sprayer based on LiDAR scanning. *Trans. Chin. Soc. Agric. Eng.* **2020**, *36*, 89–95.
9. Min, T.; Jin, B.B.; Jiang, Q.L. Variable rate fertilization control system for liquid fertilizer based on genetic algorithm. *Trans. Chin. Soc. Agric. Eng.* **2021**, *37*, 21–30.
10. Zhuang, T.F.; Yang, X.J.; Dong, X.; Zhang, T.; Yan, H.R.; Sun, X. Research status and development trend of large self-propelled sprayer booms. *Trans. Chin. Soc. Agric. Mach.* **2018**, *49* (Suppl. S1), 189–198.
11. Chang, C.Y.; Hong, W.L.; Jin, H.; Guibin, C.H.; Caiyun, L.U.; Qingjie, W.A. Design and experiment of high-frequency intermittent fertilizer supply system based on PID algorithm. *Trans. Chin. Soc. Agric. Mach.* **2020**, *51*, 45–53.
12. Li, T.; Jia, C.L.; Jing, L.W.; Zhang, B.; Lu, J. Design and experiment of bypass fertilizer-type water and fertilizer integrated automatic fertilizer applicator. *Water Sav. Irrig.* **2018**, *11*, 98–102.
13. Yi, Y.; Feng, F.Z.; Xue, F.S. Design of self-propelled remote control spraying vehicle. *Agric. Equip. Veh. Eng.* **2021**, *59*, 46–49.
14. Xiu, Y.X.; Bin, Z.; Ze, L.Z.; Shuran, S.; Zhen, L.; Hong, T.; Huang, H. Design and experiment of variable liquid fertilizer applicator for deep-fertilization based on ZigBee technology. *J. Drain. Irrig. Mach. Eng.* **2020**, *38*, 318–324.
15. Saeys, W.; Engelen, K.; Ramon, H.; Anthonis, J. An automatic depth control system for shallow manure injection, Part 1: Modelling of the depth control system. *Biosyst. Eng.* **2007**, *98*, 146–154. [CrossRef]
16. Guang, K.Z. Design and Experiment of Liquid Fertilizer Control System. Master's Thesis, Northwest A&F University, Xianyang, China, 2018.
17. Ji, C.Z.; Fan, F.M.; Ping, Z.; Shou, H.; Wen, J. Variable rate control and fertilization system of liquid fertilizer applicator based on electronic control unit. *Soybean Sci.* **2019**, *38*, 111–117.
18. Qian, Z.; Zheng, Y.W.; Yu, B.Z.; Lei, Z.; Wei, J.; Haoran, W. Rapid-response PID control technology based on generalized regression neural network for multi-user water distribution of irrigation system head. *Trans. Chin. Soc. Agric. Eng.* **2020**, *36*, 103–109.
19. Bi, P.; Zheng, J. Study on Application of Grey Prediction Fuzzy PID Control in Water and Fertilizer Precision Irrigation. In Proceedings of the 2014 IEEE International Conference on Computer and Information Technology (CIT), Xi'an, China, 11–13 September 2014; pp. 789–791.
20. Wu, Y.; Li, L.; Li, S.; Wang, H.; Zhang, M.; Sun, H.; Sygrimis, N.; Li, M. Optimal control algorithm of fertigation system in greenhouse based on EC model. *Int. J. Agric. Biol. Eng.* **2019**, *12*, 118–125. [CrossRef]
21. Zhou, R.; Zhang, L.; Fu, C.; Wang, H.; Meng, Z.; Du, C.; Shan, Y.; Bu, H. Fuzzy Neural Network PID Strategy Based on PSO Optimization for pH Control of Water and Fertilizer Integration. *Appl. Sci.* **2022**, *12*, 7383. [CrossRef]
22. Le, Y.; Xiao, L.W.; Chun, L.Z. Correction of measurement deviation of turbine flowmeter. *Energy Conserv.* **2021**, *2*, 173–175.
23. Xiu, L.Y. Error analysis in electromagnetic flowmeter. *Ind. Des.* **2017**, *10*, 141–142.
24. Kang, X.Y.; Wei, W.; Bai, T.L. Discussion on the flow test method and development trend of the liquid flowing in the pipeline. *China Met. Equip. Manuf. Technol.* **2021**, *56*, 116–119.
25. Jie, C. The application of PWM control in uniform and precise spraying for sprayer. *J. Agric. Mech. Res.* **2021**, *43*, 22–26.

Article

A Method of Modern Standardized Apple Orchard Flowering Monitoring Based on S-YOLO

Xinzhu Zhou [1], Guoxiang Sun [1,2,*], Naimin Xu [1], Xiaolei Zhang [1], Jiaqi Cai [1], Yunpeng Yuan [1] and Yinfeng Huang [1]

1 College of Engineering, Nanjing Agricultural University, Nanjing 210031, China
2 Jiangsu Province Engineering Lab for Modern Facilities of Agriculture Technology & Equipment, Nanjing 210031, China
* Correspondence: sguoxiang@njau.edu.cn; Tel.: +86-158-5076-0301

Abstract: Monitoring fruit tree flowering information in the open world is more crucial than in the research-oriented environment for managing agricultural production to increase yield and quality. This work presents a transformer-based flowering period monitoring approach in an open world in order to better monitor the whole blooming time of modern standardized orchards utilizing IoT technologies. This study takes images of flowering apple trees captured at a distance in the open world as the research object, extends the dataset by introducing the Slicing Aided Hyper Inference (SAHI) algorithm, and establishes an S-YOLO apple flower detection model by substituting the YOLOX backbone network with Swin Transformer-tiny. The experimental results show that S-YOLO outperformed YOLOX-s in the detection accuracy of the four blooming states by 7.94%, 8.05%, 3.49%, and 6.96%. It also outperformed YOLOX-s by 10.00%, 9.10%, 13.10%, and 7.20% for mAP_{ALL}, mAP_{S}, mAP_{M}, and mAP_{L}, respectively. By increasing the width and depth of the network model, the accuracy of the larger S-YOLO was 88.18%, 88.95%, 89.50%, and 91.95% for each flowering state and 39.00%, 32.10%, 50.60%, and 64.30% for each type of mAP, respectively. The results show that the transformer-based method of monitoring the apple flower growth stage utilized S-YOLO to achieve the apple flower count, percentage analysis, peak flowering time determination, and flowering intensity quantification. The method can be applied to remotely monitor flowering information and estimate flowering intensity in modern standard orchards based on IoT technology, which is important for developing fruit digital production management technology and equipment and guiding orchard production management.

Keywords: intelligent agriculture; IoT technology; apple flowering monitoring; open world; swin transformer; SAHI

Citation: Zhou, X.; Sun, G.; Xu, N.; Zhang, X.; Cai, J.; Yuan, Y.; Huang, Y. A Method of Modern Standardized Apple Orchard Flowering Monitoring Based on S-YOLO. *Agriculture* **2023**, *13*, 380. https://doi.org/10.3390/agriculture13020380

Academic Editors: Cheng Shen, Zhong Tang and Maohua Xiao

Received: 8 January 2023
Revised: 31 January 2023
Accepted: 2 February 2023
Published: 4 February 2023

1. Introduction

The quantitative gathering of information on the growing status of fruit trees using current technology facilitates the digital management of orchard production and enhances the precision of orchard production management [1]. Flowering information monitoring is one of the basic techniques for digital orchard management, and it is extensively utilized for orchard flower thinning, pest and disease control, and other management operations. Pruning and intercutting are required to obtain more significant economic returns in the apple-growing industry [2]. In the early stages of apple growth, proper flower and fruitlet thinning may increase the fruit weight per fruit and the blooming yield [3]. Current flower thinning methods mainly include manual thinning [4], chemical thinning [5–7], and mechanical thinning [8].

Traditional flowering monitoring is achieved based on human observations of specific fruit trees at specific times. That is, experts go into the orchard to randomly select a few fruit trees and estimate the flowering state with the eye. After comprehensive consideration, the overall flowering state of the orchard is obtained. Thinning after 28 days of bloom is ideal for obtaining larger, high-quality Fuji apples [3,9]. However, modern standardized orchards often have large areas and variability in the flowering times of fruit trees in different regions. It is difficult to dynamically adjust orchard flower thinning time and measures for specific fruit tree flowering information, affecting the efficiency and accuracy of modern standardized orchard flower thinning management decisions. Consequently, there is an urgent need for a method that can monitor the various growth stages of apple flowers and quantify the flowering intensity, establishing the groundwork for real-time monitoring utilizing Internet of Things (IoT) technology.

The current apple blossom monitoring methods have not been effectively studied, and most of them have been carried out in simple experimental environments, i.e., with suitable light, shooting angle, and shooting distance, achieving close to 100% detection results. Using closer imaging distances, a study on stamens in fully open flowers [10] disregarded the predictive effect of early buds and semi-open flowers for fully open flowers and found they were only appropriate for close detection. Other studies that have grouped all stages of flowers into one category for detection, even for flower clusters [11,12], have ignored the interaction effects between the flowers at different growth stages and could not accurately monitor the complete flowering process. The division of flowers into three stages of detection [13] ignored the end-flowering stage as a marker of the end of the flowering stage for determining the flowering stage of fruit trees. Other studies have divided apple flowers into 6–8 stages for detection [14,15], devising even more categories. However, the similarity between flowers at different growth stages elevates the cost of data annotation and the inability to count the number after clustering detection in the same category.

Images obtained from closer distances with a high proportion of apple blossom pixels at various stages are simpler to recognize and perform better. However, with the development of IoT monitoring devices, research based on high-resolution images of whole fruit trees acquired at a distance has become an inevitable trend. Current studies have obtained more complete images due to the long imaging distance, such as vehicle-based [16] and uncrewed aerial vehicles [17,18]. However, the tiny area of individual flowers makes the variability between flowers at different growth stages low, and only fully open flowers or even flower clusters are recognized as detection objects. In addition, none of the above studies have observed the entire growth cycle of apple blossoms or tested the models under different weather conditions. The detection models obtained may not be suitable for a wide range of weather conditions. The key to the above problem is that the current detection algorithm cannot effectively detect tiny pixel flowers in high-resolution apple tree images, let alone monitor the complete growth process in complex weather.

Convolutional neural networks are the standard model in computer vision. The related models are categorized into two groups based on whether they directly implement the classification and localization process: the Faster RCNN [19–21] series of two-stage algorithms, the SSD [22], and the YOLO series [23,24] of one-stage algorithms. The current apple flower detection algorithms are primarily separated into mask-based semantic segmentation and box-based object detection. Studies using semantic segmentation algorithms in apple blossom detection have included DeepLab-ResNet [11], Mask R-CNN [12,13], and fully convolutional neural networks [16]. Although these algorithms can segment flowers, they cannot count the number of flowers and are less effective in detecting large aggregations of flowers. The box-based apple flower detection algorithm can count the number of flowers and enable further data analysis. In work using this type of algorithm, the YOLO family, especially YOLO v4 [25], has been widely improved and achieved better detection results [10,15,26,27]. However, these studies have only examined flowers at certain times, lacking the monitoring of the whole flowering process and quantitative analysis of the flowers. Therefore, a method is needed that can accurately detect tiny apple blossoms in high-resolution images and enable multi-stage flower monitoring in the open world.

As the most advanced model in the widely used YOLO family, YOLOX [28] has superior detection performance and has been effectively implemented for similar intensive detection applications [29,30]. Although CNN models, including YOLOX, have a long history of success in target detection using translation invariance and local correlation, CNN has a restricted field of vision, making it challenging to gather global information. In contrast, Transformer does not have translational invariance and local correlation but can capture long-range dependencies. So Vision Transformer [31] performs better than pure convolutional models for large datasets, especially when massive datasets can be obtained through IoT technology.

Since the introduction of Vision Transformer, many works have tried combining CNN and Transformer to motivate the network to inherit the advantages of CNN and Transformer and retain the global and local features to the maximum extent. As a landmark work, Swin Transformer [32], with shifted windows as a prominent feature, was created. With self-attention at its foundation, Swin Transformer gathers global contextual information to establish long-range dependency on targets and extract more robust features, demonstrating the potential to replace traditional convolutional networks as the new backbone network in computer vision.

In order to achieve information monitoring of the complete flowering process using IoT technology, research based on high-resolution images of apple trees taken in complex weather is essential. However, such images are not only tough to obtain, but also the typical characteristics, such as a complex and changeable environment, a tiny proportion of flower pixels, and hazy texture and color detail information, provide obstacles for flower detection and monitoring. This study took high-resolution images of apple blossoms at the complete growth stage in the open world as the research object and used the Slicing Aided Hyper Inference (SAHI) algorithm to generate mixed datasets containing global and local information. Then an S-YOLO model was designed based on Swin Transformer, achieving the accurate detection of apple blossoms at four growth stages. An analysis model for the number and number share of apple blossoms at each stage was established, realizing the flowering intensity and flowering monitoring of the orchard or even specific fruit trees. This work gives further theoretical and technological support for monitoring orchard flowering growth using IoT technology.

2. Materials and Methods

2.1. Experimental Design

As shown in Figure 1, the specific workflow of the apple flowering stage monitoring method is as follows: Step 1: obtain images of complete apple trees in the experimental area of the orchard at the flowering stage. Step 2: label the obtained images according to bud, half-open, fully open, and end-open using labelimg, and obtain the labeled image file. Step 3: slice the annotated image using the SAHI algorithm and blend it with the original image. Step 4: build the S-YOLO model and use the blended dataset for model training and validation. Step 5: use the trained model for apple blossom detection. Step 6: perform data analysis on the detection results. Step 7: implement flowering intensity estimation and flowering period monitoring.

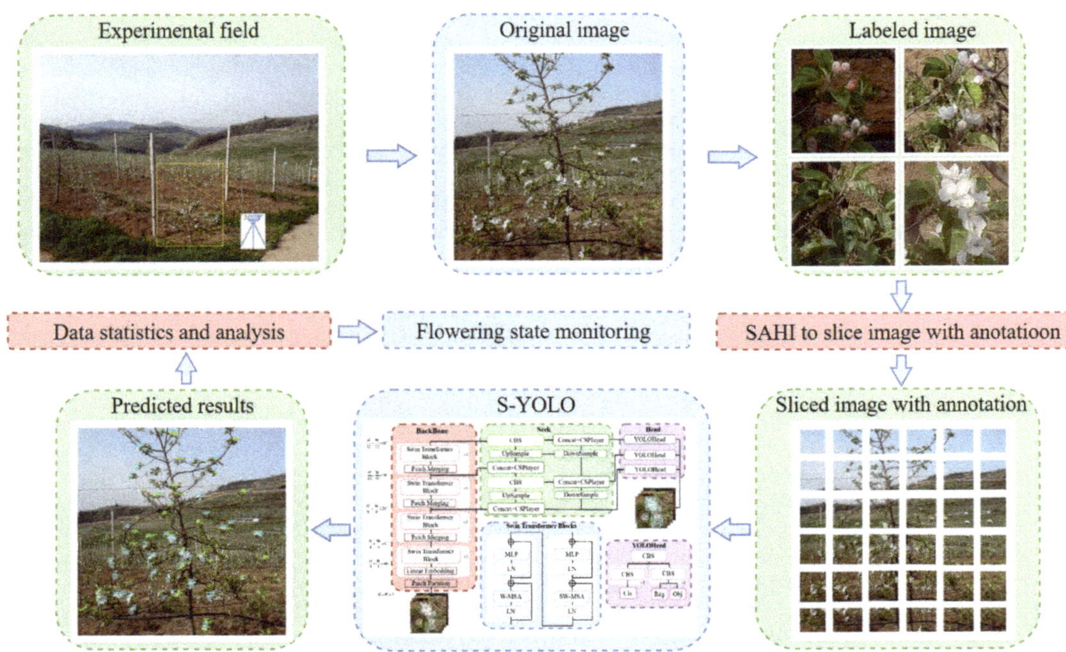

Figure 1. Workflow for monitoring the flowering state of modern standardized apple orchards.

2.2. Image Data Acquisition

The images utilized in this research were captured between 3 April 2020 and 16 April 2020, in the First Asian Alpine Orchard in Shihe Township, Lingbao City, Henan Province (E 111°4′2.6976″, N 34°27′1.44″), while the whole orchard was in the first flowering stage. The subjects were 115 red Fuji apples from 5-year-old trees in rows 4.2 m apart with 2.1 m between the plants. To simulate IoT devices to acquire images of apple trees, the investigator stood 2.5 to 3 m away from the tree's roots and used a mobile phone to capture photographs of the blooming stage of the apple trees.

The camera was operated from 10:00 am to 12:00 pm on the 13 days of shooting, and 115 images were obtained daily. A total of 1494 images of 3000 × 3000 pixels of apple trees under different weather conditions, including sunny, cloudy, and rainy days, were obtained in this study. Table 1 presents the precise weather information.

Table 1. Weather on various dates of shooting.

Date	Weather	Light Intensity	Temperature	Windy
0403	Overcast	Weak	8–16°	No
0404	Sunny	Strong	8–18°	Yes
0405	Sunny	Strong	9–20°	Yes
0406	Sunny	Strong	9–20°	Yes
0407	Sunny	Strong	9–20°	Yes
0408	Sunny	Hazy	11–25°	No
0409	Sunny	Normal	8–24°	Yes
0410	Light rain	Weak	5–10°	Yes
0411	Light rain	Weak	5–15°	Yes
0412	Sunny	Strong	5–20°	Yes
0413	Sunny	Strong	9–23°	No
0415	Sunny	Normal	12–25°	No

The flowering status of apple trees is divided into four stages: the first flowering stage, the middle flowering stage, the full flowering stage, and the last flowering stage, which correspond to the four growth statuses of apple flowers. The first stage of the apple flower is the bud (Figure 2a, red arrow), and they become half-open at the second stage (Figure 2a, green arrow and Figure 2b, green arrow) when the buds swell into white or light pink blooms that look like balloons. Once the petals have unfurled in the bud, the flower enters the fully open stage (Figure 2c, blue arrow), and the end-open stage is when the petals drop completely (Figure 2d, purple arrow). This experiment labeled 39,980 instances of the four flowering stages on 317 3000 × 3000 pixels images of Fuji apple trees using labelimg (Table 2), fulfilling the criterion that at least 3000 to 4000 instances of each class must be labeled in complicated agricultural contexts, as suggested in [27].

Figure 2. Flower images of apple tree at four flowering stages (300 × 300 pixels). (**a**) First flowering stage; (**b**) middle flowering stage; (**c**) full flowering stage; (**d**) last flowering stage.

Table 2. Unprocessed annotation data information.

Class	Bud	Half-Open	Fully Open	End-Open
Number	11,865	10,885	12,288	4942
Aspect ratio (pixels)	17.08:17.23	22.32:22.30	49.36:48.61	27.68:27.13
Average area (pixels^2)	318.86	541.71	2586.45	798.06
Area ratio (%)	0.0035	0.0060	0.0287	0.0088

Table 2 displays the raw annotation category information, including the number of annotations, the aspect ratio of various flowers, and the ratio of the area occupied in the image. The number of flower annotations in the first three growth stages was close, except for the last bloom. The aspect ratio is the ratio of the labeled flower length to the width, with the same category of flowers possessing a length-to-width pixel ratio near 1:1 and an average pixel size ratio among the different flower stages of 17:22:49:27. Therefore, the order of the size of the average area is fully open > half-open > end-open > bud. The fully open stage has the highest number of annotations and largest average area. However, its ratio in the image was less than 0.03 %, and the other three flowering stages made up less than 0.01 %, making accurate flower detection more challenging.

From the overall view of the annotation information, the image data utilized in the experiment and the annotated data were of high quality, which provided practical support for the training and validation of the model. This dataset has the following characteristics:

- The dataset used for the experiment covered a range of weather conditions, the apple tree's growth postures, and the complete flowering process.
- The original image was a high-resolution image of 3000 × 3000 pixels, where the vast majority of flowers were almost always smaller than 50 × 50 pixels.
- The flowers at each growth stage were manually labeled with sufficient numbers and fineness. All factors affecting apple flower detection, such as biometric features, gestures, shadows, and light, were considered at the human level.

2.3. Slicing Using SAHI

The high-resolution apple tree images used in this study contained a large and dense number of tiny pixel flowers. The higher resolution made directly inputting the photos into the network to extract features too computationally costly. However, reducing the resolution would result in losing information on details related to the flowers.

Multiple solutions have been developed to address the problem of small, dense objects in high-resolution photographs. The traditional method of filling and then segmenting images [6] and the method of copying and enhancing [33] images after oversampling require segmenting a large number of annotations, which results in a large number of features being altered to the point of being incompatible with the original dataset. Enlarging the target region [34] can enrich small object features, but it will add additional computational volume and is challenging to adapt to the demand for detection speed in some agricultural fields. To preserve image detail information and reduce the model calculation costs, the SAHI slicing algorithm was used to increase the model detection accuracy.

The SAHI [35] algorithm is a slicing-assisted inference approach for object detection models that perform inference by cropping the images and performing inference on them. The most notable benefit of SAHI is that it can be used in any object detection inference method, considerably enhancing the detection level of small targets while just linearly lengthening the computation time in slices. The SAHI algorithm effectively increases the precision of YOLO series detection [36].

Considering the efficacy of SAHI in small object detection applications, the SAHI algorithm was applied with a 20% coverage to the dataset utilized in this study, yielding 640 × 640 pixel images (Figure 3).

Figure 3. The process of SAHI splitting the original image.

The SAHI algorithm divides the original image of 3000 × 3000 pixels into 640 × 640 pixels, which can be directly fed into the network, eliminating the computational overhead of huge images without scaling, and minimizing the loss of detail information due to resizing by approximately five times. The dataset created by combining the original and sliced images is guaranteed to contain large images (3000 × 3000) with high semantic information and small images (640 × 640) with detailed local information. Additionally, there is a 20% overlap area between the sliced images which can also facilitate information fusion.

2.4. S-YOLO Detection Model

2.4.1. Model Construction

Swin Transformer can better capture global semantic information than traditional convolutional neural networks and can better fuse global and local information and extract more powerful features. To reduce the number of model parameters to almost the same level as that of the pure convolutional backbone network to ensure the fairness of the experimental results, Swin Transformer-tiny was substituted for the YOLOX backbone network to generate the S-YOLO model. The overall architecture of the S-YOLO is shown in Figure 4.

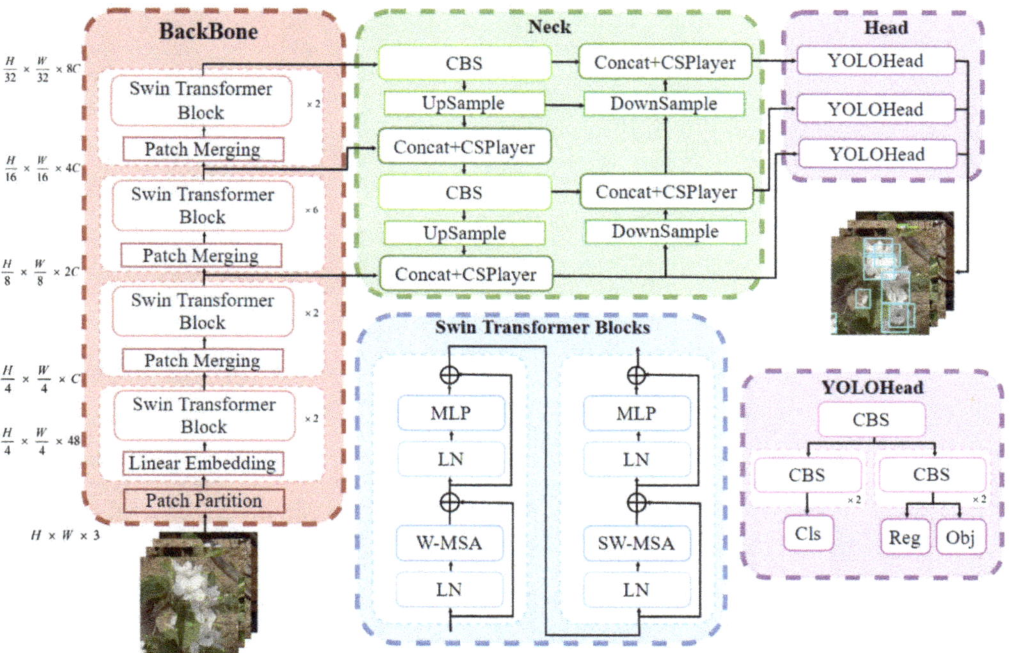

Figure 4. S-YOLO network structure.

S-YOLO is separated into four distinct size models based on the varying depths and widths of the neck and head channels. The corresponding depth and width ratios for different versions of the model are S-YOLO-tiny (S-YOLO-t: 0.33, 0.375), S-YOLO-small (S-YOLO-s: 0.33, 0.50), S-YOLO-middle (S-YOLO-m: 1.00, 1.00), and S-YOLO-large (S-YOLO-l: 1.33, 1.20).

The backbone network consists of five parts. The first part is the patch partition, and the last four parts are composed of two consecutive Swin Transformer blocks, where patch partition and linear embedding are equal to Patch Merging, which plays the role of downsampling. Swin Transformer substituted the conventional multi-head self-attention (MSA) module with the window-based multi-head self-attention (W-MSA) module and the shifted window-based multi-head self-attention (SW-MSA) module. Each Swin Transformer block consists of three components linked by a residual structure: a LayerNorm (LN) layer, a W-MSA/SW-MSA module, and a multilayer perceptron (MLP) with two completely connected layers and GELU nonlinearity. Using alternating transformations of W-MSA and SW-MSA to conduct all attention actions in a given window, the Swin Transformer block minimizes the computing volume of the model.

Detail information, such as color and texture, is crucial for flower detection. Therefore, retaining and extracting as much shallow information as possible becomes a vital consideration while building a feature extraction network. The neck part used a PAnet structure [37] to accomplish a twofold sampling of the features and to improve the network's capacity to fuse features. The CBS module, which comprises a convolutional layer (Conv), a batch normalization layer (BN), and an activation function called SiLU, is the primary module for feature extraction in the S-YOLO neck and head sections. The CSPlayer layer, using the idea of residuals, consists of two parallel CBS modules and multiple residual units in a series, which will play a role in PAnet for better image features and fusion capability extraction.

The design of the detecting head module is centered on efficiently employing the characteristics gathered from global and local information. The YOLOHead part employs decoupled heads with quicker convergence and greater precision. The decoupled head is controlled via CBS modules with varying channel counts and partitioned into classification and regression subnetworks. The classification subnetwork calculates the probability of detecting flowers belonging to distinct classes (Cls) of flower labels. In contrast, the regression subnetwork predicts the feature points' classes (Obj) and positions (Reg). Combining three sets of YOLOheads designed to detect flowers of various sizes produced a considerable number of suggestion boxes, and the anchor-free SimOTA algorithm provided the final detection results.

The hybrid dataset generated using SAHI was fed into the S-YOLO network for training and validation after data enhancement via the Mosaic [25] and MixUp [38] algorithms.

2.4.2. Model Training and Validation Environment

The mixed dataset was divided into a training set, a validation set, and a test set, at a ratio of 64:16:20. The training and validation sets were resized to 640 × 640 pixels for input to the network for training and validation. The test set was used for the network assessment and testing. The model training process was carried out using the Ubuntu 18.04 Cloud operating system (Cuda 11.0, Cudnn 8.0.4, Python 3.8, Pytorch 1.8, 4 × NVIDIA RTX 3090). The model assessment process was performed using the Windows 11 operating system (Cuda 11.3, Cudnn 8.2.1, Python 3.8, Pytorch 1.10.2, 1 × NVIDIA RTX 1650). During the experiments, the freezing and unfreezing training process was conducted with 100:200 epochs, with primary batch sizes of 128:32 and 16:16 when the SAHI algorithm was not utilized. The other critical hyperparameters are listed in Table 3.

Table 3. The primary hyperparameters of the model training process.

Hyperparameter	Value
Initial learning rate	0.01
Minimum learning rate	0.0001
Optimizer	sgd
Momentum	0.937
Weight decay	0.0005
Learn rate decay type	cos

2.4.3. Evaluation Indicators

Precision (P), recall (R), average precision (AP), and different types of mean average precision (*mAP*) are the main indicators used to assess the efficacy of the model for detecting apple flowers. The experimental outcome measures can be presented by calculating various combinations of positions within the confusion matrix: true positive (TP) is the accurate forecast for positive samples, true negative (TN) is the correct prediction for negative samples, false positive (FP) is the erroneous prognosis for positive samples, and false negative (FN) is the incorrect prediction for negative samples. In addition, the precision–recall (P–R) curve forms corresponding points between the horizontal axis, representing recall (R), and the vertical axis, representing precision (P) (with an IoU threshold equal to 0.5). Different intersections over the union (IoU) values were obtained by setting a certain degree of overlap between the prediction and the ground truth. Other specific metrics were calculated as follows:

AP: Average precision for a single category (IoU threshold from 0.5 to 0.95 in steps of 0.05), including bud, half-open, fully open, and end-open apple flowers;
mAP_{ALL}: Mean average precision of apple flowers of the four stages (all pixels);

mAP_S: mAP for small objects whose area is smaller than 322 pixels;
mAP_M: mAP for medium objects whose area is between 322 pixels and 962 pixels;
mAP_L: mAP for large objects whose area is bigger than 962 pixels.

Calculating the *P*, *R*, *AP*, and mAP for a given IoU threshold is defined in Equations (1)–(4).

$$P = \frac{TP}{TP + FP} \tag{1}$$

$$R = \frac{TP}{TP + FN} \tag{2}$$

$$AP = \sum_{k=1}^{N} \max_{\widetilde{k} \geq k} \left[P(\widetilde{k}), P(k) \right] \left[R(\widetilde{k}) - R(k) \right] \tag{3}$$

$$mAP = \frac{\sum_{i=1}^{M} AP_i}{M} \tag{4}$$

where k and $\widetilde{k}$ represent the point serial numbers before and after interpolation; N is the number of images under a category; M is the number of categories; i is the category label; and $P(k)$ and $R(k)$ are the precision and recall of the kth point. AP_i is the average precision of class i.

2.5. Apple Flowering Monitoring

In this experiment, the trained model was applied to 1494 images of apple trees taken at different times to detect flowers at different stages. The prediction of each apple tree image using S-YOLO-s to obtain the boxes corresponding to the four stages of flowers in each image and accumulating the number of boxes in the same category in all images on the same date obtained the total number of flowers in the four stages. On this basis, the total number of apple blossoms at each stage was divided by the number of images to obtain the average number of flowers at each stage in each image.

It is possible to determine the relative number share by evaluating the proportional relationships between the various stages of flowers within a single image. When the percentage of flowers at a particular stage in an image surpasses fifty percent, the fruit tree is deemed at the corresponding flowering stage. The day with the highest number of apple blossoms at a particular stage of the growth cycle is the peak time for apple blossoms at that stage. Among all images of fruit trees under a specific date, the average number of flowers in the four stages can be used to determine the flower proportion, and thus, the overall flowering status of the orchard. The percentages of fully open flowers correspond to the flowering intensities from 0 to 100, and this precise quantitative index will provide data support for flower-thinning decisions.

3. Results and Discussion

3.1. Image Slice Results

The flower images were sliced using the SAHI algorithm and combined with the original images to create a hybrid dataset (Table 4). The relevant feature changes were recorded before and after the slicing (Table 5).

Table 4. Annotated data information after slicing using the SAHI algorithm.

Class	Bud	Half-Open	Fully Open	End-Open
Number	32,060	29,402	34,703	13,648
Aspect ratio (pixels)	17.55:17.70	22.64:22.68	48.51:47.85	27.97:27.46
Average area (pixels^2)	334.50	557.75	2485.52	813.22
Area ratio (%)	0.0542	0.0894	0.3934	0.1312

Table 5. Changes in annotated data information after SAHI algorithm slicing.

Class	Bud (%)	Half-Open (%)	Fully Open (%)	End-Open (%)
Number	+170.21	+170.11	+182.41	+176.16
Aspect ratio	+2.76: +2.73	+1.43: +1.70	−1.72: −1.56	+1.05: +1.27
Average area	+4.90	+2.96	−3.90	+1.90
Area ratio	+1428.91	+1385.91	+1268.89	+1379.58

Following the SAHI algorithm slice, the percentages of flowers in each category grew by 170.21%, 170.11%, 182.41%, and 176.16%, respectively, resulting in 109,813 high-quality labeled data (a 150% increase) to the network. While allowing for a higher batch size for training, the changes in the aspect ratio and average area were within 3.00% and 5.00%, and the corresponding change in the area ratio was at least 1268.89%. In addition, the fully open stage had the largest average single flower area, which was 7.43 times bigger than the bud stage, 4.46 times bigger than the half-open stage, and 3.06 times bigger than the end-open stage. The high area of full blooms prompted the SAHI algorithm to split the fully open flowers that were at the boundary of the cut area into multiple parts and to consider them as newly fully open. This split came at the cost of a 3.90% reduction in the average areas, prompting the most significant increase in the number of fully open flowers. Fully open was the only flower growth stage with a negative average area increase.

3.2. Flower Detection with S-YOLO

3.2.1. Comparison with YOLOX-s

In the COCO dataset, the superiority of YOLOX over sophisticated models, such as PPYOLO [39], YOLOv3, and EfficientDet [40], was established [28]. YOLOv4 was proven to have greater precision and *mAP* than Faster R-CNN and SSD 300 for detecting apple pistils [10]. Consequently, the comparison experiment portion of this study was performed on the original YOLOX-s model and the modified S-YOLO-s, and the pertinent data were collected (Table 6).

Table 6. Comparison of the effects of YOLOX-s and S-YOLO-s on flowering detection in apples.

Model	P-Bud [1] (%)	P-Half-Open (%)	P-Fully Open (%)	P-End-Open (%)	mAP_{ALL} (%)	mAP_S (%)	mAP_M (%)	mAP_L (%)
YOLOX-s [2]	79.19	81.27	85.21	83.90	27.40	21.40	36.00	58.90
S-YOLO-s	87.13	89.32	88.70	90.86	37.40	30.50	49.10	66.10

[1] Represents the precision of the bud, the same as below. [2] Note: The SAHI algorithm was used in S-YOLO by default and YOLOX by contrast.

Figure 5a shows the loss curves during the training of the S-YOLO-s model. After the 100th epoch, the backbone started to thaw, and the losses of the training set and the validation set fell. The validation set loss was higher than the training set after the 170th epoch, and when the decreasing trend was smaller than the training set, the model started to overfit. Figure 5b shows the P–R curves with IoU = 3. The AP of the fully open flowers was significantly higher than the other three stages, and the differences among the AP values of the three stages were not significant. This phenomenon was significantly correlated with the higher pixel proportion of the fully open flowers.

The precision of S-YOLO-s was enhanced by 7.94%, 8.05%, 3.49%, and 6.96%, and different types of *mAP* by 10.00%, 9.10%, 13.10%, and 7.20%, respectively, compared to YOLOX-s at each flowering stage. By comparing the experimental results, it was found that the improved model's apple flower detection precision was significantly higher than the original model and indirectly higher than EfficientDet, Faster R-CNN, and SSD 300. Therefore, it is practical and feasible for the model to accurately detect apple flowers with high resolution. The detection results of YOLOX-s and S-YOLO-s for the apple tree blossoming images under four typical weather conditions of overcast, sunnier, foggy, and sunny days (Figure 6a,b) provide further proof of the model's superiority.

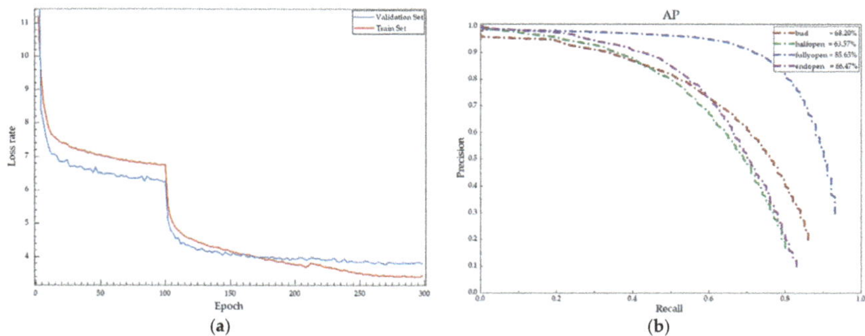

(a) (b)

Figure 5. Loss and P–R curve. (**a**) Loss curve of S-YOLO-s. (**b**) P–R curves with mixed datasets (IoU = 0.3).

(**a**)

(**b**)

(**c**)

(**d**)

Figure 6. Detection of four growth stages of apple blossoms with red, green, blue, and purple boxes using different models. (**a**) YOLOX-s detection results; (**b**) S-YOLO-s detection results; (**c**) YOLOX-l (add SAHI) detection results; (**d**) S-YOLO-l detection results.

3.2.2. The Results of Different Versions of Models

The results of the comparison experiments demonstrate that the S-YOLO model performed better in detecting apple flowers at various growth stages. An ablation experiment was designed to gain a deeper understanding of how the two improvements of adding the slicing algorithm and replacing the backbone network contributed to the results and to determine how both improvements could be used more effectively to produce better results.

Table 7 shows the results of the ablation experiments. Using YOLO-s as the baseline, the SAHI algorithm improved the precision by 0.70%, 0.04%, 1.04%, and 3.38% for each flowering stage and by 6.70%, 5.50%, 7.20%, and 7.70% for the different types of *mAP*, respectively. The results indicate that the SAHI algorithm can successfully enhance the detection impact of the model, and the degree of enhancement was proportional to the object's pixel size. After replacing the original backbone network with Swin Transformer-tiny, the model parameters took up 35.79 MB, the FLOPs took up 95.57 GB, the precision improved by 7.24%, 8.01%, 2.45%, and 3.58%, and the *mAP* improved by 3.30%, 3.60%, 5.90%, and −0.50%. After rebuilding the backbone, the results indicate that S-YOLO was more sensitive to detecting small objects of varying length and breadth. The promotion of detection precision by using Swin Transformer as a backbone network was negatively correlated with the object size, and the enhancement of the *mAP* showed an inverted U-shape with flower size, which led to negative growth of the mAP_L.

Table 7. Results of ablation experiments.

Model	P-Bud (%)	P-Half-Open (%)	P-Fully Open (%)	P-End-Open (%)	mAP_{ALL} (%)	$mAPs$ (%)	$mAPm$ (%)	$mAPl$ (%)	Parameters (M)	FLOPs (G)
YOLOX-s	79.19	81.27	85.21	83.90	27.40	21.40	36.00	58.90	8.94	26.64
YOLOX-s (+SAHI)	79.89	81.31	86.25	87.28	34.10	26.90	43.20	**66.60**	8.94	26.64
S-YOLO-s	87.13	**89.32**	88.70	90.86	37.40	30.50	49.10	66.10	35.79	95.57
S-YOLO-t	79.38	80.50	83.02	84.02	32.40	25.60	41.50	57.10	30.80	80.58
S-YOLO-m	81.95	83.48	86.96	86.86	35.10	28.20	45.40	65.70	45.89	135.35
S-YOLO-l	**88.18**	88.59	**89.50**	**91.95**	**39.00**	**32.10**	**50.60**	64.30	51.37	157.68
Swin-S	82.35	84.02	85.60	87.37	34.50	27.60	44.10	65.70	57.07	165.84

While replacing various backbones and detection heads extended the model, the detection effect exhibited a non-linear correlation with the model size. Different sizes of S-YOLO were obtained by adjusting the channel depth and width variation of YOLOX while maintaining increased data using SAHI. The *mAP* values from S-YOLO-t to S-YOLO-l were 32.40%, 37.40%, 35.10%, and 39.00%. This phenomenon, where the *mAP* did not grow as the model grew, resulted from the uncoordinated channel change between the network's backbone and neck. Swin-S replaced the YOLOX-s backbone with Swin Transformer-small. Although the number of parameters and FLOPs of the Swin-S model were higher than S-YOLO-l, the precision and *mAP* were smaller than S-YOLO-s. Therefore, the appropriate ratio of the number of structural parameters and the channel variation are necessary factors for the S-YOLO variant to achieve higher detection results.

In summary, S-YOLO-s outperformed the original model in detecting each flower stage at a high resolution, which resulted from the combined effect of the SAHI algorithm increasing the percentage of flower pixels while keeping the image features unchanged and the Swin Transformer being used as the backbone network. The high-resolution local information provided by SAHI without scaling was fed to the network along with the global information of the scaled original image. This information was fused via the Swin Transformer and subsequently fully used by the network, prompting the model to produce state-of-the-art experimental results. Moreover, S-YOLO is exceptionally sensitive to the size of the detected object, and larger detected objects will get a minor boost or even negative growth in the *mAP* compared to other objects of smaller length and width. Ablation experiments revealed the superiority of the S-YOLO performance and illustrated that appropriate channel depth variation and balanced parameter scaling will provide better results than arbitrarily expanding the model. This experiment provides data to support the replacement of the larger Swin Transformer as the backbone to obtain greater experimental results.

3.2.3. Comparing the Effects of Other Measures

The detection results on the mixed dataset (Table 8) show that the improved S-YOLO-l was slightly less effective than YOLOX-l with the comparable sizes of parameters and FLOPs on the mixed dataset, but this does not mean that the improvement of the model was unsuccessful. The dataset used for model evaluation in this experiment consisted of a 640 × 640 pixel image after slicing and the original 3000 × 3000 pixel image. However, the actual detection object in the natural environment should be 3000 × 3000 pixels, so the data in the test dataset were replaced with the raw dataset and the experiment was conducted again to obtain the new results.

Table 8. Comparison results of YOLOX-l and S-YOLO-l with similar parameters (54.15 M and 51.37 M) on mixed and raw datasets (all using SAHI).

Model	Datasets	P-Bud (%)	P-Half-Open (%)	P-Fully Open (%)	P-End-Open (%)	mAP_{ALL} (%)	mAP_S (%)	mAP_M (%)	mAP_L (%)
S-YOLO-s	Mixed	87.13	**89.32**	88.70	90.86	37.40	30.50	49.10	66.10
YOLOX-l	Mixed	85.27	88.66	**92.02**	**92.92**	**41.70**	**35.60**	**55.40**	**74.60**
S-YOLO-l	Mixed	**88.18**	88.59	89.50	91.95	39.00	32.10	50.60	64.30
YOLOX-l	$Raw_{20\%}$ *	81.00	85.80	90.30	90.00	34.30	28.30	48.20	71.80
S-YOLO-l	$Raw_{20\%}$	84.30	87.78	90.56	91.88	34.40	28.40	48.20	70.60

* The proportion of the test set used for all datasets was 20%.

The results in Table 8 show that S-YOLO-l significantly outperformed YOLOX-l in precision and achieved better results in all types of *mAP* after adjusting the test set percentage to 20%. Therefore, the improved model still outperformed the original model in the high-resolution task of detecting the growth stage of apple blossoms, even when the gain from the SAHI algorithm was ignored. In addition, if the SAHI algorithm is not used, YOLOX-l will not be trained properly due to the low number of input images. Notably, the backbone network utilized in this investigation was Swin Transformer-tiny. With enough computing resources, labeled data, and disregarding the negative impact of a bigger model, a bigger S-YOLO network with the proper parameter scaling and number of channels would produce superior detection results.

Figure 6c,d shows the detection results of YOLOX-l (add SAHI) and S-YOLO-l for the apple tree flowering images under four typical weather conditions: cloudy, more sunny, foggy, and sunny days. The image presentation results indicate that YOLOX-l, with the inclusion of the SAHI algorithm, may have performed marginally better than S-YOLOX-l at several detection locations. Notably, the initial annotation volume utilized in the experiment was 39,980, which grew to 109,813 after slicing using the SAHI technique and mixing with the original dataset, but was significantly less than the COCO dataset. Therefore, S-YOLO-l cannot fully exploit its strengths. With the development of IoT technologies, the issue of a too-small training dataset will be resolved, and it will also be possible to use bigger Swin Transformer models to obtain better detection results. By observing Table 8 and Figure 6, it is possible to infer that the enhanced S-YOLO-l beat YOLOX-l in an identical situation using the SAHI method but that the existing quantity of data did not cause S-YOLO to demonstrate an overwhelming advantage.

3.3. *Apple Flowering Monitoring Results*

A total of 1494 fruit tree images (3000 × 3000 pixels) were fed into S-YOLO-s for flower prediction to obtain the number variation and proportional variation under a time series (Figure 7b). Except for the decline in the number of buds caused by the late shooting date, the number of flowers in all three categories showed arching characteristics in line with the natural pattern for time variation. The average flower density at the peak of each stage was

55.687 on 3 April, 47.565 on 6 April, 118.183 on 9 April, and 17.522 flowers/tree on 15 April (Figure 7a) corresponding to 75.7%, 46.7%, 82.26%, and 49.58% of all flowers, respectively (Figure 7b). The flowering intensities of the orchard on different dates were 1.13%, 3.94%, 19.54%, 37.34%, 57.57%, 72.18%, 82.26%, 81.59%, 75.74%, 77.04%, 76.59%, 47.83%, and 12.18%. The experimental results show that the orchard was in the first flowering stage on 3 April, the middle flowering stage on 5 April, the full flowering stage on 7 April, and the last flowering stage on 15 April. In addition, based on the trend of the number of end-open flowers, it was predicted that only these would remain in the orchard from 17 to 18 April, which means that the orchard would be in the last flowering stage completely.

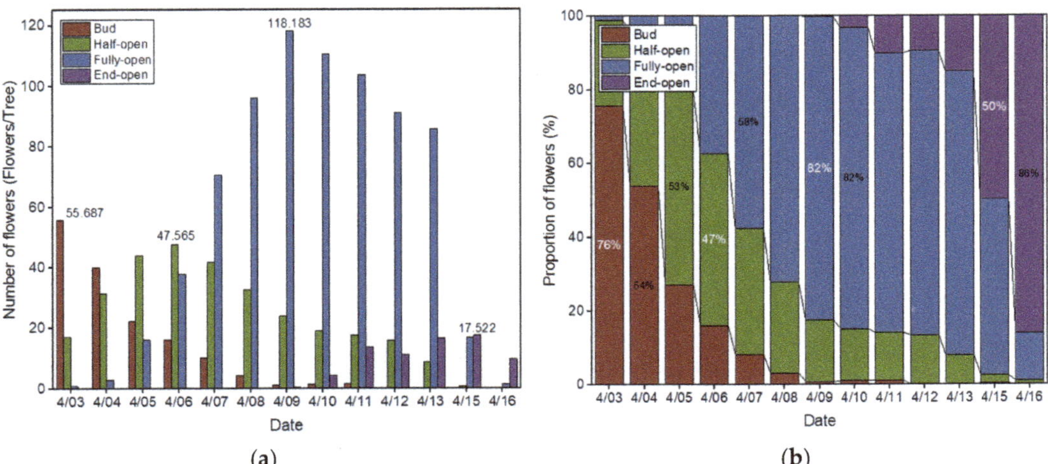

Figure 7. Flowering stage changes. (**a**) Changes in the daily number of flowers at each stage; (**b**) changes in the daily percentage of flowers at each stage.

This experiment demonstrated that with the help of an S-YOLO-s high-performance detector, it was possible to obtain time-series changes in the number and proportion of flowers. At the same time, it was possible to achieve the daily flowering intensity estimation and flower peak time determination at each stage, and finally, flowering information monitoring was realized. Notably, the results mentioned above were achieved by counting the detection results of single fruit trees; thus, S-YOLO-s is adequate for identifying the blooming phases of single fruit trees.

4. Conclusions

This study proposes a Transformer-based apple flowering monitoring method for monitoring the whole flower growth process of full fruit trees in the open world. In this work, the non-pure convolutional S-YOLO was model used to detect the four growth stages of apple blossoms accurately and to analyze the changes in the numbers and percentages of blossoms at each growth stage in order to estimate the peak flowering time and flowering intensity and to complete the monitoring process. The main conclusions are as follows.

1. Based on the combination of YOLOX and Swin Transformer, the SAHI algorithm was added to form the S-YOLO model. S-YOLO-s improved the precision compared to the original YOLOX-s by 7.94%, 8.05%, 3.49%, and 6.96% for the four flowering states and by 10.00%, 9.10%, 13.10%, and 7.20% for the mAP_{ALL}, mAP_{S}, mAP_{M}, and mAP_{L}, respectively. S-YOLO-l resulted in 88.18%, 88.95%, 89.50%, and 91.95% precision at each flowering state and 39.00%, 32.10%, 50.60%, and 64.30% for each type of mAP, respectively. Without considering the SAHI algorithm boost, the non-pure convolutional S-YOLO-l model slightly outperformed the YOLOX-l model with similar parameters and FLOPs in the original dataset, with improvements of 3.30%, 1.98%,

0.26%, and 1.88% in detection precision. In addition, using a bigger Swin Transformer as the backbone, designing an appropriate percentage of structural parameters, and collecting more training data may have resulted in improved experimental outcomes.
2. The SAHI algorithm made the object-detected aspect ratio and average area vary between 3.00% and 5.00%, respectively, while increasing the image area ratio by 1250%. The SAHI algorithm increased the number of annotations of flowers in the four growth stages by 170.20%, 170.11%, 182.41%, and 176.16%, respectively, and the total amount of annotated data increased by 150% to 109,813, providing more quality data for the model training process. The experimental results show that the SAHI algorithm improved the precision by 0.70%, 0.04%, 1.04%, 3.38%, and the *mAP* by 6.70%, 5.50%, 7.20%, 7.70% for each flowering stage, respectively, and the larger the object detected, the more the detection effect was improved.
3. Using the results of S-YOLO, the quantity and percentage of apple flowers and the flowering intensity were estimated daily for each stage of the orchard during the flowering period, and the peak time was identified. The average flower density at the peak of each stage was 55.687 on 3 April, 47.565 on 6 April, 118.183 on 9 April, and 17.522 flowers/tree on 15 April, corresponding to 75.7%, 46.7%, 82.26% and 49.58% of all flowers. On the various dates, the flowering intensities of the orchard were 1.13 %, 3.94 %, 19.54 %, 37.34 %, 57.57 %, 72.18 %, 82.26 %, 81.59 %, 75.74 %, 77.04 %, 76.59 %, 47.83 %, and 12.18 %. In addition, the orchard was at its first flowering stage on 3 April, its middle flowering stage on 5 April, its full flowering stage on 7 April, and its last flowering stage on 15 April.

The apple flower monitoring method proposed in this study is applicable to orchard environments in the open world. Based on the detection of four stages of tiny flowers in complete fruit tree images, the quantitative analysis of data and the assessment of blossom intensity were realized, and then the flower information monitoring was realized. It is important to note that the existence of diverse viewing angles, illumination fluctuations, occlusions, uncertain stances, low pixel ratio, complicated backdrops, etc., makes it challenging for models trained on the source dataset to achieve high performance. This method establishes the foundation for the proper use of IoT technology for the remote monitoring of flowering information in modern orchards.

Author Contributions: Conceptualization, X.Z. (Xinzhu Zhou); methodology, X.Z. (Xinzhu Zhou) and G.S.; software, X.Z. (Xinzhu Zhou) and X.Z. (Xiaolei Zhang); validation, X.Z. (Xinzhu Zhou) and N.X.; formal analysis, X.Z. (Xinzhu Zhou), Y.Y. and J.C.; investigation, X.Z. (Xinzhu Zhou), J.C. and Y.H.; writing—original draft preparation, X.Z. (Xinzhu Zhou), G.S. and N.X.; writing—review and editing, X.Z. (Xinzhu Zhou), Y.Y. and J.C.; project administration, G.S.; funding acquisition, G.S. All authors have read and agreed to the published version of the manuscript.

Funding: This work was supported by the key R&D Program of Jiangsu Province (No. BE2022363), High-end Foreign Experts Recruitment Plan of China (No. G2021145009L), Jiangsu agricultural science and technology Innovation Fund (No. CX(22)3097), and Jiangsu agricultural science and technology Innovation Fund (No. CX(21)2006).

Institutional Review Board Statement: Not applicable.

Data Availability Statement: The data supporting this study's findings are available from the corresponding author upon reasonable request.

Acknowledgments: The author would like to thank the editors and reviewers for their comments on improving the quality of this work and MDPI for their English language revisions.

Conflicts of Interest: The authors declare no conflict of interest.

References

1. Zhou, G.; Xia, X. Present Situation and Development Prospect of the Digital Orchard Technology. In *China's e-Science Blue Book 2020*; Springer: Berlin/Heidelberg, Germany, 2021; pp. 443–458.
2. Link, H. Significance of Flower and Fruit Thinning on Fruit Quality. *Plant Growth Regul.* **2000**, *31*, 17–26. [CrossRef]

3. Iwanami, H.; Moriya-Tanaka, Y.; Honda, C.; Hanada, T.; Wada, M. A Model for Representing the Relationships among Crop Load, Timing of Thinning, Flower Bud Formation, and Fruit Weight in Apples. *Sci. Hortic.* **2018**, *242*, 181–187. [CrossRef]
4. Bound, S.A. Precision Crop Load Management of Apple (*Malus* x *Domestica* Borkh.) without Chemicals. *Horticulturae* **2018**, *5*, 3. [CrossRef]
5. Peck, G.M.; Combs, L.D.; DeLong, C.; Yoder, K.S. Precision Apple Flower Thinning Using Organically Approved Chemicals. In Proceedings of the International Symposium on Innovation in Integrated and Organic Horticulture (INNOHORT), Avignon, France, 8–12 June 2015; pp. 47–52.
6. Farjon, G.; Krikeb, O.; Hillel, A.B.; Alchanatis, V. Detection and Counting of Flowers on Apple Trees for Better Chemical Thinning Decisions. *Precis. Agric.* **2020**, *21*, 503–521. [CrossRef]
7. Nautiyal, P.; Sharma, U.; Singh, V.; Goswami, S.; Agrawal, K.; Krishali, V.; Bisht, R.; Mittal, H.; Mehta, R.; Pokhriyal, A. Fruit Thinning: Purpose, Methods & Role of Plant Growth Regulators. *Pharma Innov. J.* **2022**, *11*, 1500–1504.
8. Solomakhin, A.A.; Blanke, M.M. Mechanical Flower Thinning Improves the Fruit Quality of Apples. *J. Sci. Food Agric.* **2010**, *90*, 735–741. [CrossRef] [PubMed]
9. Koike, H.; Tamai, H.; Ono, T.; Shigehara, I. Influence of Time of Thinning on Yield, Fruit Quality and Return Flowering of'Fuji'apple. *J. Am. Pomol. Soc.* **2003**, *57*, 169.
10. Wu, D.; Lv, S.; Jiang, M.; Song, H. Using Channel Pruning-Based YOLO v4 Deep Learning Algorithm for the Real-Time and Accurate Detection of Apple Flowers in Natural Environments. *Comput. Electron. Agric.* **2020**, *178*, 105742. [CrossRef]
11. Sun, K.; Wang, X.; Liu, S.; Liu, C. Apple, Peach, and Pear Flower Detection Using Semantic Segmentation Network and Shape Constraint Level Set. *Comput. Electron. Agric.* **2021**, *185*, 106150. [CrossRef]
12. Bhattarai, U.; Bhusal, S.; Majeed, Y.; Karkee, M. Automatic Blossom Detection in Apple Trees Using Deep Learning. *IFAC-PapersOnLine* **2020**, *53*, 15810–15815. [CrossRef]
13. Tian, Y.; Yang, G.; Wang, Z.; Li, E.; Liang, Z. Instance Segmentation of Apple Flowers Using the Improved Mask R–CNN Model. *Biosyst. Eng.* **2020**, *193*, 264–278. [CrossRef]
14. Wang, X.A.; Tang, J.; Whitty, M. DeepPhenology: Estimation of Apple Flower Phenology Distributions Based on Deep Learning. *Comput. Electron. Agric.* **2021**, *185*, 106123. [CrossRef]
15. Yuan, W.; Choi, D. UAV-Based Heating Requirement Determination for Frost Management in Apple Orchard. *Remote Sens.* **2021**, *13*, 273. [CrossRef]
16. Wang, X.A.; Tang, J.; Whitty, M. Side-View Apple Flower Mapping Using Edge-Based Fully Convolutional Networks for Variable Rate Chemical Thinning. *Comput. Electron. Agric.* **2020**, *178*, 105673. [CrossRef]
17. Piani, M.; Bortolotti, G.; Manfrini, L. Apple Orchard Flower Clusters Density Mapping by Unmanned Aerial Vehicle RGB Acquisitions. In Proceedings of the 2021 IEEE International Workshop on Metrology for Agriculture and Forestry (MetroAgriFor), Trento-Bolzano, Italy, 3–5 November 2021; IEEE: Piscataway, NJ, USA, 2021; pp. 92–96.
18. Zhang, C.; Mouton, C.; Valente, J.; Kooistra, L.; van Ooteghem, R.; de Hoog, D.; van Dalfsen, P.; Frans de Jong, P. Automatic Flower Cluster Estimation in Apple Orchards Using Aerial and Ground Based Point Clouds. *Biosyst. Eng.* **2022**, *221*, 164–180. [CrossRef]
19. Girshick, R.; Donahue, J.; Darrell, T.; Malik, J. Rich Feature Hierarchies for Accurate Object Detection and Semantic Segmentation. In Proceedings of the IEEE Conference on Computer Vision and Pattern Recognition, Columbus, OH, USA, 23–28 June 2014; pp. 580–587.
20. Ren, S.; He, K.; Girshick, R.; Sun, J. Faster R-Cnn: Towards Real-Time Object Detection with Region Proposal Networks. *Adv. Neural Inf. Process. Syst.* **2015**, *39*, 1137–1149. [CrossRef]
21. Girshick, R. Fast R-Cnn. In Proceedings of the IEEE International Conference on Computer Vision, Las Condes, Chile, 7–13 December 2015; pp. 1440–1448.
22. Liu, W.; Anguelov, D.; Erhan, D.; Szegedy, C.; Reed, S.; Fu, C.-Y.; Berg, A.C. Ssd: Single Shot Multibox Detector. In Proceedings of the European Conference on Computer Vision, Amsterdam, The Netherlands, 11–14 October 2016; Springer: Berlin/Heidelberg, Germany, 2016; pp. 21–37.
23. Redmon, J.; Divvala, S.; Girshick, R.; Farhadi, A. You Only Look Once: Unified, Real-Time Object Detection. In Proceedings of the IEEE Conference on Computer Vision and Pattern Recognition, Las Vegas, NV, USA, 27–30 June 2016; pp. 779–788.
24. Redmon, J.; Farhadi, A. Yolov3: An Incremental Improvement. *arXiv* **2018**, arXiv:1804.02767.
25. Bochkovskiy, A.; Wang, C.-Y.; Liao, H.-Y.M. Yolov4: Optimal Speed and Accuracy of Object Detection. *arXiv* **2020**, arXiv:2004.10934.
26. Guofang, C.; Zhaoying, C.; Yuliang, W.; Jinxing, W.; Guoqiang, F.; Hanqing, L. Research on Detection Method of Apple Flower Based on Data-Enhanced Deep Learning. *J. Chin. Agric. Mech.* **2022**, *43*, 148. [CrossRef]
27. Yuan, W.; Choi, D.; Bolkas, D.; Heinemann, P.H.; He, L. Sensitivity Examination of YOLOv4 Regarding Test Image Distortion and Training Dataset Attribute for Apple Flower Bud Classification. *Int. J. Remote Sens.* **2022**, *43*, 3106–3130. [CrossRef]
28. Ge, Z.; Liu, S.; Wang, F.; Li, Z.; Sun, J. Yolox: Exceeding Yolo Series in 2021. *arXiv* **2021**, arXiv:2107.08430.
29. Zhaosheng, Y.; Tao, L.; Tianle, Y.; Chengxin, J.; Chengming, S. Rapid Detection of Wheat Ears in Orthophotos From Unmanned Aerial Vehicles in Fields Based on YOLOX. *Front. Plant Sci.* **2022**, *13*, 851245. [CrossRef]
30. Zhang, Y.; Zhang, W.; Yu, J.; He, L.; Chen, J.; He, Y. Complete and Accurate Holly Fruits Counting Using YOLOX Object Detection. *Comput. Electron. Agric.* **2022**, *198*, 107062. [CrossRef]

31. Dosovitskiy, A.; Beyer, L.; Kolesnikov, A.; Weissenborn, D.; Zhai, X.; Unterthiner, T.; Dehghani, M.; Minderer, M.; Heigold, G.; Gelly, S. An Image Is Worth 16x16 Words: Transformers for Image Recognition at Scale. *arXiv* **2020**, arXiv:2010.11929.
32. Liu, Z.; Lin, Y.; Cao, Y.; Hu, H.; Wei, Y.; Zhang, Z.; Lin, S.; Guo, B. Swin Transformer: Hierarchical Vision Transformer Using Shifted Windows. In Proceedings of the IEEE/CVF International Conference on Computer Vision, Montreal, BC, Canada, 11–17 October 2021; pp. 10012–10022.
33. Kisantal, M.; Wojna, Z.; Murawski, J.; Naruniec, J.; Cho, K. Augmentation for Small Object Detection. *arXiv* **2019**, arXiv:1902.07296.
34. Chen, Z.; Wu, K.; Li, Y.; Wang, M.; Li, W. SSD-MSN: An Improved Multi-Scale Object Detection Network Based on SSD. *IEEE Access* **2019**, *7*, 80622–80632. [CrossRef]
35. Akyon, F.C.; Altinuc, S.O.; Temizel, A. Slicing Aided Hyper Inference and Fine-Tuning for Small Object Detection. *arXiv* **2022**, arXiv:2202.06934.
36. Keles, M.C.; Salmanoglu, B.; Guzel, M.S.; Gursoy, B.; Bostanci, G.E. Evaluation of YOLO Models with Sliced Inference for Small Object Detection. *arXiv* **2022**, arXiv:2203.04799.
37. Liu, S.; Qi, L.; Qin, H.; Shi, J.; Jia, J. Path Aggregation Network for Instance Segmentation. In Proceedings of the IEEE Conference on Computer Vision and Pattern Recognition, Salt Lake City, UT, USA, 18–23 June 2018; pp. 8759–8768.
38. Zhang, H.; Cisse, M.; Dauphin, Y.N.; Lopez-Paz, D. Mixup: Beyond Empirical Risk Minimization. *arXiv* **2017**, arXiv:1710.09412.
39. Long, X.; Deng, K.; Wang, G.; Zhang, Y.; Dang, Q.; Gao, Y.; Shen, H.; Ren, J.; Han, S.; Ding, E. PP-YOLO: An Effective and Efficient Implementation of Object Detector. *arXiv* **2020**, arXiv:2007.12099.
40. Tan, M.; Pang, R.; Le, Q.V. Efficientdet: Scalable and Efficient Object Detection. In Proceedings of the IEEE/CVF Conference on Computer Vision and Pattern Recognition, Seattle, WA, USA, 13–19 June 2020; pp. 10781–10790.

 agriculture

Article

Research on the Measurement Method of Feeding Rate in Silage Harvester Based on Components Power Data

Fengzhu Wang [1,2], Jizhong Wang [1,2], Yuxi Ji [1,2], Bo Zhao [1,2], Yangchun Liu [1,2], Hanlu Jiang [1,2] and Wenhua Mao [1,2,*]

[1] State Key Laboratory of Soil Plant Machine System Technology, Chinese Academy of Agricultural Mechanization Sciences Group Co., Ltd., Beijing 100083, China
[2] Chinese Academy of Agricultural Mechanization Sciences Group Co., Ltd., Beijing 100083, China
* Correspondence: maowenhua@caams.org.cn or mwh-924@163.com

Abstract: For existing problems, such as the complex interactions between a crop and a machine, the measuring difficulty and the limited measurement precision of the feeding quantity within the corn silage harvester, a method of feeding rate measurement based on key conditions data, working data cleaning, and multiple variate regression is proposed. Non-destructive rotation speed, rotation torque, and power consumption sensors are designed for the key mechanical components. The data conditions, such as rotating speed, rotating torque, power consumption, hydraulic pressure, and hydraulic flow for the key operation of parts including cutting, feeding, shredding, and throwing are monitored and collected in real-time during field harvesting. The working data are screened and preprocessed, and the Mann-Kendall boundary extraction algorithm is applied, as is multiple component time lag correction analysis, and the Grubbs exception detection method. Based on a Pearson correlation analysis results, one-factor and multiple-factor regression models are respectively developed to achieve an accurate measurement of the corn feeding rate. The field validation tests show that the working data boundary extraction results among the load-stabilizing components such as shredding roller and throwing blower are highly reliable, with a correct rate of 100%. The power monitoring data of the shredding roller and throwing blowers are significantly correlated with the crop feeding rate, with a max correlation coefficient of 0.97. The determination coefficient of the single-factor feeding rate model based on the shredding roller reaches 0.94, and the maximum absolute error of the multi-factor feeding rate model is 0.58 kg/s. The maximum relative error is ±5.84%, providing technical and data support for the automatic measuring and intelligent tuning of the feeding quantity in a silage harvester.

Keywords: feeding rate; silage harvester; power consumption; multi-variate fusion; Mann-Kendall; data screening

Citation: Wang, F.; Wang, J.; Ji, Y.; Zhao, B.; Liu, Y.; Jiang, H.; Mao, W. Research on the Measurement Method of Feeding Rate in Silage Harvester Based on Components Power Data. *Agriculture* **2023**, *13*, 391. https://doi.org/10.3390/agriculture13020391

Academic Editor: Massimo Cecchini

Received: 15 January 2023
Revised: 1 February 2023
Accepted: 4 February 2023
Published: 7 February 2023

1. Introduction

The feeding rate is an important factor affecting the overall performance of a silage harvester. The increase in the feeding rate can significantly improve the harvesting efficiency of a silage harvester [1]. However, if the feeding rate is too high, on the one hand, maize plants will be fed too fast, which may easily cause blockage and failure in the working part, especially the header, and throwing components in the harvester [2]. In this way, the harvesting progress is affected by the follow-up manual fault clearing. On the other hand, due to the excessive feeding volume, the whole machine will be in overloaded harvesting conditions for a long time, which may cause fatigue damage to the working parts and engine [3], thus reducing the overall machine service life. Conversely, when the feeding rate is too small, apart from directly reducing the harvesting efficiency, it may affect the shredding performance of the silage harvester, resulting in the average length of silage cutting or the uniformity of the cutting length not meeting the expected requirements. The ideal field operation state of the silage harvester is to maintain a stable and optimal feeding rate within the rated range on the premise of ensuring a sufficient chopping performance [4].

Currently, scholars have carried out a significant amount of research on the detection method of feeding rate in forage harvesters [5–7]. According to the detection principle, this can mainly be divided into mass flow type and volume flow type. The mass flow method mainly realizes the online detection of the feeding rate through indirect modeling by monitoring the machine's working conditions, such as speed, torque, power, pressure, capacitance, etc. [8–11]. This method has been extensively studied on grain combine harvesters and has achieved good field verification results [12–14]. However, for silage harvesters, which use whole-plant intake, the material composition and machine-crop interactions are more complex [15], and there is a lack of sufficient research on airborne applications, the existing monitoring system for a forage harvester mainly realizes the measurement of conventional parameters, such as location, rotation speed, cutter height, engine data, feeding metal, blade sharpness, etc. Most of the research applications focus on the display and early warning for the machine status, while the relationship among monitoring parameters is relatively independent, and there is a need for a better feeding rate monitoring method based on correlation analysis and multiple-parameter fusion modeling [16,17]. The volume flow detection is mainly detected by the opening of the feed inlet, without considering the influence of moisture content and material density on the model, in which repeated calibration is required [18], and further research is required. According to the model's methodology, as for the mathematical relationship between the power consumption and feeding rate for a crop combine harvester, the primary, quadratic, and exponential regression models are frequently applied, and it was shown that the measurement between feeding rate and working torque or working power is better modeled by using the primary linear regression model [19].

In consideration of the shortcomings in the existing research and the demands in feed rate monitoring, this paper proposes a measurement method based on multi-power data fusion regression for a maize silage harvester. The mechanism principle and operation process of the silage harvesting machine is analyzed, and strain resistance principles and pressure flow principles are used to design real-time power monitoring sensors for key operation components. Highly reliable pre-processing algorithms for dynamic field data are studied, a one-variable linear model and multiple-variable fusion regression model based on a correlation analysis is established to measure the feeding rate, and field harvesting experiments are carried out to validate the relevant sensors, processing algorithms, and measurement models.

2. Materials and Methods

2.1. Feeding Rate Factors Analysis

The feeding rate of a silage harvester is the total quality of silage fed into the silage harvester per second. During the harvesting operation, the silage feeding rate is influenced by many factors, which can be divided into antecedent factors and subsequent factors according to the time before and after the crop is fed into the machine. The antecedent factors include the harvesting speed, crop density, feeding width, plant moisture content, stubble height, etc. The subsequent factors include the relevant operating condition parameters of the silage harvester, such as the speed, torque, and power of the key working components. The antecedent factors are largely dependent on crop characteristics and vary considerably between land areas [20]; it is difficult to obtain precise data to support the accurate measurement of the feeding rate in field experiments. The subsequent factors can directly characterize the working load of the silage harvester along with the material feeding, and can dynamically respond to the feeding rate change in real time. Therefore, this paper constructs a detection model to automatically obtain the feeding rate through the subsequent power data of multiple components of the maize silage machine.

2.2. *Power Measuring Principle*

2.2.1. Parameter Symbol Definition

Variable symbols are defined as follows:
P_r—the power consumption of the mechanical-driven part, kW
n_r—the rotation speed of the mechanical-driven part, r/min
v_r—the rotation linear velocity of the mechanical-driven part, m/s
T_r—the rotation torque of the mechanical-driven part, N.m
F—the rotation force of the mechanical-driven part, r/min
R—the rotation radius of the mechanical-driven part, m
P_h—the power consumption of the hydraulic-driven part, kW
q—the output flow of the hydraulic pump, L/min
P_o—the outlet pressure of the hydraulic pump, MPa
P_i—the inlet pressure of the hydraulic pump, Mpa
q—the output flow of the hydraulic pump, L/min
P_o—the outlet pressure of the hydraulic pump, MPa
P_i—the inlet pressure of the hydraulic pump, MPa
V_0—the rated flow of the hydraulic pump, L/min
η_v—the volumetric efficiency of the hydraulic pump
n_0—the engine speed, r/min
k—the transmission ratio the hydraulic pump and the engine

2.2.2. Measuring Principle Analysis

According to the machine power roadmap of the maize silage harvester, the power sources of key operation parts include both mechanical drive and hydraulic drive.

For mechanically-driven working parts, the operating power can be acquired by measuring the rotational speed and rotating torque. For rotation torque measurement, the principle of torsion resistance strain gauge is used, and the strain gauge is attached to the rotating elastic shaft to form a strain bridge detection circuit, which outputs a frequency signal proportional to the torsion value, and thus the working torque is obtained. For rotation speed measurement, a speed sensor is mounted on the fixed part of the torque sensor, and an encoder with multiple detection positions is installed on the rotating part of the torque sensor. When the encoder rotates with the working shaft, the sensor outputs a series of speed pulse signals, and the working speed is measured. According to the definition of mechanical torque, the torque can be calculated by Formula (1).

$$T_r = FR \tag{1}$$

The relationship between linear speed and rotational speed is shown as follows:

$$v_r = \frac{n_r}{60} 2\pi R \tag{2}$$

The power consumption of the mechanical rotating parts is then given as follows:

$$P_r = \frac{1}{1000} F v_r = \frac{1}{1000} \times \frac{T_r}{R} \times \frac{n_r}{60} \times 2\pi R = \frac{n_r T_r}{9549.3} \tag{3}$$

For hydraulic-driven working parts, the operating power can be obtained by the differential pressure between the outlet and inlet and the hydraulic flow of the hydraulic pump, and the power of the hydraulic parts is calculated as follows.

$$P_h = \frac{(p_o - p_i)}{60} q = \frac{n_0 k V_0 \eta_v}{60} (p_o - p_i) \tag{4}$$

2.3. *Power Sensor Design*

On the basis of not changing the original body structure, the torque and speed sensors are designed to meet the power detection requirements of different installation locations in the silage harvester.

2.3.1. Header Power Sensor

The header is driven by a spline shaft. Under normal operating conditions, the corn silage harvester works in the way of full-width without row-control. Theoretically, it is believed that the operating power of the left header and the right header are equal. Considering the installation space on the header shaft, the original T-shaped driving shaft on the right side of the header is selected as the installation position of the sensors. Two sensors are designed to measure the dynamic power of the header in the shaft-mounting type, where the power sensor of the inside shaft is used to measure the power at the small and fixed rotary cutter disk, while the power sensor of the outside shaft is applied to measure the dynamic power of the large folding and rotary cutter disk, as shown in Figure 1.

Figure 1. Physical view of header power sensor. 1: power sensor of inside shaft 2: T-type drive shaft 3: power sensor of outside shaft.

2.3.2. Shredding Power Sensor

The shredding rollers are driven by the pulley. The original drive pulley is made of cast iron, and its elastic properties cannot meet the requirements of accurate torque measurement. According to the mechanical dimensions of the original cast iron pulley, 40CrNiMoA high-strength alloy steel material with better inherent elastic properties is selected to design a pulley-type shredding torque and speed sensor, as shown in Figure 2.

Figure 2. Physical view of shredding power sensor. 1: elastic pulley; 2: mounting base; 3: torque sensor; 4: speed sensor.

The sensor mainly consists of three parts: an elastic pulley, a torque-sensing component, and a speed-sensing component. The elastic pulley is utilized to replace the original cast iron pulley, providing the mechanical connection and power transmission for shredding rollers. The torque sensor component includes a resistive torsion strain bridge, a signal acquisition circuit, a signal output circuit, etc. The speed sensor component includes an inductive speed sensor mounted with tiny round screw threads, a speed probe mounting hole, and a speed measuring panel with 30 detection points equally distributed along the circumference. The inductive speed sensor is fixed on the sensor shell at a distance of 1.0 mm from the speed panel, and while as the disc rotates, the speed sensor continuously outputs pulses.

2.3.3. Blowing Power Sensor

The throwing blower is driven by the pulley. The pulley-type throwing blower torque and speed sensor are designed according to the original mechanical dimensions. The torque measurement is implemented in a similar structure as the shredding rollers. As for the speed detection, due to the small size of the throwing pulley, the outward structure of the speed panel is designed to achieve real-time acquisition of speed pulse signals, while 30 measuring points are evenly arranged along the end face of the pulley, as shown in Figure 3.

Figure 3. Physical view of blowing power sensor. 1: elastic pulley; 2: velocimetric panel; 3: mounting base; 4: torque sensor; 5: speed sensor.

2.3.4. Power Sensor Accuracy

The YD322 oil pressure sensor with ±0.1% precision produced by Xian Yunyi Instrument Co., Ltd., Xi'an, China. is applied in the hydraulic power measurement. While the PY-M5S1.5N-L2M hall-type speed sensor with ±0.1% precision manufactured by Delan Electronics Co., Ltd., Dongguan, China. and the HX-900 Rotary Torque Sensor produced by Sanhe Yanjiao Huaxin Mechanical & Electrical Co., Ltd., Langfang, China. is used in the mechanical power measurement. Meanwhile, the CX8525 torque calibrator is used for static torque calibration and measurement, where the loading torque was set to 0%, ±20%, ±40%, ±60%, and ±100% of the full scale of the sensor by electromagnetically controlling the standard loading weights, and it was determined through testing that the static error of the torque sensors was lower than ±0.3%.

3. Algorithms and Models

3.1. Data Pre-Processing Algorithm

The field data acquisition system collects real-time machine-state data continuously. In order to avoid the interference of invalid data to the power-feeding model, only the data collected during the silage harvesting period should be selected as the original data input of the model.

3.1.1. Effective Data Filtering

The shredding roller and throwing blower are the key operating components of the silage harvester, which have more than 60% of the power consumption of the whole machine and have more significant data variation characteristics. Therefore, the monitoring data of these two components are used as the screening basis, and combined with the time-lag analysis model based on material flow, the screening of all measurement data is realized. The main process is as follows.

(1) Define a collection of data categories, D = $\{d_1, d_2, d_3, d_4\}$ = {1, 2, 3, 0}. Where d_1, d_2, d_3, and d_4 represent pre-acceleration data, field harvesting data, harvesting stopped data, and non-experimental suspended data, respectively.

(2) Preliminary classification is carried out depending on the rotation speed of the shredding roller, where the experimental and non-experimental data segments are obtained.

$$D_i = \begin{cases} 0 & n_c < 0.5n_{ce} \\ 1,2,3 & n_c \geq 0.5n_{ce} \end{cases} \tag{5}$$

where D_i is the data category of the i_{th} sampling. n_c *and* n_{ce} are the real-time shredding speed and rated shredding speed, respectively.

(3) The Mann-Kendall algorithm is used to detect the boundary points of field harvesting data [21]. Mann-Kendall is a non-parametric time-series rank test method which has good applicability to field data with anomalous interference and unknown sample distribution. However, it is not appropriate for multiple mutation detection. Thus, the experiment samples in the original torque detection sequence are segmented into two data subsets, which are independently used to detect the loaded mutation boundary points and the unloaded mutation boundary points. In the case of a subset $X = \{x_1, x_2, x_3 \ldots, x_n\}$ with n torque detection sequences, first, a forward rank sequence S_k is constructed by forwarding traversal and calculation of the cumulative count of torque values at the i_{th} time greater than that at the j_{th} time in the sequence, as shown in Formula (6).

$$\begin{aligned} S_k &= \sum_{i=1}^{k} p_k \ (k=2,3,\cdots,n) \\ p_k &= \sum_{j=1}^{k} m_{kj} \ (j=1,2,\cdots,k) \\ m_{kj} &= \begin{cases} 1, x_k > x_j \\ 0, x_k \leq x_j \end{cases} \end{aligned} \tag{6}$$

Secondly, based on the sequence mean $E(S_k)$ and sequence variance $var(S_k)$, the forwarding statistics sequence UF_k is calculated by Formula (7).

$$\begin{aligned} UF_k &= \begin{cases} 0 & (k=1) \\ \frac{(S_k - E(S_k))}{\sqrt{var(S_k)}} & (k=2,3,\cdots,n) \end{cases} \\ E(S_k) &= \frac{k(k-1)}{4} \\ var(S_k) &= \frac{k(k-1)(2k+5)}{72} \end{aligned} \tag{7}$$

Thirdly, the reverse statistic sequence UB_k is calculated by reconstructing the reverse detection sequence $X_B = \{x_n, x_{n-1}, \ldots, x_1\}$, repeating the above traversal process, and taking the negative. Then, under the assumption that the statistic follows the standard normal distribution [22], the solution in Equation (8) under the $U_{0.05}$ significance level constraint is the rising or falling boundary of the effective data segment.

$$\begin{cases} UF_k(i) - UB_k(i) = 0 \\ s.t. UF_k(i) \in U_{0.05}, i = 2,3,\cdots n \end{cases} \tag{8}$$

(4) Through the steps above, the data screening for shedding rollers and throwing blowers can be achieved. However, as for header and feeding parts, the data variation characteristics are not sufficient to enable precise boundary extraction. Thus, we locate the harvesting data segment of these two parts by the material's time-delay model in the silage harvester, as shown in Figure 4.

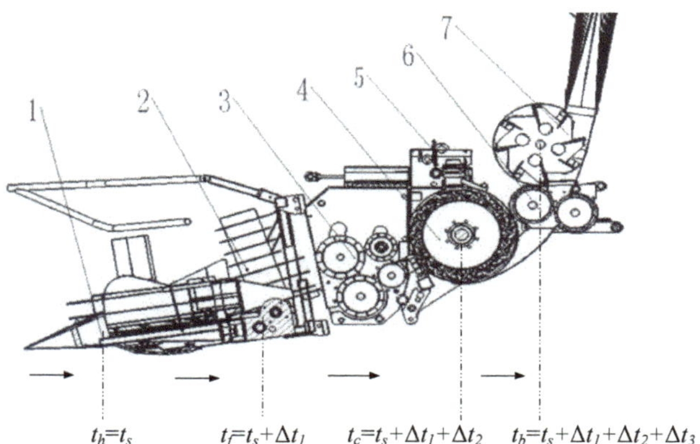

Figure 4. Time lag analysis of silage flow in the harvester. 1: header; 2: conveying roller; 3: feeding rollers; 4: shedding rollers; 5: self-sharpening unit; 6: self-sharpening unit; 7: kernel-processing unit.

As the silage is fed into the harvester at the time t_h, there is almost no time delay for header torque change; After the collection of the header, the silage then reaches the feeding conveying rollers at the time t_f, while a time lag of Δt_1 exists for the feeding power. After the silage is compacted and conveyed by the feeding rollers, it arrives at the shedding rollers at the time t_c, and there is a time lag of (Δt_1 + Δt_2) for the shredding torque change. Finally, after the crop is shredded, it reaches the blower at the time tb, while the blower load data has a time delay of (Δt_1 + Δt_2 + Δt_3). Δt_1 and Δt_2 are calculated by Formula (9).

$$\begin{aligned} \Delta t_1 + \Delta t_2 &= \frac{S_1}{120\pi n_h R_h} + \frac{S_2}{120\pi n_f R_f} = \frac{S_1 n_f R_f + S_2 n_h R_h}{120\pi n_f n_h R_f R_h} \\ \Delta t_2 &= \frac{S_2}{120\pi n_f R_f} \end{aligned} \tag{9}$$

Thus, the effective data segments of the header, feeding unit, shredding rollers, and throwing blows are X_h, X_f, X_c, and X_b respectively, as shown in the Formula (10).

$$\begin{aligned} X_h &= [x(t_{cs} - \Delta t_1 - \Delta t_2), x(t_{cs} - \Delta t_1 - \Delta t_2 + T), \ldots, x(t_{cs} - \Delta t_1 - \Delta t_2 + iT), \ldots, x(t_{ce} - \Delta t_1 - \Delta t_2)] \\ X_f &= [x(t_{cs} - \Delta t_2), x(t_{cs} - \Delta t_2 + T), \ldots, x(t_{cs} - \Delta t_2 + iT), \ldots, x(t_{ce} - \Delta t_2)] \\ X_c &= [x(t_{cs}), x(t_{cs} + T), \ldots, x(t_{cs} + iT), \ldots, x(t_{ce})] \\ X_b &= [x(t_{bs}), x(t_{bs} + T), \ldots, x(t_{bs} + iT), \ldots, x(t_{be})] \end{aligned} \tag{10}$$

where S_1 and S_2 are the moving distance of the crop at the header and the feeder, respectively. R_h and R_f are the radii of the header cutting disk and the upper feed roller, respectively. n_h and n_f are the rotation speed of the header and the upper feed roller, respectively. t_{cs} and t_{ce} denote the rising edge and descending edge of the filtered data in the shedder. t_{bs} and t_{be} express the rising edge and descending edge of the filtered data in the blower.

3.1.2. Exception Data Handling

There are inevitably some anomalous data among the field monitoring samples. Anomalous data include outliers value and the missing value. The outliers are generally evaluated based on the statistical principle. In this study, the Grubbs criterion is applied to verify data and reject outliers one by one with a confidence level of 0.05, while the Grubbs criterion has a more rigorous result [23,24], particularly suitable for this application, within a sample size of 25 to 185 [25]. Taking the screened shedding unit data $X_c = [x(t_{cs}), x(t_{cs}+T), \ldots, x(t_{cs}+iT), \ldots, x(t_{cs}+pT)]$ as an example, the process of data exception handling is as follows.

(1) the sample mean and standard deviation are calculated by the Bessel Formula (11).

$$\begin{cases} \overline{x} = \sum_{i=1}^{p} x(t_{cs} + iT) \\ \sigma = \sqrt{\frac{1}{p-1} \sum_{i=1}^{p} \left(x(t_{cs} + iT) - \overline{x} \right)^2} \end{cases} \tag{11}$$

(2) Reorganize the samples in order of numerical size $\{x(t)^1, x(t)^2 \ldots, x(t)^p\}$, where $x(t)^1 \leq x(t)^2 \leq \ldots x(t)^p$. The residual error of suspicious outliers vi is calculated by the Formula (12).

$$v_i = \max\{(|x(t)^1 - (\overline{x})|), (|x(t)^p - (\overline{x})|)\} \tag{12}$$

(3) Compare the suspicious residual error vi with the critical value of Grubbs coefficient G_0. If Formula (13) is satisfied, the data is determined to be an outlier and removed from the data sequence. The new sequence is then retested until there are no outliers in the data series.

$$\frac{v_i}{\sigma} > G_0(n, \alpha = 0.05) \tag{13}$$

(4) Stuff the rejected outliers and missing values by the nearest neighbor method [26], as shown in Formula(14).

$$x_i = \frac{1}{2}x_{i-1} + \frac{1}{2}x_{i+1} \tag{14}$$

3.2. *Power Feeding Rate Model*

3.2.1. Single-Variable Regression Model

Taking the real-time power consumption of each key operating component in the maize silage harvesting process as the input of the model, a one-dimensional linear regression model between the single power data and the feeding quantity is established, as shown in Formula (15).

$$\begin{cases} q_m = a + bx + \varepsilon \\ \varepsilon \in N(0, \sigma^2) \end{cases} \tag{15}$$

where q_m is the feeding rate of the silage harvester. x represents the power of a key component. a and b are model regression parameters. ε is residual in the single regression.

3.2.2. Multi-Variable Regression Model

Through preferentially selecting the main factors affecting the feeding rate by correlation analysis in the silage harvester, the least squares multiple regression model [27] between multi-power data and feeding quantity is developed to achieve multi-parameter calibration and fused feeding rate detection, as shown in Formula (16).

$$\begin{cases} q_m = b_0 + b_1x_1 + \cdots + b_2x_i + \cdots + b_nx_n + \varepsilon_m \\ \varepsilon_m \in N(0, \sigma^2) \end{cases} \tag{16}$$

where $b_0, b_1, \ldots$, and b_n are model regression parameters, ε_m is the residual of the multi-regression.

4. Results and Discussion

To verify the availability of the feeding rate detection model, harvesting tests are carried out in Rizhao, Shandong Province with a self-propelled silage harvester. First, the real-time power monitoring sensors are integrated into the test prototype, covering both the mechanical driving parts and hydraulic driving parts. Secondly, by adjusting the harvesting speed between 1 km/h and 4 km/h, seven groups of field experiments are designed and conducted under different feeding rates, where condition monitoring data under different operating status is simultaneously collected and stored. Thirdly, the raw monitoring data is preprocessed by the combination of the Mann-Kendall data filtering algorithm and the

Grubbs exception handling algorithm. Finally, the univariate and multivariate model based on the power data is applied to realize the feeding rate measurement.

4.1. Field Experiment Design

Field experiments are realized group by group, as shown in Figure 5, while the harvested crops are upright maize at the wax ripeness stage with a moisture content of 65–70%.

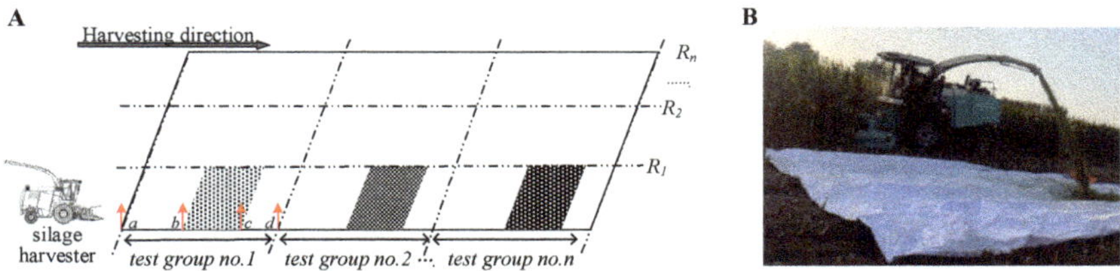

Figure 5. The experiment process of harvesting in the field(where R_1, R_2, ... , and R_n are harvesting crop rows); (**A**) Grouping schematic; (**B**) Picture of the field test.

As shown in Figure 5A, during each group, the silage harvester is adjusted to a predetermined harvesting velocity under the rated working state in the pre-acceleration area (the area between position a and position b in test group no.1). After that, the normal harvesting progress is carried out in the harvesting area (the area between position b and position c in test group no.1) according to the setting velocity and cutting height of 150 mm, while simultaneously all of the silage materials thrown from the outlet of the throwing cylinder were collected manually. Reaching point c, the harvester was parked and shut down in the stopping area (the area between position c and position d in test group no.1). As shown in Figure 5B, the raw material harvested and processed by the silage harvester all falls on the aggregate canvas. After manual collection, it is weighed with a platform scale with an accuracy of 0.01 kg to obtain the cumulative mass of the harvested material. After that, the true value of the feeding rate is calculated with the harvesting time.

4.2. Test Results Analysis

4.2.1. Data Pre-Processing Results

First, the Mann-Kendall algorithm is used to screen the effective data segments. The detection results of the mutation boundary on both sides of the data area are shown in Figure 6. It can be seen that the algorithm can accurately capture the rising and falling edge of the torque monitoring value. Moreover, the data fluctuations related to the dynamic load changes during the harvesting have less influence on the mutation boundary detection.

Meanwhile, the power data of all components are subjected to effective boundary detection, and the test results are shown in Table 1. The boundary extraction accuracy of the shedding data and blowing data are 100%, while that of the inside header, outside header, and feeding part are 64.3%, 71.4%, and 85.7%, respectively. The reason for the analysis is that the header and the feeding part are the front-end mechanisms in plant feeding, which have the characteristics of strong operating vibration, frequency data fluctuations, and small amplitude changes between different working conditions; the accuracy of edge detection is lower compared with the rear-end mechanism. Considering that, the time-delay model described in Formula (10) is applied to achieve data filtering for front-end mechanisms.

Then, through Grubbs Criterion anomaly detection and neighborhood interpolation, the processing results of field test data are finally obtained, as shown in Table 2.

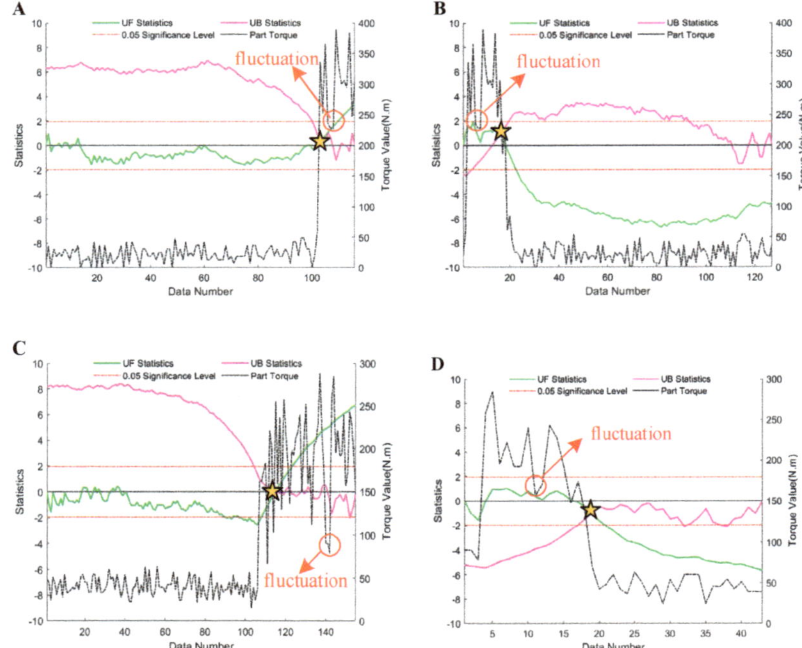

Figure 6. Results of data filtering; (**A**) The rising boundary of the shredding data; (**B**) The descending boundary of the shredding data; (**C**) The rising boundary of the blowing data; (**D**) The descending boundary of the blowing data.

Table 1. Results of boundary detection for data filtering.

Part Name	Rising Detection (Number)		Descending Detection (Number)		Accuracy Rate (%)
	Real	Detected	Real	Detected	
Inside header	7	5	7	4	64.3
Outside header	7	5	7	5	71.4
Feeding part	7	6	7	6	85.7
Shredding rollers	7	7	7	7	100
Throwing blower	7	7	7	7	100

Table 2. Pre-processed data of field harvesting.

Group No.	Velocity /m·s^{-1}	Feeding Rate /kg·s^{-1}	Inside Header Power/kW	Outside Header Power/kW	Feeding Power /kW	Shredding Power /kW	Throwing Blower /kW
1	2.38	11.69	1.37	0.71	17.26	24.3	34.56
2	1.70	9.15	1.3	0.55	16.55	18.83	29.07
3	3.06	17.30	1.7	0.8	19.62	35.03	51.43
4	2.04	11.41	1.74	0.74	15.87	25.49	31.29
5	2.23	12.42	1.68	0.72	14.71	26.14	33.25
6	2.50	14.88	1.47	0.71	15.78	35.67	40.94
7	3.29	18.48	1.79	0.87	17.52	39.26	43.06

4.2.2. Correlation Analysis Results

Correlation analysis between the monitored variables and feeding rate is performed based on the field trial data, and the correlation R is calculated by Formula (17) through the Pearson correlation coefficient [28].

$$R = \frac{\sum_{i=1}^{n} x_i y_i - \frac{\sum_{i=1}^{n} x_i \sum_{i=1}^{n} y_i}{n}}{\sqrt{\left(\sum_{i=1}^{n} x_i^2 - \frac{\left(\sum_{i=1}^{n} x_i\right)^2}{n}\right)\left(\sum_{i=1}^{n} y_i^2 - \frac{\left(\sum_{i=1}^{n} y_i\right)^2}{n}\right)}} \quad (17)$$

where X = $\{x_1, x_2 \ldots, x_n\}$ and Y = $\{y_1, y_2 \ldots, y_n\}$ are the sampled data sets of the two monitored variables respectively; R is the correlation coefficient between the variable X and Y; and n is the sample set size.

The heat map of the correlation among variables is shown in Figure 7. The results show that the influence of each operating part on the feeding rate, in descending order, is the shredding rollers, the throwing blower, the outside header, the inside header, and the hydraulic feeding part. The correlation coefficient of the outside header, the shredding rollers, and the throwing blower relative to the feeding rate is 0.87, 0.97, and 0.90, respectively. All of the above correlation coefficients are greater than 0.85, presenting a strong correlation relationship with the feeding rate. The correlation coefficient of the inside header and the feeding part relative to the feeding rate is 0.63 and 0.54, respectively, showing a weak correlation with the feeding rate.

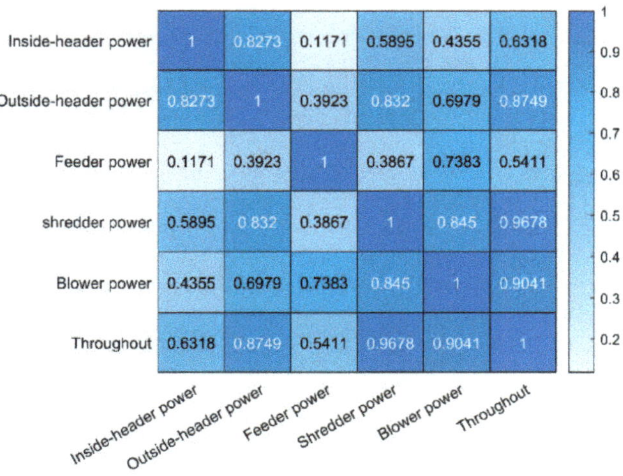

Figure 7. Heatmap of the variable correlation coefficient.

4.2.3. Feeding Rate Measurement Results

(1) single-variable regression results

The univariate linear regression models between the feeding rate and the power data of key components in maize silage harvester are respectively established by using the single factor analysis method, and the model results are shown in Figure 8. Along with the increase in feeding rate, the power of each working part shows an increasing trend. According to the modeling performance, they are ranked into shredding roller, throwing blower, outside-header, inside-header, and hydraulic feeding part power, which are consistent with the correlation analysis results. The linear relationship between the shredding power and the feeding rate is the most significant, with a model coefficient of determination R^2 of 0.94, while the coefficient of determination R^2 for the feeding rate model based on the blowing power is 0.82. Therefore, if a single-factor measurement is considered, a shredding-power-based feeding rate model can achieve better detection results.

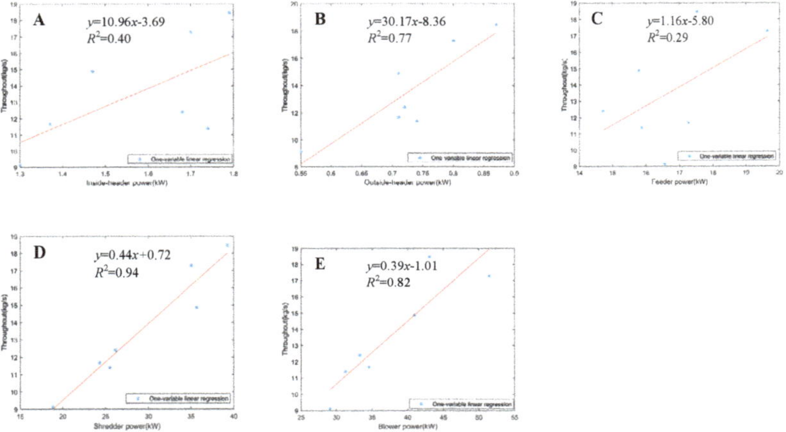

Figure 8. Results of the single-variable regression model; (**A**) The model between inside-header power and feeding rate; (**B**) The model between outside-header power and feeding rate; (**C**) The model between feeding power and feeding rate; (**D**) The model between shredding power and feeding rate; (**E**) The model between blowing power and feeding rate.

(2) multiple-variable regression results

The first three factors that affect the feeding rate, including the inside-header power, the shredding power, and the blowing power, are selected as model inputs to develop a multivariate least squares regression model for the detection of the feeding rate. The result is shown in Formula (18), where the parameter b_0 is −4.0326 within the confidence interval of [−12.0817, 4.0165], the parameter b_1 is 7.9867 within the confidence interval of [−8.5042, 24.4776], the parameter b_2 is 0.2353 within the confidence interval [−0.0564, 0.5271], and the parameter b_3 is 0.1315 within the confidence interval [−0.0817, 0.3446]. While the model coefficient of determination R^2 is 0.9792, the model statistic F is 73.8684, and the significance factor P is 0.00035 (far less than 0.05). The multiple-variable model is reliable.

$$q_m = -4.0326 + 7.9867 \cdot p_2 + 0.2353 \cdot p_4 + 0.1315 \cdot p_5 \quad (18)$$

where p_2, p_4, and p_5 are the power of the outside header, shredding rollers, and blower, respectively.

Furthermore, a residual analysis is made for the model and the results are shown in Figure 9. The residual value of the data in the samples is close to zero, while the confidence interval of the residual includes zero and there are no exceptions, indicating that the model can better fit the original data.

Meanwhile, the accuracy of the multiple regression feeding rate model is analyzed, and the results are shown in Table 3. The maximum absolute error of the model is 0.58 kg/s, and the maximum relative error is ±5.84%, which meets the requirements of the field feeding rate measurement and evaluation. Furthermore, the influence factors of feeding rate measurement accuracy are analyzed. There are primarily two aspects that contribute to the detection error. Firstly, as the compositions of maize forage plants in the machine vary, the working power of key components changes dynamically during harvesting, and the feeding rate measurement is modeled based on the average value of the power consumption, which introduces detection errors. Secondly, although the variation range in the crop moisture content during the field harvesting is slight with a range of 65%~70%, the crop moisture content has a direct influence on chopping power consumption, which is the core parameter of feeding rate detection, and thus increases the measurement error.

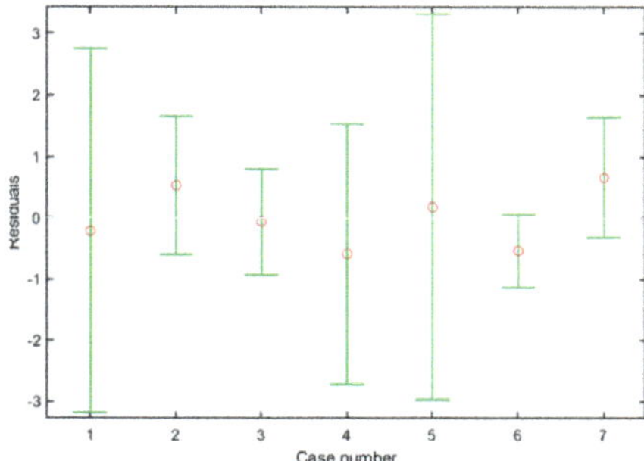

Figure 9. The plot of residual case order.

Table 3. Results of the multi-variable regression model error analysis.

Group No.	Actual Feeding Rate /kg·s^{-1}	Detected Feeding Rate /kg·s^{-1}	Absolute Error /kg·s^{-1}	Relative Error /%
1	9.15	8.61	−0.53	−5.84
2	11.41	11.99	0.58	5.12
3	11.69	11.90	0.21	1.79
4	12.42	12.24	−0.18	−1.45
5	14.88	15.41	0.53	3.58
6	17.30	17.36	0.06	0.36
7	18.48	17.82	−0.67	−3.60

5. Conclusions

A feed rate measurement method based on multi-component power monitoring, effective data automatic screening, and multivariate data fusion regression was studied in this paper, and field experiments were carried out to validate the model. The conclusions are as follows:

(1) The Mann-Kendall algorithm is applied to realize the automatic boundary extraction in the condition data filtering, which is highly applicable in the shredding rollers and throwing blower with a stable load. The data screening and pre-processing for the whole condition data are then achieved effectively in combination with the material-flow time lag correction model.

(2) The Pearson correlation analysis shows that the correlation coefficients between the feeding rate and the power data of the outside header, throwing blower, and shedding roller are all higher than 0.85, indicating a strong relation, which can be used to build a multivariate fusion feeding rate model.

(3) The multivariate least squares feeding rate model is obtained with excellent significance verification results. The maximum absolute error is 0.58 kg/s, while the maximum relative error is ±5.84%.

(4) As for the feeding rate measurement means of the maize silage harvester, if a single parameter methodology is considered, it is proposed to take shredding power as the input of the model, which provides good measurement accuracy. If a multiple conditions parameter methodology is used, the model based on the fusion of cutting power, shredding power, and blowing power can achieve a better detection result. Furthermore, there is still much room for the improvement of model accuracy and robustness by integrating crop characteristics.

Author Contributions: Conceptualization, B.Z. and Y.L.; methodology, F.W. and H.J.; software, F.W. and Y.L.; validation, F.W. and Y.J.; formal analysis, Y.J.; investigation, J.W.; resources, W.M.; writing—original draft preparation, F.W.; writing—review and editing, F.W., B.Z., Y.L. and W.M.; visualization, H.J. and J.W.; supervision, Y.J.; project administration, W.M.; funding acquisition, B.Z. and W.M. All authors have read and agreed to the published version of the manuscript.

Funding: This research is supported by the National Key Research and Development China Project (2020YFB1709603).

Institutional Review Board Statement: Not applicable.

Data Availability Statement: The original data supporting the study in this article are included in the article; further inquiries can be directed to the corresponding author.

Acknowledgments: The authors thank Shandong Wuzheng Group Co., Ltd. and the Experimental Base of Chinese Academy of Agricultural Mechanization Sciences Group Co., Ltd. for providing the experimental equipment.

Conflicts of Interest: The authors declare that the research was conducted in the absence of any commercial or financial relationships that could be construed as a potential conflict of interest.

References

1. Santos, A.P.M.d.; Santos, E.M.; Araújo, G.G.L.d.; Oliveira, J.S.d.; Zanine, A.d.M.; Pinho, R.M.A.; Cruz, G.F.d.L.; Ferreira, D.d.J.; Perazzo, A.F.; Pereira, D.M.; et al. Effect of Inoculation with Preactivated Lactobacillus Buchneri and Urea on Fermentative Profile, Aerobic Stability and Nutritive Value in Corn Silage. *Agriculture* **2020**, *10*, 335. [CrossRef]
2. Zhou, X.; Xu, X.; Zhang, J.; Wang, L.; Wang, D.; Zhang, P. Fault diagnosis of silage harvester based on a modified random forest. *Inf. Process. Agric.* **2022**. [CrossRef]
3. Stellmach, S.; Braun, L.M.; Wächter, M.; Esderts, A.; Diekhaus, S. On load assumptions for self-propelled forage harvesters. *Int. J. Fatigue* **2021**, *147*, 106114. [CrossRef]
4. Kayad, A.; Paraforos, D.S.; Marinello, F.; Fountas, S. Latest Advances in Sensor Applications in Agriculture. *Agriculture* **2020**, *10*, 362. [CrossRef]
5. Dale Maughan, J.; K Mathanker, S.; E Grift, T.; Christopher Hansen, A. *Yield Monitoring and Mapping Systems for Hay and Forage Harvesting: A Review*; American Society of Agricultural and Biological Engineers: St. Joseph, MI, USA, 2012.
6. Mohsenimanesh, A.; Nieuwenhof, P.; Necsulescu, D.-S.; Laguë, C. Monitoring a Hydraulically-Driven Feed Roll System with Sensors on aPrototype Pull-Type Forage Harvester. *Appl. Eng. Agric.* **2017**, *33*, 23–30. [CrossRef]
7. Ferraretto, L.F.; Shaver, R.D.; Luck, B.D. Silage review: Recent advances and future technologies for whole-plant and fractionated corn silage harvesting. *J. Dairy Sci.* **2018**, *101*, 3937–3951. [CrossRef]
8. Kumhala, F.; Prosek, V.; Kroulik, M. Capacitive sensor for chopped maize throughput measurement. *Comput. Electron. Agric.* **2010**, *70*, 234–238. [CrossRef]
9. Maharlouie, M.M.; Kamgar, S.; Loghavi, M. Field evaluation and comparison of two silage corn mass flow rate sensors developed for yield monitoring. *Int. J. Agric.* **2013**, *3*, 730.
10. Lisowski, A.; Klonowski, J.; Sypuła, M.; Chlebowski, J.; Kostyra, K.; Nowakowski, T.; Strużyk, A.; Świętochowski, A.; Dąbrowska, M.; Mieszkalski, L.; et al. Energy of feeding and chopping of biomass processing in the working units of forage harvester and energy balance of methane production from selected energy plants species. *Biomass Bioenergy* **2019**, *128*, 105301. [CrossRef]
11. Chengxiao, F.; Zhigang, L.; YanWei, Y.; Bo, Z.; Yangchun, L.; Liming, Z. on-line detection of mass flow rate of self-propelled corn silage harvester based on capacitance method. *J. Chin. Agric. Mech.* **2020**, *41*, 137–142. [CrossRef]
12. Shenghua, Z.; Guozhong, Z.; Shijie, Z.; Jianwei, F.; Gan, X.; Anwer, M. Designing a Real-time Feed Measurement System for Horizontal Axial Flow Threshing Drum Based on Thin Film Sensor. *J. Huazhong Agric. Univ.* **2020**, *39*, 160–169. [CrossRef]
13. Sun, Y.; Liu, R.; Zhang, M.; Li, M.; Zhang, Z.; Li, H. Design of feed rate monitoring system and estimation method for yield distribution information on combine harvester. *Comput. Electron. Agric.* **2022**, *201*, 107322. [CrossRef]
14. Zhang, Y.; Yin, Y.; Meng, Z.; Chen, D.; Qin, W.; Wang, Q.; Dai, D. Development and testing of a grain combine harvester throughput monitoring system. *Comput. Electron. Agric.* **2022**, *200*, 107253. [CrossRef]
15. Lin, J. Mathematical Model for Improved Mass Flow Estimation in the Feeder Housing of a Forage Harvester. Master's Thesis, University of Illinois at Urbana-Champaign, Urbana, IL, USA, 2012.
16. Yingshuai, J. Development of Intelligent Silage Monitoring System and Silage Operation Device. Master's Thesis, University of Jinan, Jinan, China, 2020.
17. Siebald, H.; Hensel, O.; Beneke, F.; Merbach, L.; Walther, C.; Kirchner, S.M.; Huster, J. Real-time acoustic monitoring of cutting blade sharpness in agricultural machinery. *IEEE/ASME Trans. Mechatron.* **2017**, *22*, 2411–2419. [CrossRef]
18. Worek, F.; Thurner, S. Yield measurement of wilted forage and silage maize with forage harvesters. In *Precision Agriculture'21*; Wageningen Academic Publishers: Wageningen, The Netherlands, 2021; pp. 628–637.

19. Yawei, Z. Mechanisms and Control Strategies Research on Threshing and Seperaing Quility of Combine Harvester. Ph.D. Thesis, China Agricultural University, Beijing, China, 2018.
20. Pallottino, F.; Antonucci, F.; Costa, C.; Bisaglia, C.; Figorilli, S.; Menesatti, P. Optoelectronic proximal sensing vehicle-mounted technologies in precision agriculture: A review. *Comput. Electron. Agric.* **2019**, *162*, 859–873. [CrossRef]
21. Chen, X.; Li, Y.; Yao, N.; Liu, D.L.; Javed, T.; Liu, C.; Liu, F. Impacts of multi-timescale SPEI and SMDI variations on winter wheat yields. *Agric. Syst.* **2020**, *185*, 102955. [CrossRef]
22. Li, J.; He, S.; Wang, J.; Ma, W.; Ye, H. Investigating the spatiotemporal changes and driving factors of nighttime light patterns in RCEP Countries based on remote sensed satellite images. *J. Clean. Prod.* **2022**, *359*, 131944. [CrossRef]
23. Chen, Z.; Xu, K.; Wei, J.; Dong, G. Voltage fault detection for lithium-ion battery pack using local outlier factor. *Measurement* **2019**, *146*, 544–556. [CrossRef]
24. Torres, A.B.B.; da Rocha, A.R.; Coelho da Silva, T.L.; de Souza, J.N.; Gondim, R.S. Multilevel data fusion for the internet of things in smart agriculture. *Comput. Electron. Agric.* **2020**, *171*, 105309. [CrossRef]
25. Xiong, Y.; WU, X. The Generalizing Application of Four Judging Criterions for Gross Errors. *Phys. Exp. Coll.* **2010**, *23*, 66–68. [CrossRef]
26. Zhenqian, Z.; Cheng, P.; Yifan, S.; Renjie, L.; Man, Z.; Han, L. Signal Analysis and Processing of Combine Harvester Feedrate Monitoring System. *Trans. Chin. Soc. Agric. Mach.* **2019**, *50*, 73–78.
27. Choi, J.-H.; Park, S.H.; Jung, D.-H.; Park, Y.J.; Yang, J.-S.; Park, J.-E.; Lee, H.; Kim, S.M. Hyperspectral Imaging-Based Multiple Predicting Models for Functional Component Contents in Brassica juncea. *Agriculture* **2022**, *12*, 1515. [CrossRef]
28. Djordjević, B.; Mane, A.S.; Krmac, E. Analysis of dependency and importance of key indicators for railway sustainability monitoring: A new integrated approach with DEA and Pearson correlation. *Res. Transp. Bus. Manag.* **2021**, *41*, 100650. [CrossRef]

Article

Numerical Simulation of the Picking Process of Supernormal Jujube Branches

Ren Zhang [1], Guofeng Wang [1,2], Wei Wang [1,*], Dezhi Ren [1], Yuanjuan Gong [1], Xiang Yue [1], Junming Hou [1] and Mengmeng Yang [1]

[1] College of Engineering, Shenyang Agricultural University, 120 Dongling Road, Shenhe District, Shenyang 110161, China

[2] Key Laboratory of Liaoning Province for Clean Combustion Power Generation and Heat-Supply Technology, Shenyang Institute of Engineering, Shenyang 110136, China

* Correspondence: syww@syau.edu.cn; Tel.: +86-13-898-179-216

Abstract: This paper elaborates on a digital simulation study on supernormal particle flow used to investigate and analyze the process of picking up jujube branches, which was a meaningful attempt to search for accurate and effective advanced numerical analogy methods in the agricultural field. In this paper, the meshless technology based on the element-free Galerkin method was used for the first time to present the effects of particle size, particle number and particle acting force on the movement of irregular particles, and the influence of the gear rotation speed, the feeding amount, and the jujube branch size on the movement behavior as well as the picking rate. It can describe not only the particles' dynamic movement in the process of picking up jujube twigs, such as feeding, collision, throwing and rolling, but also the effect of the quality and shape caused by the particle size, which in turn affects the surface force of particles and interparticle acting force, thereby affecting the weight function in the analytical solution, the total feeding amount and the effect of the acting force resulting from the particles' contact, roll and collision caused by gear rotation. The findings reveal that the digital simulation, based on the meshless Galerkin technology and Rocky software, is effective in dealing with issues related to supernormal particle flow. By eliminating the influence of geometric shapes on calculations, the method boasts an effective solution to the movement problems of irregularly shaped particles, which would be further applied in the agriculture field.

Keywords: meshless Galerkin method; supernormal particles; jujube branches; picking mechanism; numerical model

Citation: Zhang, R.; Wang, G.; Wang, W.; Ren, D.; Gong, Y.; Yue, X.; Hou, J.; Yang, M. Numerical Simulation of the Picking Process of Supernormal Jujube Branches. *Agriculture* **2023**, *13*, 408. https://doi.org/10.3390/agriculture13020408

Academic Editor: Massimo Cecchini

Received: 20 December 2022
Revised: 24 January 2023
Accepted: 25 January 2023
Published: 9 February 2023

1. Introduction

Supernormal particles, particles with irregular geometric shapes whose length is at least several times the diameter, cannot be simplified into a spherical-shape mathematical model. The flow of supernormal particles has the characteristics of anisotropy and involves complex forces, so conventional methods for non-spherical particles such as equivalent diameter, shape factor and specific surface area are not suitable for supernormal particle problems to be solved in the agricultural field [1–3]. Taking the problem of jujube branch picking as an example, since jujube pruning stubs have typical characteristics such as jujube thorns, crotches, sizeable stem-to-length ratio, and high degree of irregularity, the jujube branch picking process is commonly seen as a specific abnormal particle flow problem. Exploring advanced and accurate simulation methods is the key to solving the complex flow problem of agricultural supernormal particles [4–7].

Presently, research on spherical particles, non-spherical particles and slender particles is extensive in the agricultural field, with different algorithms of particle movement being widely applied in various types of research. A two-stage dust removal device for straw carbonization flue gas has been designed by Xin et al. [8] and simulated the separation efficiency of dust particles using the DPM model to test the relationship between the particle size and the separation efficiency of the cyclone separator. The separation of grains and short stalks under the action of airflow using the CFD-DEM coupling method has been simulated by Jiang et al. [9], which solved the problem of the indoor settlement of imperfect grains reasonably. The flow characteristics of straw-like slender particles in the flow field has been researched by Cai et al. [10], and a two-way coupling relationship between the slender particles and the flow field has been constructed. Since there are many crops with highly irregular geometric characteristics in the agricultural field, it is urgent to explore an appropriate method to cope with the flow characteristics of such supernormal particles.

There are plenty of calculation methods for particle flow, such as the finite element method based on Lagrange (Euler), the lattice-Boltzmann method and the immersed boundary method [11–13]. The finite element method based on Lagrange (Euler) involves thinking of particles as point sources and then using the Lagrangian method to calculate particle motion, solve fluid interaction with N-S equation, and employ phase interaction to describe the law of particle force and the interaction between particles and walls [14,15]. The lattice-Boltzmann method uses the Boltzmann equation of microscopic statistical mechanics to calculate, and it provides a simplified dynamic model for the microscopic particles in the system. The macroscopic physical quantities of particles (such as density, velocity, etc.) are obtained through mathematical statistical methods, which meet the macroscopic fluid dynamics equations, so this method boasts relatively prominent advantages in describing fluid effects [16,17]. The embedded boundary method proposed by Fadlun et al. uses linear velocity interpolation on the interface between the wall and the fluid to obtain the velocity of the calculated point on the interface and to calculate the interaction force between the fluid and the embedded boundary of the solid wall [18].

Research has shown that a new DEM numerical solution method based on the meshless Galerkin method is being studied by scholars [19–21]. The method uses the weight function and the kernel function to solve the mechanical properties of any point in the region to solve the particle motion problem eventually [22–24]. Instead of mesh generation, it only needs the node information. In addition, it has the advantages of high accuracy, fast convergence, and convenient post-processing. Finally, the meshless Galerkin method eliminates the influence of geometry on the flow to the largest extent; emphasizes the interaction between different particles, between particles and walls, and between particles and fluids; and provides a theoretical basis and mathematical method for solving supernormal particle flow problems.

In summary, the movement of supernormal particle flow is a widespread phenomenon in the agricultural multiphase flow field. However, the understanding of supernormal particle flow in related studies is neither systematic nor in depth, such as that of the mechanical picking of jujube branches, hairy vetch, and straw. The meshless Galerkin method was first used to simulate the motion of supernormal particles, whose numerical results could further reveal the influence of the main flow factors (particle size, particle number and particle force) on the jujube branch. Based on numerical analysis and experimental results, the relationship between sizes of jujube branches, the total feeding number, the rotation speed and shape of the gears and the picking rate of jujube branches were clarified. This research provides a necessary theoretical basis for solving the problem of supernormal particle flow in agricultural machinery engineering, which boasts extensive use prospects and important academic value in the mechanical picking of crop residues.

2. Materials and Methods

2.1. Jujube Branches Model

The jujube branch samples were taken from the jujube garden of the 11th regiment of Alar city, the first division of Tarim, Xinjiang. The cultivation mode was dwarfing culture and compact planting, and the tree was 1 year old, as shown in Figure 1a. The projection method was used to obtain three-dimensional coordinates of critical nodes of the jujube branch, and then the curvature and length of each jujube branch were measured. After modeling, the jujube branch model was produced, as shown in Figure 1b.

(a)

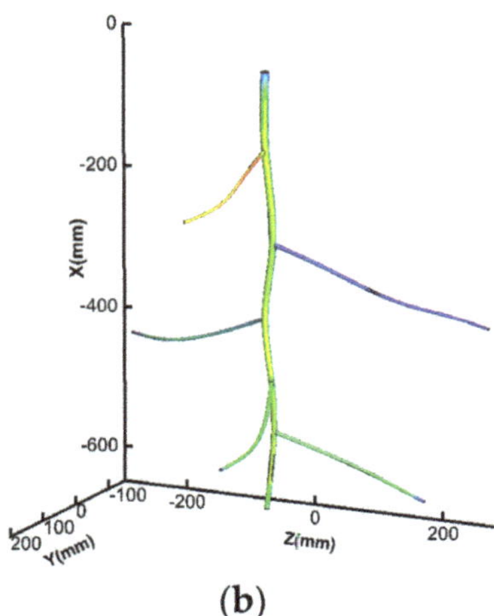

(b)

Figure 1. Jujube branch model. (**a**) Physical map of jujube branches to be picked. (**b**) Single jujube branch numerical model.

To study the effect of particle size on flow, the selected jujube branches were statistically analyzed to determine the value range of jujube branch size. Test design factors were defined based on the X-direction size of the jujube branch, and the three-level design value of the experiment was determined at the same time. Combined with the randomness principle of the orthogonal experiment, three jujube branch sizes were defined as size I, size II, and size III. The specific jujube branch sizes are shown in Table 1. This physical model does not have the characteristics of spherical particles and can be regarded as a complex supernormal particle.

Table 1. Sizes and parameters of jujube branch.

Jujube Branch Size	Size in X Dimension (mm)	Size in Y Dimension (mm)	Size in Z Dimension (mm)	Surface Area (m^2)	Volume (10^{-5} m^3)	Weight (kg)	Maximum Screening Size (mm)
Size I	640	370	550	0.038	6.79	0.095	459
Size II	800	463	688	0.059	13.26	0.185	573
Size III	480	278	412	0.021	2.86	0.040	344

2.2. *Jujube Picking Mechanism*

Figure 2 shows a three-dimensional model of the jujube picking equipment. The equipment is mainly composed of five parts: plane and convey mechanisms, a pair of roller picking mechanisms, crush mechanisms, screen mechanisms and walk mechanisms. The tractor was used as the traction power to realize the field picking process, with the jujube picking equipment connecter to the traction frame. In the process of forward movement, the jujube was gathered in front of the shovelling teeth by the rolling brush, and the shovelling teeth were used to transport the jujube branches to the upper and lower gear components to complete the levelling and transportation of the jujube. The jujube could be collected under the action of upper and lower gears. The collected jujube was crushed by the crushing hammer of the crushing device, and the crushing process of the jujube was realized. Through the sieve plates, the crushed straw residue was evenly sprinkled back onto the field. The research emphasis of the equipment is to improve the picking rate, which can not only improve the clogging problem of the collecting system, but also reduce the power consumption; therefore, the research on picking rate is of great significance to the equipment used for picking up branches.

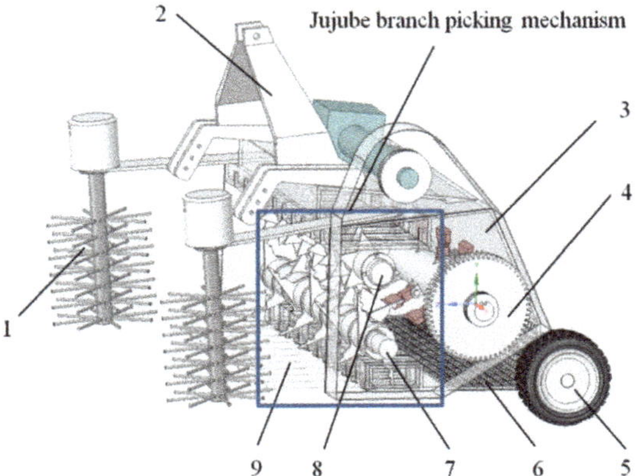

Figure 2. Three-dimensional model of the jujube picking equipment. 1. Rolling brush 2. Traction frame 3. Device shell 4. Crushing mechanism 5. Walking wheels 6. Sieve plates 7. Lower gear assembly 8. Upper gear assembly 9. Collect shovelling teeth.

To obtain the reliable data of the picking rate of the equipment, the numerical simulation model was simplified according to the structure of the test bench, as shown in Figure 3. The rotation direction of the lower gear shaft was opposite to the forward direction, so that the picking mechanism produced a sufficient picking process, the lifting distance and time of the material were extended, and finally, the jujube branches were fully separated. Jujube picking can be regarded as a kind of supernormal particle movement, in which the study of jujube branch size can be regarded as the study of particle size, and the study of total feeding amount can be regarded as the study of the effects of particle number on supernormal particle flow. What is more, the gear will collide with the jujube branch during the rotation, which will affect the speed and direction of the jujube branch. Therefore, the research on the speed of the gear can be regarded as the study of the influence of external force on supernormal particle flow. Due to the frequent occurrence of missing branches during the picking process, the picking rate is an important indicator of jujube picking. The picking rate of jujube branches is shown in Formula (1).

$$Q_1 = \frac{C_1}{S_1} \times 100\% \quad (1)$$

where Q_1 is picking rate, C_1 is the number of jujube branches successfully picked up into the collection box by the jujube picking agency, and S_1 is the total number of jujube branches fed in the picking test.

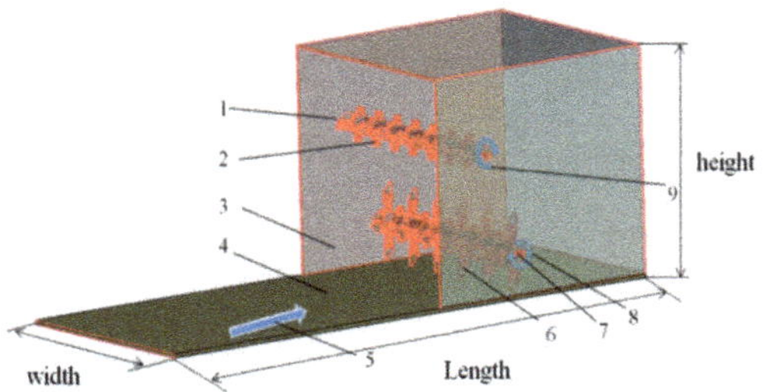

Figure 3. Simplified numerical model of jujube branch picking mechanism. 1. Upper gear shaft 2. Upper gear 3. Device shell 4. Conveyor 5. Movement direction of conveyer belt 6. Lower gear 7. Lower gear shaft 8. Counterclockwise rotation 9. Clockwise rotation.

2.3. Numerical Methods

For the first time, the jujube branch is used as a supernormal particle model, and the meshless Galerkin method was first used to simulate the motion of supernormal particles. This method is to eliminate the influence of the shape of the jujube branch on the flow. Compared with the traditional CFD method, the meshless Galerkin method takes the advantage of not considering the force on the interface between the particles and the flow field, and the effect of particle movement on the mesh deformation of the fluid [25–29]. The study found that the meshless Galerkin method mainly constructs an approximate weight function for the field function according to the node, which is a basic feature of the method. The meshless method includes two key steps: (1) constructing an approximate weight function for the field function; and (2) solving the meshless discretization of partial differential equations.

The jujube branches were transported, entangled, collided, escaped and collected during the mechanical picking process. Among them, the particle size determines the particle weight, the particle force, particle movement speed, and displacement value in the field function. The number of particles affects the relative position between particles, which in turn affects the force between particles and particle movement speed. The gear rotation speed directly affects the force between particles and walls, resulting in changes in particle movement speed and particle displacement. Therefore, the field functions to be calculated are the force, the velocity and the displacement values; the force of the particles is the volume force and the surface force, including the collision force between the particles, the collision contact force, and the rolling force of the particles. The particle movement speed and displacement values were calculated using the current particle position, while speed and time data were used to calculate the relevant data of the particle in the next time step. The specific field function equations are shown in Formulas (2)–(4).

$$m_i\frac{dv_i}{dt} = \sum F_{net} = \sum F_{body} + \sum F_{surface} = \sum F_{body} + \sum F_{contact} + \sum F_{rolling} \quad (2)$$

$$v_{new} = v_{old} + \int_t^{t+\Delta t} \frac{\sum F_{net}}{m} dt \quad (3)$$

$$x_{new} = x_{old} + \int_t^{t+\Delta t} v_{new} dt \quad (4)$$

where m_i is the particle mass at the node x_i, kg·m^{-3}, F is the particle force, N, v_i is the particle velocity at the node x_i, m·s^{-1}, x is the particle displacement at the node x_i, m.

The general expression of the equivalent integral form of the differential equation system based on the meshless Galerkin method to solve each node in the domain was calculated using Formula (5).

$$\int_{\Omega_j} wA\left[\sum_{i=1}^{n} N_i(x)u_i\right] d\Omega_j + \int_{\Gamma_j} \overline{w}B\left[\sum_{i=1}^{n} N_i(x)u_i\right] d\Gamma_j = 0 \quad (5)$$

where Ω represents the solution domain of the problem, Γ represents the boundary of the solution domain (including displacement boundary and force boundary), A and B are the differential operators of independent variables (such as spatial coordinates, time coordinates, etc.). Functions w and w are test functions (or weight functions); $N_i(x)$ is the shape function of the approximate function $u(x)$; u_i is the value of the field function $u(x)$ to be solved at node x_i.

The approximate function $u(x)$ is constructed by solving a set of discrete points $x_i = (i = 1, 2, \cdots, N)$ in the domain by solving a set of discrete u_i known in the function $u(x)$, and the expression of the global approximation function $u^h(x)$ at the solution point x_i is:

$$u^h(x,\overline{x}) = \sum_{i=1}^{m} p_i(\overline{x})a_i(x) = p^T(\overline{x})a(x) \tag{6}$$

where $\overline{x}$ is the coordinate of each node in the solution domain app, $p_i(\overline{x})$ is the basis function, m is the number of basis function, and $a_i(x)$ is the undetermined coefficient.

The primary and secondary basis functions in a two-dimensional space can be expressed as:

$$p^T(\overline{x}) = [1, x, y], m = 3 \tag{7}$$

$$p^T(\overline{x}) = \left[1, x, y, x^2, xy, y^2\right], m = 6 \tag{8}$$

The sum of the global approximation function $u^h(x)$ and weight squared error at the solution point is x_i is:

$$J = \sum_{j}^{N} w_j \left[\sum_{i=1}^{m} p_i(\overline{x})a_i(x) - u(x_i)\right]^2 \tag{9}$$

Using the principle of least squares method to solve the undetermined coefficient $a_i(x)$ when J =0, we can obtain:

$$\sum_{j}^{m}\left[\sum_{i=1}^{N} w_i p_i(x_i) p_j(x_j)\right] a_i(x) = \left[\sum_{i=1}^{N} w_i p_i(x_i)\right] u(x_i) \tag{10}$$

The undetermined coefficient $a_i(x)$ can be obtained from the above formula

$$a(x) = A^{-1}Bu \tag{11}$$

It can be obtained by integrating Equations (6) and (11)

$$u^h(x,\overline{x}) = p^T(\overline{x})A^{-1}Bu = N(x,\overline{x})u \tag{12}$$

The shape function can be obtained as follows:

$$N(x,\overline{x}) = p^T(\overline{x})A^{-1}B \tag{13}$$

The weight function type obtained by the Rocky software is the Gaussian exponential weight function, and its expression is:

$$w_i = \begin{cases} \frac{e^{-(d_i/c_i)^{2k}} - e^{-(r_i/c_i)^{2k}}}{1 - e^{-(r_i/c_i)^{2k}}} & , 0 \leq d_i \leq r_i \\ 0 & , d_i \leq r_i \end{cases} \tag{14}$$

where d_i is the distance from the node x_i to the field point x, m, c_i is the constant of the control function shape, and r_i is the solution domain of the weight function w_i, which is the radius of the node x_i, m.

Under the simulation control of the above equations, jujube branches' movement phenomena, such as transportation, entanglement, collision, escape and collection, occur. Particle size determines the particle mass, which in turn affects particle force, particle movement speed, and the displacement value in the field function. The number of particles affects the position between particles, as well as the force between particles and the speed of particle movement. The rotation speed of the gear directly affects the force between the particles and wall surface, and then affects the particle movement speed and particle displacement.

2.4. Simulation Parameters

Solidworks software was used to draw jujube branch particles and the simplified model of the jujube branch picking mechanism. A full-scale three-dimensional model was imported into the Rocky software, of which the computational domain size was 1.8 m (length) × 1.3 m (width) × 1.1 m (height). Firstly, the left side of the model was defined as the entrance of the jujube branch, the translation speed of conveyor belt was set to 0.42 m/s, and the rotation speed and direction of upper and lower gears were given. Secondly, according to the physical parameters of jujube particles, gears, and conveyor belt, the interaction force between particles and gear mechanism, as well as between particles and conveying device, were set. Then, the jujube particle model was input, and the total number of imported particles was determined. Finally, the total calculation time and time step were defined as 20 s and 0.05 s, respectively. Table 2 shows the initial boundary conditions and mechanical parameters of the picking mechanism and jujube branch particles required for the simulation. The design parameters of the prototype machine were taken as reference, and the simulation and experimental data of Ning Xinjie and Almeida E et al. were used to define the physical parameters of jujube branches.

Table 2. Material parameters used in simulation.

Parameters	Values
Translation speed of conveyor belt (m/s)	0.42
Rotating speed of upper and lower gears (r/min)	100, 120 or 150
Total number of particles	16, 23 or 30
Particle shape	I, II or III
Grain density of jujube branch (kg·m^{-3})	837
Poisson's ratio of particles	0.5
Shear modulus of particles (Pa)	1.0×10^9
Coefficient of restitution between particles	0.5
Coefficient of static friction between particles	0.45
Coefficient of dynamic friction between particles	0.08
Density of the gear mechanism (steel, kg·m^{-3})	7800
Poisson's ratio of gear mechanism	0.3
Shear modulus of gear mechanism (Pa)	7×10^{10}
Coefficient of static friction between particles and gear shifting mechanism	0.35
Coefficient of dynamic friction between particles and gear shifting mechanism	0.04
Restitution coefficient of particles and gear shifting mechanism	0.5
Conveying device (rubber, kg·m^{-3})	1400
Poisson's ratio of conveying device	0.3
Shear modulus of conveying device (Pa)	1×10^{11}
Coefficient of static friction between particles and conveying device	0.9
Coefficient of dynamic friction between particles and conveying device	0.1
Restitution coefficient of particles and conveying device	0.5
Calculation of total time (s)	20
Time step (s)	0.05

3. Results and Discussions

3.1. Numerical Analysis of Movement during Jujube Branch Picking

Figure 4 shows the dynamic analysis results obtained by the Rocky software. The simulation time is 20 s; the filling time of jujube branches varied from 0 s to 16 s with an average speed of one jujube branch per second. Jujube branches were conveyed, entangled, collided, escaped, and collected on the conveying device and picking mechanism. Figure 4a shows the state of jujube branches after the filling time of 4 s. Meanwhile, four jujube branches were piled up on the conveyor belt, and the first jujube branch was rolled sideways by the action of the gear shift. Figure 4b shows the state of jujube branches at 10 s, when the counter-roller gear starts to pick up the first jujube branch. Then, under the influence of gear and adjacent jujube branches, jujube branches close to the gear tooth begin to collide and roll, which results in the escape of jujube branches. Figure 4c shows that jujube branches within the dotted line were leaving the test bed under the impact of collision when the time was 13.9 s. Under the working condition of this analysis, a total of three jujube branches fell off from the test bed, all of which slipped from the edge of the conveyor belt after collision and winding. Figure 4d shows that one jujube branch was successfully collected at 16.9 s. When jujube branches were regarded as supernormal particles, the meshless Galerkin method (MGM) could be used to describe the dynamic process of jujube branch movements, such as transportation, collision, and tumbling.

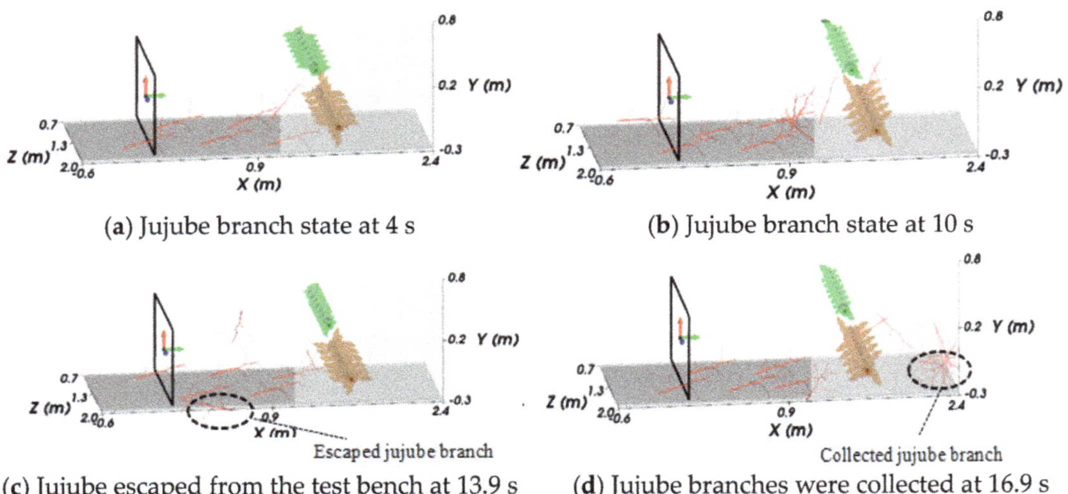

Figure 4. Dynamic picking process of jujube branches.

3.2. Influence of Jujube Branch Size on Picking Rate

In order to study the relationship between the size and picking rate of jujube branches, a numerical analysis was carried out on the movement law and picking rate changes for three sizes of jujube branches in the picking progress.

Figure 5a shows the movement of the simulation at the time point of 10.65 s when a total of 16 jujube branches with a size of 480 mm were fed. It can be found that the feeding was more uniform because the number of jujube branches was relatively small and there was little interference among each other. At the same time, due to the smaller size of the jujube branch, the weight was correspondingly lighter. Further study of the dynamic process showed that only winding and rolling occurred after the collision of jujube branches, not flinging. It is concluded that when the jujube branch was lighter, the force caused by collision can be reduced, the distance that the jujube branch was pushed away by the gear was shorter, and there was less rolling, which is beneficial to the final collection. Figure 5b shows the jujube branch's movement when jujube branch size was 640 mm and fed with a total of 23 jujube branches. As shown in the figure, due to the increase in the size of the jujube branches, the force between the jujube branch and the shifting teeth was more potent, resulting in throwing off the jujube branch. In addition, the jujube branch and the shifting teeth were also entangled due to the increase in twig size, leading to the decrease in the picking rate. Figure 5c shows the movement of jujube branches when the size of jujube branch was 800 mm and the total amount of jujube branch feeding was 30. As the size of jujube branches increased, the mass of a single particle also increased, but the amount of the dialing power of the dial tooth was limited; it was difficult for the dial tooth to dial the jujube branch, causing the jujube branches to pile up on each other. However, the jujube branches were transported by the conveyor belt and entered between the two gears, and then they were finally collected after throwing out the equipment under the agitation of the gears. The simulation results showed that large-size jujube branches were helpful for collection due to the influence of weight and size. The impact of the collision caused by the shifting of the teeth only causes the jujube branch to roll, and the rolling jujube branch will have an entanglement and rolling influence on the subsequent jujube branch.

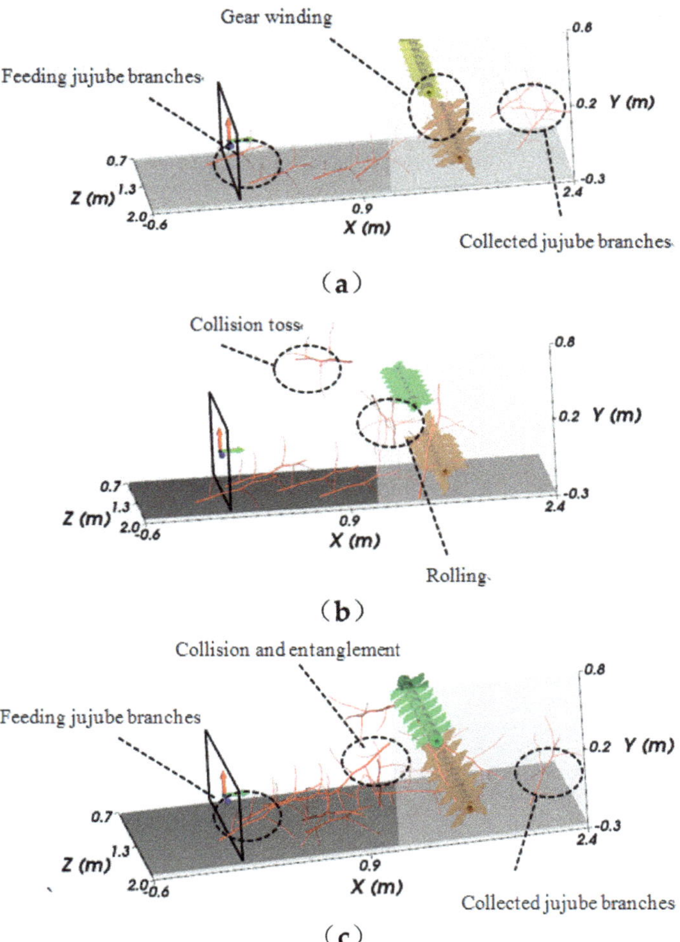

Figure 5. Movement behavior analysis of jujube branches in the picking mechanism. (**a**) State of jujube branches with a size of 480 mm when the time point is 10.65 s. (**b**) State of jujube branches with a size of 640 mm when the time point is 12.9 s. (**c**) Collection state of jujube branches with a size of 800 mm when the time point is 13.65 s.

In this research, the influence of the size of jujube branches on the picking up rate of jujube branches was obtained by single-factor analysis with experimental and numerical results (Figure 6). It can be found from Figure 6 that the picking rate of the test result was higher than that of the numerical simulation, the curve of the test result has relatively smaller fluctuation, and the numerical result has a deviation between the two values at the size of 640 mm. Further research on the number of jujube branches found that when the size is 640 mm, the number of jujube branches collected in the numerical result is 13 and the test result is 15, with the deviation between the two being 2.

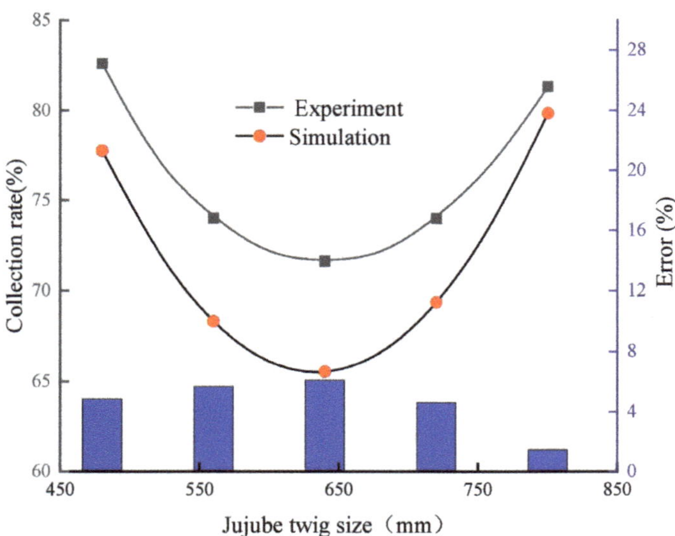

Figure 6. Relationship between jujube branch size and picking rate.

The experimental and numerical results show that the influence of jujube branch size on the picking is parabolic distribution. When the size of jujube branches is 480 mm and 800 mm, the picking rate of jujube branches can reach 75%. Combined with the analysis of the field function equation, it can be found that the particle size of the jujube branch is one of the important factors affecting the particle movement. According to the field function equation, the particle size determines the mass m_i of a single particle and affects the volume force F_{body} of the particle. In addition, the particle size also influences the contact force $F_{contact}$ and tumbling force $F_{rolling}$. At the same time, according to the weight function equation, it can be found that the particle size is related to the distance di from the node x_i of the weight function to the field point x and the radius r_i of the support domain. The change in the particle size will inevitably affect the value of the weight function.

3.3. Influence of the Total Number of Jujube Branches on Picking Rate

When the feeding number was 16, 23, or 30, the influence of the total feeding number on the picking rate was analyzed. Figure 7 shows the relationship between the numbers of jujube branches collected at the exit of the picking mechanism over time. The figure shows that the jujube branch collection process was irregular because the collision among jujube branches and between jujube branches and the wall affect the movement of the jujube branches. In the process of t = (0,10 s), the number of jujube branches collected was smaller, and with time, the number of jujube branches increases. Figure 7a shows the process of t = (14 s, 20 s): the collection number and time were relatively uniform, indicating that the collection effect was better. The numerical calculation result of the picking rate under this working condition was 75%. Figure 7b shows that during the t = (13 s, 20 s) process, the distribution of the number of collections was not uniform, as shown in Figure 7a, indicating that the jujube branches interfere with each other during the collection process. The numerical calculation result of the picking rate under this working condition was 56.6%. In Figure 7c, the feeding number of jujube branches was the most, and the interaction among jujube branches was the most intense. When t = 9 s, this influence begins to take effect in the collection process.

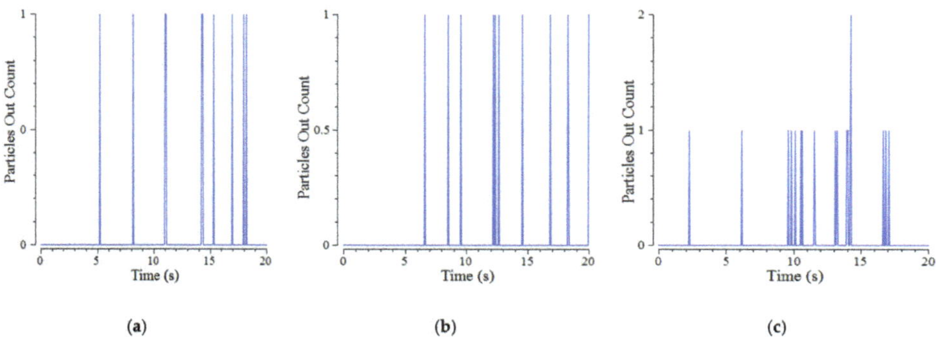

Figure 7. Relationship between the numbers of jujube branches collected over time. (**a**) Feed 16 jujube branches. (**b**) Feed 23 jujube branches. (**c**) Feed 30 jujube branches.

Figure 8 shows the effect of the total feeding number on the movement of jujube branches. From Figure 8a, it can be found that the three adjacent jujube branches at the entrance were relatively far apart, which reduces the entanglement and collision of the jujube branches in the tooth setting area. With the increase in the number of particles, the distance among the three adjacent jujube branches at the entrance became smaller, and the collision and entanglement of the jujube branches in the area of the teeth increased, as shown in Figure 8b. When the number of jujube branches increased to 30, the jujube branches overlapped and were arranged in a staggered manner (Figure 8c), and collision and entanglement may occur during the movement.

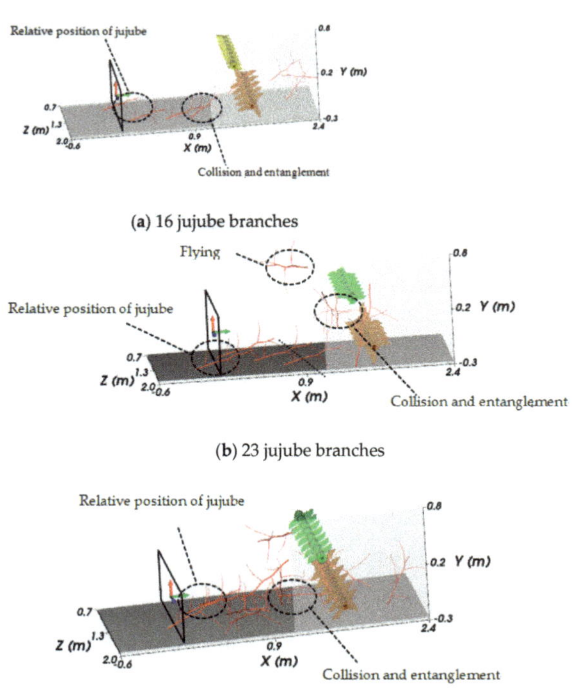

Figure 8. Effect of the total number of jujube branches fed on the movement position of jujube branches.

As the total feeding number increases, both the value and the test results show a trend of first declining and then rising, as shown in Figure 9. The increase in jujube branches leads to more serious interference between jujube branches, and the probability of entanglement collision and jujube branch escape increases. When the number reaches a certain level, the entanglement and collision of jujube branches will increase the picking rate of jujube branches.

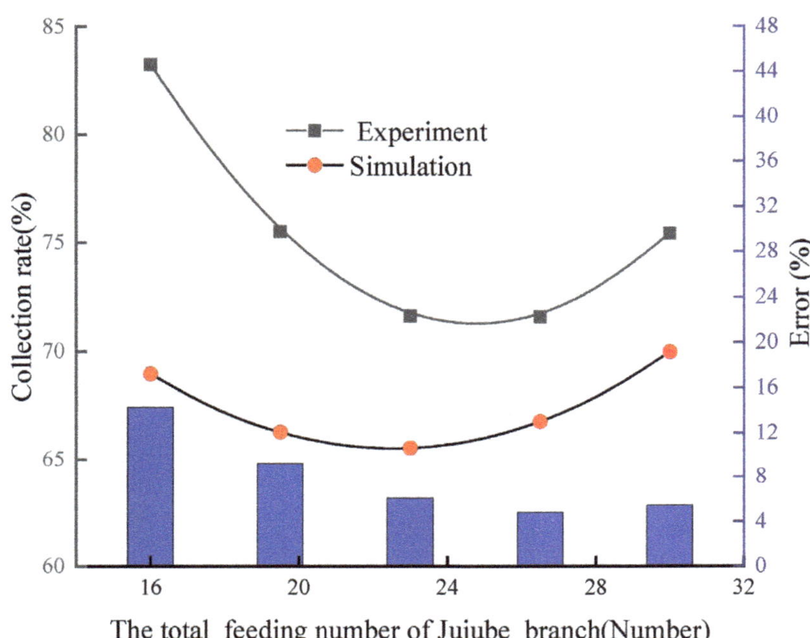

Figure 9. Relationship between the total number of jujube branches and the picking rate.

The total feeding number of jujube branches does not affect the quality and geometric properties of a single particle, nor does it affect the distance d_i from the weight function node x_i to the field point x and the support domain radius r_i. According to the field function equation, the total number of feeds affects the contact force $F_{contact}$ between particles, and influences the collisions between different particles, as well as between particles and walls. Compared to the size of jujube branches, the total feeding number affects the picking rate of jujube branches less.

3.4. *Influence of Gear Rotation Speed on Picking Rate*

The jujube branch picking mechanism makes collisions among jujube branches as well as collisions between jujube branches and the wall surface through the dialing of the teeth, which form a granular force. At the same time, with the appearance of tumbling, collision, entanglement, and throwing, the picking rate changes. When the gear rotation speed was 100 r/min, the collision and force between the jujube branch and the gear may be small, and the entanglement and throwing phenomenon may be rarer (Figure 10a). When the gear rotation speed increases to 120 r/min, the force of the gear on the jujube branch increases, and collisions and throws become more frequent (Figure 10b). When the rotation speed reaches 150 r/min, the force of the shifting tooth is further strengthened, and the number of jujube branches that are shifted away increases (Figure 10c).

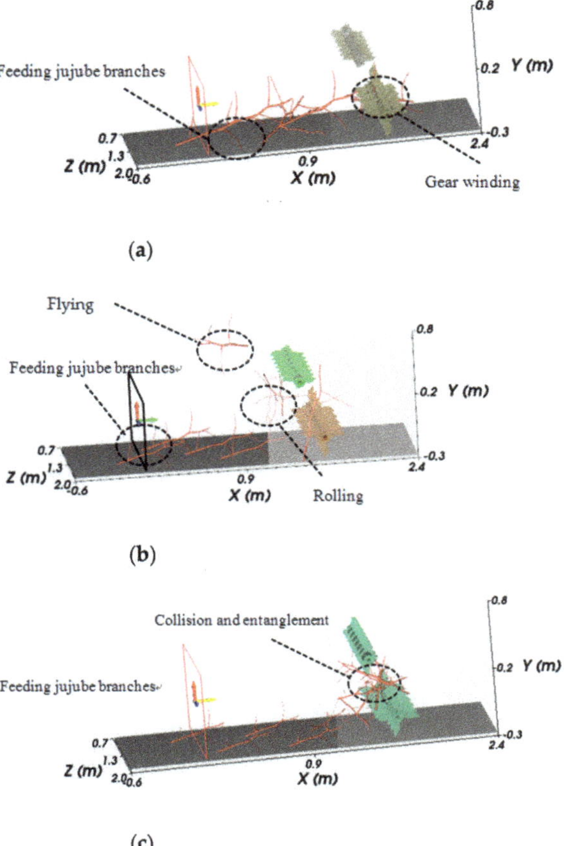

Figure 10. Influence of the gear rotation speed on the movement of jujube branches. (**a**) Gear rotation speed is 100 r/min. (**b**) Gear rotation speed is 120 r/min. (**c**) Gear rotation speed is 150 r/min.

It can be found from Figure 11 that with the increase in gear rotation speed, the collision effect between particles becomes stronger and the particle pickup rate decreases.

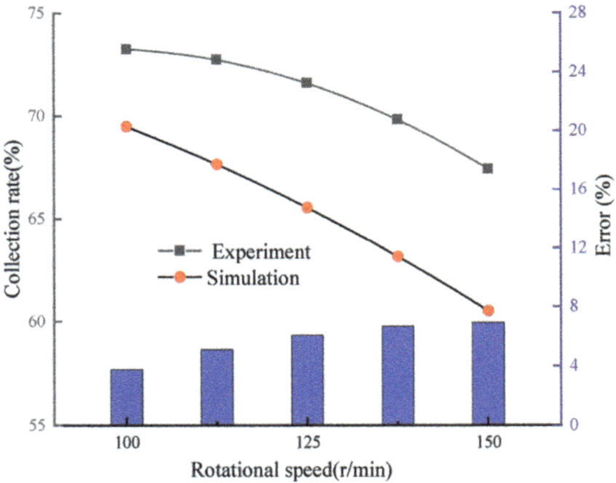

Figure 11. Relationship between gear speed and picking rate.

Numerical simulation values and test results show that the increase in the rotation speed of the gear makes the interaction between the particles more serious and reduces the picking rate of jujube branches. Compared with the size of the jujube branch and the total number of feeds, the speed of the gear shift affects the particle contact force and tumbling force less than the size of the jujube branch and more than the total number of feeds.

3.5. Orthogonal Test Verification

Figure 12 shows an experimental prototype of the jujube picking mechanism, which is mainly composed of a pair of roller picking mechanisms, an electromagnetic speed regulating motor, a conveyor belt mechanism, a chain drive system, a collection box, and a JD1 electromagnetic speed regulating motor inverter. The jujube branch samples were placed on the conveying device, which moved steadily under the drive of the frequency conversion motor. The jujube branch samples were transported to the picking area of the pair-roller picking mechanism, and the picking mechanism performed a pair-roller picking movement to pick up the jujube branch samples on the conveyor belt and move them to the collection box.

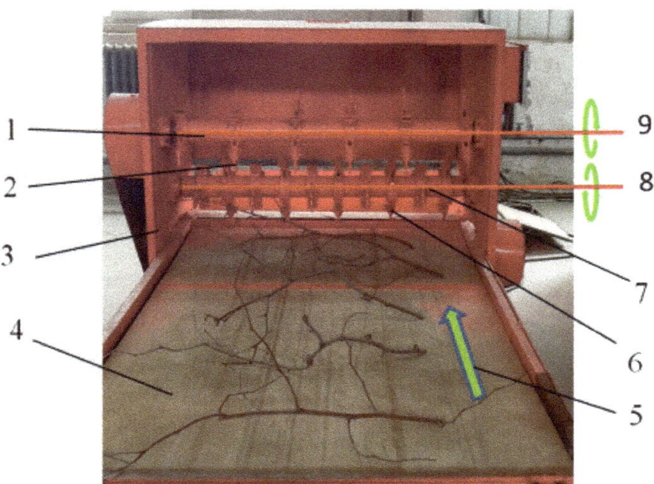

Figure 12. Prototype of jujube branch picking mechanism. 1. Upper gear shaft 2. Upper gear 3. Device shell 4. Conveyor 5. Movement direction of conveyer belt 6. Lower gear 7. Lower gear shaft 8. Counterclockwise rotation 9. Clockwise rotation.

The jujube picking mechanism test device was used for orthogonal test verification. Each variable parameter was set, among which the size of the jujube branches was 480 mm, 640 mm and 800 mm; the total number of feedings was 16, 23 and 30; and the rotation speed of the gear was 100 r/min, 120 r/min, and 150 r/min. The test plan L9 (3^4) was selected, and the orthogonal test was performed according to the randomness principle of the orthogonal test, as shown in Table 3.

Table 3. Influence factors and levels of test.

Factors / Level	A Total Number of Feeding (Branch)	B Size of Jujube Branch (mm)	C Rotation Speed of the Gear (r/min)
1	16	640	150
2	23	800	100
3	30	480	120

As shown in Table 4, numerical simulation values and test results showed that the size of jujube branches was an important factor affecting the picking rate. The picking rate of jujube branches was the worst when the size of jujube branches was 640 mm. When the size of jujube branches was 800 mm and 480 mm, the picking rate of jujube branches was relatively higher. From the numerical value and the test results, the picking rate was highest when the jujube branch size was 480 mm, which indicated that the jujube branches with a smaller size were easier to be collected.

Table 4. Experimental design and comparative analysis with simulation results.

Test Number		Factors				Test Results	Numerical Results
		A Total Number of Feeding (branch)	B Size of Jujube Branch (mm)	Test Error	C Rotation Speed of the Gear (r/min)	Picking Rate (%)	
1		1	1	1	1	75	62.5
2		1	2	2	2	81.3	75
3		1	3	3	3	81.3	75
4		2	1	2	3	65.2	56.6
5		2	2	3	1	82.6	73.9
6		2	3	1	2	82.6	73.9
7		3	1	3	2	60	56.7
8		3	2	1	3	76.7	70
9		3	3	2	1	86.7	73.3
Test analysis	k1	237.6	200.2	234.3	244.3		
	k3	230.4	240.6	233.2	223.9		
	k2	223.4	250.6	223.9	223.2		
	R	14.2	50.4	10.4	21.1		
	Optimal order of factors			B > C > A $B_3C_1A_1$			
Numerical analysis	k1	212.5	175.8	206.4	209.7		
	k3	204.4	218.9	204.9	205.6		
	k2	200	222.2	205.6	201.6		
	R	12.5	46.4	1.5	8.1		
	Optimal order of factors			B > A > C $B_3C_1A_1$			

The extreme difference between the gear rotation speed and the total number of jujube branch feeds was relatively close, and the effect on the picking rate was basically the same. From the perspective of the optimal plan, the test results were consistent with the numerical results. The optimal plan was $B_3C_1A_1$, which means that when the jujube branch size was 480 mm, the total number of jujube branches was 16 and the gear rotation speed was 150 r/min, the picking rate was the highest. The test results proved the reliability of the numerical results, indicating that the use of numerical methods based on the meshless Galerkin method can solve the problem related to supernormal particle flow in the process of picking up jujube branches.

4. Discussion

The numerical analysis method provides an important research method for the research and development of the pick-up equipment of jujube branches. The method can effectively simulate the pick-up process of jujube branches and obtain the important factors. It is not difficult to find out that the core of the influence on the picking rate is the collision and winding of jujube branches. In addition to the above factors, the shape of the gear is also an important factor. Using the optimal total feeding number, the size of jujube branches and the rotation speed of the gear, the shape of the gear can be further simulated and tested, and its influence on the picking rate can be obtained. Figure 13 shows three gear shapes (the unit of size is mm), where gear 2 is the initial design gear and gear 1 and 3 are the modified shapes.

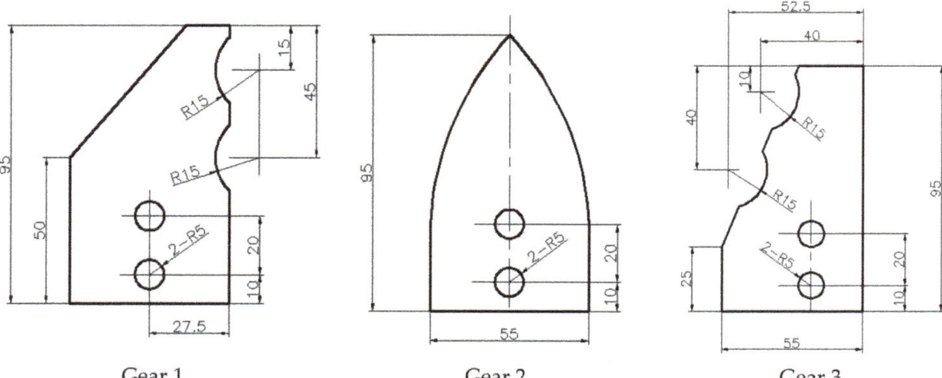

Figure 13. Gear Profile.

The collection situation of the jujube branch picking process could be obtained quickly using the numerical simulation method. Figure 14 shows the gear 3 collection of jujube branches. In the process of 12~20 s, when the interval of collecting jujube branches was relatively stable, the number of collected jujube branches was almost the same, with 21 jujube branches out of the total input of 24 being collected and the collection efficiency being 87.5%.

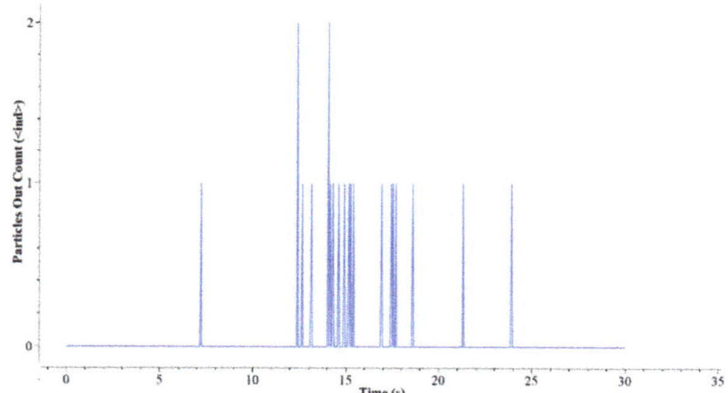

Figure 14. Collection of jujube branches on gear 3.

The relevant experimental research on the gear profile was carried out, and Table 5 shows the parameters involved in the experiment. The experimental results under the influence of a single factor of tooth profile are obtained by repeatedly testing different gears five times and by taking the average value of five data. The experimental data show that the picking rate is 83.3% when gear 3 is used, which is close to the numerical results.

Table 5. Experimental design on single-factor effect test of tooth profile.

Factors	Height of Gear (cm)	Gear Pitch (cm)	Rotation Speed (r/min)	Number of Feeding	Pick Up a Number	Picking Rate
Gear1	13.5	10	150	24	17	70.8%
Gear2	13.5	10	150	24	15	62.5%
Gear3	13.5	10	150	24	20	83.3%

5. Conclusions

The meshless Galerkin method (MGM) is an effective solution for problems related to supernormal and irregular particles. Without gridding the supernormal particles in the computational domain, the dynamic movement process of supernormal particles such as transportation, collision, throwing, and tumbling can be described only by constructing approximate weight functions and solving partial differential equations with the field function (particle force, particle movement velocity, and displacement value). In this paper, the meshless Galerkin method (MGM) and the method of combining numerical simulation and experimental verification are employed to study the influence of jujube branch size (particle size), the total number of feeds (number of particles), and the gear rotation speed and shape (particle force) on the picking rate of jujube branches.

(1) The numerical and experimental results show that the size of jujube branches not only affected the entire field function, but also influenced the weight function, which was an important factor affecting the picking rate of jujube branches. The influence curve of jujube branch size on the picking rate presented a parabolic distribution. When the jujube branch size was 480 mm and 800 mm, the jujube picking rate could reach a peak of 75%.

(2) The total feeding number and the rotation speed of the gear also have an effect on the picking rate of jujube branches. These two factors will affect the force between the particles, the particles and the wall, and the force of the particles in the field function, causing the jujube branches to roll, collide, entangle, and flutter.

(3) With the increase in the number of particles, the picking rate presented a parabolic distribution, with the minimum trough values being 65.5 and 71.1% and the maximum peak values being 69.9% and 83.2%, respectively. The rotation speed of the gear is in a negative correlation with the picking rate, which means the smaller the speed is, the higher the collection rate of jujube branches will be, with the value and the maximum picking rate in the experiment being 69.5% and 73.2%, respectively.

Author Contributions: Formal analysis, J.H.; Funding acquisition, W.W.; Investigation, X.Y.; Project administration, W.W.; Software, D.R.; Supervision, W.W.; Validation, Y.G.; Writing—original draft, R.Z. and G.W.; Writing—review and editing, W.W. and M.Y. All authors have read and agreed to the published version of the manuscript.

Funding: This research was funded by National Nature Science Foundation of China, grant number 32171900.

Institutional Review Board Statement: Not applicable.

Data Availability Statement: Not applicable.

Acknowledgments: The authors would like to thank the research team members for their contributions to this work.

Conflicts of Interest: The authors declare no conflict of interest.

References

1. Ma, Z.; Li, Y.; Xu, L.; Chen, J.; Zhao, Z.; Tang, Z. Dispersion and migration of agricultural particles in a variable-amplitude screen box based on the discrete element method. *Comput. Electron. Agric.* **2017**, *142*, 173–180. [CrossRef]
2. Jebahi, M.; Dau, F.; Charles, J. Multiscale Modeling of Complex Dynamic Problems: An Overview and Recent Developments. *Arch. Comput. Methods Eng.* **2016**, *23*, 101–138. [CrossRef]
3. Tchen, C. Mean Value and Correlation Problem Connected with the Motion of Small Particles Suspended in a Turbulent Fluid. Ph.D. Dissertation, Delft Univerity, Martinus Nijhoff, Hague, The Netherlands, 1947.
4. Kafashan, J.; Wiącek, J.; Abd Rahman, N.; Gan, J. Two-dimensional particle shapes modelling for DEM simulations in engineering: A review. *Granul. Matter* **2019**, *21*. [CrossRef]
5. Ito, S.; Nagatani, T.; Saegusa, T. Volatile jam and flow fluctuation in counter flow of slender particles. *Physica A* **2007**, *373*, 672–682.
6. Dosta, M.; Antonyuk, S.; Broeckel, U. DEM simulations of amorphous irregular shaped micrometer-sized titania agglomerates at compression. *Adv. Powder Technol.* **2015**, *26*, 767–777.

7. Fanyi, L. *Discrete Element Modelling of the Wheat Parpticles and Short Straw in Cleaning Devices. PhD Thesis;* Northwest A and F University: Shanxi, China, 2018.
8. Xin, J.M.; Chen, T.Y.; Meng, J.; Wu, L.Y.; Jiao, J.K.; Zhang, Q.; Liu, C.H.; Song, Y.Q.; Ren, W.T. Design and numerical simulation of two-stage device for dust removal from flue gas of straw carboniz. *J. Huazhong Agric. Univ.* **2018**, *37*, 100–107.
9. Jiang, E.; Sun, Z.; Pan, Z.; Wang, L. Numerical simulation based on CFD-DEM and experiment of grain moving laws in inertia separation chamber. *Trans. Chin. Soc. Agric. Mach.* **2014**, *45*, 117–122.
10. Cai, J.; Zhao, X.B.; Geng, F. Numerical study on the fluidization characteristics of slender particles with gas/solid two-way coupling. *Acta Energ. Sol. Sin.* **2018**, *21*, 3169–3177.
11. Guoxiang, S.; Yongbo, L.; Xiaochan, W.; Yu, J. CFD simulation and experiment of distribution characteristics for droplet of knapsack sprayer. *Trans. Chin. Soc. Agric. Eng.* **2012**, *28*, 73–79.
12. Han, D.; Zhang, D.; Yang, L.; Li, K.; Zhang, T.; Wang, Y.; Cui, T. EDEM-CFD simulation and experiment of working performance of inside-filling air-blowing seed metering device in maize. *Trans. Chin. Soc. Agric. Eng.* **2017**, *33*, 23–31.
13. Du, J.; Hu, G.M.; Fang, Z.Q. Accelerating CFD-DEM simulation of dilute pneumatic conveying with bends. *Int. J. Fluid Mach. Syst.* **2015**, *8*, 84–93. [CrossRef]
14. Hua, L.; Meina, Z.; Wenqing, Y. Optimization of airflow field on air-and-screen cleaning device based on CFD. *Trans. Chin. Soc. Agric. Mach.* **2013**, *44*, 12–16.
15. Tsuji, Y.; Kawaguchi, T.; Tanaka, T. Discrete particle simulation of two-dimensional fluidized bed. *Powder Technol.* **1993**, *77*, 79–87. [CrossRef]
16. Han, H.F. Numerical Study on Gas-Solid Two-Phase Flow Based on the Lattice Boltzmann Method: Direct Numerical Simulation and Efficient Implementation. Ph.D. Thesis, Huazhong University of Science and Technology, Wuhan, China, 2013.
17. Chen, L.; Mo, C.; Wang, L.; Cui, H. Direct numerical simulation of the self-propelled Janus particle: Use of grid-refined fluctuating lattice Boltzmann method. *Microfluid. Nanofluidics* **2019**, *23*, 73.1–73.16.
18. Wang, Z.L. Fully Resolved Direct Numerical Simulation Technique and Its Application in Multiphase Flows. Ph.D. Thesis, Zhejiang University, Zhejiang, China, 2010.
19. Thomas, S. Meshless methods: An overview and recent developments. *Comput. Methods Appl. Mech. Eng.* **2005**, *139*, 3–47.
20. Zhang, X. Application of Meshless Local Petrov-Galerkin Method in Large Deformation Problems. Ph.D. Thesis, Tsinghua University, Beijing, China, 2006.
21. Zeng, H.Q. Parallel algorithms and parallel implementation of meshless numerical simulation. Ph.D. Thesis, University of Science and Technology of China, Hefei, China, 2006.
22. Pan, X.F.; Zhang, X.; Lu M, W. Meshless Galerkin least-squares method. *Comput. Mech.* **2005**, *35*, 182–189.
23. Keck, R.; Hietel, D. A projection technique for incompressible flow in the meshless finite volume particle method. *Adv. Comput. Math.* **2005**, *23*, 143–169.
24. Zhang, Q.; Banerjee, U. Numerical integration in Galerkin meshless methods, applied to elliptic Neumann problem with non-constant coefficients. *Adv. Comput. Math.* **2012**, *37*, 453–492. [CrossRef]
25. Xu, Z.; Chun, F.Z.; Wan, M.Z.; Yang, F. Discrete element simulation and its validation on vibration and deformation of railway ballast. *Yantu Lixue/Rock Soil Mech.* **2017**, *38*, 1481–1488.
26. Xinjie, N.; Chengqian, J.; Xiang, Y. Research status and development trend of air and screen cleaning device for cereal combine harvesters. *J. Chin. Agric. Mechmination* **2018**, *39*, 9–14.
27. Zhang, F.Q. The Application of Smoothed Particle Hydrodynamics Method in Structure Analysis. Master's Thesis, Nanjing University of Aeronautics and Astronautics, Nanjing, China, 2007.
28. de Almeida, E.; Spogis, N.; Taranto, O.P.; Silva, M.A. Theoretical study of pneumatic separation of sugarcane bagasse particles. *Biomass Bioenergy* **2019**, *127*, 105256.
29. Almeida, E. Exploratory Study of Rocky/Fluent Software Coupling-Case: Pneumatic Separation of Sugarcane Bagasse Particles. Ph.D. Thesis, University of Campinas, Campinas, Brazil, 2018.

 agriculture

Article

Research on Fault Diagnosis of HMCVT Shift Hydraulic System Based on Optimized BPNN and CNN

Jiabo Wang [1,2], Zhixiong Lu [1], Guangming Wang [3], Ghulam Hussain [4], Shanhu Zhao [5], Haijun Zhang [1] and Maohua Xiao [1,*]

1 College of Engineering, Nanjing Agricultural University, No. 40 Dianjiangtai Road, Pukou District, Nanjing 210031, China
2 College of Mechanical and Electrical Engineering, Jiangsu Vocational College of Agriculture and Forestry, No. 19 Wenchang East Road, Jurong 212400, China
3 College of Mechanical and Electronic Engineering, Shandong Agricultural University, No. 61 Daizong Street, Taishan District, Taian 271018, China
4 Ghulam Ishaq Khan Institute of Engineering Sciences and Technology, Tarbela Road, District Swabi, Khyber Pakhtoon Khwa, Topi 23460, Pakistan
5 Jiangsu Yueda Intelligent Agricultural Equipment Co., Ltd., No. 9 Nenjiang Road, Economic and Technological Development Zone, Yancheng 224100, China
* Correspondence: xiaomaohua@njau.edu.cn; Tel.: +86-139-5175-6153

Abstract: There are some problems in the shifting process of hydraulic CVT, such as irregularity, low stability and high failure rate. In this paper, the BP neural network and convolutional neural network are used for fault diagnosis of the HMCVT hydraulic system. Firstly, through experiments, 120 groups of pressure and flow data under normal and four typical fault modes were obtained and preprocessed; they were divided into 80 groups of training samples and 40 groups of test samples via random extraction, using the BP neural network model and convolutional neural network model for fault classification. The results show that compared with BP, PSO-BP and other models, the fault diagnosis rate of the BAS-BP neural network model can reach 92.5%, and the average diagnosis accuracy rate of the convolutional neural network can reach 97.5%, which can be effectively applied to the fault diagnosis of the HMCVT hydraulic system and provide some reference for the shifting reliability of hydraulic CVT.

Keywords: HMCVT; fault diagnosis; BP algorithm; CNN; attribute reduction

Citation: Wang, J.; Lu, Z.; Wang, G.; Hussain, G.; Zhao, S.; Zhang, H.; Xiao, M. Research on Fault Diagnosis of HMCVT Shift Hydraulic System Based on Optimized BPNN and CNN. *Agriculture* **2023**, *13*, 461. https://doi.org/10.3390/agriculture13020461

Academic Editor: Filipe Neves Dos Santos

Received: 12 January 2023
Revised: 3 February 2023
Accepted: 14 February 2023
Published: 15 February 2023

1. Introduction

Hydro-mechanical continuously variable transmission (HMCVT) [1–3] is highly automated, and its shifting process is completely carried out under its transmission control unit (TCU) [4,5]. The fault of the position clutch or hydraulic control system will have a great impact on its shifting quality [6,7]. Therefore, in order to discover potential faults in time and improve the reliability of the shifting operation, the TCU needs to perform real-time fault monitoring. However, in the current fault diagnosis related to it, most of the research directions are mechanical traditional gearboxes, and few are specifically aimed at HMCVT. With the wide application of HMCVT, improving the reliability of the shifting process will become the direction of rapid development in the future [8,9].

The structure of the transmission system of HMCVT is complex, but overall it can be divided into mechanical systems and hydraulic systems. Mechanical system faults are mainly gear faults, and the current fault diagnosis methods for mechanical systems are relatively mature, such as wavelet analysis [10,11], support vector machine [12], hidden Markov model [13], etc. Hydraulic system faults include pump motor hydraulic system failures and clutch hydraulic system failures, which can be identified by analyzing pressure, flow, power and other data. Wang Guangming et al. studied gearbox speed ratio control and proposed a hydraulic system fault diagnosis method based on the Fisher criterion kernel method for its clutch [14]; Grover Zurita et al. proposed a multi-channel deep support vector classification method for gearbox fault diagnosis [15]; Lin Ruilin et al. proposed the application of a robust residual support vector machine in fault diagnosis and realized the leakage fault diagnosis of the electro-hydraulic servo system [16] and Han Zhengze studied the fault diagnosis method of the rack rail hydraulic system, and constructed the fault diagnosis rules of the pilot system based on the fault tree method [17].

From the aspect of BP neural network optimization, particle swarm optimization (PSO), as a random search algorithm based on population, has been applied to BP neural network optimization because of its high accuracy and fast convergence. Zou Lan and others used the PSO algorithm to optimize the SOMBP neural network prediction model, and the recognition rate of the optimized model increased from 90% to 95% [18]. However, although the model recognition rate of the PSO algorithm's optimization has been greatly improved [19], the PSO algorithm also has some defects such as slow network convergence speed and it being easy to fall into the local optimum with the increase in iteration times, which means it is difficult to meet the use requirements. Therefore, PSO still has a lot of room for improvement [20,21].

In recent years, many scholars have conducted a series of studies on the intelligent diagnosis method of hydraulic system faults. Additionally, the BP neural network [22–24] and convolutional neural network [25,26] are popular among them. In order to make up for the shortcomings of previous research, the BP neural network optimization model and the convolutional neural network model are applied to the fault diagnosis of the HMCVT shift hydraulic system in this paper, and the classification results are compared.

2. Construction of HMCVT Shift Hydraulic System Test Platform

The structure of the test platform of the HMCVT shift hydraulic system is shown in Figure 1. In the hydraulic oil circuit system, the external motor, 1, drives the oil pump, 2, to supply oil to the system, and the overflow valve, 4, adjusts the pressure of the oil circuit into the clutch; the five electromagnetic directional valves, 8, are all installed on the integrated valve plate, 9, to control the pressure oil of the wet clutch to be turned on or off to realize the shifting operation; the oil flow of the clutch is controlled and regulated by the flow speed regulating valve, 7; at the same time, in order to detect the oil pressure fluctuations that occur during the engagement and disengagement of each clutch, a pressure sensor, 6, is installed in each clutch oil circuit; the flow sensor, 5, installed in the main oil passage is used to detect changes in hydraulic oil. Table 1 shows the relevant parameters of the main components in the clutch oil circuit control system. The HMCVT overall bench test system is shown in Figure 2. To increase the objectivity and comparability of the data calculation, when calculating statistics, the starting point is the first sampling point after the controller sends the segment change command, and the end point is the sampling point with a time of 1 s from the start point. That is, the length of the data set is measured in time and the scale is 1 s.

Table 1. Main components of the test bench for the shift hydraulic system.

Number	Part Name	Model	Main Performance Parameters
1	Asynchronous motor	JO2-22-4	Rated speed: 1450 r/min
2	Vane pump	YB1-6.3	Displacement: 6.3 mL/r
3	Relief valve	2FRM5/10QB	Adjustment range: 0~10 L/min
4	Pressure sensor	NS-F	Measuring range: 0~10 MPa Output signal: 0~5 V
5	Flow sensors	LWGB-4 Turbine flow sensor	Measuring range: 0~0.4 m^3/h Output signal: 4~20 mA

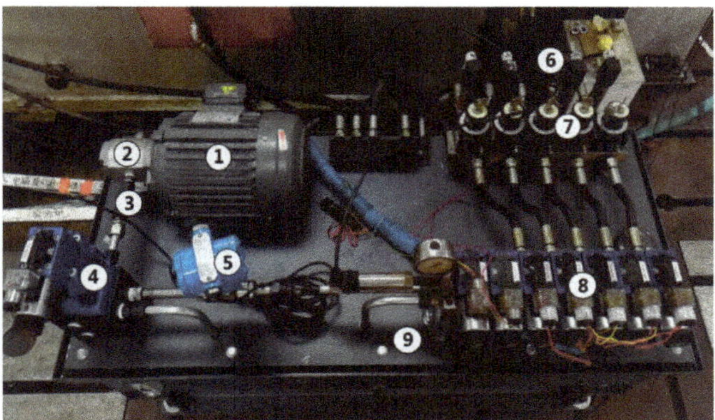

Figure 1. HMCVT shift hydraulic control system. 1. Motor. 2. Oil pump. 3. Check valve. 4. Relief valve. 5. Flow sensors. 6. Pressure sensor. 7. Control valve. 8. Electromagnetic valve. 9. Integrated valve plate.

Figure 2. HMCVT overall test bench. 1. Diesel engines. 2. Front speed torque meter. 3. Transmission. 4. Hydraulic fixed motor. 5. Hydraulic variable pump. 6. Rear speed torque meter. 7. Loader. 8. Measurement and control platform. 9. Shift hydraulic control system.

3. Obtaining Test of Fault Sample of HMCVT Shift Hydraulic System

3.1. Typical Failure Type

According to the operation record of the HMCVT test platform, the fault of the shift hydraulic system of the HMCVT can be defined as five types of modes to be identified, namely:

(1) Normal mode (Fault Type 1): all of the parameters of the shift hydraulic system are in the normal working range, and there is no abnormality in the shifting process.
(2) The clutch piston is stuck (Fault Type 2): The clutch will always be in a different degree of slipping state. The lighter can change the segment but cannot load (the slip will stall as soon as it is loaded), and the heavy will directly burn the clutch.
(3) The seal ring at the rotary joint is damaged (Fault Type 3): Local internal leakage occurs in the oil passage, but this leakage occurs inside the oil passage. When the oil passage is filled for the first time, the oil pressure is difficult to establish. As the oil channel is filled with hydraulic oil, its influence on the establishment of the shift section hydraulic pressure is no longer significant, and this fault has an impact on the shift section quality of the transmission.
(4) The outlet oil passage of the governor valve is blocked (Fault Type 4): The oil filling flow of the clutch is reduced, and the sliding time between the main and driven friction plates is extended. This not only deteriorates the quality of the shift, but also increases the risk of power interruption and clutch burnout. This fault mode has a gradual characteristic and is not easy to detect.

(5) Leakage of the branch of the oil pipeline (Fault Type 5): the fault generally occurs at the place of the pipe joint of the branch oil pipeline, and a partial external leakage occurs in the oil pipeline.

The oil pressure may be established normally, but the steady-state pressure of the clutch branch oil circuit is reduced, and it will cause air to mix in, causing cavitation and vibration of the hydraulic oil during the circulation process. This situation also reduces the shift section quality of the transmission.

Changes in the parameters of the hydraulic system can have a greater impact on the quality of the change section. Therefore, if the fault is not diagnosed in time, when this change reaches a certain level, it will inevitably have a significant impact on the shift section quality, thereby affecting driving comfort and threatening driving safety.

3.2. Fault Simulation and Data Obtained

Before the test operation, the standard of the experimental parameters needs to be determined first. After a number of optimization experiments on the shift section of the continuously variable transmission, it was found that when the oil pressure is 4 MPa and the flow rate is 5 L/min, the overall performance of the shift section is the best. Therefore, all of the following experiments were conducted under this parameter index.

Among the aforementioned fault modes, the T1 mode does not require special processing, and can directly collect oil pressure and flow data; the T2 mode can be simulated when the clutch is in the state of oil drain disconnection, by filling the joint gap between the clutch's main and driven shafts with sandpaper. At this time, the clutch piston was completely unable to extend and was forced to be stuck in place; the T3 mode can be simulated by installing seal rings with different degrees of wear on the rotary joint; T4 can be used to conduct simulation tests while reducing the opening of the governor valve. Additionally, because the characteristic component of the T4 mode is the gradual fault, the flow level can be controlled within the range of 0~4 L/min, while ensuring the interval of 1 L/min for the test. The T5 mode can be simulated by unscrewing the branch pipe joint.

The data to be measured in the test are the flow data of the clutch main oil circuit and the pressure data of the clutch branch oil circuit. Due to the strict proportional relationship between the observed value and the actual value of the original data, the original data can be directly identified without conversion. The data recording cycle of the data and programs obtained through the high-speed data acquisition system is 16 ms. The test was conducted 120 times in total, 24 times for each fault simulation test. In addition, we considered that the original data points of the flow and pressure of the hydraulic system were huge in the shifting process, so its characteristic attributes were calculated based on the following six statistics:

$$X_f = \sqrt{\frac{1}{N_f}\sum_{i=1}^{N} x_{fi}^2} \tag{1}$$

$$C_f = \frac{max\left(\left|x_{fi}\right|\right)}{X} \tag{2}$$

$$K_f = \frac{\sum_{i=1}^{N} x_{fi}^4}{N_f X^4} \tag{3}$$

$$I_f = \frac{max\left(\left|x_{fi}\right|\right)}{\frac{1}{N_f}\sum_{i=1}^{N}\left|x_{fi}\right|} \tag{4}$$

$$S_f = \frac{X}{\frac{1}{N_f}\sum_{i=1}^{N}\left|x_{fi}\right|} \tag{5}$$

$$X_p = \sqrt{\frac{1}{N_p}\sum_{i=1}^{N} x_{pi}^2} \quad (6)$$

In the formula, X_f, C_f, K_f, I_f and S_f are flow statistics, respectively, representing the root mean square value of flow, peak factor, kurtosis factor, pulse factor and form factor during the transition period; X_p is the pressure statistic, which represents the root mean square pressure of the pressure during the transition; x_{fi} and x_{pi} represent the data of the i-th sampling point of flow and pressure, respectively, and N_f and N_p represent the total number of data sampling points of flow and pressure, respectively. After the attribute calculation, a sample set of 120 fault data characteristic attributes was obtained. Randomly, we set 80 of them as training samples and 40 of them as test samples.

4. BP Method for FAULT Diagnosis of HMCVT Shift Hydraulic System

4.1. Fault Diagnosis of Shift Hydraulic System Based on BP Neural Network

The BP neural network is mainly composed of an input layer. The output layer and hidden layer are composed of three parts, in which the number of input layers is determined by the eigenvalue of fault data and the number of output layers is determined by the fault diagnosis result. In the data collection, this paper collects the flow and pressure fault signals of the HMCVT hydraulic system as characteristic values, and takes the corresponding working state as output characteristics. The structure of the neural network is constructed according to the actual training results. Therefore, the number of neurons in the input layer of the neural network is set to six, the number of neurons in the output layer is one and the number of hidden layer neurons is as follows.

$$N = 2n + 1 \quad (7)$$

$$N = \sqrt{n + m} + \alpha \quad (8)$$

$$N = \log_2 n \quad (9)$$

In the formula, N—the number of neurons in the hidden layer; n—the number of neurons in the input layer; m—the number of neurons in the output layer; α—the constant from 1 to 10. After many neural network trainings, the number of neurons in the hidden layer of this paper was selected as 10. Its network structure is shown in Figure 3.

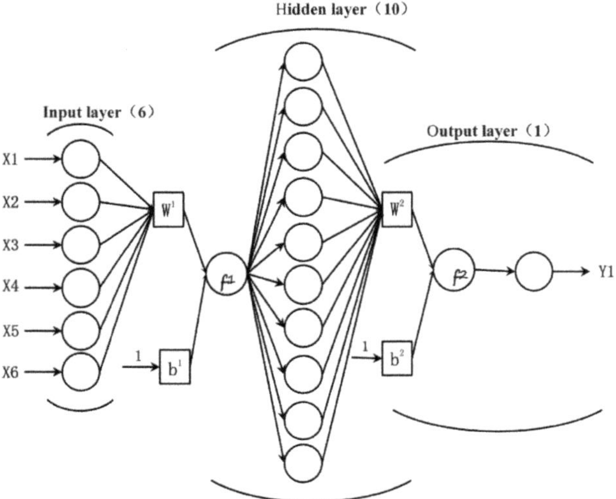

Figure 3. The topology of the three-layer neural network.

The BP neural network established in this paper comprises the following: one input layer, one hidden layer and one output layer. The BP neural network established in this paper comprises the following: one input layer, one hidden layer and one output layer, in which the number of neurons in the input layer is six, the number of neurons in the hidden layer is ten and the number of neurons in the output layer is one. The transfer function of the hidden layer uses the tansig function, and the output layer uses the purelin function. We inputted 40 groups of test samples into the unoptimized BP neural network model, and the resulting diagnosis results of the test samples are shown in Figure 4.

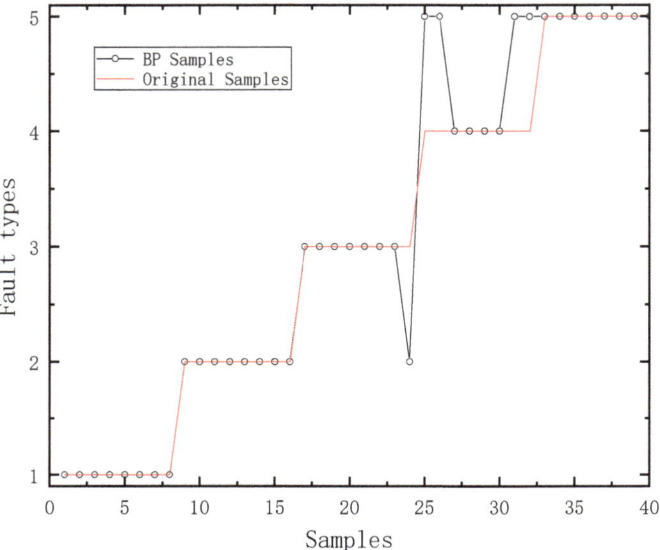

Figure 4. The effect of fault diagnosis of the unoptimized BP neural network test sample.

It can be seen from Figure 4 that the unoptimized BP neural network model has a good recognition ability for the normal mode (Fault Type 1), the clutch piston being stuck (Fault Type 2) and branch oil pipeline joint leakage (Fault Type 5).

4.2. Fault Diagnosis of Shift Hydraulic System Based on PSO-BP Neural Network

Number of population particles and population particle number: The determination of particle number mainly depends on the complexity of fault type. If the total number of particles is small, it is not conducive to the overall optimization, and if the total number of particles is large, it will increase the calculation of the population. According to relevant data, generally, when the number of particles is maintained in the range of 20–40, the optimization result will be relatively good. If the problem is very complicated, the number of particles can be increased to more than 100 [27]. Aiming at the problem of fault diagnosis of the shift hydraulic system, this paper sets the number of population particles to 20.

The dimension of the particle: The value of the problem can be determined by the dimension of the problem. According to the fault diagnosis and data characteristics of the hydraulic system, the selected dimension is 81 in this paper.

The range of the particles: From the characteristics of the optimized problem, different change intervals for each dimension can be determined. According to the characteristics of the flow and pressure data of the shift hydraulic system, the range selected in this paper is $(-5, 5)$.

Maximum speed V_{max}: In general, the range of particles will be represented by V_{max}, which is an important basis for determining the maximum distance that particles can move in each iteration. According to the characteristics of the flow and pressure data of the shift hydraulic system, the range is selected as (−1, 1).

Learning factor c: generally, the learning factors c1 and c2 take the value of two.

Forty groups of test samples were inputted into the BP neural network optimized by particle swarm, and the resulting diagnosis results of the test samples are shown in Figure 5. The BP neural network optimized by the particle swarm has a strong ability to recognize the five fault modes of the shift hydraulic system. It mainly has deviations in the identification process of seal ring damage (Fault Type 3) and oil passage blockage fault (Fault Type 4), but it has a more obvious improvement in fault recognition compared to the unoptimized BP neural network model.

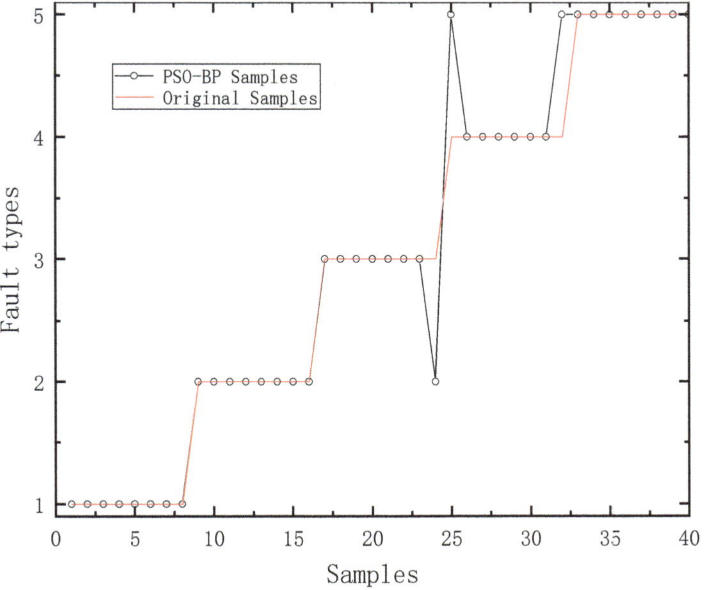

Figure 5. The effect of fault diagnosis of the PSO-BP neural network test sample.

4.3. Fault Diagnosis of Shift Hydraulic System Based on BAS-BP Neural Network Model

Beetle antennae search (BAS) is more effective than particle swarm optimization. Because the beetle antenna search can accurately find the expected target without specific function form and gradient information, it is applied to various optimization models to improve the efficiency of fault diagnosis [28].

Combining the beetle antennae search algorithm with the neural network, the global search capability of the beetle antennae search algorithm was used to optimize the initial weights and thresholds of the neural network. Moreover, compared with the particle swarm algorithm, the beetle antennae search algorithm is much simpler, because it only requires one beetle, which greatly reduces the amount of calculation. Its specific process roadmap is shown in Figure 6.

The beetle antennae search algorithm only has two parameters that need to be set, namely, the distance, d0, between the two whiskers and the ratio constant, c, between the step size and the distance of the two whiskers. In this paper, d0 = 1 and c = 5.

When applying the BP neural network model optimized by the beetle antennae search algorithm to the fault diagnosis of the shift hydraulic system, the fitness curve of samples is shown in Figure 7.

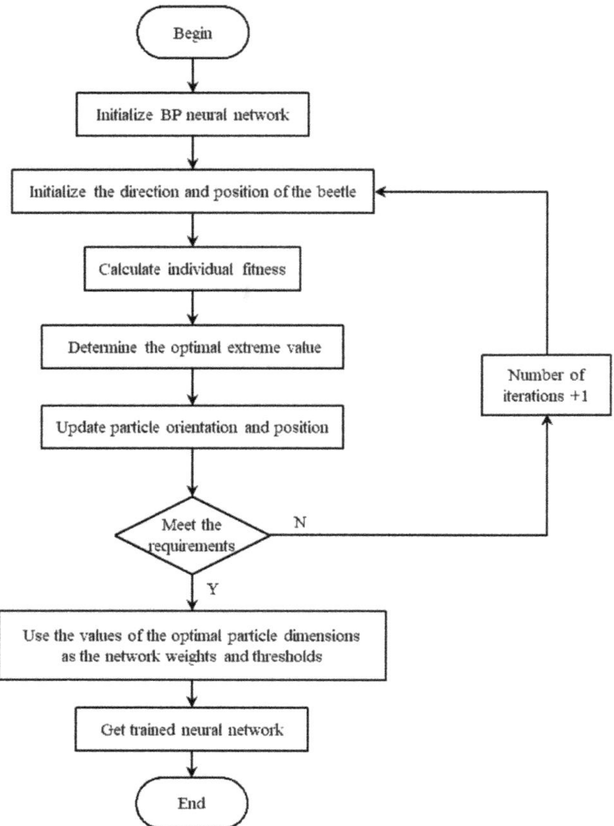

Figure 6. BAS-BP neural network algorithm flowchart.

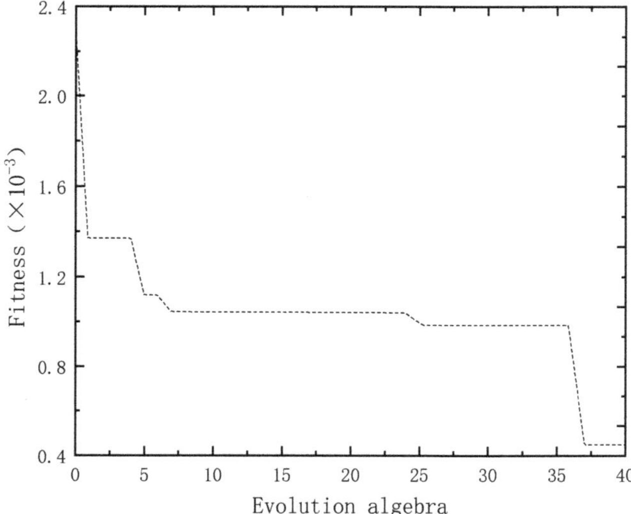

Figure 7. Fitness curve of BAS-BP neural network.

It can be seen from Figure 7 that when using the beetle antennae search algorithm to optimize the BP neural network, the optimization speed of its fitness value is slow, which is mainly caused by its lesser parameters and lesser calculation amount. Although the optimization speed of the beetle antennae search algorithm must be slow, its optimization calculation process is faster.

Forty groups of test samples were inputted into the BP neural network optimized by the beetle antennae search algorithm, and the resulting diagnosis results of the test samples are shown in Figure 8.

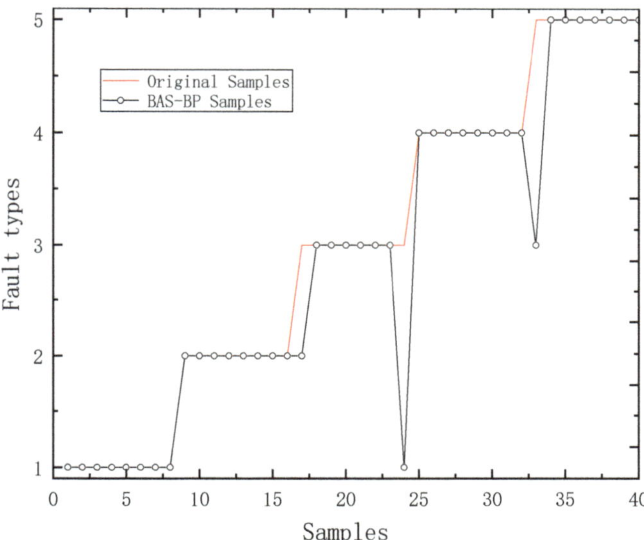

Figure 8. The effect of fault diagnosis of the BAS-BP neural network test sample.

It can be seen from Figure 8 that the BP neural network model optimized by the beetle antennae search algorithm must have the strongest ability to recognize the five fault modes of the shift hydraulic system. Compared with the unoptimized BP neural network model and the particle-swarm-optimized BP neural network model, the fault recognition accuracy rate of the oil channel blockage fault (Fault Type 4) can reach 100%. However, it has a large deviation in the process of identifying the seal ring fault (Fault Type 3). Its recognition accuracy rate for pipeline joint leakage (Fault Type 5) is no higher than that of the unoptimized BP neural network model and the particle-swarm-optimized BP neural network model.

From the above analysis, it can be seen that the optimized BP neural network model is better than the unoptimized BP neural network model for fault recognition. Additionally, the BAS-BP neural network model has the strongest ability to identify the fault of the oil channel blockage fault T4, which is conducive to further analysis to determine and eliminate the fault.

The comparison of the fault diagnoses of the test samples of the three neural network models is shown in Figure 9 and the test sample fault diagnosis correct rate of the three neural network models is shown in Table 2.

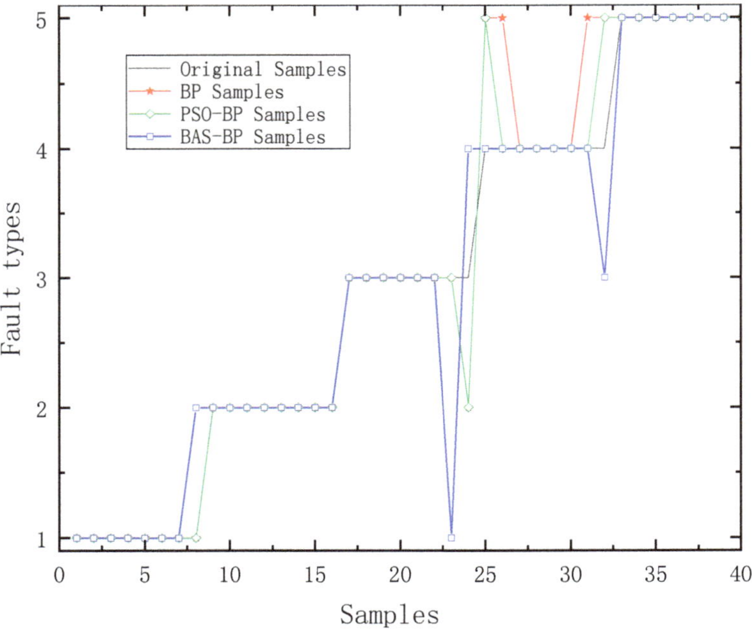

Figure 9. Contrast diagram of fault diagnoses for BP, PSO-BP and BAS-BP.

Table 2. The accuracy of test samples fault diagnosis.

Models / Modes	T1	T2	T3	T4	T5	Correct Rate
BP	100%	100%	87.5%	50%	100%	87.5%
PSO-BP	100%	100%	87.5%	75%	100%	92.5%
BAS-BP	100%	100%	75%	100%	87.5%	92.5%

5. CNN Method for Fault Diagnosis of HMCVT Shift Hydraulic System

5.1. Convolutional Neural Network Overview

The convolutional neural network (CNN) is a hierarchical neural network. Its advantage is that the network model is simple, and the image can be directly used as the input of the network, which reduces the workload. These characteristics make the convolutional neural network have obvious advantages in identifying two-dimensional graphics [29].

5.1.1. Convolutional Layer

The convolution layer uses the convolution kernel to convolute the input image. Additionally, the activation function is used to extract the texture features of the image to enhance the features. The convolution operation can be expressed as:

$$x_j^l = f\left(\sum_{i \in M_j} x_i^{l-1} * k_{ij}^l + b_j\right) \tag{10}$$

where l represents the current layer number, k_{ij} is the weight matrix of the convolution kernel and M_j represents the set of input feature maps; b_j is an offset term corresponding to each feature in the convolution layer.

5.1.2. Pooling Layer

The pooling layer is also called a sub-sampling layer. It is usually located after the convolution layer. Using the sub-sampling function can reduce redundant features, further avoid overfitting and reduce network parameters. The mathematical model can be described as:

$$x_j^l = f\left(\beta_j^l \; down\left(x_j^l - 1 + b_j^l\right)\right) \tag{11}$$

where down(.) represents the sub-sampling function. Generally, this function sums each different nxn block in the input image, so that the output image is n times smaller in both spatial dimensions. Each output map has its own multiplication bias β and addition bias b.

5.1.3. Fully Connected Layer

The fully connected layer is located at the end of the convolutional neural network and it is used to calculate the output of the entire network. After the downsampling is completed, many corresponding feature maps can be obtained. At this time, all of its pixels must be arranged in columns to form a feature vector, and then all of them are connected to the output layer to serve as a fully connected layer. Additionally, Softmax was used as the classifier.

5.1.4. Convolutional Neural Network Structure

Convolutional neural networks can determine the basic parameters of the corresponding structure of the network by analyzing specific problems. Figure 10 shows the structure of a basic convolutional neural network. Its overall architecture is similar to the convolutional neural network model corresponding to the CNN code in Deep Learning Toolbox.

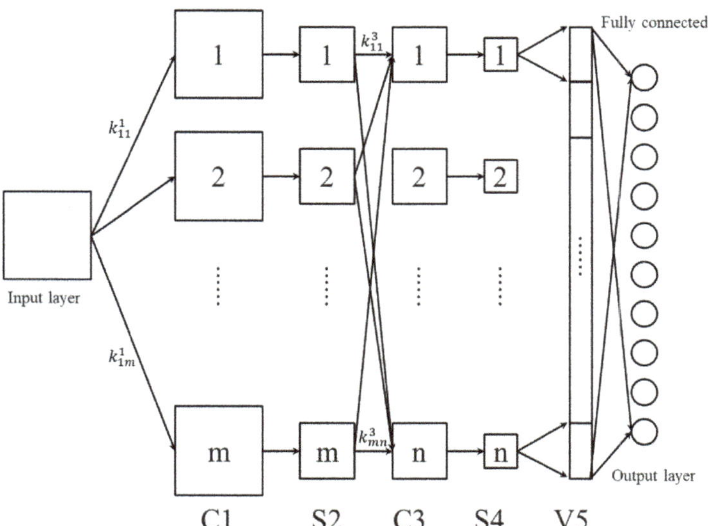

Figure 10. Convolution neural network structure.

It can be clearly seen from Figure 10 that the number of input layers, output layers and feature vector layers is one, and the number of convolutional layers and downsampling layers is two. They connect and cooperate with each other to construct a complete network. In this process, when a pixel value of $p \times p$ is used as an input layer signal, m convolution kernels of type $k_1 \times k_1$ start to perform the corresponding convolution operation with a step size of one. Additionally, through the activation function $m\ (p - k_1 + 1) \times (p - k_1 + 1)$, feature maps are obtained to form the convolution layer $C1$. Then, it completes the entire pooling process through the pooling area of model size $c \times c$, so as to obtain the sampling layer $S2$. At this time, there are still m feature maps, and the side length becomes $(p - k_1 + 1)/c$. Next, the $S2$ layer and the n convolution kernels of all models of $k_1 \times k_1$ each start to perform the corresponding convolution operation, and then use the activation function to obtain the convolution layer C3. After completing the pooling process again, the sampling layer $S4$ can be obtained. Then, arrange all of its n feature maps in columns to obtain the feature vector layer $V5$ that was originally required. $V5$ is fully connected to the output layer to obtain the output.

5.2. Attribute Reduction Based on Rough Set

The aforementioned six sets of sample attributes X_f, C_f, K_f, I_f, S_f and X_p have different abilities to distinguish the fault mode. In order to reduce unnecessary attribute calculation, on the premise of ensuring the correct rate of fault diagnosis, the attributes that contribute less to fault diagnosis can be deleted. This paper is based on the rough set theory [30–32] for attribute reduction.

For a fixed decision system $S = (U, C \cup D, V, f)$, once there is $B_1 \subseteq B_2 \subseteq C$, then there is $|Pos_{B_1}(D)| \leq |Pos_{B_2}(D)| \leq |Pos_C(D)|$, while $|\gamma_{B_1}(D)| \leq |\gamma_{B_2}(D)| \leq |\gamma_C(D)|$, that is, the dependence formula intuitively shows that the dependence degree and the attribute set have an inverse proportional relationship. Additionally, the calculation method based on attribute dependency is as follows [33,34]:

(1) Calculation method in the case of deleting attributes: for a fixed decision system $DS = (U, C \cup D, V, f)$, $\forall B \subseteq C$; once the attribute $a \in B$, then the calculation method of the importance of the conditional attribute a for the decision attribute D is as follows:

$$Sig(a, B, D) = \gamma_B(D) - \gamma_{B-\{a\}}(D) \tag{12}$$

According to the formula, it can be seen that when the attribute is removed from the condition attribute set, the weakened dependence of the decision attribute on the condition attribute can be used as the basis for the meaning of the attribute to the decision attribute.

(2) Calculation method in the case of adding attributes: for a fixed decision system DS, once the attribute $a \in C$, but $a \notin B$, the expression of the importance of the conditional attribute a relative to the conditional attribute set B for the decision attribute D is as follows:

$$Sig(a, B, D) = \gamma_{B\cup\{a\}}(D) - \gamma_B(D) \tag{13}$$

In the same way, when the attributes in the conditional attribute set are greatly added, the size of the increased dependence of the decision attribute on the condition attribute can be used as the basis for the meaning of the attribute to the decision attribute.

In this paper, the attribute reduction in the neighborhood rough set was applied to the fault diagnosis of the shift hydraulic system, and the attribute reduction in the training sample set collected by the experiment was performed.

After calculation, the attribute importance values of the characteristic attributes are shown in Table 3. Obviously, after attribute reduction, mode T1 retains one characteristic attribute of X_p; mode T2 retains two characteristic attributes of X_f and X_p; mode T3 retains two feature attributes of X_f and X_p; mode T4 retains three feature attributes of X_f, S_f and X_p and mode T5 retains two feature attributes of S_f and X_p.

Table 3. Calculated results of attributes' importance values.

Modes	X_f	C_f	K_f	I_f	S_f	X_p
T1	0	0	0	0	0	1
T2	0.0278	0	0	0	0	0.9722
T3	0.2222	0	0	0	0	0.7778
T4	0.0217	0	0	0	0.3478	0.6304
T5	0	0	0	0	0.4324	0.5676

As can be seen from Table 3, among the six characteristic attributes of the five failure modes, after the attribute reduction, the attribute importance of the root mean square value of the pressure X_p exceeds 50%, which are common and indispensable characteristic attributes in the fault modes. That is to say, compared with the flow data, the pressure data has the greatest influence on the recognition of the fault mode of the shift hydraulic system. Therefore, on the premise of ensuring a high accuracy of fault diagnosis, it can be considered to use only the original pressure data of the shift hydraulic system as the input to train the convolutional neural network model, and then to obtain a fault diagnosis model with an ideal effect.

5.3. Fault Diagnosis Results Based on CNN and Neighborhood Rough Set

Convolutional neural network training generally requires a large number of data sets, and the input datum is a two-dimensional image, so this paper uses the translation transformation method to amplify the data of 1000 original pressure data in each of the five modes obtained by the experiment. That is to say, in each mode, 400 data are continuously taken as a group for each of the 1000 data at intervals of five to generate a 20 × 20 size two-dimensional grayscale image in PNG format. After translation transformation, 120 grayscale images were generated in each fault mode, 80 groups were set as training samples and 40 groups were set as test samples using a random method. The two-dimensional grayscale image under the five fault modes is shown in Figure 11.

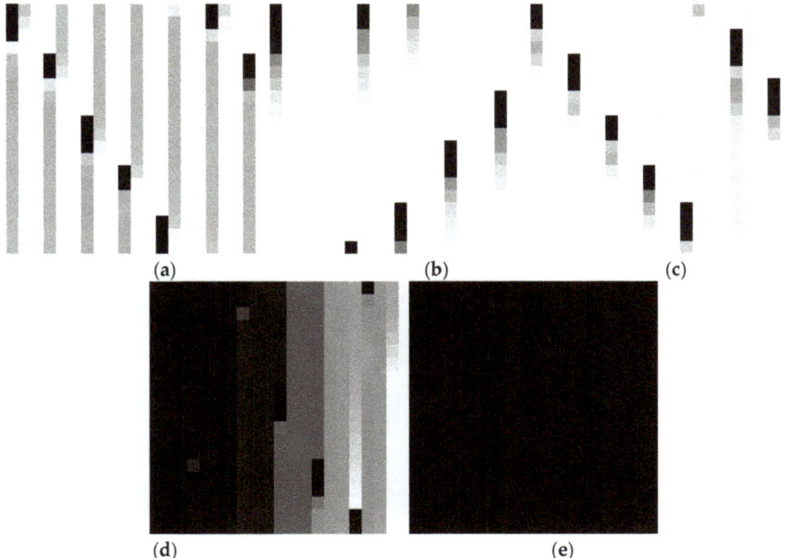

Figure 11. Two-dimensional gray image under five types of fault modes. (**a**) Normal mode T1, (**b**) spring fault T2, (**c**) seal ring fault T3, (**d**) blockage of oil passage T4 and (**e**) cavitation seal T5.

This paper used an eight-layer neural network, an input layer, three convolutional layers, two pooling layers, a fully connected layer and an output layer to establish a convolutional neural network model. We used Softmax as the classifier to obtain the classification results and fault diagnosis accuracy. The specific flowchart is shown in Figure 12.

In Figure 12, the fault signal is the oil pressure signal during the shifting process collected by the clutch branch oil pressure sensor. We transformed the oil pressure signal data to generate a two-dimensional gray image and adjusted the two-dimensional image using a size of 20 × 20 as the input training sample. After training the established convolutional neural network model, we inputted the test samples to obtain the sample classification and its fault recognition rate via the Softmax classifier. The diagnosis results are shown in Figure 13.

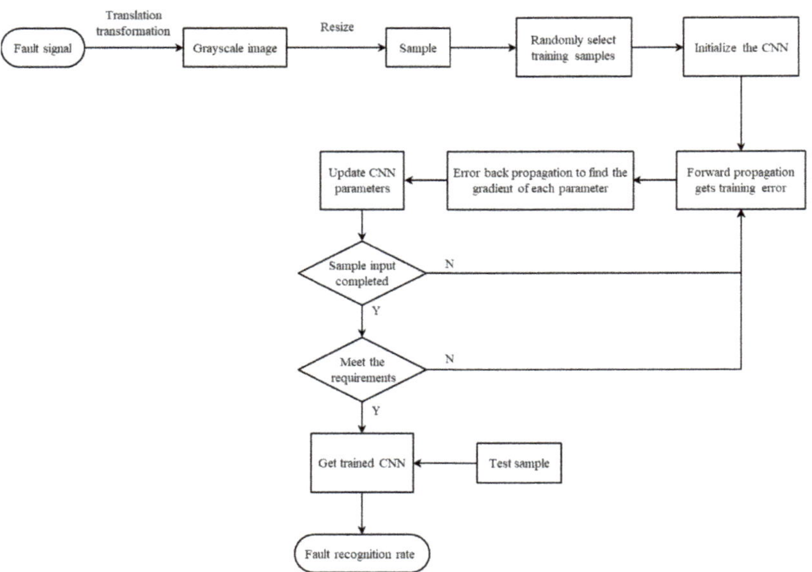

Figure 12. Flowchart of fault diagnosis for shift hydraulic system.

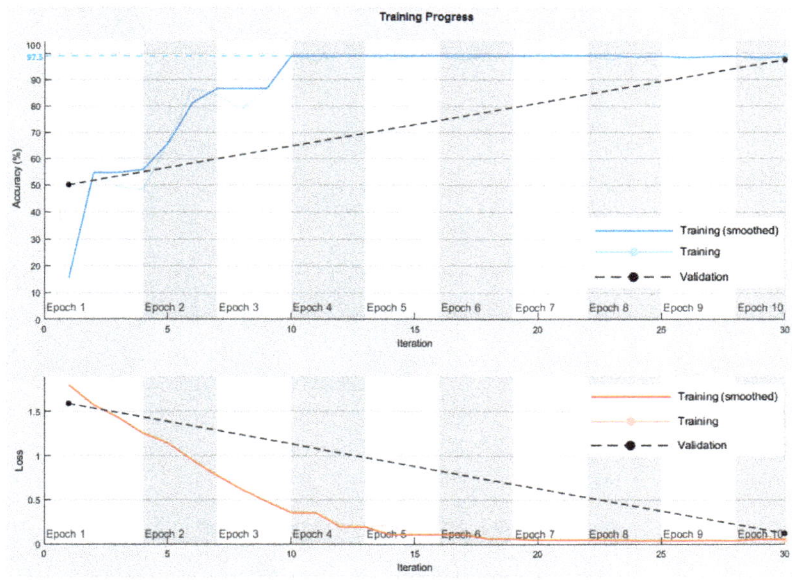

Figure 13. Fault diagnosis effect diagram of shift hydraulic system.

It can be seen from Figure 13, compared with the shallow optimized BP neural network model, that the hydraulic system fault diagnosis model based on the convolutional neural network can make the average diagnosis accuracy rate of each failure mode reach 97.5% with few iterations. This not only shows that it is feasible to use the pressure data of the hydraulic system as the only input parameter to identify the fault mode of the HMCVT shift hydraulic system, but also it is proved that the convolutional neural network as a deep learning algorithm is significantly better than the shallow optimized BP neural algorithm in the application effect of fault diagnosis of the shift hydraulic system.

6. Conclusions

Aiming at the problems existing in the BP neural network and convolutional neural network, the BP neural network optimization model and convolutional neural network model were established, respectively. They were applied to the fault diagnosis of the shifting hydraulic system of hydraulic continuously variable transmission. The results show the following:

(1) Various types of faults have greater separability after nonlinear transformation by the BP network, the overall similarity is low and the model fault classification effect is good.
(2) The optimized BP neural network model is better than the unoptimized BP neural network model for fault recognition. Its average diagnostic accuracy rate reached 92.5%.
(3) The BAS-BP neural network model has the strongest ability to identify the fault of the oil channel blockage fault T4, which is conducive to further analysis to determine and eliminate the fault.
(4) The experiment shows that the pressure data has the greatest influence on the recognition of the fault mode of the shift hydraulic system. It is feasible to use the pressure data of the hydraulic system as the only input parameter to identify the fault mode of the HMCVT shift hydraulic system.
(5) The convolutional neural network under the same test conditions is significantly better than the shallow optimized BP neural network in the application of the fault diagnosis of the shift hydraulic system, and its fault diagnosis accuracy can reach 97.5%.

For the follow-up research, there is still room for further improvement. In the optimization of the BP network, a better algorithm model can be proposed. In the selection of data sets, more complex and fault data can be collected, and the comprehensive verification can be carried out through the simulation and experimental data sets to improve the reliability of the algorithm. In the data feature extraction, more optimized feature extraction can be carried out to improve the accuracy of fault diagnosis.

Author Contributions: Conceptualization J.W., Z.L. and G.W.; formal analysis, Z.L. and G.W.; investigation, J.W., G.H. and S.Z.; resources, Z.L., S.Z. and M.X.; data curation, J.W. and G.W.; writing—original draft preparation, J.W.; writing—review and editing, J.W.; visualization, J.W.; supervision, Z.L., G.W., H.Z., G.H., S.Z. and M.X.; project administration, H.Z. and M.X.; funding acquisition, M.X. All authors have read and agreed to the published version of the manuscript.

Funding: The work described in this paper was fully supported by the National Key R&D Program (2022YFD2001204), Basic Scientific Research Business Expenses of Central Universities (XUEKEN2022015), Jiangsu Agricultural Science and Technology Independent Innovation Fund (CX (22) 3101), Jiangsu International Science and Technology Cooperation Project (BZ2021007) and Jiangsu Modern Agricultural Machinery Equipment and Technology Demonstration and Promotion Project (NJ2021-06).

Institutional Review Board Statement: Not applicable.

Informed Consent Statement: Not applicable.

Data Availability Statement: The data presented in this study are available upon request from the first author at (gerab.wang@jsafc.edu.cn).

Conflicts of Interest: The authors declare no conflict of interest.

References

1. Lei, X.; Cai, Z.; Zhang, M.; Ma, W. Analyses of Transmission Error of Mechanical Shift Gear in HMCVT. *China Mech. Eng.* **2015**, *26*, 3253–3259.
2. Xia, Y.; Sun, D. Characteristic analysis on a new hydro-mechanical continuously variable transmission system. *Mech. Mach. Theory* **2018**, *126*, 457–467. [CrossRef]
3. Geng, G.; Zhou, K.; Xiao, M.; Zhang, H.; Wang, G. Research on starting process of hydro-mechanical continuously variable transmission for tractor. *J. Nanjing Agric. Univ.* **2016**, *39*, 332–340.
4. Macor, A.; Rossetti, A. Optimization of hydro-mechanical power split transmissions. *Mech. Mach. Theory* **2011**, *46*, 1901–1919. [CrossRef]
5. Hu, J.B.; Wei, C.; Du, J.Y. A Study on the Speed Ratio Follow-Up Control System of Hydro-Mechanical Transmission. *Trans. Beijing Inst. Technol.* **2008**, *28*, 481–485.
6. Wang, G.; Song, Y.; Wang, J.; Xiao, M.; Cao, Y.; Chen, W.; Wang, J. Shift quality of tractors fitted with hydrostatic power split CVT during starting. *Biosyst. Eng.* **2020**, *196*, 183–201. [CrossRef]
7. Kim, D.C.; Kim, K.U.; Park, Y.J.; Huh, J.Y. Analysis of shifting performance of power shuttle transmission. *J. Terramech.* **2007**, *44*, 111–122. [CrossRef]
8. Zhang, X.; Tai, J.; Wang, G.; Zhang, H.; Wang, C.; Zhong, C. Hydraulic fault diagnosis of hydro-mechanical continuously variable transmission in shift based on BP method. *J. Chin. Agric. Mech.* **2016**, *37*, 133–139. [CrossRef]
9. Ma, B. Study on Hydraulic System Fault Diagnosis of Hydro-Mechanical Continuously Variable Transmission. Nanjing Agricultural University, Nanjing, China, 2015.
10. Xue, L.; Jiang, H.; Zhao, Y.; Wang, J.; Wang, G.; Xiao, M. Fault diagnosis of wet clutch control system of tractor hydrostatic power split continuously variable transmission. *Comput. Electron. Agric.* **2022**, *194*, 106778. [CrossRef]
11. Li, Y.; Liang, X.; Xu, M.; Huang, W. Early fault feature extraction of rolling bearing based on ICD and tunable Q-factor wavelet transform. *Mech. Syst. Signal Process.* **2017**, *86*, 204–223. [CrossRef]
12. Zhang, X.; Zhao, J.; Li, H.; Ni, X.; Sun, F. Compound fault diagnosis for gearbox based on NIC-DWT-WOASVM. *J. Vib. Shock.* **2020**, *39*, 146–151+164.
13. Xia, L. Research on Fault Diagnosis and Related Algorithm Based on Hidden Markov Model. Ph.D. thesis, Huazhong University of Science and Technology, Wuhan, China, 2014.
14. Wang, G.; Zhang, X.; Zhu, S.; Zhang, H.; Shi, L.; Zhong, C. Hydraulic failure diagnosis of tractor hydro-mechanical continuously variable transmission in shifting process. *Trans. Chin. Soc. Agric. Eng.* **2015**, *31*, 25–34.
15. Li, C.; Sanchez, R.-V.; Zurita, G.; Cerrada, M.; Cabrera, D.; Vásquez, R.E. Multimodal deep support vector classification with homologous features and its application to gearbox fault diagnosis. *Neurocomputing* **2015**, *168*, 119–127. [CrossRef]
16. Lin, R.; Ming, T.; Wu, J. Leakage Faults Detection and Isolation for Electro-hydraulic Servo System. *Mach. Tool Hydraul.* **2012**, *40*, 162–166+169.
17. Han, Z. Research on Hydraulic System Fault Diagnosis Method of Rack-Rail Car. M.D. thesis, Dalian University of Technology, Dalian, China, 2013.
18. Zou, L. Research on Shield Machine Fault Diagnosis Method Based on SOM-BP Neural Network. M.D. thesis, Xi'an University of Technology, Xi'an, China, 2018.
19. Xin, H. Research on Switch Fault Diagnosis System Based on PSO-BP. M.D. thesis, Lanzhou Jiaotong University, Lanzhou, China, 2020.
20. Imtiaz, T.; Khan, B.H.; Khanam, N. Fast and improved PSO (FIPSO)-based deterministic and adaptive MPPT technique under partial shading conditions. *IET Renew. Power Gener.* **2020**, *14*, 3164–3171. [CrossRef]
21. Peng-cheng, Y.; Song-hang, S.; Chao-yin, Z.; Xiao-fei, Z. Research on classification of coal mine water source by improved BP neural network algorithm. *Spectrosc. Spectr. Anal.* **2021**, *41*, 2288–2293.
22. Jarusek, R.; Volna, E.; Kotyrba, M. Photomontage detection using steganography technique based on a neural network. *Neural Netw.* **2019**, *116*, 150–165. [CrossRef]
23. Su, L.; Zha, Z.; Lu, X.; Shi, T.; Liao, G. Using BP network for ultrasonic inspection of flip chip solder joints. *Mech. Syst. Signal Process.* **2013**, *34*, 183–190. [CrossRef]
24. Song, Z.; Geng, D.; Su, C.; Liu, Y. Vibration prediction of a hydro-power house base on IFA-BPNN. *J. Vib. Shock.* **2017**, *36*, 64–69.
25. Gong, W.F.; Chen, H.; Zhang, Z.H.; Zhang, M.L.; Guan, C.; Wang, X. Intelligent fault diagnosis for rolling bearing based on improved convolutional neural network. *J. Vib. Eng.* **2020**, *33*, 400–413.
26. Chandrasekhar, V.; Lin, J.; Morère, O.; Goh, H.; Veillard, A. A practical guide to CNNs and Fisher Vectors for image instance retrieval. *Signal Process.* **2016**, *128*, 426–439. [CrossRef]
27. Jing, Y.; Sun, A.; Zhang, M. Fault diagnosis of three-phase bridge rectifier circuit based on particle swarm optimization. *J. Comput. Appl.* **2019**, *39*, 60–64.
28. Jiang, X.; Li, S. BAS: Beetle Antennae Search Algorithm for Optimization Problems. *Int. J. Robot. Control* **2018**, *1*, 1–5. [CrossRef]
29. Qu, J.; Yu, L.; Yuan, T.; Tian, Y.; Gao, F. A hierarchical intelligent fault diagnosis algorithm based on convolutional neural network. *Control Decis.* **2019**, *34*, 2619–2626.
30. Pan, R.; Wang, X.; Yi, C.; Zhang, Z.; Fan, Y.; Bao, W. Multi-objective optimization method for thresholds learning and neighborhood computing in a neighborhood based decision-theoretic rough set model. *Neurocomputing* **2017**, *266*, 619–630. [CrossRef]

31. Zhang, N. Research of Attribute Reduction Algorithm Based on Neighborhood Rough Set. M.D. thesis, Hunan University, Changsha, China, 2017.
32. Deng, D.; Li, Y.; Huang, H. Concept Drift and Attribute Reduction From the Viewpoint of F-rough Sets. *Acta Autom. Sin.* **2018**, *44*, 1781–1789.
33. Yao, S.; Xu, F.; Wu, Z.; Chen, J.; Wang, J.; Wang, W. Non-monotonic attribute reduction based on neighborhood rough mutual information entropy. *Control Decis.* **2019**, *34*, 353–361.
34. Hu, Q.; Yu, D.; Xie, Z. Numerical Attribute Reduction Based on Neighborhood Granulation and Rough Approximation. *J. Softw.* **2008**, *19*, 640–649. [CrossRef]

 agriculture

Article

An Application of Artificial Neural Network for Predicting Threshing Performance in a Flexible Threshing Device

Lan Ma [1,2,3], Fangping Xie [1,3,*], Dawei Liu [1,3], Xiushan Wang [1,3] and Zhanfeng Zhang [4]

1 College of Mechanical and Electrical Engineering, Hunan Agricultural University, Changsha 410102, China; malan@caas.cn (L.M.); liudawei@hunau.edu.cn (D.L.); hnwxs@hunau.edu.cn (X.W.)
2 Institute of Bast Fiber Crops, Chinese Academy of Agricultural Sciences, Changsha 410205, China
3 Hunan Key Laboratory of Intelligent Agricultural Machinery Corporation, Changsha 410102, China
4 Changsha Zichen Technology Development Co., Ltd., Changsha 410221, China
* Correspondence: hunanxie2002@163.com

Abstract: Rice is a widely cultivated food crop worldwide, and threshing is one of the most important operations of combine harvesters in grain production. It is a complex, nonlinear, multi-parameter physical process. The flexible threshing device has unique advantages in reducing the grain damage rate and has already been one of the major concerns in engineering design. Using the measured test database of the flexible threshing test bench, the rotation speed of the threshing cylinder (RS), threshing clearance of the concave sieve (TC), separation clearance of the concave sieve (SC), and feeding quantity (FQ) are used as the input layer. In contrast, the crushing rate (Y_P), impurity rate of the threshed material (Y_Z), and loss rate (Y_S) are used in the output layer. A 4-5-3-3 artificial neural network (ANN) model, with a backpropagation learning algorithm, was developed to predict the threshing performance of the flexible threshing device. Next, we explored the degree to which the inputs affect the outputs. The results showed that the R of the threshing performance model validation set in the hidden layer reached 0.980, and the root mean square error ($RMSE$) and the average absolute error (MAE) were less than 0.139 and 0.153, respectively. The built neural network model predicted the performance of the flexible threshing device, and the regression determination coefficient R^2 between the prediction data and the experimental data was 0.953. The results showed revealed that the data combined with the ANN method is an effective approach for predicting the threshing performance of the flexible threshing device in rice. Moreover, the sensitivity analysis showed that RS, TC, and SC were crucial factors influencing the performance of the flexible threshing device, with an average relative importance of 15.00%, 14.89%, and 14.32%, respectively. FQ had the least effect on threshing performance, with an average threshing relative importance of 11.65%. Our findings can be leveraged to optimize the threshing performance of future flexible threshing devices.

Keywords: rice; flexible threshing cylinder; artificial neural network; threshing clearance of concave sieve; separating clearance of concave sieve; feeding quantity; threshing performance

Citation: Ma, L.; Xie, F.; Liu, D.; Wang, X.; Zhang, Z. An Application of Artificial Neural Network for Predicting Threshing Performance in a Flexible Threshing Device. *Agriculture* **2023**, *13*, 788. https://doi.org/10.3390/agriculture13040788

Academic Editors: Massimo Cecchini, Cheng Shen, Zhong Tang and Maohua Xiao

Received: 19 March 2023
Revised: 23 March 2023
Accepted: 27 March 2023
Published: 29 March 2023

1. Introduction

Rice is one of the four main staple food crops in China, with a perennial planting area of 30 million hectares [1]. Mechanized rice production relies heavily on the harvest process as an essential step. Threshing is a key link in the rice harvesting process; it is a complex, nonlinear, and uncertain process, with several influencing parameters and large nonlinearity [2,3]. The impact of threshing on rice determines how much grain is lost during the harvest and processing stages. Double cropping rice in southern China has a short harvesting duration. The performance parameters of the threshing and separation device directly affect the operation quality of the combined rice harvester, i.e., the core working component. The longitudinal axial threshing device is characterized by long threshing time, smooth threshing process, good adaptability, and relatively soft threshing effect, and it is broadly used in combined harvesters [4]. Researchers in agricultural mechanization are interested in the flexible threshing tooth due to its lower impact force and rate of damage to the cracked grains compared to its rigid counterpart [5]. For this reason, it is suitable for increasing the synthesis benefit in grain production [6]. Several scholars have studied the application of flexible materials in agricultural engineering. In 1972, Duane L et al. [7] designed a self-made collision test device to analyze the effects of corn grain velocity, collision surface material, collision angle, and other parameters on the extent of grain collision damage. One study found that when the impact surface was polyurethane, the damage degree of the grain was one-fifth of that when the impact surface was steel, and one-sixth of that when the impact surface was concrete. This is an inaugural study focusing on the effect of flexible materials on grain, demonstrating the benefit of flexible materials in reducing grain damage degree. Shi Qingxiang et al. [8] performed a comparative study on the flexible and rigid threshing elements, demonstrating that flexible threshing with flexible teeth made of flexible materials can extend the threshing time and reduce grain breakage with feasible flexible threshing. Xie Fangping et al. [9] utilized polyurethane plastic cylindrical strips as the teeth of flexible threshing rods to conduct a dynamic analysis of the threshing of flexible rod teeth. Consequently, they found that the indexes of flexible threshing, for instance, non-removal rate and impurity rate, were similar to those of rigid rod teeth threshing, and the crushing rate was significantly lower than that of rigid rod teeth threshing. Ren Xuguang et al. [10] analyzed the threshing process of rice using the conservation law of capacity and noted that it is conducive to rice threshing when the flexible teeth periodically hit the ear of rice, and a resonance response occurred. Su Yuan et al. [11] modified the conventional Q235 carbon steel teeth into nitrile rubber composite nail teeth and polyurethane rubber nail teeth. The test found better grain removal performance of nitrile rubber composite nail teeth than that of polyurethane rubber nail teeth and traditional carbon steel nail teeth. Geng Duanyang et al. [12] designed a cross-axial flow flexible corn threshing device. To realize flexible and low-damage threshing of corn ears, the threshing element combined a structure of flexible nail teeth and an elastic short grain rod. Li Yibo et al. [13] performed a bench test to explore the effect of composite nail teeth of different outer materials on the threshing performance and self-wear resistance of corn ear. The results showed that the rubber composite nail teeth had the best comprehensive effects in threshing and self-anti-fraying performance, the breakage rate of maize was lower compared with that of traditional carbon steel nail teeth, and the non-threshing rate of maize was similar to that of traditional carbon steel nail teeth, thus meeting the conditions of technical specifications for threshing quality evaluation of maize harvester. Fu Jun et al. [14] established a rigid–flexible coupled wheat threshing arch tooth. Under similar operating conditions, the damage rate of the rigid-flexible coupled arch tooth was significantly reduced, unlike that of the standard arch tooth, with significant loss reduction and threshing effect. Qian Zhenjie et al. [15] introduced the increase and decrease constraint strategy to establish a multi-friction dynamic model of flexible threshing teeth on grains. As a consequence, it was observed that the continuous normal striking force and repeated minor tangential kneading force of flexible teeth on grains combined to reduce the grain damage rate. Reports on the longitudinal axial flow threshing cylinder with a hollow core and flexible rod teeth used in rice threshing are limited. Flexible threshing can reduce the crushing rate of rice grains and, thus, developing a comprehensive and accurate evaluation model of flexible threshing has important theoretical value and practical significance.

In recent years, the artificial neural network (ANN) has achieved desired performance and high accuracy in predicting laboratory data because of its capacity to describe non-linear systems. As a result, it is widely applied in the fields of mathematics, engineering, medicine, economy, environment, and agriculture [16], particularly where some traditional modeling methods have failed [17]. Artificial neural network technology has been utilized in harvester systems by some researchers [18,19]. Nonetheless, few studies have been conducted on the threshing performance of a flexible threshing device using artificial neural networks. Due to the uncertainty of the threshing condition and the complexity of the factors affecting the threshing device, the threshing performance prediction is a nonlinear problem affected by multiple factors. Nevertheless, the BP neural network is a nonlinear dynamic system [20,21] with powerful nonlinear [22] and generalization capacity and can identify complex relationships among the data [23]. Herein, parameters [24] affecting the performance of the threshing device and threshing performance indicators [25,26] were based on the parameters reported by several studies.

In the laboratory-based flexible threshing bench test, the rotated speed of the threshing cylinder, threshing clearance of the concave sieve, and separating clearance of the concave, as well as feeding quantity, were selected as the inputs of the model based on the BP neural network. The neural network model was established between inputs and their threshing characteristic of the crushing rate, impurity rate of threshed material, and entrainment loss rate. Further, the threshing performance index was predicted under different parameters. The objectives of this study included: (1) Determining the feasibility of artificial neural network technology in predicting the threshing performance of the flexible threshing device and providing executable procedures for an artificial neural network model for practical application; (2) Investigating the effect of artificial neural network geometry and some internal parameters on model performance; (3) Exploring the relative significance of factors influencing threshing performance through sensitivity analysis.

2. Materials and Methods

2.1. Test Materials and Equipment

The plots with basically similar crop growth rates were selected as the experimental sampling area. The rice variety tested was Xiangzaoxian 24. Table 1 shows the main material characteristics of the rice. The rice flexible threshing test was conducted in the Agricultural Machinery Engineering Training Center of Hunan Agricultural University from July 11 to 18, 2022. Figure 1 shows the test equipment, and Table 2 shows the equipment parameters.

Table 1. Main physical characteristic parameters of harvesting rice.

Rice Varieties	Plant Height/mm	Panicle Length/mm	Middle Stem Diameter/mm	Middle Stem Wall Thickness/mm	Number of Shoots per Ear	Number of Grains per Ear	Thousand-Grain Mass/g	Stem Moisture Content/%	Grain Moisture Content /%	Yield per Unit Area /kg·hm^{-2}	Ratio of Grass to Grain
Xiangzaoxian No. 24	833 ± 64	182 ± 12	32.18 ± 0.3	0.4 ± 0.1	12 ± 1.6	110 ± 22.9	30.02 ± 1.0	55.68 ± 4.8	22.42 ± 0.8	6230	1:(0.83 ± 0.1)

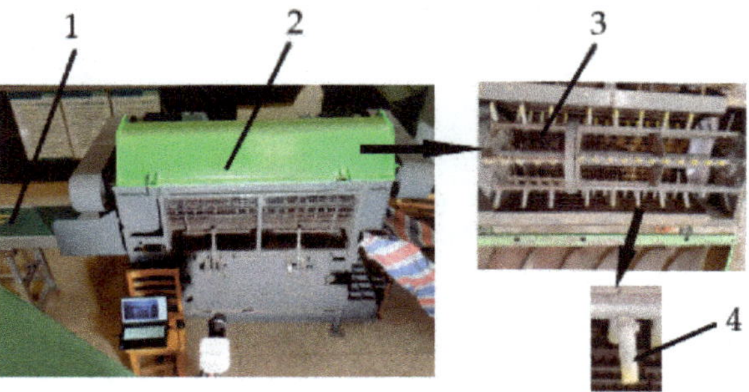

Figure 1. Flexible threshing experiment device. 1. Feeding device 2. Threshing device 3. Threshing cylinder 4. Flexible threshing teeth.

Table 2. Parameter table of flexible threshing device.

Parameters	Values
The total length of the cylinder/mm	1935
Threshing cylinder diameter/mm	620
Cylinder speed/($r \cdot min^{-1}$)	400–1500
Threshing clearance of concave sieve/mm	0–60
Separating clearance of concave Sieve/mm	0–60
Feeding rate/($kg \cdot s^{-1}$)	0.5–5

2.2. *Test Method*

The test was conducted following GB/T 5262—2008 and GB/T 5982—2005.

The thousand-grain quality was determined using the national standard method to explore grain and stem water content and according to the GB 5519-85 "grain, oil test thousand grain weight determination method".

The plots with similar crop growth were selected as the sample areas. Rice plants were artificially fed uniformly into the longitudinal axial flow threshing drum. In the multi-factor experiment, the material of each group weighed 10 kg. Three parallel tests were performed using similar parameter combinations, and the average value was taken. The performance evaluation indexes of the system were categorized into the crushing rate, impurity rate of threshed material (impurity rate for short), and entrainment loss rate. The mix that was threshed was collected in the receiving box located under the adaptable threshing mechanism. The mix released from the end of the cylinder was accumulated with the help of a tarpaulin attached to it. After each parallel test, the crushing rate and impurity rate of the threshing system were calculated using the mix, which was discharged into the receiving box. The mixture discharged onto the tarpaulin attached to the end of the cylinder was analyzed to determine the entrainment loss rate. The calculation formulas of the crushing, impurity, and entrainment loss rates are, respectively:

$$Y_P = \frac{W_P}{W_X} \times 100\% \tag{1}$$

$$Y_Z = \frac{W_{XZ}}{W_{Xh}} \times 100\% \tag{2}$$

$$Y_S = \frac{W_W}{W} \times 100\% \tag{3}$$

where Y_P is the crushing rate, %; W_P is the mass of crushed grains in the sample, g; W_X represents the total grain weight in the sample, g; Y_Z is the impurity rate of threshed material, %; W_{XZ} is the impurity mass in the extruded sample, g; W_{Xh} is the total mass of extruded samples, g; Y_S is the entrainment loss rate, %; W_W is the grain mass discharged from the tail of the drum, g; W is the grain weight of each group of test extracts, g.

2.3. *Building the ANN Model*

2.3.1. Development of Neural Network Model

One of the most commonly used neural network models is the BP neural network, which utilizes the BP algorithm. Even the most complex nonlinear relationship completely approximates it. The information is dispersed and stored in the neurons of the network. The computation is extremely fast due to parallel processing. Since neural networks are self-learning and adaptive, they can deal with uncertain or unknown systems. This system is excellent when simultaneously processing both quantitative and qualitative information. It can coordinate a wide range of input information relations and is, thus, ideal for fusion and multimedia applications. A well-trained artificial neural network can function as a predictive model for a specific application, which is a data processing system inspired by biological neural systems. The predictive power of an ANN is derived from training on experimental data, which is then validated using independent data. Artificial neural networks can relearn and adapt to improve their performance by updating data availability [27]. The structure and operation of ANNs have been described by numerous authors [28]. The modeling used in feedforward neural networks for prediction was designed to capture the correlation between the historical model inputs and their corresponding outputs. This is accomplished by repeatedly feeding the model examples of input/output relationships and adjusting the model coefficients (i.e., connection weights) to minimize the error function between the historical output and the model-predicted outputs.

This article follows the procedure of the artificial neural network model as described by Maier and Dandy [29]. They include determining model inputs and outputs, dividing and preprocessing available data, selecting an appropriate network architecture, optimizing connection weights (training), setting stopping criteria, and validating the model. A typical algorithm flow diagram is shown in Figure 2. In this work, all calculations and programming were executed in MATLAB (R2016a, 9.0.0.341360). The data used to calibrate and validate the neural network model were obtained from the bench field measurements of the flexible threshing experiment device and the corresponding information on the feeding amount and material characteristics. The data cover a wide range of variation in different operating parameters types and threshing properties. The database comprises a total of 25 individual cases. The statistics of the input and output parameters used for the artificial neural networks are shown in Table 3. Figure 3 is a database of all the threshing performance metrics for the ANN.

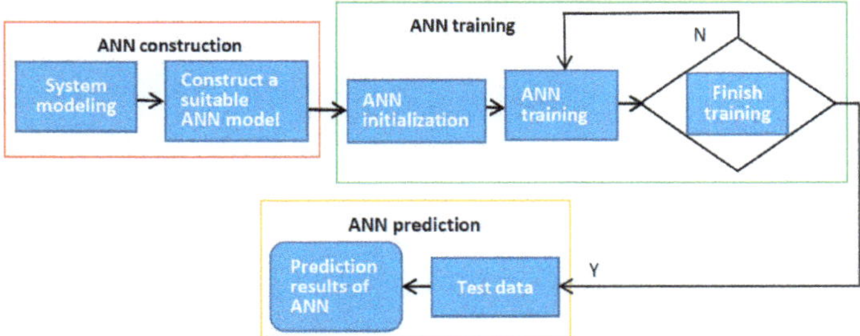

Figure 2. The artificial neural network algorithm flow chart.

Table 3. Statistical criteria of the input parameters and performance attributes (output parameters) used in the ANN model.

Parameters		Statistical Criteria					
		Minimum	Maximum	Average	Standard Deviation	Median	Variance
Training set	inputs	1	800	190.4500	300.8486	25	9.0154×10^4
	outputs	0.0490	1.1960	0.4142	0.4074	0.2070	0.1660
Validation set	inputs	2	800	183.3750	291.1664	30	8.4778×10^4
	outputs	0.0490	1.1960	0.3984	0.4231	0.1920	0.1790
Testing set	inputs	1.5	800	197.7750	311.6720	35	9.7139×10^4
	outputs	0.04093	0.9970	0.3995	0.4093	0.1920	0.1675

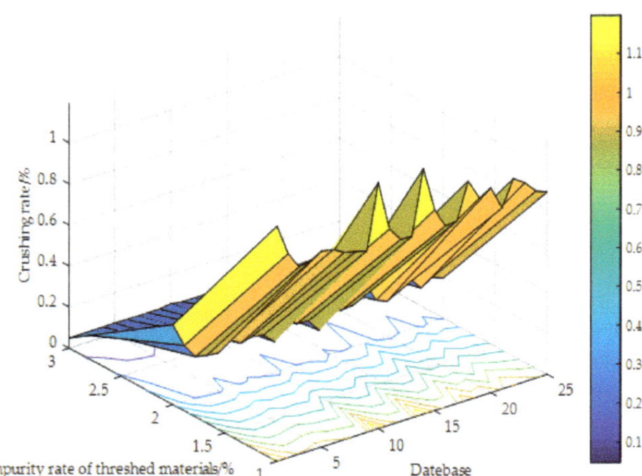

Figure 3. The threshing performance index database for the artificial neural network.

2.3.2. Model Inputs and Outputs

A thorough comprehension of the determinants of threshing performance is required to obtain accurate threshing performance prediction. The rotational speed of the cylinder is closely associated with the performance of the thresher because high-speed results in cracking of the grain, and low-speed leads to unthreshed grain. Moreover, the threshing clearance and separating clearance of the concave sieve significantly influence the threshing performance. Furthermore, the size of the feeding quantity is closely correlated to the threshing characteristics [30].

The primary factors affecting threshing performance include the rotational speed of the cylinder, threshing clearance of the concave sieve, separating clearance of the concave sieve, and feeding quantity. Other factors include total threshing power consumption and grain moisture content that contribute to a lesser degree, thus considered secondary. Grain moisture content was excluded in this work since the tests were conducted under specific moisture content conditions during the harvest period.

The aforementioned factors, i.e., the rotational speed of cylinder (RS), threshing clearance of concave sieve (TC), separating clearance of concave sieve (SC), and feeding quantity (FQ), were introduced to the ANN as the model input variables. On the other hand, the crushing rate (Y_P), impurity rate of threshed materials (Y_Z), and entrainment loss rate (Y_S) were the output variables. Sensitivity analysis was conducted on the trained network to identify the input variables with the most significant impact on threshing performance predictions.

Sensitivity analysis (Figure 4) was based on the validation set, where the input variable was RS, TC, SC, and FQ. To normalize the input variables, the value of the input variable was first changed, the trained network was introduced, the maximum and minimum output values were recorded, the difference between the maximum and the minimum value was computed, the difference to the maximum value was calculated, before finally taking the mean of all ratios as the sensitivity of the classification variable. Lastly, the sensitivity size was compared to establish the sensitivity of each categorical variable to the output variable. The sensitivity analysis results will be discussed later.

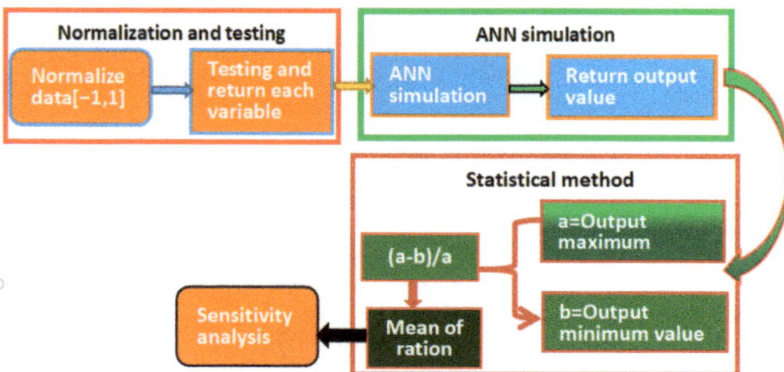

Figure 4. The sensitivity analysis flow.

2.3.3. Data Division and Preprocessing

The database was randomly divided into three sets, i.e., training, testing, and validation. A training set was used to construct the neural network model, whereas an independent validation set was used to estimate model performance in the deployed environment [31]. In total, 60% of the data were used for training, 20% for testing, and 20% for validation. Table 4 shows the orthogonal test for different levels as well as the data ranges used for the ANN model variables.

Table 4. Orthogonal test factor level table used for Artificial Neural Network Model Variables.

Factors / Levels	Rotational Speed of Cylinder, RS (r/min)	Threshing Clearance of Concave Sieve, TC (mm)	Separating Clearance of Concave Sieve, SC (mm)	Feeding Quantity, FQ (kg/s)
1	600	15	15	1.0
2	650	20	25	1.5
3	700	25	35	2.0
4	750	30	45	2.5
5	800	35	55	3.0

Notably, it is critical to preprocess the data into an appropriate format before applying it to the ANN. Preprocessing the data by scaling is crucial in ensuring that all variables receive equal attention during training. The output variables must be scaled to commensurate with the limits of the transfer functions used in the output layer. Although scaling the input variables is not necessary, it is often recommended [32]. Here, the input and output variables were scaled between −1.0 and 1.0, as the purelin sigmoidal transfer function was used in the output layer.

2.3.4. Model Architecture

Determining the network architecture is one of the crucial and challenging tasks in the development of ANN models because it requires the selection of several hidden layers and the number of nodes in each of these.

The number of model inputs and outputs restricts the number of nodes in the input and output layers. The input layer of the ANN model developed in this work had four nodes, one for each of the model inputs (i.e., a rotational speed of cylinder (RS), threshing clearance of concave sieve (TC), separating clearance of concave sieve (SC), feeding quantity (FQ)). On the other hand, the output layer had three nodes (i.e., crushing rate (Y_P), impurity rate of threshed materials (Y_Z), and entrainment loss rate (Y_S)) representing the measured value of threshing performance.

Figure 5 shows the basic elements of an artificial neuron. Artificial neurons mainly comprise weight bias and activation functions. The BP neural network is the most popular and widely used artificial neural network architecture [33]. It involves an input layer, one or more hidden layers, and an output layer. Evidence suggests that a network with a threshold, at least one S-shaped hidden layer, and a linear input layer can approximate any rational number [34]. Mathematical expressions and interpretations of artificial neural networks can be referred to in reference [35].

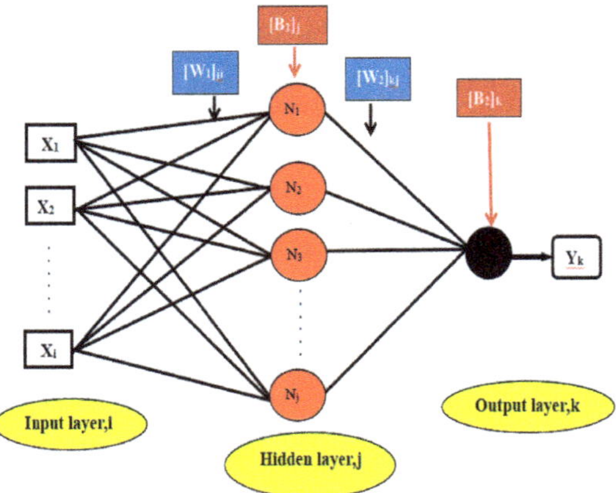

Figure 5. Schematic diagram of an artificial neural network.

The activation function introduces nonlinearity into the neural network, making it more powerful than the linear transformation. The Levenberg–Marquardt algorithm is the most commonly used multi-layer perception training algorithm. It is a gradient descent technique [36] used to reduce the error of specific training patterns. The network was built using the Levenberg–Marquardt backpropagation technique. Tansig is a common nonlinear activation function for nodes in the hidden layer. Figure 6 depicts the architecture of the artificial neural network system described in this paper. W is a weight matrix for the hidden and output layers, and N_{ij} is a node that computes a weighted sum of its inputs and passes the sum through a soft nonlinearity or activity function.

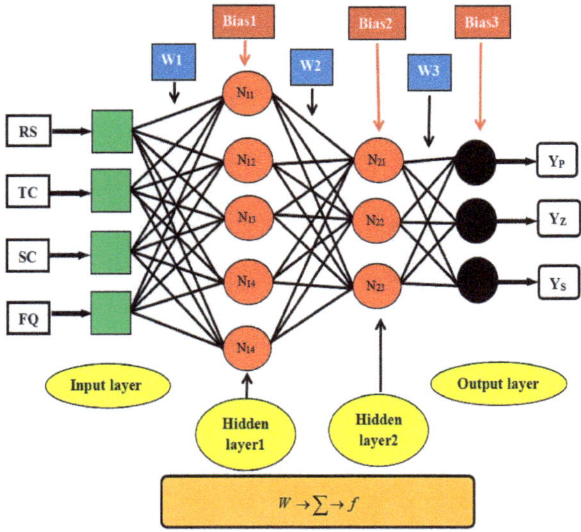

Figure 6. The artificial neural network architecture of threshing performance.

2.3.5. Weight Optimization

"Training" or "learning" is the process of optimizing the connection weights. The goal is to identify a global solution to what is typically a highly nonlinear optimization problem. The method most commonly used for finding the optimum weight combination for feedforward neural networks is the backpropagation algorithm [37], which is based on first-order gradient descent. Feedforward networks trained with the backpropagation algorithm have been applied successfully to numerous agricultural engineering problems [38,39], hence, also used in this work.

2.3.6. Stopping Criteria and ANN Model Validation

The following are conditions for the neural network to stop: 1. Meet the accuracy requirements; 2. Complete the maximum number of iterations.

Backpropagation works by minimizing a cost function. The mean squared error (MSE) is the most common cost function.

Validation data were used to validate the performance of the trained model once the training phase of the model was completed. Additionally, the validation set was used to determine the optimum number of hidden layer nodes and the optimum internal parameters (learning rate, momentum, and initial weights). The MSE was used to validate the performance of the ANN in terms of the different number of hidden layer nodes according to Equation (4).

$$MSE = \frac{\sum_{i=1}^{m} (y_i - \hat{y}_i)^2}{m} \tag{4}$$

The evaluation parameters metrics of root mean square error ($RMSE$) [40], correlation coefficient (R), and mean absolute error (MAE) were utilized to assess the performance of the models by comparing the target and output values of networks.

$$RMSE = \sqrt{\frac{\sum_{i=1}^{m} (y_i - \hat{y}_i)^2}{m}} \tag{5}$$

$$R = \sqrt{\frac{\left(\sum_{i=1}^{m} (y_i - \overline{y})(\hat{y}_i - \overline{\hat{y}})\right)^2}{\sum_{i=1}^{m} (y_i - \overline{y})^2 \bullet \sum_{i=1}^{m} (\hat{y}_i - \overline{\hat{y}})^2}} \tag{6}$$

$$MAE = \frac{1}{m}\sum_{i=1}^{m} |\mathrm{y_i} - \hat{y}_i| \tag{7}$$

The $RMSE$, R, and MAE values were calculated in all stages: training; validating; and testing. Where y_i, $\hat{y}_i$ are the observed value and predicted values, $\overline{y}$, $\overline{\hat{y}}$ are the average observed and predicted values, and m is the total number of points in each dataset, respectively. Using this parameter aids in selecting the best structure and network and provides the possibility of understanding the proximity of the model.

After model construction, the variable parameters of the experimental trials were entered as the new input model, and the actual results were compared with the model. Microsoft Excel 2016 software was used to analyze the correlation coefficient between the actual results and the output of the neural network model.

3. Results

3.1. Evaluation of the Number of Hidden Layer Nodes

The BP network has a varied number of nodes in the hidden layer, and the hidden nodes affect the error of the output connected neurons [41]. If the number of neurons in the hidden layer is too small, the network's ability to learn is limited, resulting in the need for more training to decrease its fault tolerance. On the other hand, network iterations will increase with too many neurons, thereby extending the training time of the network, and reducing the generalization capacity of the network, resulting in a decrease in predictive ability. The optimal number of nodes needs to be explored to confirm the effect of different nodes on network performance. In practical situations, the number of nodes in the hidden layer is selected by first determining the approximate range of the number of nodes using the empirical formula before using the step-wise test strategy to establish the best number of nodes with the smallest error by training and comparing the networks with different neurons. The best number of hidden layer nodes can be derived from the following formula [42,43]:

$$l = \sqrt{(m+n)} + a \tag{8}$$

where l represents the number of neurons in the hidden layer, n denotes the number of neurons in the input layer, m is the number of neurons in the output layer, a is the constant, and $1 < a < 10$. According to this formula, the value range of the hidden layer nodes of the network was 4–12, and the performance of the artificial neural network under different numbers of nodes is shown in Figure 7. When the number of hidden layers was 5, the minimum MSE was 0.00080796, indicating superior model performance.

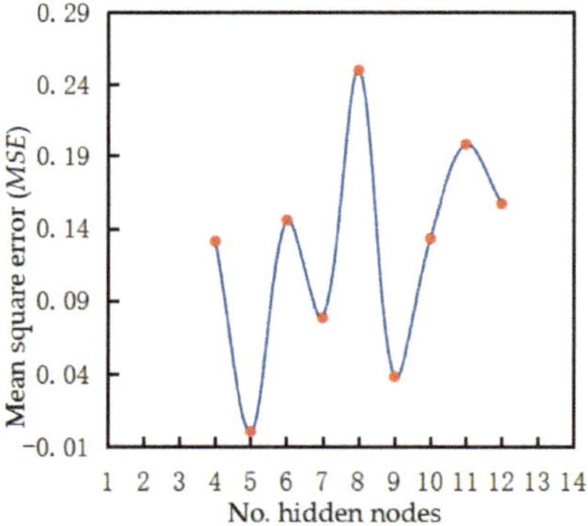

Figure 7. Performance of artificial neural network models with different hidden layer nodes (learning rate = 0.1 and training goal = 0.001).

Table 5 summarizes the predictive performance of the optimal neural network. The findings showed a validation set of $R = 0.979$, $RMSE$ of 0.138, and MAE of 0.153. The ANN model with a 4-5-3-3 structure performed effectively. Table 5 further shows the results of the model, which were generally consistent with those obtained during training and testing, indicating that the model can generalize within the range of data used for training.

Table 5. Artificial Neural Network Results.

Dataset	*R*	*RMSE*	*MAE*
Training set	0.97596	0.079148	0.14100
Validation set	0.97981	0.13823	0.15260
Testing set	0.99041	0.086466	0.13543

Based on the data shown in Figure 8, the error curves of the model training sample, the corrected sample, and the test sample were well correlated. The curve trend slowly decreased, indicating that the network was trained on the training data. To avoid overfitting with the validation data, the MSE between the initial fitting and validation will become smaller and smaller, but as the network begins to overfit the training data, the *MSE* will become larger. In the default setting, the training ends when the validation error is added six consecutive times, and the best performance is obtained from the lowest validation error period (drawing circle). Finally, the obtained best artificial neural network parameters are shown in Table 6. Figure 9 shows the training state of the model training phase.

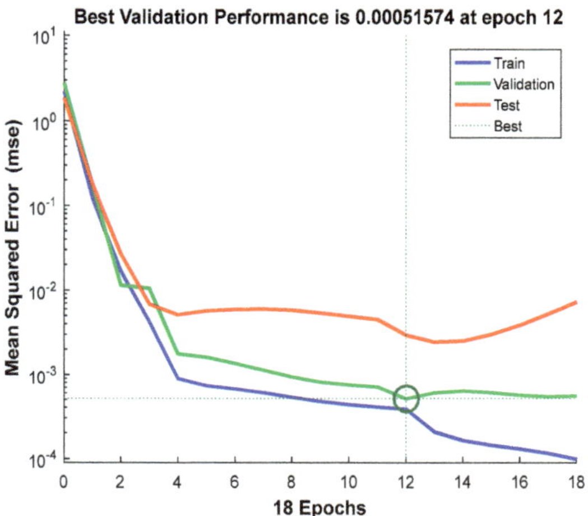

Figure 8. The neural network training performance (epoch 18, validation stop).

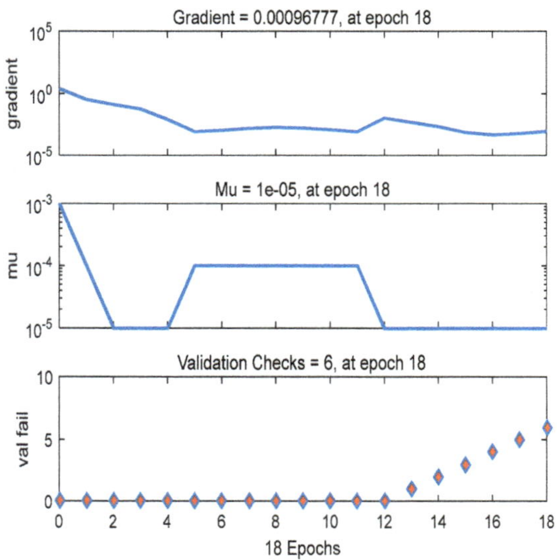

Figure 9. The neural network training state (epoch 18, validation stop).

Table 6. Optimum ANN parameters for the design of the model.

Sr.No.	Parameter	Description
1	No. of input nodes	Varying from 1 to 25 in the cascaded training procedure
2	No. of output nodes	3
3	No. of hidden layers	2
4	No. of neurons in the hidden layer (Hn)	5–3
5	Training rule	Levenberg-Marquardt (LM)
6	Activation function	Sigmoid
7	Network type	Feed-forward (FF)
8	Training method	Backpropagation algorithm

3.2. Evaluation of Prediction Results

The regression curves for assessing the accuracy of the ANN estimation are shown in Figure 10. Estimates of the threshing performance of the ANN were evaluated by regression analysis between the predicted and experimental data. To validate the ANN model, we applied the estimation and regression methods. The regression value for the threshing characteristics was calculated as 0.9525. Figure 10 displays the optimal curve resulting from multiple iterations of the R^2 curve.

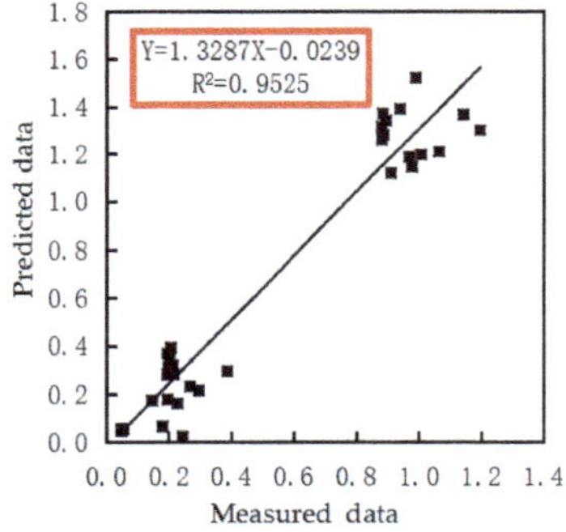

Figure 10. The predicted and the measured values of threshing performance.

3.3. Sensitivity Analysis

Sensitivity analysis was performed to examine the sensitivity of the various factors influencing the threshing characteristics, and Table 7 shows the effects of the various input factors. As shown, different input variable values, i.e., the size of different sensitivity, reflected the effect of the input variable on the output variable. RS affected the predicted threshing performance when the network values had distinct input variables. However, the relative importance of the remaining input variables varied based on changes in the input variables. RS was the most important input in all trials followed by TC, SC, and FQ. Sensitivity analysis revealed that RS, TC, and SC were the most vital factors affecting threshing performance, with an average relative importance of 15.00%, 14.89%, and 14.32%, respectively. The results further showed that FQ had a minimal effect on threshing performance, with an average relative importance of 11.65%.

Table 7. Sensitivity Analyses of the Relative Importance of Artificial Neural Network Input Variables.

Trial No.	Relative Importance for Input Variables			
	RS	TC	SC	FQ
1	0.2417	0.1083	0.1208	0.1013
2	0.1831	0.1575	0.1468	0.086
3	0.1426	0.1278	0.1892	0.0828
4	0.0507	0.2191	0.1274	0.1805
Average	0.1500	0.1489	0.1432	0.1165

4. Discussion

Threshing is one of the most critical operations of combine harvesters during grain production, which is a complex, nonlinear, multi-parameter physical process. The working performance index of the threshing device has a significant on the separation, cleaning, and other parts and the working quality of the whole machine and has always been one of the main concerns of the engineered design. A flexible threshing device has the advantage of reducing the crushing rate of rice grain. Therefore, a comprehensive and accurate design of a flexible threshing performance evaluation model has important theoretical value and practical significance. In this study, the BP artificial neural network was used to model the threshing performance factors based on four factors: RS, TC, SC, and FQ. Determining the optimal network architecture is related to the number of hidden layers and neurons. The optimum network geometry was found to be 4-5-3-3 by evaluating different number of hidden layer nodes in this study. The performance of the ANN model was verified by comparing the predicted dataset with the experimental results (measured data). The sensitivity analysis performed for the described ANN model indicated that four working variables of the flexible threshing device had the greatest contribution to threshing performance attributes compared to FQ. These results can guide the optimal design of a flexible threshing cylinder to achieve the maximum performance of the device.

5. Conclusions

This study analyzed different numbers of hidden layer nodes and found that when the number of hidden layer nodes was five, the minimum MSE was 0.00080796, indicating that the model performed well. The results indicated that backpropagation neural networks could predict the threshing performance of the flexible threshing device with an acceptable degree of accuracy ($R = 0.980$, $RMSE = 0.138$, $MAE = 0.153$). The built neural network model prediction predicted the performance of the flexible threshing device well. The regression determination coefficient R^2 between the predicted and experimental data was 0.953, indicating that the predicted data of the built neural network model was in good agreement with the experimental data. The ANN method is an effective method for predicting the threshing performance of flexible threshing devices in rice. The established artificial neural network model exhibited stable prediction of the threshing performance of the flexible threshing device during operation. The sensitivity analysis revealed that RS, TC, and SC are important factors affecting the performance of the flexible threshing device, with an average relative importance of 15.00%, 14.89%, and 14.32%, respectively. FQ had the least impact on threshing performance, with an average threshing relative importance of 11.65%. These results can guide the optimal design of flexible threshing cylinders and improve the performance of the flexible threshing device.

Author Contributions: Conceptualization, L.M. and F.X.; methodology, L.M.; software, L.M.; validation, L.M., F.X. and D.L.; formal analysis, L.M.; investigation, D.L.; resources, X.W. and Z.Z.; data curation, X.W. and D.L.; writing—original draft preparation, L.M.; writing—review and editing, L.M., F.X. and Z.Z.; visualization, L.M.; supervision, F.X.; project administration, D.L.; funding acquisition, F.X. All authors have read and agreed to the published version of the manuscript.

Funding: This research was funded by Hunan High-Tech Industry Technology Leading Plan Project (Science and Technology Research category) (2020NK2002); Hunan Agricultural Machinery Equipment and Technology Innovation Research and Development Project (Xiangcai Agricultural Index (2021) No.47); and Hunan Agricultural Machinery Equipment and Technology Innovation Research and Development Project (Xiangcai Agricultural Index [2020] No.107).

Institutional Review Board Statement: Not applicable.

Informed Consent Statement: Not applicable.

Data Availability Statement: The data presented in this study are available on request from the corresponding author.

Conflicts of Interest: The authors declare no conflict of interest.

References

1. Han, B.; Meng, F.C.; Liang, L.N.; Yang, Y.N.; Wang, B. Numerical simulation and experiment on performance of supplying seeds mechanism of directional precision seeding device for japonica rice. *Trans. CSAE* **2016**, *32*, 22–29. (In Chinese)
2. Craessaerts, G.; Saeys, W.; Missotten, B.; De Baerdemaeker, J. A genetic input selection methodology for identification of the cleaning process on a combine harvester. Part I: Selection of relevant input variables for identification of the sieve losses. *Biosyst. Eng.* **2007**, *98*, 166–175. [CrossRef]
3. Maertens, K.; De Baerdemaeker, J. Design of a virtual combine harvester. *Math. Comput. Simul.* **2004**, *65*, 49–57. [CrossRef]
4. Liang, Z.W.; Li, Y.M.; Baerdemaeker, J.; Xu, L.Z.; Saeys, W. Development and testing of a multi-duct cleaning device for tangential-longitudinal flow rice combine harvesters. *Biosyst. Eng.* **2019**, *182*, 95–106. [CrossRef]
5. Shi, Q.X.; Liu, S.D.; Ji, J.T.; Fu, R.X.; Ni, C.G. Research on seed-control feed and soft threshing for rice. *Trans. Chin. Soc. Agric. Mach.* **1996**, *27*, 41–46. (In Chinese)
6. Qian, Z.J.; Jin, C.Q.; Zhang, D. Multiple frictional impact dynamics of threshing process between flexible tooth and grain kernel. *Comput. Electron. Agric.* **2017**, *141*, 276–285. [CrossRef]
7. Keller, D.L. Corn kernel damage due to high velocity impact. *Trans. ASAE* **1972**, *15*, 330–332. [CrossRef]
8. Shi, Q.X.; Liu, S.D.; Ji, J.T.; Fu, R.X.; Ni, C.G. Studies on the mechanism of speed-controlled feeding and soft threshing. *Trans. CSAE* **1996**, *12*, 177–180. (In Chinese)
9. Xie, F.P.; Luo, X.W.; Lu, X.Y.; Sun, S.L.; Ren, S.G.; Tang, C.Z. Threshing principle of flexible pole-teeth roller for paddy rice. *Trans. CSAE* **2009**, *25*, 110–114. (In Chinese)
10. Ren, S.G.; Xie, F.P.; Luo, X.W.; Sun, S.L. Analysis and test of power consumption in paddy threshing using flexible and rigid teeth. *Trans. CSAE* **2013**, *29*, 12–18. (In Chinese)
11. Su, Y.; Liu, H.; Xu, Y.; Cui, T.; Qu, Z.; Zhang, D.X. Optimization and experiment of spike-tooth elements of axial flow corn threshing device. *Trans. Chin. Soc. Agric. Mach.* **2018**, *49*, 258–265. (In Chinese)
12. Geng, D.Y.; He, K.; Wang, Q.; Jin, C.Q.; Zhang, G.H.; Liu, X.F. Design and experiment on transverse axial flow flexible threshing device for corn. *Trans. Chin. Soc. Agric. Mach.* **2019**, *50*, 101–108. (In Chinese)
13. Li, Y.B.; Jiang, J.J.; Xu, Y.; Cui, T.; Su, Y.; Qiao, M.M. Preparation and threshing performance tests of rubber composite nail teeth under maize with high moisture content. *Trans. Chin. Soc. Agric. Mach.* **2020**, *51*, 158–167. (In Chinese)
14. Fu, J.; Zhang, Y.C.; Cheng, C.; Chen, Z.; Tang, X.L.; Ren, L.Q. Design and experiment of bow tooth of rigid flexible coupling for wheat threshing. *J. Jilin Univ. (Eng. Technol. Ed.)* **2020**, *50*, 730–738. (In Chinese)
15. Qin, Z.J.; Jin, C.Q.; Yuang, W.S.; Ni, Y.L.; Zhang, G.Y. Frictional impact dynamics model of threshing process between flexible teeth and grains. *J. Jilin Univ. (Eng. Technol. Ed.)* **2021**, *51*, 1121–1130. (In Chinese)
16. Safa, M.; Samarasinghe, S. Determination and modelling of energy consumption in wheat production using neural networks: "A case study in Canterbury province, New Zealand". *Energy* **2011**, *36*, 5140–5147. [CrossRef]
17. Hertz, J. Introduction to the theory of neural computation. *Am. J. Phys.* **1994**, *62*, 668. [CrossRef]
18. Mirzazadeh, A.; Abdollahpour, H.; Mahmoudi, A.; Bukat, A.R. Intelligent modeling of material separation in combine harvester's thresher by ANN. *Int. J. Agric. Crop Sci.* **2012**, *4*, 1767–1777.
19. Gundoshmian, T.M.; Ardabili, S.; Mosavi, A.; Várkonyi-Kóczy, A.R. Prediction of combine harvester performance using hybrid machine learning modeling and response surface methodology. In *Engineering for Sustainable Future, Proceedings of the 18th International Conference on Global Research and Education Inter-Academia–2019, Budapest, Hungary, 4–7 September 2019*; Springer International Publishing: New York, NY, USA, 2020; pp. 345–360.
20. Hornik, K.; Stinchcome, M.; White, H. Multilayer feedford networks are universal approximaters. *Neural Netw.* **1989**, *2*, 359–379. [CrossRef]
21. Therneau, T.M.; Grambsch, P.M.; Fleming, T.R. Martingale-based residuals for survival models. *Biometrika* **1990**, *77*, 147–160. [CrossRef]
22. Jahirul, M.; Rahman, S.; Masjuki, H.; Kalam, M.; Rashid, M. Application of artificial neural networks (ANN) for prediction the performance of a dual fuel internal combustion engine. *HKIE Trans.* **2009**, *16*, 14–20. [CrossRef]

23. Cirak, B.; Demirtas, S. An application of artificial neural network for predicting engine torque in a biodiesel engine. *Am. J. Energy Res.* **2014**, *2*, 74–80. [CrossRef]
24. Shpokas, L. Research of Grain Damage Caused by High-Performance Combine Harvesters. *Mot. Power Ind. Agric.* **2007**, *9*, 168–177.
25. Maertens, K.; Ramon, H.; De Baerdemaeker, J. An on-the-go monitoring algorithm for separation processes in combine harvesters. *Comput. Electron. Agric.* **2004**, *43*, 197–207. [CrossRef]
26. Miu, P.I.; Kutzbach, H.D. Modeling and simulation of grain threshing and separation in threshing units-Part I. *Comput. Electron. Agric.* **2008**, *60*, 96–104. [CrossRef]
27. Hertz, J.A.; Krogh, A.S.; Palmer, R.G. *Introduction to the Theory of Neural Computation*; Addison-Wesley Publishing Company: Boston, MA, USA, 1991; pp. 115–156.
28. Fausett, L.V. *Fundamentals of Neural Networks: Architectures, Algorithms, and Applications*; Prentice Hall: Upper Saddle River, NJ, USA, 1994.
29. Maier, H.R.; Dandy, G.C. *Applications of Artificial Neural Networks to Forecasting of Surface Water Quality Variables: Issues, Applications and Challenges*; Artificial Neural Networks in, Hydrology; Govindaraju, R.S., Rao, A.R., Eds.; Kluwer: Dordrecht, The Netherlands, 2000; pp. 287–309.
30. Špokas, L.; Steponavičius, D.; Petkevičius, S. Impact of technological parameters of threshing apparatus on grain damage. *Agron. Res.* **2008**, *6*, 367–376.
31. Twomey, J.M.; Smith, A.E. *Validation and Verification Artificial Neural Networks for Civil Engineers: Fundamentals and Applications*; Kartam, N., Flood, I., Garrett, J.H., Eds.; ASCE: New York, NY, USA, 1997; pp. 44–64.
32. Masters, T. *Practical Neural Network Recipes in C++*; Academic Press Professional, Inc.: San Diego, CA, USA, 1993.
33. Rumelhart, D.E.; Hinton, G.E.; Williams, R.J. Learning representations by back propagating errors. *Nature* **1986**, *323*, 533–536. [CrossRef]
34. Han, L.Q. *Theory, Design, and Application of Artificial Neural Networks*, 2nd ed.; Chemical Industry Press: Beijing, China, 2007.
35. Haykin, S. *Neural Networks: A Comprehensive Foundation*; Prentice Hall International Inc.: Upper Saddle River, NJ, USA, 1999.
36. Eyercioglu, O.; Kanca, E.; Pala, M.; Ozbay, E. Prediction of martensite and austenite start temperatures of the Fe-based shape memory alloys by artificial neural networks. *J. Mater. Process. Technol.* **2008**, *200*, 146–152. [CrossRef]
37. Rumelhart, D.E.; Hinton, G.E.; Williams, R.J. *Learning Internal Representation by Error Propagation*; Parallel Distributed Processing; Rumelhart, D.E., McClelland, J.L., Eds.; MIT Press: Cambridge, MA, USA, 1986; Volume 1, Chapter 8.
38. Zhao, Z.; Qin, F.; Tian, C.J.; Yang, S.X. Monitoring method of total seed mass in a vibrating tray using artificial neural network. *Sensors* **2018**, *18*, 3659. [CrossRef]
39. Kosari-Moghaddam, A.; Rohani, A.; Kosari-Moghaddam, L.; Esmaeipour-Troujeni, M. Developing a Radial Basis Function Neural Networks to Predict the Working Days for Tillage Operation in Crop Production. *Int. J. Agric. Manag. Dev. (IJAMD)* **2018**, *9*, 119–133.
40. Al-Dosary, N.M.N.; Aboukarima, A.M.; Al-Hamed, S.A. Evaluation of Artificial Neural Network to Model Performance Attributes of a Mechanization Unit (Tractor-Chisel Plow) under Different Working Variables. *Agriculture* **2022**, *12*, 840. [CrossRef]
41. Jinchuan, K.; Xinzhe, L. Empirical analysis of optimal hidden neurons in neural network modeling for stock prediction. In Proceedings of the Pacific-Asia Workshop on Computational Intelligence and Industrial Application, Wuhan, China, 19–20 December 2008; pp. 828–832.
42. Feisi Technology Product Research and Development Center. *Neural Network Theory and the MATLAB7 Implementation*; Electronic Industry Press: Beijing, China, 2005.
43. Fu, H.X.; Zhao, H. *MATLAB Neural Network Application Design*; Machinery Press: Beijing, China, 2010; pp. 83–97.

Article

Experimental Research for Digging and Inverting of Upright Peanuts by Digger-Inverter

Haiyang Shen [1,2], Hongguang Yang [1], Qimin Gao [1,2], Fengwei Gu [1], Zhichao Hu [1,2,*], Feng Wu [1], Youqing Chen [1] and Mingzhu Cao [1]

1 Nanjing Institute of Agricultural Mechanization, Ministry of Agriculture and Rural Affairs, Nanjing 210014, China; 82101221060@caas.cn (H.S.); yanghongguang@caas.cn (H.Y.); qmgao93@163.com (Q.G.); wufeng@caas.cn (F.W.); chenyouqing@caas.cn (Y.C.); caomingzhu@caas.cn (M.C.)
2 Graduate School of Chinese Academy of Agricultural Sciences, Beijing 100081, China
* Correspondence: huzhichao@caas.cn

Abstract: In order to quickly dry the peanut pods and effectively reduce the mildew on peanut pods in rainy weather, this paper analyzed the research status of peanut digging and inverting technology in China and abroad, combined with the peanut two-stage harvesting mode. The orthogonal experiment was carried out by using the peanut digger-inverter to study the different forms of upright type, taking the rate of vines inverting, the rate of buried pods, and the rate of fallen pods as the evaluation indexes, and taking the traveling speed of the tractor A, the line speed of the conveyor chain B, and the line speed of the inverting roller C as the experimental factors. The results showed that, in the states of unpressed vines and pressed vines, the order of influence of the peanut digger-inverter on the evaluation indices was A > C > B. The optimal horizontal combination of unpressed vines was $A_2C_3B_2$, when the traveling speed of the tractor is 1.06 m/s, the line speed of the inverting roller is 2.12 m/s, and the line speed of the conveyor chain is 1.02 m/s; the rate of vines inverting was 71.07%, the rate of buried pods was 0.2%, and the rate of fallen pods was 0.22%. Under the condition of vines pressing, the best horizontal combination is $A_2C_2B_2$, when the travelling speed of the tractor is 1.01 m/s, the line speed of the inverting roller is 1.88 m/s, and the line speed of the conveyor chain is 1.02 m/s; the rate of vines inverting was 74.29%, the rate of buried pods was 0.14%, and the rate of fallen pods was 0.33%. The paired *t*-test was carried out by the peanut digger-inverter under the state of pressed and unpressed vines. There was little difference in the influence of each factor on the rate of fallen pods and the rate of buried pods, but there was a significant difference in the influence on the rate of vines inverting. The rate of inversion of vines under the state of pressed vines was higher than that under the state of unpressed vines. The research results will provide certain technical support for the late optimization of the peanut digger-inverter and create a hardware foundation for the intelligence and information harvesting of peanuts.

Keywords: agricultural machinery; peanuts; digger-inverter; orthogonal experiment; paired *t*-test; unpressed vines; pressed vines

Citation: Shen, H.; Yang, H.; Gao, Q.; Gu, F.; Hu, Z.; Wu, F.; Chen, Y.; Cao, M. Experimental Research for Digging and Inverting of Upright Peanuts by Digger-Inverter. *Agriculture* **2023**, *13*, 847. https://doi.org/10.3390/agriculture13040847

Academic Editor: Alessio Cappelli

Received: 16 March 2023
Revised: 31 March 2023
Accepted: 6 April 2023
Published: 10 April 2023

1. Introduction

Peanut (*Arachis hypogea* L.) is an annual leguminous herb in the rose order, also known as "everlasting pod", "mud-bean", "crocus bean", and so on [1]. Peanut is the fourth largest oil crop in the world [2–4], and it is also one of the most vital oil crops and economic crops that play a crucial role in ensuring domestic edible oil supply and diversifying food consumption in China [5–7]. According to the statistics of the Food and Agriculture Organization (FAO) of the United Nations and the National Bureau of Statistics of China, in 2020, the planting area for peanuts in the world was 3.16×10^7 ha with a total yield of 5.36×10^7 t, and the planting area for peanuts in China was 4.62×10^6 ha with a total yield of 1.81×10^7 t. The area and total yield were ranked second and first in the world, accounting for 14.63% and 33.64% of the global peanut area and yield, respectively [8,9].

The research and development of Chinese peanut production machinery began in the 1960s. After half a century of effort, all the links in peanut mechanization production have made great progress [10–13]. In 2016, the mechanization rates of the three main links of peanut cultivation, sowing, and harvesting were 72.61%, 43.1%, and 33.91%, respectively, and the comprehensive mechanization rate was 52.14% [14]. At present, there are two main operating modes in China's peanut harvesting machinery, which are two-stage harvest and combine harvest [15]. When the two-stage harvester operates, the digging harvester places peanut plant pods in the field facing up, which is conducive to the rapid drying of pods, and it can effectively reduce the occurrence of peanut pod mold in rainy weather [16,17]. Therefore, whether it is abroad or in China, there is a demand for peanut digging harvesting techniques with vines inverting functions. At present, most of the peanut-digging harvesting machines in countries such as the United States have the function of inverting the vines [18]. In China, the two-stage rewarding areas are mainly produced in Huanghuaihai Northeast and Northwest, while the peanuts of Huanghuaihai Marine District are also harvested. It is often harvested in the same season, and the technical demand for digging and harvesting with the function of the vines inverting is the most urgent.

The peanut digger-inverter mainly has an inverting wheel and a curve-type inverting rod mechanism, which can realize the peanut plant pod facing up and orderly inverting laying. In the United States, there are five major agricultural machinery manufacturing companies, including Armadas Industries (AMADAS), Suffolk, VA, USA; Kelley Manufacturing Company (KMC), Tifton, GA, USA; Colombo North America Inc. (COLOMBO), Adel, GA, USA; Ferguson Manufacturing Company (FERGUSON), Suffolk, VA, USA; and Pearman Corporation (PEARMAN), Tifton, GA, USA. They have produced peanut harvesters with a peanut vine inverting function, and their vines inverting mechanisms are similar in structure and have been widely used in the world, but it does not apply to Chinese upright peanuts [19–23]. For example, in 2011, the relevant machine from KMC Company in the United States was introduced into Xinjiang. The peanut varieties planted in Xinjiang are upright and mulched, resulting in poor vine inversion, serious mulching film entanglement, high pod fall and loss rates, and other problems [24]. In combination with the common planting of upright peanut varieties in the main producing areas of China, they generally have the characteristics of an upright and compact plant type, few branches, and a higher main stem. Zhengzhou Xechuang Mechanical and Electrical Equipment Co., Ltd., Zhengzhou, China, independently developed Hongtian 4HS-2 peanut harvester; Heishan County Jianguo Agricultural Machinery Machinery Co., Ltd., Suzhou, China, independently developed Jianguo brand 2H-1 peanut harvester; Shandong Jiarun Heavy Industry Machinery Co., Ltd., Linyi, China, independently developed supply chain peanut harvester, Ministry of Agriculture and Rural Affairs of Nanjing Institute of Agricultural Mechanization independently developed elevator chain peanut harvester; peanut digging and harvesting machines with neat laying functions have been widely used in Shandong, the Northeast, and other major peanut producing areas [25–27]. These machines can only achieve the orderly laying of peanuts and cannot prevent the vines from inverting. On the basis of referring to the harvesting technology of the United States, China has also developed some similar peanut digging and harvesting machines with the function of inverting vines. For example, Mechanical Equipment Research Institute of Xinjiang Academy of Agricultural Reclamation Sciences [28], Zhengyang Chuangxin Machinery Co., Ltd., Zhumadian, China [29], Henan Shifeng Machinery Manufacturing Co., Ltd., Luoyang, China [30], Xinjiang Agricultural University [31], Shenyang Agricultural University [32], Henan Polytechnic University [16], etc., these research results are only reflected in literature and have not been applied in real production practice. Therefore, the optimization of the working parameters of the peanut digger-inverter is helpful to promote the rapid industrialization of the machine.

In view of the current research status of peanut digging and inverting technology in China and abroad, in order to quickly dry the pods and effectively reduce the peanut pod sticking mildew in rainy weather, this paper, combined with the peanut two-stage harvesting mode, used the peanut digger-inverter to conduct experimental research on upright type peanuts under two conditions of unpressed vines and pressed vines, and found the parameters of the peanut digger-inverter suitable for Chinese peanut varieties. It will provide certain technical support for the later design of the peanut digger-inverter and create the hardware foundation for peanut intelligence and information harvesting.

2. Materials and Methods

2.1. Agronomic Process Based on Peanut Two-Stage Harvesting Operation Mode

Two-stage harvesting is the operation process of using more than two machines to complete the digging of peanuts, the separation of pod soil, laying and drying, picking up pods, and other working conditions. There are mainly three operation modes: first, after the peanuts are dug and laid by the digging and harvesting machine, they are artificially dried in the field, and finally they are collected and harvested together, namely "Digging–Separation of soil from peanut pods–Laying–Field drying–Tiling–Pickup combined harvesting." Second, after the digging and laying of peanuts by the digging and harvesting machine, the peanuts are picked up manually and taken to the drying field and other places for centralized drying. Finally, the dried pods are picked by the pod picking machine, namely: "Digging–Separation of soil from peanut pods–Laying–Manual pickup–Dry in the sun or pile up-Pick the dried pod." Three is the peanut after digging and harvesting by machine in a digging shop, artificial pick-up after the use of machinery, or artificial fresh pod picking, which is "Digging–Separation of soil from peanut pods–Laying–Manual pickup–Pick pod from fresh plants" [33,34]. The agronomic process of a specific peanut under the two-stage operation mode is shown in Figure 1.

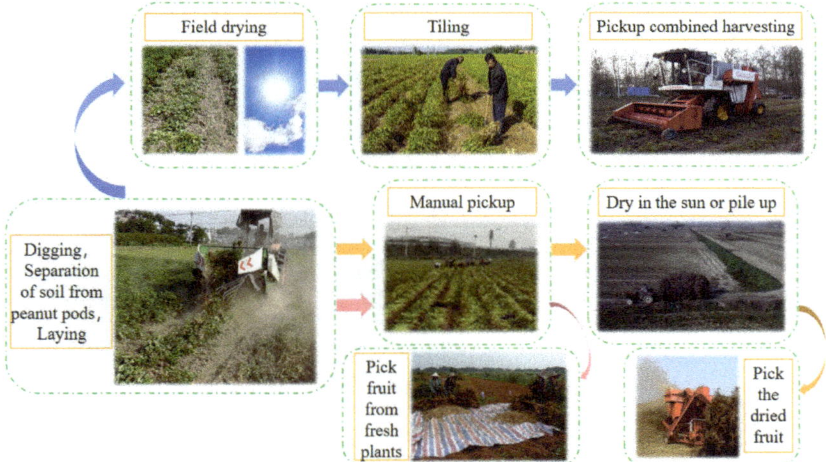

Figure 1. Agronomic process of peanut in two-stage operation mode.

2.2. Design of Overall Structure and Principle

2.2.1. Overall Structure

Nanjing Institute of Agricultural Mechanization, Ministry of Agriculture and Rural Affairs, and Henan Nongyouwang Agricultural Equipment Technology Co., Ltd., Zhumadian, China. have jointly developed a kind of peanut digger-inverter, which is mainly composed of a depth limiting roller, V-shaped digging shovel, vines guide rod, frame, loop conveyor chain, harrow teeth, inverting roller, inverting rod, transmission assembly, folding plate, and rear wheel, etc., with its basic structure shown in Figure 2. The peanut digger-inverter developed can complete the digging and harvesting, soil separation, vine transportation, vine inverting, and laying of peanuts in one go. During the operation of the peanut digger-inverter, driven by the tractor, the digging shovel first breaks ridges and digs, and the soil is separated through the loop conveyor chain, and finally the inverting roller and inverting rod are spread in the field. The main structural parameters and technical parameters of the machine are shown in Table 1.

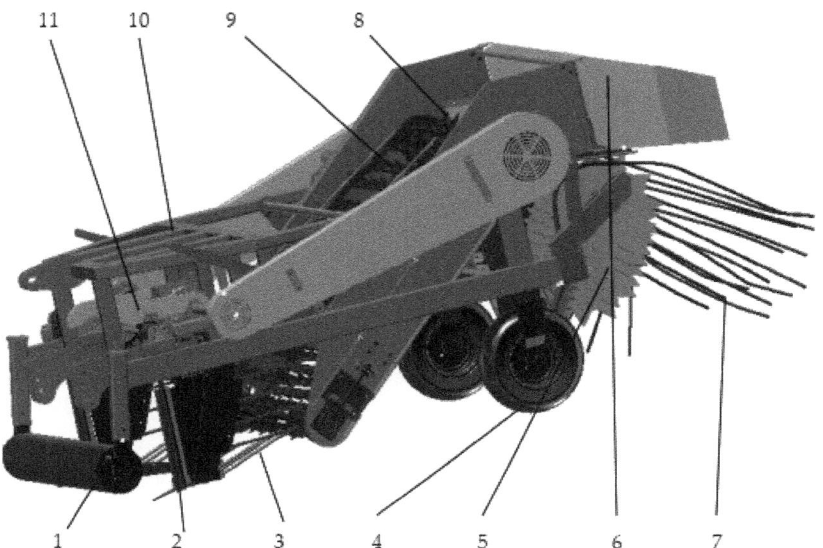

Figure 2. Peanut digger-inverter: 1. Depth limiting roller 2. V-shaped digging shovel 3. Vines guide rod 4. Rear wheel 5. inverting roller 6. Collecting plate 7. Inverting rod 8. Harrow teeth 9. Back loop conveyor chain 10. Frame 11. Transmission assembly.

Table 1. Structural parameters and technical parameters of peanut digger-inverter.

Parameters	Design Value
Type	Suspension-type
Machine dimensions: size (Length × width × height)/(mm × mm × mm)	3500 × 2100 × 1550
Total weight/kg	1500
Most suitable row spacing/mm	800
Numbers of ridge	Double ridge
Working width/mm	1800
Digging depth/mm	≤250
Travelling speed/(m/s)	0.7~1.3
Efficiency/(ha/h)	0.45~0.84

2.2.2. Working Principle

The working principle of the peanut digger-inverter is as follows: With the advance of the machine, the peanut vine moves backward along the vine's guide rod and is supported by the toothed conveying rod. The peanut vines keep the top up and the root down on the vine's guide rod and conveying rod, and the peanut vines continue to be transported backward to the vine inverting roller. Under the rotating action of the vine inverting roller, it is handed over to the vine inverting rod. Under the combined action of inverting rod and inverting plate, the left and right rows of peanuts within the width of a pair of digging and shoveling operations support and cooperate with each other, reaching the state of rooting up.

2.3. Experimental Instruments and Conditions

2.3.1. Experimental Instruments

During the experiment, the instruments needed mainly include: a tractor (Kubota, M954-k), a peanut digger-inverter (as shown in Figure 3), an XJP-02A speed digital display instrument (measuring range 1~9999 r/min, accuracy ±0.02%), a toolbox (including all tools for disassembling parts of the peanut vine inverting and harvesting machine), a tape measure (100 m), an ICS465 electronic platform scale (measuring range: 50 kg, accuracy: 0.02 kg), a marker, label paper, a woven bag, a spade, a multifunctional electronic stopwatch, etc.

Figure 3. Peanut digger-inverter experimental machine.

2.3.2. Experimental Conditions

From 31 July 2022 to 2 August 2022, the peanut digger-inverter experiment was conducted in Baishi Village, Tangjiang Town, Nankang District, and Ganzhou City, Jiangxi Province. In the peanut production experimental base of the Ganzhou Academy of Agricultural Sciences, Jiangxi Province, the experiment of the peanut digger-inverter was carried out. The experimental land size reached 0.5 ha, the terrain was flat, the soil was sticky, and the peanut variety was "Osmanthus 73". This kind of peanut is an upright type. The planting mode was single ridge and double row, and the ridge spacing was 90 cm. During the experiment, the peanut was fully mature, and the average plant height was 54 cm, which met the basic requirements for peanut harvest.

2.4. Experimental Factors, Indicators, and Methods

2.4.1. Experimental Factors

In the field experiment, there are many factors affecting the peanut digger-inverter, among which the traveling speed of the tractor is the most important factor. The traveling speed of the tractor will not only affect the working efficiency of the whole machine but also affect the working effect of the peanuts because the tractor is the most uncontrollable factor in the whole experiment process, which is also the biggest change. At the same time, driving the tractor too fast will lead to poor peanut planting and vine inversion. According to a literature review and calculation, the traveling speed range of tractors is 0.7~1.3 m/s [16,31,32]. The rotation speed of the loop conveyor chain is also a major factor; too fast rotation will lead to peanut straw dispersion, and too slow rotation will lead to peanut straw congestion, so take the line speed of the conveyor chain as an experimental factor. The known rotation radius is 0.08 m, the maximum rotation speed is determined to be 135 r/min, and the minimum rotation speed is 100 r/min. Through a literature review and calculation, it can be known that the line speed of the conveyor chain range is 0.84~1.13 m/s [16,31,32]. In order to ensure the effect of peanut vines inverting, it is required that the absolute velocity of peanut vines landing be zero, so the rotation speed of the peanut inverting roll is a major experimental factor. Given that the radius of the inverting roll is 0.3 m, the maximum rotation speed is determined to be 70 r/min, and the minimum rotation speed is 50 r/min. Through literature review and calculation, it is known that the line speed of the inverting roller range is 1.57~2.12 m/s [16,31,32]. Therefore, the traveling speed of the tractor, the line speed of the conveyor chain, and the line speed of the inverting roller are determined as the main factors affecting the operation index.

The experimental scheme used in this paper is an orthogonal experiment with three factors and three levels. For the three test factors, the traveling speed of the tractor is A, the line speed of the conveyor chain is B, and the line speed of the inverting roller is C. The experimental research is carried out in the two states of pressed and unpressed vines. Experimental factors and levels are shown in Table 2. Orthogonal table L_9 (3^4) is established [35].

Table 2. Test factors and levels.

Factors	Levels		
	1	2	3
Traveling speed of the tractor A/(m/s)	0.7~0.9	0.9~1.1	1.1~1.3
Line speed of the conveyor chain B/(m/s)	0.84	1.02	1.13
Line speed of the inverting roller C/(m/s)	1.57	1.88	2.12

Note: The travelling speed of the tractor is not the same value every time measured, so the values of the three levels are range values.

2.4.2. Experimental Indexes

According to DB34/T534-2022 Anhui Provincial Standard "Technical Specification for Mechanized Peanut Harvesting" [36] and DG/T077-2019 agricultural machinery extension identification outline "Peanut Harvesting" [37], the travelling speed of the tractor and the rate of buried pods, the rate of fallen pods, and the rate of vines inverting of peanuts were measured in the field experiment of the peanut digger-inverter machine.

(1) Measurement of traveling speed of tractor

The length of the measuring area is 30 m, the rated speed of the tractor engine can be ensured (and the speed of the rear power output shaft can be ensured), the suitable working gear is selected for full operation, a stroke is measured, and the time through the measuring area is recorded. The operating speed is calculated according to Formula (1):

$$V = L/T \tag{1}$$

where:

V—The traveling speed of the tractor, expressed in meters per second (m/s);
L—The length of the entire test area, expressed in meters (m);
T—Time to pass through the test area, expressed in seconds (s).

(2) Determination of the rate of buried pods, the rate of fallen pods, and the rate of vines inverting in the peanut digger-inverter

In the measuring area, three equal distances were taken, each of which was 3 m long and had one working width. All the pods falling on the ground and buried in the soil layer (after the removal of naturally falling pods) were collected in each plot and called quality. The rates of buried pods, fallen pods, and vines inverting were calculated according to Formulas (2)–(4), and the average value of the three plots was taken as the evaluation index.

$$P_m = M_m/M_x \times 100\% \tag{2}$$

$$P_s = M_s/M_x \times 100\% \tag{3}$$

$$P_f = l/L \times 100\% \tag{4}$$

where:

P_m—The rate of buried pods, the weight of the pod buried in the soil layer in the plot divided by the total weight of the pod in the whole plot (%);
M_m—The pod mass (excluding naturally fallen pods) buried in the soil layer in the plot, expressed in grams (g);
M_x—The total pod weight of crops in the plot, expressed in grams (g);
P_s—The rate of fallen pods, the mass of pods dropped in the cell divided by the total mass of pods in the whole cell (%);
M_s—Mass of pods dropped in the plot, expressed in grams (g);
P_f—The rate of vines inverting; the number of peanut vines with no pod in the community divided by the total number of peanut vines in the community (%);
l—The number of peanut vines without pods on the ground after peanut harvest in the community; unit is the number of vines;
L—Total number of peanut vines in the community; unit is the number of vines.

2.4.3. Experimental Methods

The self-developed peanut digger-inverter was driven by the tractor to carry out field experiments according to the working width of the machine; each experiment is 2 ridges and 4 rows. The theoretical traveling speed of the tractor was controlled by the gear position, and the actual operating speed was calculated at the end of each measuring area. The line speed of the conveyor chain and the line speed of the inverting roller are controlled by the rotating speed of the conveying shaft, and the orthogonal experiment between unpressed vines and pressed vines was carried out on the designed orthogonal experiment scheme. The range, variance, and comprehensive analysis methods were used to analyze the orthogonal experimental results, and the optimal working parameter combination was obtained [35,38–40]. The paired *t*-test method was used to test the correlation between the two states of pressed vines and unpressed vines, and the best working effect was obtained under the state of pressed vines and unpressed vines [35,38]. The field experiment process and effect are shown in Figure 4.

(**a**)

(**b**)

Figure 4. *Cont.*

(c)

(d)

(e)

Figure 4. Field experiment flow of peanut digger-inverter: (**a**) The process of field experiment of peanut digger-inverter. (**b**) The effect of peanut digger-inverter after field operation. (**c**) Random 3 m sampling of a stroke after peanut operation. (**d**) The sampled peanuts were picked manually. (**e**) The buried pod, fallen pod, and total pod weight of peanuts shall be weighed after operation.

There are two states in the experimental object of the peanut digger-inverter, namely unpressed vines and pressed vines. The unpressed vines refer to the upright peanut growing naturally, while the pressed vines refer to the two ridges of peanut vines falling to the middle of the ridges by artificial or mechanical means to reach the morphology of a trailing peanut. Through the comparison of these two ways of experimenting, it can be seen that the operation effect of the peanut digger-inverter harvester is better under the condition of pressed vines than unpressed vines.

3. Results

3.1. Results of Unpressed Vines Experiment

According to the experimental method, the experimental results of the peanut digger-inverter under the condition of unpressed vines are shown in Table 3, and the range analysis and variance analysis are shown in Figure 5 and Table 4.

Table 3. Experimental design and results of unpressed vines.

Test Number	Travelling Speed of the Tractor A/(m/s)	Line Speed of the Conveyor Chain B/(m/s)	Line Speed of the Inverting Roller C/(m/s)	Rate of Vines Inverting P_f/(%)	Rate of Buried Pods P_m/(%)	Rate of Fallen Pods P_s/(%)
1	1 (0.89)	1 (0.84)	1 (1.57)	55.56	0	0.08
2	1 (0.86)	2 (1.02)	2 (1.88)	68.06	0.18	0.35
3	1 (0.84)	3 (1.13)	3 (2.12)	60	0.21	0.19
4	2 (1.04)	1 (0.84)	2 (1.88)	64.29	0.15	0.72
5	2 (1.01)	2 (1.02)	3 (2.12)	69.7	0.13	0.23
6	2 (1.05)	3 (1.13)	1 (1.57)	62.5	0.12	0.44
7	3 (1.25)	1 (0.84)	3 (2.12)	65.31	0.77	0.77
8	3 (1.29)	2 (1.02)	1 (1.57)	72.73	1.32	0.54
9	3 (1.13)	3 (1.13)	2 (1.88)	73.44	1.08	1.45

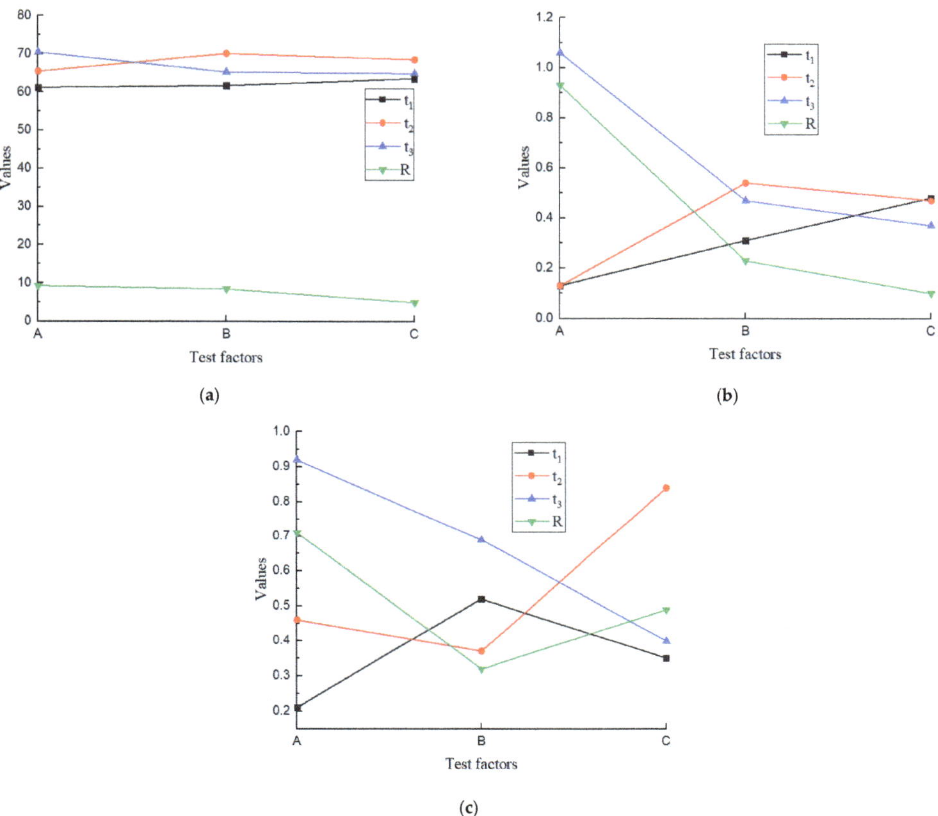

Figure 5. Range analysis of unpressed vines experiment results; (**a**) P_m, (**b**) P_s, (**c**) P_f.

Table 4. Experimental variance analysis of unpressed vines.

Indexes	Variance Source	Sum of Squares	Free Degree	F-Value	*p*-Value	Significance
P_f	A	129.613	2	95.98	0.01	**
	B	107.724	2	79.771	0.012	**
	C	39.891	2	29.539	0.033	**
	Error	1.35	2			
	T	278.579	8			
P_m	A	1.711	2	25.141	0.038	*
	B	0.088	2	1.294	0.436	
	C	0.022	2	0.326	0.754	
	Error	0.068	2			
	T	1.89	8			
P_s	A	0.783	2	47.567	0.021	**
	B	0.154	2	9.34	0.097	*
	C	0.435	2	26.433	0.036	**
	Error	0.016	2			
	T	1.389	8			

(P_f) R square = 0.995, (P_m) R square = 0.964, (P_s) R square = 0.988. Note: The critical value of significant judgment $F_{0.01}(2,2) = 99$, $F_{0.05}(3,3) = 19$, $F_{0.1}(2,2) = 9$. * Indicates that the factors have some influence on the test index ($0.05 < p \leq 0.1$), ** indicates that the factors have a significant influence on the test index ($0.01 < p \leq 0.05$).

The variable R stands for range, and the range of a factor is the difference between the maximum and minimum values of the mean values of each level of the factor, t_1 represents the average value of experimental results at the level of 1 for each experimental factor, t_2 represents the average value of experimental results at the level of 2 for each experimental factor, and t_3 represents the average value of experimental results at the level of 3 for each experimental factor.

According to the range analysis in Figure 5a, the order of influence of all factors on the rate of vines inverting is as follows: A > B > C. According to the comprehensive comparability, the higher the average value of each factor group of the rate of vines inverting, the better the level of factor. Moreover, according to the variance results in Table 4, the impact of all factors on the rate of vines inverting is extremely significant, so the maximum horizontal combination is $A_3B_2C_2$. Similarly, according to the range analysis in Figure 5b, the order of influence of all factors on the rate of buried pods is as follows: A > B > C. From comprehensive comparability, it can be seen that the lower the average value of each factor group in the rate of buried pods, the better the level of the factor. Moreover, according to variance results in Table 4, it can be seen that the traveling speed of the tractor has a significant impact on the rate of buried pods, while the line speed of the conveyor chain and the line speed of the inverting roller have no significant impact on the rate of buried pods. At the same time, A_1 and A_2 are the same, so the minimum horizontal combination is chosen as $A_2B_1C_3$ by mechanical performance. Similarly, from the range analysis in Figure 5c, it can be seen that the order of influence of all factors on the rate of fallen pods is as follows: A > C > B. From the comprehensive comparability, it can be seen that the lower the average value of each factor group for the rate of fallen pods, the better the level of that factor. Moreover, from the variance results in Table 4, it can be seen that the influence of the traveling speed of the tractor and the line speed of the inverting roller on the rate of fallen pods is extremely significant, and the line speed of the conveyor chain has a significant effect on the rate of fallen pods, so the minimum horizontal combination is $A_1C_1B_2$.

According to the comprehensive balance method for multi-index data analysis of an orthogonal experiment, the above analysis results are summarized, and the summary table is shown in Table 5.

Table 5. Summary of results of unpressed vines.

Index	Factor Importance Order	The Best Level and the Next Best Level		
		A	B	C
P_f	A, B, C	A_3 or A_2	B_2	C_2 or C_3
P_m	A	A_1 and A_2		
P_s	A, C, B	A_1 or A_2	B_2	C_1 or C_3

As can be seen from Table 5, factor A is the most important at the level of A_2, followed by factor C at the level of C_3, and then factor B at the level of B_2. In conclusion, the best horizontal combination can be $A_2C_3B_2$, that is, the traveling speed of the tractor is 0.9~1.1 m/s, the line speed of the inverting roller is 2.12 m/s, and the line speed of the conveying chain is 1.02 m/s. In order to further verify the operation effect, three repeated experiments were carried out under the above optimal working parameters: the average traveling speed of the tractor was 1.06 m/s, the rate of vines inverting was 71.07%, the rate of buried pods was 0.2%, and the rate of fallen pods was 0.22%. The rate of buried pods and the rate of fallen pods were far less than the identification standard of the peanut harvester.

3.2. Results of Pressed Vines Experiment

According to the experimental method, the experimental results of the peanut digger-inverter under the condition of pressed vines are shown in Table 6, and the range analysis and variance analysis are shown in Figure 6 and Table 7.

Table 6. Experimental design and results of pressed vines.

Test Number	Travelling Speed of the Tractor A/(m/s)	Line Speed of the Conveyor Chain B/(m/s)	Line Speed of the Inverting Roller C/(m/s)	Rate of Vines Inverting P_f/(%)	Rate of Buried Pods P_m/(%)	Rate of Fallen Pods P_s/(%)
1	1 (0.74)	1 (0.84)	1 (1.57)	60	0.13	0.36
2	1 (0.71)	2 (1.02)	2 (1.88)	73.77	0.19	0.38
3	1 (0.82)	3 (1.13)	3 (2.12)	66.67	0.37	0.6
4	2 (0.96)	1 (0.84)	2 (1.88)	67.8	0.1	0.39
5	2 (1.01)	2 (1.02)	3 (2.12)	68.18	0.36	0.89
6	2 (1.09)	3 (1.13)	1 (1.57)	67.69	0.76	0.93
7	3 (1.26)	1 (0.84)	3 (2.12)	70.97	0.79	1.02
8	3 (1.29)	2 (1.02)	1 (1.57)	77.92	0.89	1.15
9	3 (1.15)	3 (1.13)	2 (1.88)	80.49	0.8	0.8

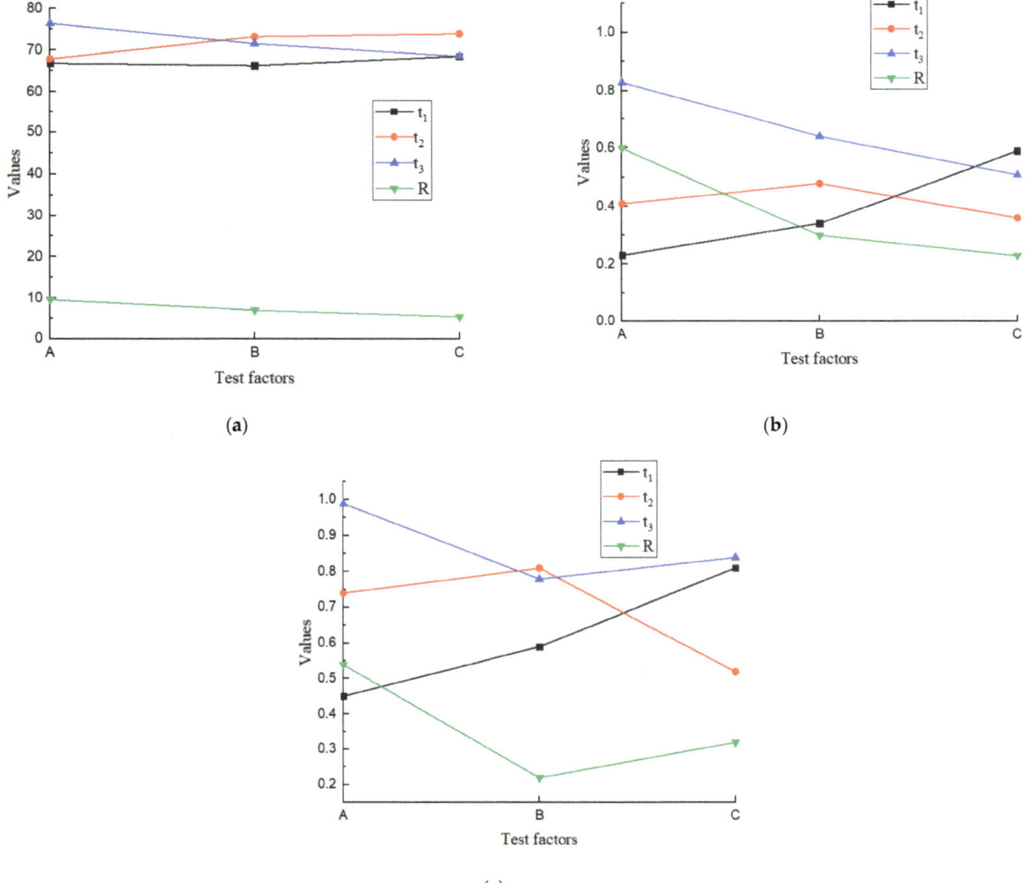

Figure 6. Range analysis of pressed vine experiment results, (**a**) P_m, (**b**) P_s, (**c**) P_f.

Table 7. Experimental variance analysis of pressed vines.

Indexes	Variance Source	Sum of Squares	Free Degree	F-Value	*p*-Value	Significance
P_m	A	167.662	2	53.935	0.018	**
	B	80.997	2	26.056	0.037	**
	C	59.376	2	19.1	0.05	**
	Error	3.109	2			
	T	311.144	8			
P_s	A	0.564	2	14.419	0.065	*
	B	0.138	2	3.538	0.22	
	C	0.081	2	2.071	0.326	
	Error	0.039	2			
	T	0.822	8			
P_f	A	0.443	2	32.556	0.03	**
	B	0.083	2	6.07	0.141	
	C	0.183	2	13.421	0.069	*
	Error	0.014	2			
	T	0.723	8			

(P_f) R square = 0.99, (P_m) R square = 0.952, (P_s) R square = 0.981. Note: The critical value of significant judgment $F_{0.01}(2,2) = 99$, $F_{0.05}(3,3) = 19$, $F_{0.1}(2,2) = 9$. * indicates that the factors have some influence on the test index ($0.05 < p \leq 0.1$), ** indicates that the factors have a significant influence on the test index ($0.01 < p \leq 0.05$).

The variable R stands for range, and the range of a factor is the difference between the maximum and minimum values of the mean values of each level of the factor, t_1 represents the average value of experimental results at the level of 1 for each experimental factor, t_2 represents the average value of experimental results at the level of 2 for each experimental factor, and t_3 represents the average value of experimental results at the level of 3 for each experimental factor.

According to the range analysis in Figure 6a, it can be seen that the order of influence of all factors on the rate of vines inverting is as follows: A > B > C. From comprehensive comparability, it can be seen that the higher the average value of each factor group, the better the level of that factor. Moreover, according to the variance results in Table 7, it can be seen that each factor has an extremely significant impact on the rate of vines inverting, so the maximum horizontal combination is $A_3B_2C_2$. Similarly, according to the range analysis in Figure 6b, it can be seen that the order of influence of all factors on the rate of buried pods is as follows: A > B > C. From the comprehensive comparability, it can be seen that the lower the average value of each factor group for the rate of buried pods, the better the factor level. Moreover, from the variance results in Table 7, it can be seen that the traveling speed of the tractor has a significant impact on the rate of buried pods. The line speed of the conveyor chain and the line speed of the inverting roller had no significant effect on the rate of buried pods, so the minimum horizontal combination was chosen as $A_1B_1C_2$. Similarly, according to the range analysis in Figure 6c, it can be seen that the order of influence of all factors on the rate of fallen pods is as follows: A > C > B. Based on comprehensive comparability, it can be seen that the smaller the average value of each factor group, the better the level of that factor. Moreover, according to variance results in Table 7, it can be seen that the influence of the traveling speed of the tractor on the rate of fallen pods is extremely significant, the influence of the line speed of the inverting roller on the rate of fallen pods is significant, and the influence of the line speed of the conveyor chain on the rate of fallen pods is insignificant, so the minimum horizontal combination is $A_1C_2B_1$.

According to the comprehensive balance method of multi-index data analysis in an orthogonal experiment, the above analysis results are summarized, and the summary table is shown in Table 8.

Table 8. Summary of results of pressed vines.

Index	Factor Importance Order	The Best Level and the Next Best Level		
		A	B	C
P_f	A, B, C	A_3 or A_2	B_2	C_2
P_m	A	A_1 or A_2		
P_s	A, C	A_1 or A_2		C_2

As can be seen from Table 8, factor A is the most important at the level of A_2, followed by factor C at the level of C_2, and then factor B at the level of B_2. In summary, the best horizontal combination can be $A_2C_2B_2$, that is, the traveling speed of the tractor is 0.9~1.1 m/s, the line speed of the inverting roller is 1.88 m/s, and the line speed of the conveyor chain is 1.02 m/s. In order to further verify the operation effect, three repeated experiments were carried out under the above optimal working parameters: the average traveling speed of the tractor was 1.01 m/s, the rate of vines inverting was 74.29%, the rate of buried pods was 0.14%, and the rate of fallen pods was 0.33%. The rate of buried pods and the rate of fallen pods were far less than the identification standard of the peanut harvester.

3.3. Results of Pairing Analysis between Unpressed Vines and Pressed Vines

The paired *t*-test was conducted for the rate of vines inverting, the rate of buried pods and the rate of fallen pods, of unpressed and pressed vines, and the significance level was 0.05. The correlation of paired samples was shown in Table 9, the test results of paired samples were shown in Table 10, and the box diagram of paired sample comparison was shown in Figure 7.

Table 9. Correlation of paired samples.

Sample Group	Correlation	*p*-Value	Significance
Traveling speed of tractor A	0.913	0.001	***
Line speed of the conveyor chain B	0.787	0.012	**
Line speed of the inverting roller C	0.281	0.464	

Note: The critical value of significant judgment $t_{0.01}(8) = 2.896$, $t_{0.05}(8) = 1.86$, $t_{0.1}(8) = 1.397$. ** indicates that the factors have a significant influence on the test index ($0.01 < p \leq 0.05$), *** indicates that the factors have a very significant influence on the test index ($p \leq 0.01$).

Table 10. Paired sample test.

Sample Group	T	df	*p*-Value	Significance
Traveling speed of tractor A	−5.479	8	0.001	***
Line speed of the conveyor chain B	−0.468	8	0.652	
Line speed of the inverting roller C	−1.326	8	0.222	

Note: The critical value of significant judgment $t_{0.01}(8) = 2.896$, $t_{0.05}(8) = 1.86$, $t_{0.1}(8) = 1.397$. *** indicates that the factors have a very significant influence on the test index ($p \leq 0.01$).

As can be seen from Table 9, the correlation between unpressed vines and pressed vines to the rate of vines inverting was 0.913, which was significant ($p = 0.001 < 0.01$), the correlation between unpressed vines and pressed vines was 0.787, and the correlation was significant ($p = 0.012 < 0.05$), and the correlation between unpressed vines and pressed vines was 0.281, and the correlation was insignificant ($p = 0.464 > 0.05$). As can be seen from Table 10, at the significance level of 0.05, the paired *t*-test was conducted for the rate of vines inverting on unpressed vines and pressed vines, and the T-value was −5.479, and the significance *p* value was 0.001 < 0.01. It can be seen that unpressed vines and pressed vines not only have a significant correlation but also a significant difference in the rate of vines inverting. According to the comparison in Figure 7a, the mean and median values of the rate of vines inverting on pressed vines are higher than those on unpressed vines. At the significance level of 0.05, the paired *t*-test was conducted for the rate of buried pods of unpressed vines and pressed vines and the rate of fallen pods of unpressed vines and pressed vines. The *p* values of the rates of buried pods of unpressed vines and pressed vines and the rates of fallen pods of unpressed vines and pressed vines were all greater than 0.05. Therefore, there was no significant difference between the rate of buried pods of unpressed vines and pressed vines and the rate of fallen pods of unpressed vines and pressed vines. According to the comparison in Figure 7b,c, there was little difference between the average rate of buried pods and the average percentage of falling pods between the unpressed and pressed vines.

As summarized, under the condition of pressed and unpressed vines, the influence of various factors on the rate of fallen pods and the rate of buried pods is small, but there is a difference in the rate at which vines invert. The rate of vines inverting peanut under the state of pressed vines is higher than that under the state of unpressed vines.

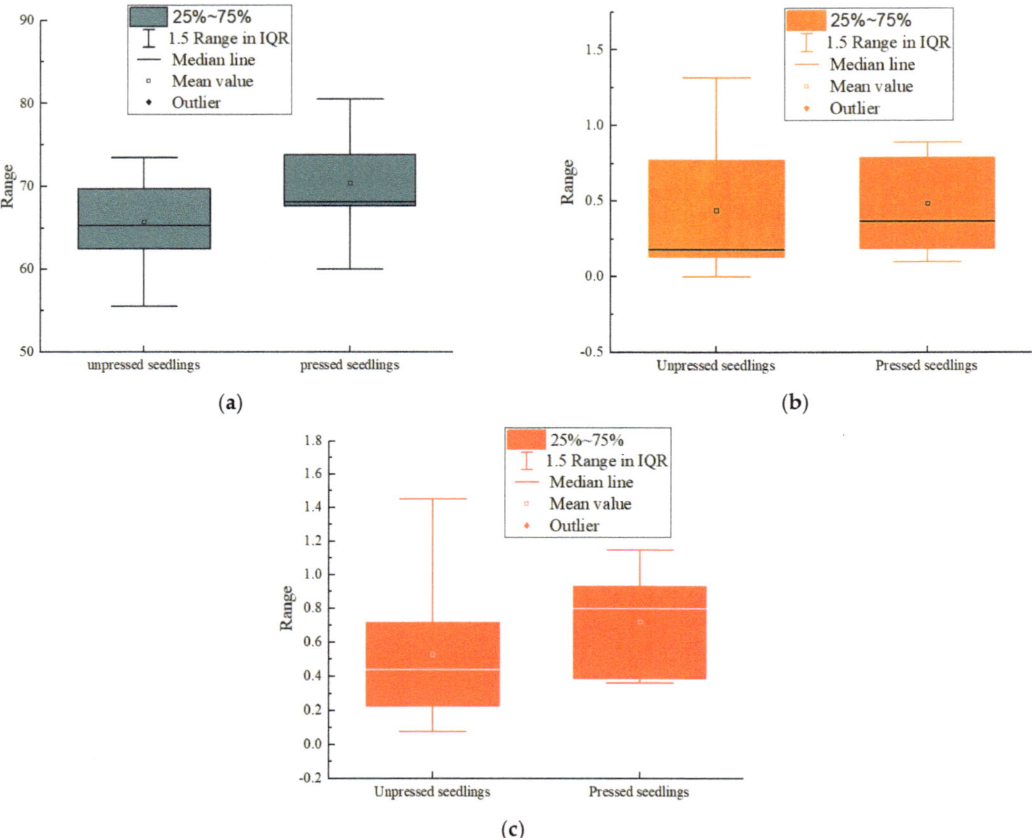

Figure 7. Box diagram of paired samples. (**a**) the rate of vines inverting; (**b**) the rate of buried pods; (**c**) the rate of fallen pods.

4. Discussion

In the process of peanut digger-inverter operation, under the state of unpressed vines and pressed vines, the influencing sequence of each factor on each index is the same, that is, the traveling speed of the tractor has the greatest influence on the rate of vines inverting, the rate of buried pods, and the rate of fallen pods of the peanut digger-inverter. This is because when the tractor's travel speed is high, there will be many peanut vines and pods on the peanut digger-inverter machine. First, it will cause blockage and serious vine accumulation, which is not conducive to the later vines flipping. Second, the peanut digging shovel will not maintain the digging depth at the beginning of the operation, which will break the peanut pod, resulting in buried pods and falling pods. Third, the pod's falling speed will also accelerate, which will cause the rigid acceleration of the peanut pod landing at that moment, resulting in a peanut falling pod. When the speed of the tractor is slow, the number of peanut vines will be small, and the peanut vines cannot stand on their hands independently, resulting in a very low rate of peanut vines inverting [16,41,42]. The line speed of the conveyor chain only has an effect on the rate of vines inverting; it has no effect on the rate of buried pods or the rate of fallen pods. This is because the line speed of the conveyor chain can only affect the number of peanut vines on the back of the inverting roller and has no effect on the peanut pod. However, the speed of the tractor, the line speed of the inverting roller is the most influential factor on the rate of vines inverting and the rate of fallen peanut pods, and it has no influence on the rate of buried pods. This is because the inverting roller is too large, which will lead to the peanut vines being thrown out, resulting in their inability to stand on their hands and the pod falling off. The inverting roller is too small, which will lead to the peanut vines not being able to turn over to the ground in time, resulting in congestion and the peanut vines not being able to turn over [28,43].

Under different states of unpressed vines and pressed vines, there are significant differences in the paired *t*-test of the rate of vines inverting for each factor of the peanut digger-inverter. This is because the lodging peanut vines are dependent, while the independent peanut vines have an extremely poor inverting effect, and the trailing and semi-trailing peanut varieties have a better inverting effect. Therefore, peanut planting row spacing should be matched with the harvester. The left and right rows of peanut vines support each other [32,44].

The relevant experimental results showed that the peanut vine inverting effect was better under the pressed vine state, and under the same conditions, the rate of vine inverting was 4.3% higher than that of unpressed vines and 3.1% higher than that of the turnover laying device of the peanut harvester. In addition, the peanut digger-inverter has a much smaller rate of fallen pods than the turnover-laying device of the peanut harvester. The data pairs for each experiment are shown in Table 11.

Table 11. Comparison of test results of performance indices.

Type of Peanut Harvesting Structure	Sources	Performance Indexes	
		Rate of Vines Inverting/(%)	Rate of Fallen Pods/(%)
The peanut digger-inverter	This study (unpressed vines)	71.07	0.22
	This study (pressed vines)	74.29	0.33
Turnover Laying Device	Ref. [32]	72	2.1

5. Conclusions

In order to quickly dry peanut pods and effectively reduce the mildew of peanut pods sticking to the ground in rainy weather, this paper analyzed the current research status of peanut digging and inverting technology in China and abroad. Combined with the two-stage harvesting mode of peanut, the orthogonal experiment was carried out with the peanut digger-inverter on the upright peanut under two states of pressed and unpressed vines. The results showed that:

(1) Through the analysis of range, variance, and the comprehensive balance method of an orthogonal experiment, it can be seen that the influence sequence of each factor of the peanut digger-inverter on the evaluation index under the condition of unpressed vines is as follows: A > C > B, the best horizontal combination is $A_2C_3B_2$, that is, the speed of the tractor is 0.9~1.1 m/s, the line speed of the conveying chain is 1.02 m/s, and the line speed of the inverting roller is 2.12 m/s. Under the horizontal combination, the experimental results were as follows: the average travelling speed of the tractor was 1.06 m/s, the rate of

vines inverting was 71.07%, the rate of buried pods was 0.2%, and the rate of fallen pods was 0.22%.

(2) Through the analysis of range, variance, and the comprehensive balance method of an orthogonal experiment, it can be seen that the influence sequence of each factor of the peanut digger-inverter on the evaluation index under the condition of pressed vines is as follows: A > C > B, the best horizontal combination is $A_2C_2B_2$, that is, the traveling speed of the tractor is 0.9~1.1 m/s, the speed of the conveying chain is 1.02 m/s, and the line speed of the inverting roller is 1.88 m/s. Under this horizontal combination, the experimental results were as follows: the average travelling speed of the tractor was 1.01 m/s, the rate of vines inverting was 74.29%, the rate of buried pods was 0.14%, and the rate of fallen pods was 0.33%.

(3) Through the paired *t*-test, it can be seen that under the state of pressed vines and unpressed vines, there is little difference in the influence of various factors on the rate of fallen pods and the rate of buried pods, but there is a difference in the influence on the rate of vines inverting. The rate of peanut vines inverting under the state of pressed vines is higher than that under the state of unpressed vines.

On the basis of the optimization of the working parameters of the peanut digger-inverter, in the future, we will develop a set of the peanut digger-inverter suitable for upright peanuts and carry out intelligence and information research on the existing basis.

Author Contributions: Conceptualization, H.S., F.G. and Z.H.; methodology, H.S. and H.Y.; software, H.S. and H.Y.; validation, H.S., H.Y. and F.W.; formal analysis, H.S. and Q.G.; investigation, Q.G.; resources, F.W.; data curation, H.S., F.W. and Y.C.; writing—original draft preparation, H.S.; writing—review and editing, M.C., Q.G. and Z.H.; visualization, Y.C.; supervision, Z.H.; project administration, M.C.; funding acquisition, F.G. All authors have read and agreed to the published version of the manuscript.

Funding: This research was funded by the National Peanut Industry Technology System, grant number CARS-13.

Institutional Review Board Statement: Not applicable.

Informed Consent Statement: Not applicable.

Data Availability Statement: The data presented in this study are available on demand from the first author at (82101221060@caas.cn).

Acknowledgments: Thanks to Ganzhou Experimental Station of the National Peanut Industry System for providing the experimental site.

Conflicts of Interest: The authors declare no conflict of interest.

References

1. Chen, Y.C. Falling peanut in Ming Dynasty, shaped like taro but not peanut in early Qing Dynasty, the first real peanut in Chongming. *Local Chron. Jiangsu* **2018**, *174*, 85–88.
2. Zhang, J.L.; Geng, Y.; Guo, F.; Li, X.G.; Wan, S.B. Research progress on the mechanism of improving peanut yield by single-seed precision sowing. *J. Integr. Agric.* **2020**, *19*, 1919–1927. [CrossRef]
3. Shi, L.; Wang, B.; Hu, Z.; Yang, H. Mechanism and Experiment of Full-Feeding Tangential-Flow Picking for Peanut Harvesting. *Agriculture* **2022**, *12*, 1448. [CrossRef]
4. Chen, Z.K.; Wu, H.C.; Zhang, Y.H.; Peng, B.L.; Gu, F.W.; Hu, Z.C. Development of automatic depth control device for semi-feeding four-row peanut combine harvester. *Trans. Chin. Soc. Agric. Eng.* **2018**, *34*, 10–18.
5. Zhao, X.S.; Ran, W.J.; Hao, J.J.; Bai, W.J.; Yang, X.L. Design and experiment of the double-seed hole vines precision seed metering device for peanuts. *Int. J. Agric. Biol. Eng.* **2022**, *15*, 107–114.
6. Xie, Y.K.; Lin, Y.W.; Li, X.Y.; Yang, H.; Han, J.H.; Shang, C.J.; Li, A.Q.; Xiao, H.W.; Lu, F.Y. Peanut drying: Effects of various drying methods on drying kinetic models, physicochemical properties, germination characteristics and microstructure. *Inf. Process. Agric.* **2022**. [CrossRef]
7. Wang, B.; Hu, Z.C.; Peng, B.L.; Zhang, Y.H.; Gu, F.W.; Shi, L.L.; Gao, X.M. Structure operation parameter optimization for elastic steel pole oscillating screen of semi-feeding four rows peanut combine harvester. *Trans. Chin. Soc. Agric. Eng.* **2017**, *33*, 20–28.
8. National Bureau of Statistics of China. *China Statistical Yearbook*; China Statistics Press: Beijing, China, 2022.

9. Food and Agriculture Organization (FAO) of the United Nations Database. 2022. Available online: http://faostat3.fao.org/download/Q/QC/E (accessed on 25 December 2022).
10. Hu, Z.C. *Study on Key Technologies of Half-Feed Peanut Combine Harvester*; Nanjing Agricultural University: Nanjing, China, 2011.
11. Chen, Z.Y.; Gao, L.X.; Chen, C.; Butts, C.L. Analysis on technology status and development of peanut harvest mechanization of China and the United States. *Trans. Chin. Soc. Agric. Mach.* **2017**, *48*, 1–21.
12. Zhang, J.S. Exploring the main points of mechanization technology of peanut production. *Shihezi Sci. Technol.* **2022**, *261*, 6–7.
13. Chen, M.D.; Zhai, X.T.; Zhang, H.; Yang, R.B.; Wang, D.W.; Shang, S.Q. Study on control strategy of the vine clamping conveying system in the peanut combine harvester. *Comput. Electron. Agric.* **2020**, *178*, 105744. [CrossRef]
14. The Comprehensive Mechanization Rate of Peanuts in China has Reached 52.14%. 2017. Available online: http//www.nongjitong.com/news/2017/42210.html (accessed on 25 December 2022).
15. Wang, B.; Gu, F.; Cao, M.; Xie, H.; Wu, F.; Peng, B.; Hu, Z. Analysis and Evaluation of the Influence of Different Drum Forms of Peanut Harvester on Pod-Pickup Quality. *Agriculture* **2022**, *12*, 769. [CrossRef]
16. Zheng, J.S. Design and experiment of peanut digging and placing machine based on two-stage harvest. *J. Agric. Mech. Res.* **2022**, *44*, 133–139.
17. Xu, T.; Shen, Y.Z.; Gao, L.X.; Zhang, X.D.; Lv, C.Y.; Liu, Z.X. Spring-finger peanut pickup mechanism based on two-stage harvest. *Trans. Chin. Soc. Agric. Eng.* **2016**, *47*, 90–97.
18. Gao, L.X.; Chen, Z.Y.; Charles, C.; Butts, C.L. Development course of peanut harvest mechanization technology of the United States and enlightenment to China. *Trans. Chin. Soc. Agric. Eng.* **2017**, *33*, 1–9.
19. AMADAS INDUSTRIES. Peanut Diggers. 2015. Available online: http://www.amadas.com/agriculture/peanuts/peanut-diggers (accessed on 25 December 2022).
20. Kelly Manufacturing Co. Vine Conditioner and Vine Lifter. 2015. Available online: http://www.Kelleymfg.com/products/peanut/vine_conditioner_lifter.aspx (accessed on 25 December 2022).
21. COLOMBO. Colombo Dump Cart 61.52.12. 2015. Available online: http://colombona.com/colombo-dump-cart-cta-61.52.12 (accessed on 25 December 2022).
22. Mike, B. *Peanut Digger and Combine Efficiency. Cooperative Extension of Colleges of Agricultural and Environmental Science*; The University of Georgia: Athens, GA, USA, 2009.
23. NC State University, BAE. Peanut Harvesting Equipment: Diggers and Combines. 2017. Available online: http://www.bae.ncsu.edu/topic/agmachine/farmequip/harvest/peanut_harvest_guide.htm (accessed on 25 December 2022).
24. Wang, L.; Wei, J.J.; Li, Y. Development of peanut's whole course mechanization in our country and the application in Xinjiang. *Chin. Agric. Sci. Bull.* **2014**, *30*, 161–168.
25. Wang, Y.B. *Development of 4HS-80 Peanut Harvester*; Henan Province, Zhengzhou Xechuang Mechanical and Electrical Equipment Co., Ltd.: Zhengzhou, China, 2014.
26. Shen, J.M. Research on 4H-1 peanut harvester. *Mech. Rural. Pastor. Areas* **2008**, *1*, 14–15.
27. Hu, Z.C.; Peng, B.L.; Xie, H.X.; Tian, L.J.; Wang, H.O.; Wu, F. Design and Experiment of Peanut Hoist Chain Harvester. *Trans. Chin. Soc. Agric. Mach.* **2008**, *11*, 220–222.
28. He, Y.C.; Tang, Z.H.; Yang, H.J.; Meng, X.J.; Qin, T.R.; Zhang, D.C. Design and experiment of 4HQ-150 peanut plucking harvester. *J. Gansu Agric. Univ.* **2018**, *53*, 180–186.
29. Wang, G.; Wang, H.Y.; Zhao, Y.; Li, Y.L. A Kind of Tilting Peanut Excavator. CN213755725U, 23 July 2021.
30. Ma, Y.; Wang, Y.J. A Kind of Peanut Excavator and Its Flipping Device. CN213427033U, 15 June 2021.
31. Guo, H.; Guo, W.H. A Small Peanut Digging and Recycling Machine. CN209151579U, 26 July 2019.
32. Gao, L.X.; Wang, D.W.; Dong, H.S. Design and Test of Turnover Laying Device of Peanut Harvester. *J. Shenyang Agric. Univ.* **2016**, *47*, 57–63.
33. Wang, S.Y.; Hu, Z.C.; Chen, Y.Q.; Wu, H.C.; Wang, Y.W.; Wu, F.; Gu, F.W. Integration of agricultural machinery and agronomy for mechanised peanut production using the vine for animal feed. *Biosyst. Eng.* **2022**, *219*, 113–152. [CrossRef]
34. Wang, B.K.; W, N.; Hu, Z.C.; Wang, H.O.; Chen, Y.Q. Experience and thought of development of peanut harvesting mechanization at home and abroad. *J. Chin. Agric. Mech.* **2011**, *4*, 6–9.
35. Mao, S.S.; Zhao, J.X.; Chen, Y. *Experimental Design*; China Statistics Press: Beijing, China, 2014.
36. *DB34/T534-2022*; Technical Specification for Mechanization of Peanut Harvesting. Local Standard of Anhui Province: Hefei, China, 2022.
37. *DG/T077-2019*; Peanut Harvest. Agricultural Machinery Extension Appraisal Outline: Beijing, China, 2019.
38. Xu, X.H.; He, M.Z. *Experimental Design and Application of Design Expert and SPSS*; Scientific Press: Beijing, China, 2010.
39. Wang, S.Y.; Hu, Z.C.; Yao, L.J.; Peng, B.L.; Wang, B.; Wang, Y.W. Simulation and parameter optimisation of pickup device for full-feed peanut combine harvester. *Comput. Electron. Agric.* **2022**, *192*, 106602. [CrossRef]
40. Yang, H.; Cao, M.; Wang, B.; Hu, Z.; Xu, H.; Wang, S.; Yu, Z. Design and Test of a Tangential-Axial Flow Picking Device for Peanut Combine Harvesting. *Agriculture* **2022**, *12*, 179. [CrossRef]
41. Tian, L.X.; Shang, S.Q.; Wang, D.W.; Shen, S.L. Development and experiment of 4HT-2 peanut strip laying harvester. *Agric. Mech. Res.* **2018**, *40*, 87–91.
42. Yu, W.J.; Yang, R.B.; Shang, S.Q.; Shi, C.; Yang, H.G. Shovel sieve modular design and test of peanut segment harvester. *J. Agric. Mech. Res.* **2016**, *38*, 163–166, 171.

43. Zhao, S.J.; Wang, D.W.; Wang, Y.Y.; Luan, G.D.; Sun, Q.W. Orderly laid the development of peanut harvester. *J. Agric. Mech. Res.* **2013**, *35*, 73–75, 79.
44. Guo, W.H.; Guo, H.; Yu, X.D.; Wang, M.C.; Xing, S.K.; Peng, B. Based on a two-part process of peanut harvest machine is designed with the test. *J. Chin. Agric. Mech.* **2020**, *9*, 34–39.

 agriculture

Article

Design and Experiment of an Underactuated Broccoli-Picking Manipulator

Huimin Xu [1], Gaohong Yu [1,2,*], Chenyu Niu [1,3], Xiong Zhao [1,2], Yimiao Wang [1] and Yijin Chen [1]

[1] College of Mechanical Engineering, Zhejiang Sci-Tech University, Hangzhou 310018, China; 202010601017@mails.zstu.edu.cn (H.X.); zhaoxiong@zstu.edu.cn (X.Z.)
[2] Key Laboratory of Transplanting Equipment and Technology of Zhejiang Province, Hangzhou 310018, China
[3] New H3C Artificial Intelligence Technologies Co., Ltd., Hangzhou 310018, China
* Correspondence: yugh@zstu.edu.cn

Abstract: Mature broccoli has large flower balls and thick stems. Therefore, manual broccoli picking is laborious and energy-consuming. However, the big spheroid vegetable-picking manipulator has a complex structure and poor enveloping effect and easily causes mechanical damage. Therefore, a broccoli flower ball-picking manipulator with a compact structure and simple control system was designed. The manipulator was smart in structure and stable in configuration when enveloped in flower balls. First, a physical damage test was carried out on broccoli according to the underactuated manipulator's design scheme. The maximum surface pressure of the flower ball was 30 N, and the maximum cutting force of the stem was 35 N. Then, kinematic analysis was completed, and the statical model of the underactuated mechanism was established. The dimension of the underactuated mechanism for each connecting rod was determined based on the damage test results and design requirements. The sizes of each connecting rod were 50 cm, 90 cm, 50 cm, 90 cm, 50 cm, 60 cm, and 65 cm. The statical model calculated the required thrust of the underactuated mechanism as 598.66–702.88 N. Then, the manipulator was simulated to verify its reliability of the manipulator. Finally, the manipulator's motion track, speed, and motor speed were determined in advance in the laboratory environment. One-hundred picking tests were carried out on mature broccoli with a 135–185 mm diameter. Results showed that the manipulator had an 84% success rate in picking and a 100% lossless rate. The fastest single harvest time in the test stand was 11.37 s when the speed of the robot arm was 3.4 m/s, and the speed of the stepper motor was 60 r/min.

Keywords: broccoli-picking; underactuated mechanism; manipulator; simulation; test

Citation: Xu, H.; Yu, G.; Niu, C.; Zhao, X.; Wang, Y.; Chen, Y. Design and Experiment of an Underactuated Broccoli-Picking Manipulator. *Agriculture* **2023**, *13*, 848. https://doi.org/10.3390/agriculture13040848

Academic Editors: Cheng Shen, Zhong Tang and Maohua Xiao

Received: 25 March 2023
Revised: 6 April 2023
Accepted: 6 April 2023
Published: 11 April 2023

1. Introduction

China is a major producer of vegetables in the world, with its production scale and export scale ranking first in the world [1]. As a vegetable favored by the world, broccoli's planting scale is increasing yearly. At present, the total planting area in China is about 100,000 hectares, and the output accounts for 50% of the global total [2,3]. With the increase in planting scale, traditional manual harvesting is time-consuming and laborious; therefore, the development of picking equipment is urgently needed [3,4].

The design of a manipulator in vegetable-picking equipment is very important and requires a stable structure, accurate picking, fast response, and minimal damage to the crop [5,6]. Developed countries began to study picking robots in the 1960s, and the research and development of vibrating, suction, shear, and other manipulators promoted the mechanization of picking equipment, but the actual efficiency was not high [7–9]. At present, fruit and vegetable picking is accomplished by designing a manipulator with a stable structure and coordinating with vision [10–12]. Kinugawa et al. [13] designed an underactuated manipulator for circular plates, which is not applicable to plates with other shapes. Xiong et al. [14] improved the flexibility of the end-effector for the difficult task of picking in an unstructured environment and carried on a two-arm cooperative robot to complete strawberry picking. However, due to the lack of flexibility of the end-effector, the success rate of picking was 70%. Arad et al. [15] developed a sweet pepper-picking robot with six degrees of freedom, but its picking efficiency could not be guaranteed owing to poor algorithm recognition. Although the research and development of picking equipment started late in China, it is developing rapidly now. Various kinds of picking equipment, such as pneumatic apple picking, hand-held tea tender shoot picking, cooperative kiwi picking, and under-driven grape picking, have emerged one after another [16–19]. The underactuated manipulator has good grasping stability, a strong envelope, and high flexibility [20–22]. Yang et al. [20] proposed a new underactuated manipulator, which has three degrees of freedom and ensures the stability of grasping but is not applicable to crops. The humanoid finger mechanism designed by Yin et al. [19] ensures flexible grasping by installing torsional springs with different stiffness coefficients at the knuckle limit, but the structural strength of the mechanism cannot be determined. Ma et al. [23] designed a Y-type underactuated manipulator according to the growth characteristics of sweet pear. Although the picking rate of the device is high, the degree of damage caused by the picking of this manipulator cannot be judged.

Broccoli flower balls are big, have an unequal diameter, and have a thick stem. Therefore, few robot hands are suitable for enveloping flower balls. At present, the mechanized harvesting equipment of cabbage uses high-horsepower tractors loaded with harvesters to complete one-time harvesting. Although highly efficient, it has no selective harvesting function [24]. Blok [25] et al. developed a broccoli image recognition system to solve the problem of selective harvesting. Although selective harvesting can be achieved, the harvesting efficiency is too low. Lapalmeagtech Co., Ltd. combined Sami4.0 [26] with a broccoli harvester to develop a fully automatic harvesting robot. Although the harvester is highly efficient, its large fuselage is not suitable for use in hilly areas. In China, Shandong Hualong Agricultural Equipment Co., Ltd. (Qingzhou, China) developed a broccoli-harvesting and loading machine, which relies on human labor to harvest broccoli and place the flower balls into the harvesting conveyor belt, but a whole mechanized harvesting operation has not been realized [27]. Yu et al. [28] designed a hanging bucket for broccoli harvesting, which can reduce the labor force. However, the actual nature of the device is similar to that of Shandong Hualong's broccoli-harvesting device, which relies on human resources. However, the broccoli-harvesting equipment of China Agricultural University [29] and Jiangsu University [30] is still in the research and development stage.

In this paper, an underactuated broccoli-picking manipulator was designed to solve problems in the manual picking of mature broccoli, such as big flower balls, thick stems, and huge time and labor consumption. The underactuated broccoli-picking manipulator will be designed to have a compact underactuated structure and a simple control system to realize the stable picking of flower balls. First, physical damage tests were carried out to determine the range of flower ball surface pressure and stem-cutting force. Then, the mechanical structure and control system of the manipulator was designed from the view of the statics of the picking mechanism. Finally, a picking test bed was set up to carry out the indoor picking test.

2. Design Scheme and Working Principle of the Underactuated Broccoli-Picking Manipulator

In this paper, 300 mature Zhejiang 'Tai Lu' series were selected for studying broccoli's biological character data statistics. The physical characteristics of suitable broccoli were analyzed, as shown in Figure 1. The suitable characteristics were a flower ball diameter Φ of 14–18 cm, a plant height h_1 of 45–50 cm, a stem diameter Φ_1 of 0.34–0.46 cm, a flower ball height h of 7–9 cm, and a harvested plant height h_3 of 12–15 cm.

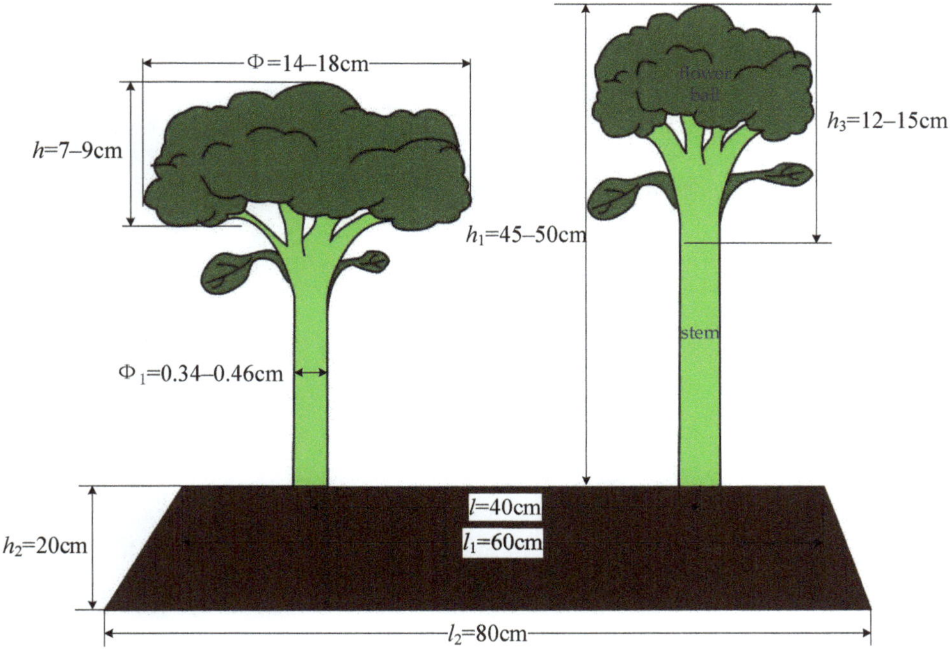

Figure 1. Planting patterns and biological characteristics of broccoli.

A single-stepper motor driven by an underdrive manipulator with two degrees of freedom, which was designed based on a compact underdrive structure, fewer drive actuators, a simple and efficient motion control system, broccoli physical characteristics, and low-loss picking agricultural requirements, can achieve the low-loss enveloping of flower balls with different diameters. The structure of the manipulator is shown in Figure 2.

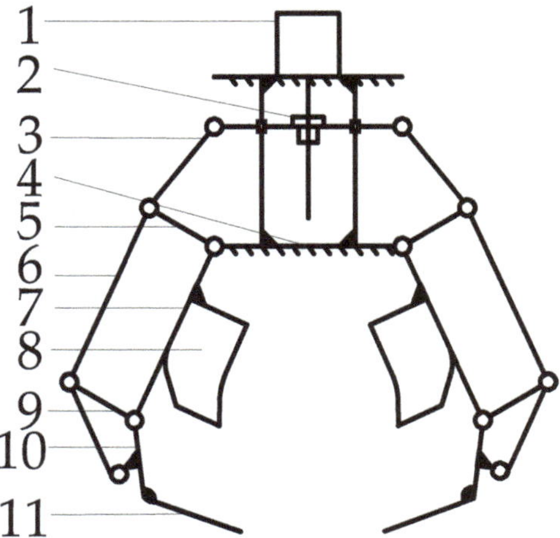

Figure 2. Schematic diagram of manipulator structure: 1 stepper motor; 2 lift platforms composed of worm-and-nut; 3 drive-rod; 4 frame bottom plates; 5, 6 linkage; 7, 9 rockers; 8 clamp guard plates; 10 swing-rod; 11 cutting-blade.

The manipulator is composed of a set of symmetrical underactuated clamping mechanisms, clamping guard plates, cutting blades, and a single-stepper motor, which is picked by a clamping–cutting method. The working principle is shown in Figure 3. The manipulator moves directly above the broccoli and enters the picking area (Figure 3a). Under the action of drive rod force, the clamp guard plates (8) envelope the flower ball (Figure 3b) while the rocker (7) remains motionless. At this time, the rocker (9) drives the swing-rod (10) on the cutting blade (11) to allow the closure of the cutting blades and complete the stem cutting (Figure 3c). Finally, the manipulator driven by the motor picks the flower ball (Figure 3d).

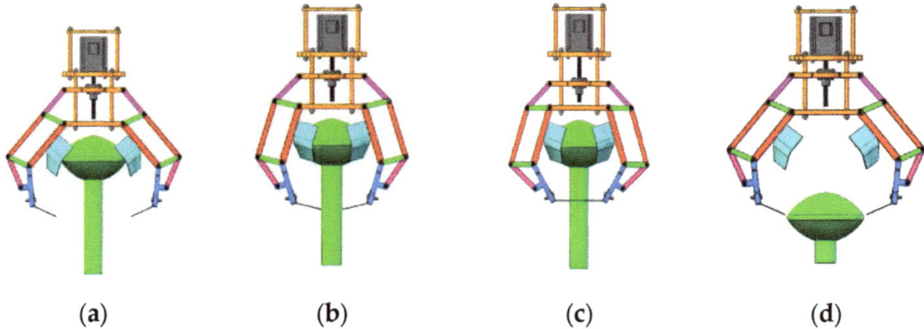

Figure 3. Working principle diagram of picking manipulator: (**a**) The manipulator enters the area to be picked, (**b**) Envelope flower ball, (**c**) Clamp–cut flower ball, and (**d**) Release flower ball.

3. Mechanism Design of the Underactuated Manipulator

3.1. Determination of the Range of Flower Ball Surface Pressure and Stem-Cutting Force

Broccoli picking involves two actions: enveloping and cutting. The enveloping action is performed by the clamping pressure of the underactuated mechanism, and stem separation is performed by the cutting tool. The maximum clamping pressure on the flower ball surface and the cutting force of the stem needs to be determined to provide a theoretical basis for the design of the rod length of the mechanical hand. The clamping action of the manipulator is composed of four contact forces: clamping pressure $F_1 = F_1'$ and cutting force $F_3 = F_3'$ (Figure 4).

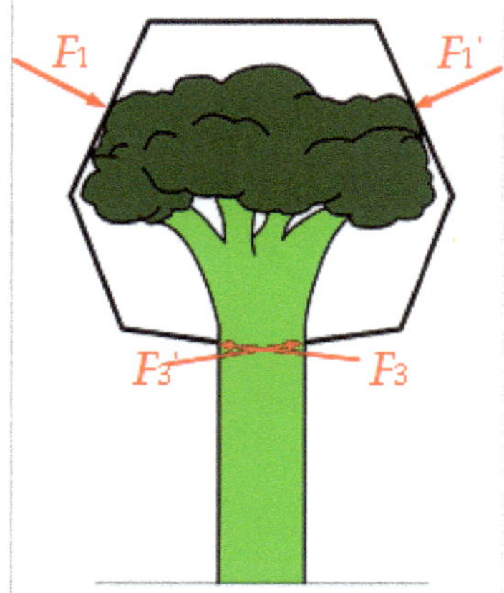

Figure 4. Clamping workspace of the manipulator.

An LDW-1 universal material test machine and cutting tool (304 stainless steel; length, 80 mm; width, 50 mm; thickness, 2 mm; tool point angle, 8°) was used. Thirty flowers with no surface damage were selected to determine the flower surface pressure and stem-cutting force, as shown in Figure 5.

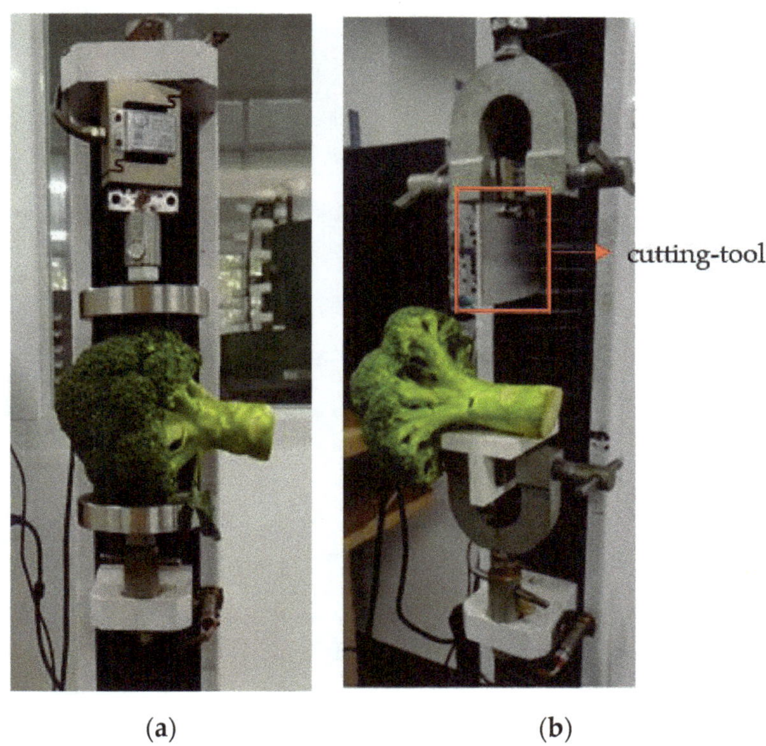

Figure 5. Broccoli physical damage test: (**a**) Flower ball pressure test; (**b**) Stem-cutting force test.

The dynamic compression mold of the universal material test machine was operated at a uniform speed of 10 mm/min. Different loads (30, 40, and 50 N) were applied to the surface of the flower ball, and the damage degree was compared and analyzed, as shown in Figure 6.

When the pressure was 30 N, the surface of the flower ball had no obvious damage. When the pressure was 40 N, small flower buds on the compressed surface of the flower ball were slightly deformed, and the small flower buds on the compressed part were damaged. When the pressure was 50 N, the small stem on the inner surface of the flower ball under pressure was deformed, and the small bud under pressure was damaged seriously. Therefore, the clamping pressure was maintained at 25–30 N, which can ensure the non-destructive enveloping of the flower ball.

The fixture of the moving platform on the test machine drove the cutting tool to move at uniform speeds of 30, 80, and 100 mm/min, and 30 groups of cutting tests were carried out. When the cutting blade was in contact with the broccoli stem, the force sensor on the test machine could detect the cutting force in real-time. The cutting force statistics are shown in Figure 7. Each dot in Figure 7 represents the maximum stem-cutting force for each set of tests. In the 30 groups of stem-cutting force, the cutting force represented by red dots was less than that of blue dots and more than that of green dots. The blue dot is the maximum cutting force of 37.28 N, the green dot is the minimum cutting force of 28.36 N, and the average cutting force is 34.89 N. According to the above cutting force characteristics, keeping the cutting force at 30 N–35 N can ensure that the stems can be cut and separated.

Figure 6. Comparison of pressure damages on broccoli surfaces at different pressures: (**a**) 30 N; (**b**) 40 N; (**c**) 50 N.

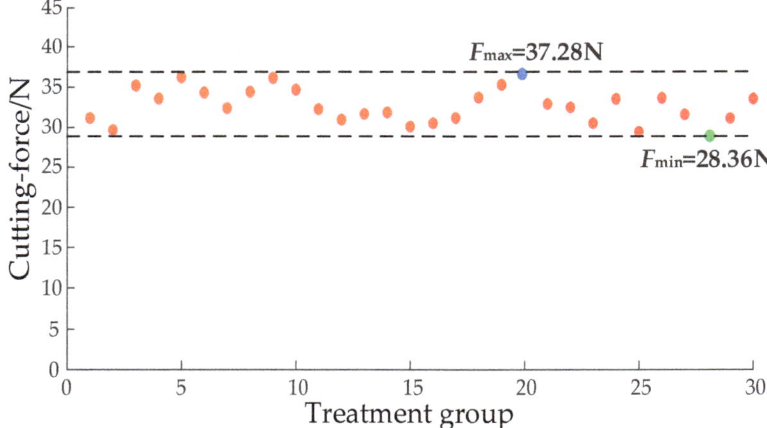

Figure 7. Stem-cutting force test statistics.

3.2. Kinematics Analysis of the Underactuated Mechanism

As shown in Figure 8, the underdrive mechanism designed in this paper is bilaterally symmetric. Therefore, one side of the mechanism was analyzed. When the underactuated mechanism was not in contact with the target object, the relative position of each linkage in the mechanism remained unchanged under the constraint of the torsional spring. The meaning of specific mathematical symbols can be found in Table 1. The specific meanings of mathematical symbols in Figure 8 can be referred to in Table 1. The red characters θ_1, θ_2, and θ_3 in Figure 8 respectively represent angle variations of rods AB, AD, and DK.

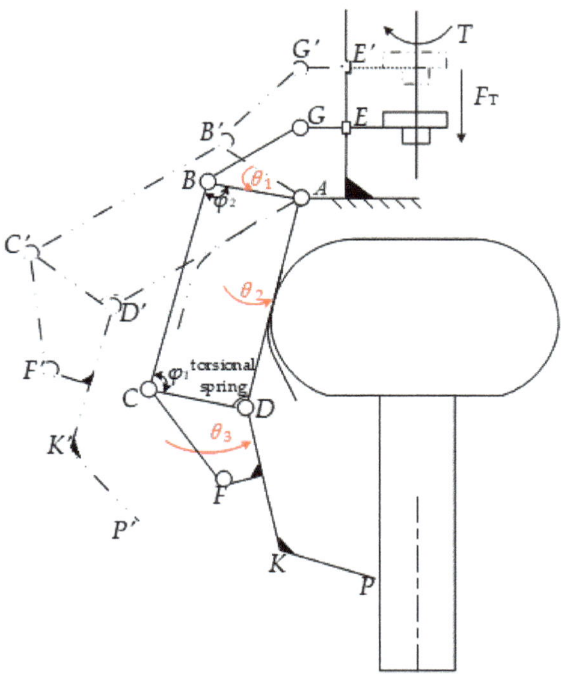

Figure 8. Motion analysis of the underactuated mechanism.

Table 1. Symbol declaration of underactuated mechanism.

Symbol	Declaration	Symbol	Declaration
a	Rod L_{AB}	b	Rod L_{BC}
c	Rod L_{CD}	d	Rod L_{AD}
g	Rod L_{BG}	l_1	Half distance of rod L_{AD}
l_2	Rod L_{DK}	α	$\angle AGB$ angle
$\theta_1, \theta_2, \theta_3$	Angle variations of rods AB, AD, and DK	$\dot{\theta}_1\dot{\theta}_2\dot{\theta}_3$	Angular velocities of rods AB, AD, and DK
ω	Matrix of angular velocities $\dot{\theta}_1$ and $\dot{\theta}_3$	v	Matrix of normal velocity vectors v_1 and v_2 at the contact point
T	Matrix of the underdrive mechanism input torque T_0 and torsional spring torque T_1	F	Matrix of normal forces F_1 and F_3
		θ	Angle between cutting force F_3 and compound force F_2
φ_1	Angle between rods L_{AB} and L_{BC}	φ_2	Angle between rods L_{AB} and L_{BC}

For the velocity at the contact points of each joint, the following formula can be obtained according to the projection theorem of the velocity of the rigid body plane motion:

$$\begin{cases} v = \begin{bmatrix} v_1 \\ v_2 \end{bmatrix} = \begin{bmatrix} l_1 & 0 \\ l_2 + d\cos\theta_3 & l_2 \end{bmatrix} \begin{bmatrix} \dot{\theta}_2 \\ \dot{\theta}_3 \end{bmatrix} \\ J_v = \begin{bmatrix} l_1 & 0 \\ l_2 + d\cos\theta_3 & l_2 \end{bmatrix} \end{cases}. \quad (1)$$

The four-bar mechanism has the following form:

$$\begin{cases} \begin{bmatrix} \dot{\theta}_2 \\ \dot{\theta}_3 \end{bmatrix} = \begin{bmatrix} 1 & -A \\ 0 & 1 \end{bmatrix} \begin{bmatrix} \dot{\theta}_1 \\ \dot{\theta}_3 \end{bmatrix} \\ J_w = \begin{bmatrix} 1 & -A \\ 0 & 1 \end{bmatrix} \end{cases}, \tag{2}$$

where A is shown in the following equations:

$$\begin{cases} A = \frac{(\sqrt{1-B}+\sin\varphi_2)\sqrt{1-C}}{\sqrt{1-B}(\sqrt{1-C}+\sin\varphi_1)} = 1 \\ B = \frac{[a^2+b^2-c^2+2cd\cos\varphi_2-d^2]^2}{4a^2b^2} = \cos^2\varphi_2 \\ C = \frac{[c^2+b^2-a^2+2ad\cos\varphi_1-d^2]^2}{4b^2c^2} = \cos^2\varphi_1 \end{cases}, \tag{3}$$

Therefore, when the slider E moves down to drive the drive rod BG to move, the underdrive mechanism rotates around base point A. When the clamping guard plates contact the surface of the flower ball, the rocker AD stops moving. The driving force drives the swing rod DK to move and closes the cutting blade to overcome the binding force of the torsional spring and complete the clamping action.

3.3. Statics Analysis of the Underactuated Mechanism

As shown in Figure 9, the unilateral clamping mechanism drives rod BG to move through spiral rotation and drives the four-bar mechanism to complete the whole moving process. The specific meanings of mathematical symbols in Figure 9 can be referred to in Table 1. The red characters F_1, F_2, and F_3 in Figure 9 respectively represent the pressure on the surface of the flower ball, the force on the DK rod, and the cutting force of the stem.

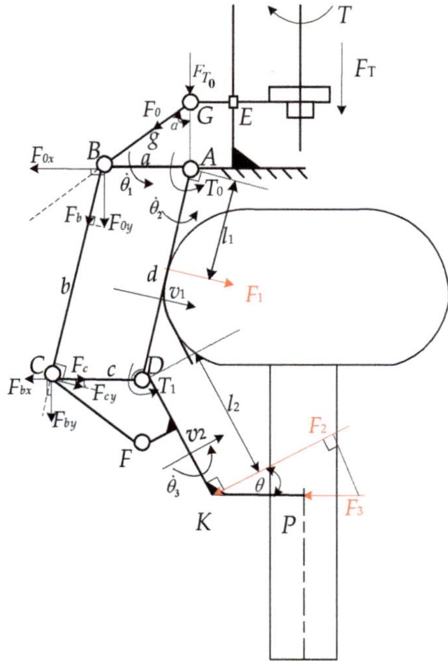

Figure 9. Statics analysis diagram of underactuated mechanism.

In Figure 9, rod *GE* is pushed down by the left thrust F_{T0} to rotate the driving rod *GB*, making the four-bar mechanism rotate around base point *A*. The total thrust F_T and the force of rod *BG* are shown in Equations (4) and (5):

$$F_T = 2F_{T_0}, \tag{4}$$

$$F_0 = F_{T_0}\cos\alpha, \tag{5}$$

The force F_b of rod *BC* is the component force F_{0y} of rod *BG* as shown in Equation (6):

$$\begin{cases} F_{0y} = F_0\sin(\frac{\pi}{2} - \alpha) \\ F_b = F_{0y}\sin\varphi_1 \end{cases}, \tag{6}$$

The underactuated mechanism approaches the flower ball under the force F_{cy} perpendicular to the rod *BC* as shown in Equation (7):

$$\begin{cases} F_c = F_{bx} \\ F_{bx} = F_b\sin\varphi_1 \\ F_{cy} = F_c\sin\varphi_1 \end{cases}. \tag{7}$$

Therefore, the torque T_0 applied by linkage *AB* around point *A* is shown in Equation (8):

$$T_0 = aF_{T_0}\cos\alpha\sin(\frac{\pi}{2} - \alpha)\cos(\alpha + \theta_1) = aF_{0y}\cos(\alpha + \theta_1), \tag{8}$$

When picking is completed, rod *AD* and cutting blade *KP* are constrained by F_1 and F_3. The mechanical model of the unilateral underdrive is [19,20]:

$$\boldsymbol{T^T\omega} = \boldsymbol{F^Tv}, \tag{9}$$

Each vector in Equation (9) can be obtained according to Figure 9.

$$\begin{cases} \boldsymbol{T^T} = \begin{bmatrix} T_0 & T_1 \end{bmatrix}^T \\ \boldsymbol{\omega^T} = \begin{bmatrix} \dot{\theta}_1 & \dot{\theta}_3 \end{bmatrix}^T \\ \boldsymbol{F^T} = \begin{bmatrix} F_1 & F_2 \end{bmatrix}^T \\ \boldsymbol{v^T} = \begin{bmatrix} v_1 & v_2 \end{bmatrix}^T \end{cases}, \tag{10}$$

In Equation (10), the torsional spring torque is $T_1 = -(k\theta_3 + T_0^1)$, k is the stiffness coefficient of the torsional spring (k = 52.85 N·mm/°), and $T_0{}^1$ is the initial torque of the torsional spring ($T_0{}^1$ = 317.1 N·mm).

According to the relation between the contact force and input torque, the contact force vector of the underdriven mechanism can be obtained using the virtual work principle as follows:

$$\boldsymbol{F} = \boldsymbol{TJ_v^{-T}J_w^{-T}}, \tag{11}$$

Clamping pressure F_1 and cutting force F_3 can be obtained from Equation (11).

$$\begin{cases} F_1 = -\frac{(1 - Al_1\cos\theta_3)T_0 + (\mathrm{d}\cos\theta_3 + l_2)T_1}{l_1 l_2} \\ F_2 = \frac{AT_0(\mathrm{d}\cos\theta_3 - l_2) + (l_1\cos\theta_3 + l_2)T_1}{l_1 l_2} \\ F_3 = \frac{AT_0(\mathrm{d}\cos\theta_3 - l_2) + (l_1\cos\theta_3 + l_2)T_1}{l_1 l_2\cos\theta} \end{cases}, \tag{12}$$

According to Equation (12), the clamping pressure F_1 and cutting force F_3 are related to T_0 and T_1, whereas θ_1, θ_2, θ_3, and other angles affect the motion position of the mechanism, as shown in Figure 8. The related parameters are listed in Table 1.

3.4. Parameter Determination of Each Rod of the Underactuated Mechanism

In this paper, a stepper motor, model 57HD5401-110, with a torque of 1.2 N·m, was selected as the driving actuator of the manipulator, together with the 4 mm lead of a TBI high-precision linear screw. The thrust was calculated by Equation (13):

$$F_r = \frac{2\pi n T_{stepper}}{S}, \tag{13}$$

The transmission efficiency n of the lead screw is generally 85–90%. Therefore, the thrust of the nut on the lead screw driven by the motor is 1601.4–1695.6 N.

When the motor is not driven, φ_2 in the unilateral underactuated mechanism is 120° and θ = 26° as shown in Figure 9. Therefore, according to the requirements of the design scheme, a torsional spring was placed in the distal knuckle D to restrict the relative rotation of the two joints. According to the picking requirements and previous design experience [31], rods AB, BG, and L_{CF} were set as 50, 50, and 65 mm, respectively. The length of the bottom plate was set to 130 mm to avoid affecting the enveloping effect of the manipulator with a short-frame bottom plate. The length and width of the cutting blade were designed to be 80 and 50 mm, respectively, to ensure that the manipulator can cut at one time. The objective function was designed to optimize the length of rod CF and determine its parameters, as shown in Equation (14).

$$\begin{cases} F_2 l_2 > T_1 \\ F_2 = F_3\cos\theta \end{cases}, \tag{14}$$

When the unilateral underactuated mechanism contacts broccoli, rod AD remains stationary, and the driving force overcomes the binding force of the torsional spring and continues to drive the swing rod DK to move. At this time, φ_2 is 102°, and the change of $\angle CDA$ at joint D is 18°. Therefore, the torsional spring torque T_1 is –1268.4 N·mm and the constraint condition is set as 30 N $\leq F_3 \leq$ 35 N. The following results were obtained according to the constraints of F_3.

$$\begin{cases} F_3 = 35\ \text{N}, l_2 > 38.81\ \text{mm} \\ F_3 = 30\ \text{N}, l_2 > 45.28\ \text{mm} \end{cases}'$$

Therefore, l_2 is rounded to 50 mm, and L_{DK} is 50 mm.

Additionally, the objective function was designed according to the design scheme and picking requirements, as shown in Equation (15), to optimize and determine the length of rod AD.

$$T_0 \geq 2F_{cy} l_1 + F_2 l_2 + T_1, \tag{15}$$

Substituting the cutting force into the above equation and setting the constraint condition to 30 N $\leq F_3 \leq$ 35 N obtained the following results:

$$\begin{cases} F_3 = 30\ \text{N}, l_1 \leq 46.44\ \text{mm} \\ F_3 = 35\ \text{N}, l_1 \leq 44.36\ \text{mm} \end{cases}.$$

The height of the plants left after cutting was ensured to be 12–15 cm, and the structure was kept compact by setting l_1 to 45 mm and rods BC and AD to 90 mm.

Substituting l_1 = 45 mm, l_2 = 50 mm, T_0 = 5357.13 N·mm, and T_1 = –1268.4 N·mm into Equation (12) yielded the following results:

$$\begin{cases} F_1 = 23.08\ \text{N} \\ F_2 = 32.57\ \text{N}. \\ F_3 = 34.89\ \text{N} \end{cases}$$

In this result, the flower ball surface pressure F_1 is 23.08 N, whereas the pressure range in the previous paper is 25–30 N. Therefore, this result is less than the pressure range and meets the requirement of low-loss picking. In addition, the cutting force F_3 in this result is 34.89 N, which conforms to the cutting force range of 30–35 N in the stem-cutting test and meets the requirements for picking.

Therefore, the parameters of each rod are shown in Table 2.

Table 2. Parameters of the connecting rods of the underactuated mechanism (unit: mm).

Symbol	L_{AB}	L_{BC}	L_{CD}	L_{AD}	L_{DK}	L_{BG}	L_{CF}
Parameter	50	90	50	90	50	60	65

3.5. Determination of Underactuated Mechanism Thrust

F_1 and F_2 in Equation (11) can be obtained by Equation (16):

$$T_0 = F_1 l_1 + (d\cos\theta_3 + l_2)F_2, \tag{16}$$

The following can be obtained from Equation (8):

$$F_{T_0} = \frac{T_0}{a\cos\alpha\sin(\frac{\pi}{2} - \alpha)}, \tag{17}$$

Substituting $l_1 = 45$ mm, $l_2 = 50$ mm, $d = 90$ mm, $\theta = 21°$, $\theta_3 = 18°$, $\alpha = 55°$, $25\ \text{N} \leq F_1 \leq 30$ N, and $25\ \text{N} \leq F_3 \leq 30$ N into Equations (16) and (17) resulted in:

$$\begin{cases} F_1 = 25\ \text{N}, F_2 = 28.01\ \text{N}, F_{T_0} = 299.28\ \text{N} \\ F_1 = 25\ \text{N}, F_2 = 32.68\ \text{N}, F_{T_0} = 337.77\ \text{N} \\ F_1 = 30\ \text{N}, F_2 = 28.01\ \text{N}, F_{T_0} = 312.95\ \text{N} \\ F_1 = 30\ \text{N}, F_2 = 32.68\ \text{N}, F_{T_0} = 351.44\ \text{N} \end{cases}.$$

The manipulator has a symmetrical structure of the manipulator; therefore, thrust F_T was calculated as shown in Equation (18):

$$F_T = 2F_{T_0}, \tag{18}$$

The required thrust F_T range of the manipulator is 598.66–702.88 N. According to the result, the thrust of the motor drive nut is greater than that required by the underactuated mechanism:

$$F_r > F_T = 2F_{T_0}. \tag{19}$$

Therefore, the thrust of the stepper motor meets the requirements of picking and the requirements of the scheme design of the underactuated mechanism.

4. Broccoli-Picking Test

4.1. Simulation Analysis of Manipulator Motion

Aluminum alloy material, 302 steel, and S304 material were imported into the manipulator in Adams software, and constraints were added to further verify the rationality of the design of the manipulator. The manipulator motion simulation was completed at a driving force of 2000 N and a moving plate speed of 5.5 mm/s. Marker points were added to the contact position between the middle of the clamping guard plates and the contact position between the cutting blades and the stem of the mechanical hand to measure the contact force of the two positions after the simulation (Figure 10), thus verifying the reasonable design of the manipulator.

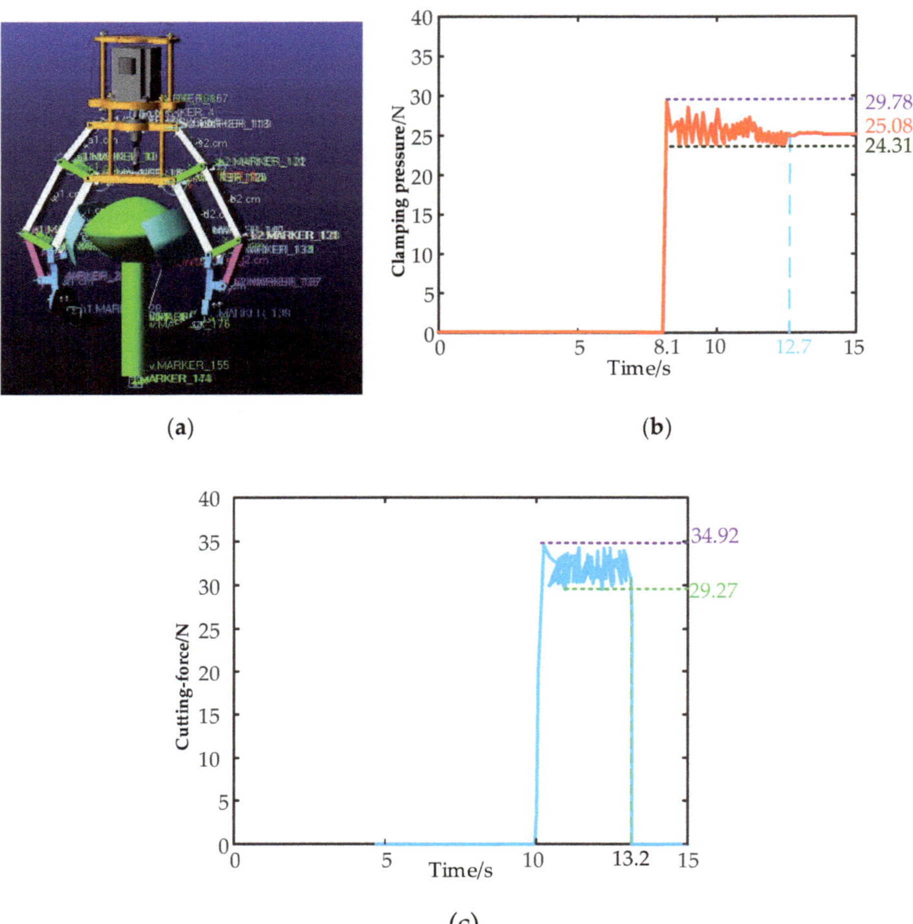

Figure 10. Manipulator motion simulation: (**a**) Manipulator motion simulation; (**b**) Clamping pressure curve; (**c**) Cutting force curve.

In Figure 10b, the red curve represented the pressure on the guard plates when the manipulator was clamping the flower ball, and the blue line represented the pressure stabilization time point. At 12.7 s, the clamping pressure tended to stabilize at 25.08 N, while purple represented the maximum pressure of 29.78 N and green represented the minimum pressure of 24.31 N. In Figure 10c, the blue curve represented the pressure on the blade when the manipulator was cutting the stem. When the cutting was finished at 13.2 s, the cutting force was 0 N. Purple represented the maximum cutting force of 34.92 N and green represented the minimum cutting force of 29.27 N. Therefore, it can be concluded from Figure 10b,c that the flower ball surface pressure and stem cutting force in the simulation test were basically consistent with the forces tested in the actual test above.

4.2. Broccoli-Picking Test Stand Set Up

According to the requirements of planting and picking, the feasibility of the picking manipulator was verified by setting up a picking test stand. Set up a picking test stand in the laboratory environment, as shown in Figure 11. The length, width, and height of the test stand support frame are 108, 100, and 120 cm, respectively, as shown in Figure 11(1). The control cabinet was placed on the upper layer of the support frame, as shown in Figure 11(2).

Figure 11(3) shows that the Siasun GCR5-910 robot arm with six degrees of freedom was hung upside down in the middle of the support frame of the test stand. The manipulator was connected to the end flange of the robot arm, as shown in Figure 11(4). A telepad with a length, width, and height of 120, 30, and 20 cm, respectively, was placed in the middle of the support frame of the test stand to simulate ridge planting conditions. Broccoli plant fixers were installed at 40 cm intervals on the telepad, as shown in Figure 11(5). Baskets with a height of 20 cm and a diameter of 30 cm were placed at the four corners horizontally from the middle point of the support frame, as shown in Figure 11(6).

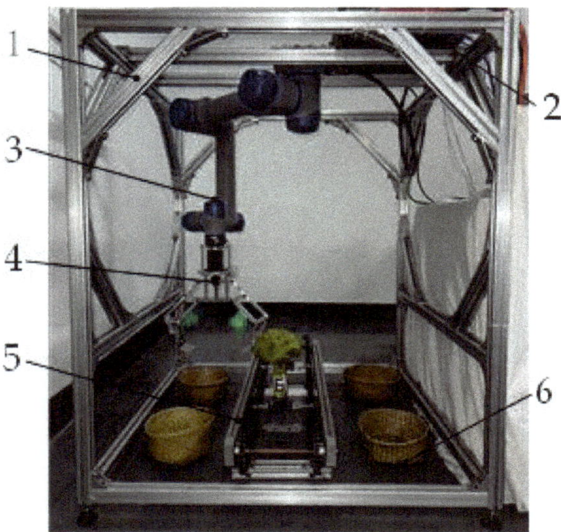

Figure 11. Broccoli-picking test stand: (1) test stand support frame; (2) control cabinet; (3) robot arm; (4) manipulator; (5) telepad; (6) baskets.

4.3. Establishment of Broccoli-Picking Control System

The control system of the picking test stand is composed of a manipulator with six degrees of freedom, a control cabinet, an underactuated manipulator, a stepper motor, and a stepper motor controller. The schematic diagram of the control system of the picking test stand is shown in Figure 12.

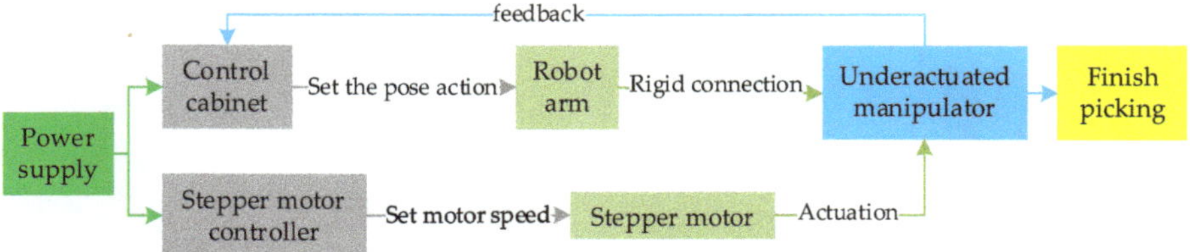

Figure 12. Diagram of the broccoli-picking test stand system.

The motion trajectory of the mechanical arm was planned in advance, and the motion path and posture of the robot arm were determined directly through the demonstrator to facilitate the determination of the position of the broccoli. In this way, the clamping and unloading time of the manipulator was set to 2.5 s, and the speed of the stepper motor was adjusted to 60 r/min. The picking test was carried out in the picking test environment without interference.

4.4. The Picking Test of Testbed

First, the power supply is turned on. After the robot arm is fully started, the manipulator is driven to the top of the broccoli plant through the control panel. The underactuated mechanism gradually approaches the plant under the motor drive (Figure 13a). The manipulator holds the guard plates and cutting blade in contact with the plant (Figure 13b). Broccoli is cut under the motor drive (Figure 13c) and reaches the top of the harvest basket under the traction of the robot arm. The manipulator opens under the drive of the motor, and the broccoli falls into the harvest baskets (Figure 13d). The process is shown in Figure 13.

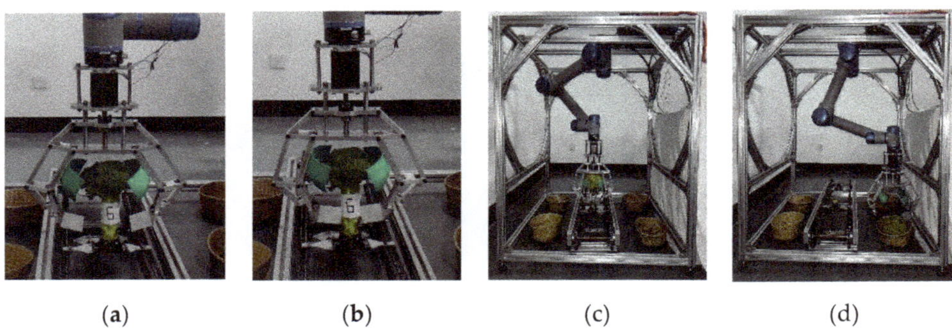

(a) (b) (c) (d)

Figure 13. Experimental picking process of broccoli: (**a**) The manipulator moves closer to the broccoli; (**b**) The manipulator contacts the broccoli; (**c**) Clamping and cutting; (**d**) Broccoli is released into the harvest basket.

5. Experiment and Analysis of Picking

5.1. Material and Method

The picking experiment was carried out on 6 November 2022. A simulated picking test was conducted under an experimental environment to verify the success rate, non-destructive rate, and picking efficiency of the mechanism. In the early stage of the experiment, the flower balls of 100 broccoli were divided into five groups with 20 broccoli each according to their diameter: 135–145, 145–155, 155–165, 165–175, and 175–185 mm. When the broccolis were brought back, their leaves had been removed; therefore, the broccolis in the experiment had no leaves; The manipulator was opened to a maximum size of 21 cm by the stepper motor drive. The broccoli was then attached to the telepad, as shown in Figure 14. Then, the test was carried out in the way described above, and these test data were recorded.

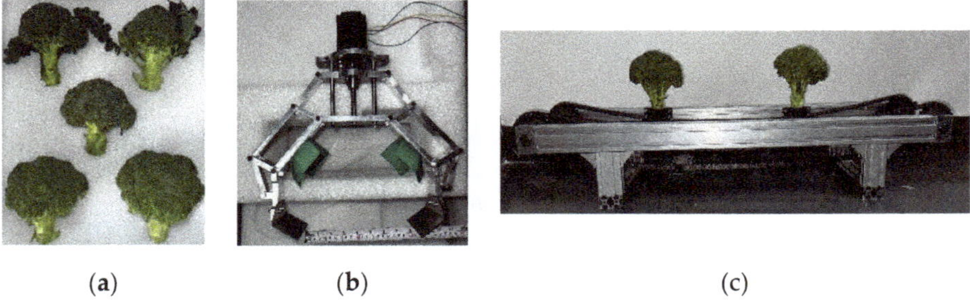

(a) (b) (c)

Figure 14. Test material: (**a**) Broccoli of different diameters; (**b**) Initial state of the manipulator; (**c**) Broccoli-picking location.

5.2. Analysis of Picking Results

The criteria for successful harvest were divided into three conditions: planting retention height, flower ball extrusion state, and stem incision state (Figure 15).

(a) (b)

Figure 15. Successful picking factors: (**a**) Clamping damage was low, and plant height was greater than 12 cm; (**b**) The incision surface was flat and non-adhesive.

Ten broccoli plants were selected from each of the five groups to determine the success rate of picking. The robot arm ran at 2.8 m/s, and the stepper motor ran at 60 r/min. As shown in Table 3, the success rate of broccoli picking can reach 84% on average.

Table 3. The success rate of picking.

Group	Flower Ball Diameter (mm)	Total Number of Picks	Number of Successful Picks	Number of Picking Failures	Picking Success Rate (%)	Number of Broccolis with Harvesting Heights Higher than 12 cm
1	135–145	10	8	2	80	9
2	145–155	10	9	1	90	10
3	155–165	10	9	1	90	9
4	165–175	10	8	2	80	10
5	175–185	10	8	2	80	8
Total		50	42	8	84	46

Afterward, 10 broccoli plants were sampled from each of the five groups. The average harvesting time of a single broccoli plant was calculated under the five different moving speeds of the robot arm. The results are shown in Table 4.

Table 4. Average harvesting time under different movement speeds.

Motion Speed of the Arm (m/s)	Average Time for the Arm to Reach the Broccoli (s)	Average Time for the Arm to Finish Picking (s)	Average Time to Release to Harvest Basket (s)	Total Time (s)	Number of Broccoli with Harvesting Height Higher than 12 cm
2.2	2.86	2.02	10.23	15.11	10
2.5	2.57	2.02	9.32	13.91	9
2.8	2.33	2.02	8.55	12.9	8
3.1	2.31	2.02	7.18	11.51	10
3.4	2.27	2.02	7.08	11.37	9

When the speed of the robot arm exceeded 3.4 m/s, the picking test bench shook violently under the action of inertia. Therefore, the fastest single flower ball harvesting speed was 11.37 s in the laboratory environment when the arm speed was 3.4 m/s, and the stepper motor speed was 60 r/min.

Mechanical damage characteristics of fruits and vegetables include static pressure, vibration, and impact damage [32]. However, the manipulator used in this paper exerts pressure on flower balls during the picking process. According to the simulation test and damage observation of 100 post-harvest flower balls, the pressure generated by the manipulator does not damage the flower ball surface. Therefore, the non-destructive rate of picking 100 broccoli flower balls was 100%.

5.3. Manipulator Characteristic Evaluation

In this paper, a symmetrical clamping and cutting style broccoli picking manipulator is designed based on the underactuated principle, which has the following characteristics.

(1) The manipulator has a simple and compact structure and simple control system, and stable configuration when picking broccoli.

(2) In the laboratory environment, the success rate of picking broccoli with the six-degree-of-freedom robot arm equipped with the manipulator was 84%, and the non-destructive rate was 100%.

(3) In the closed state, the maximum strength of the manipulator is 28 cm, the widest part is 21 cm, and the overall mass is 3.86 kg. Therefore, the size of the manipulator can be further reduced in the future to make the structure more flexible and lightweight processing. Thus applicable to other cruciferous ball-type vegetables.

6. Conclusions

(1) According to the physical characteristics of broccoli, a symmetrical double-finger underactuated manipulator was designed for picking operation by means of clamping and cutting.

(2) According to the surface pressure and stem-cutting force test of the broccoli flower ball and the analysis of kinematics and static mechanics of the underactuated manipulator, each connecting rod of the underactuated mechanism was determined to be 50 cm, 90 cm, and 50 cm, 90 cm, 50 cm, 60 cm, 65 cm, and the driving force of the manipulator was 598.66 N–702.88 N. The manipulator has the ability to pick broccoli with a diameter of 135 mm–185 mm. The reliability of the manipulator was verified by the simulation analysis of the manipulator motion.

(3) According to the requirements of broccoli planting and picking, picking test in a broccoli-picking test stand through the laboratory environment with 100 broccolis proved that the manipulator had a picking success rate of 84% and a lossless rate of 100% when the speed of the arm was 3.4 m/s, and the speed of the stepper motor was 60 r/min. The fastest single harvest time of the test stand was 11.37 s.

Author Contributions: Original draft, investigation, resources, validation, review and editing, H.X., C.N. and Y.C.; Supervision, review, and editing, funding acquisition, G.Y.; Methodology, data curation, software, visualization, conceptualization, C.N. and H.X.; Project administration, G.Y. and X.Z.; Formal analysis, Y.W. and H.X. All authors have read and agreed to the published version of the manuscript.

Funding: This research was funded by the Key Research and Development Project of the Science and Technology Department of Zhejiang Province, China, grant number 2021C02021; National key research and development program, China, (No. 2022YFD2001800).

Institutional Review Board Statement: Not applicable.

Informed Consent Statement: Not applicable.

Data Availability Statement: Not applicable.

Acknowledgments: Key Research and Development Project of the Science and Technology Department of Zhejiang Province, China, (No. 2021C02021); National key research and development program, China, (No. 2022YFD2001800).

Conflicts of Interest: The authors declare no conflict of interest.

References

1. Aloo, S.O.; Ofosu, F.K.; Kilonzi, S.M.; Shabbir, U.; Oh, D.H. Edible plant sprouts: Health benefits, trends, and opportunities for novel exploration. *Nutrients* **2021**, *13*, 2882. [CrossRef] [PubMed]
2. Li, Z.S.; Liu, Y.M.; Fang, Z.Y.; Yang, L.M.; Zhuang, M.; Zhang, Y.Y.; Lv, H.H.; Wang, Y. Development status, existing problems, and coping strategies of the Chinese broccoli industry. *China Veg.* **2019**, *4*, 1–5. (In Chinese) [CrossRef]
3. Vrochidou, E.; Tsakalidou, V.N.; Kalathas, I.; Gkrimpizis, T.; Pachidis, T.; Kaburlasos, V.G. An Overview of End Effectors in Agricultural Robotic Harvesting Systems. *Agriculture* **2022**, *12*, 1240. [CrossRef]
4. Biswas, N.; Aslekar, A. Improving Agricultural Productivity: Use of Automation and Robotics. In Proceedings of the 2022 International Conference on Decision Aid Sciences and Applications, Chiangrai, Thailand, 23–25 March 2022. [CrossRef]
5. Zhang, B.; Xie, Y.; Zhou, J.; Wang, K.; Zhang, Z. State-of-the-Art robotic grippers, grasping and control strategies, as well as their applications in agricultural robots: A review. *Comput. Electron. Agric.* **2020**, *177*, 105694. [CrossRef]
6. Wang, Z.; Xun, Y.; Wang, Y.; Yang, Q. Review of smart robots for fruit and vegetable picking in agriculture. *Int. J. Agric. Biol. Eng.* **2022**, *15*, 33–54. [CrossRef]
7. Mochiyama, H.; Gunji, M.; Niiyama, R. Ostrich-Inspired Soft Robotics: A Flexible Bipedal Manipulator for Aggressive Physical Interaction. *J. Robot. Mechatron.* **2022**, *34*, 212–218. [CrossRef]
8. Hota, R.K.; Kumar, C.S. Effect of design parameters on strong and immobilizing grasps with an underactuated robotic hand. *Robotica* **2022**, *40*, 3769–3785. [CrossRef]
9. Birrell, S.; Hughes, J.; Cai, J.Y.; Iida, F. A field-tested robotic harvesting system for iceberg lettuce. *J. Field Robot.* **2020**, *37*, 225–245. [CrossRef]
10. He, T.; Aslam, S.; Tong, Z.; Seo, J. Scooping manipulation via motion control with a two-fingered gripper and its application to bin picking. *IEEE Robot. Autom. Lett.* **2021**, *6*, 6394–6401. [CrossRef]
11. Gao, G.; Chapman, J.; Matsunaga, S.; Mariyama, T.; MacDonald, B.; Liarokapis, M. A Dexterous, Reconfigurable, Adaptive Robot Hand Combining Anthropo-morphic and Interdigitated Configurations. In Proceedings of the 2021 IEEE/RSJ International Conference on Intelligent Robots and Systems, Prague, Czech Republic, 27 September–1 October 2021. [CrossRef]
12. Williams, H.A.; Jones, M.H.; Nejati, M.; Seabright, M.J.; Bell, J.; Penhall, N.D.; Barnett, J.J.; Duke, M.D.; Scarfe, A.J.; Ahn, H.S.; et al. Robotic kiwifruit harvesting using machine vision, convolutional neural networks, and robotic arms. *Biosyst. Eng.* **2019**, *181*, 140–156. [CrossRef]
13. Kinugawa, J.; Suzuki, H.; Terayama, J.; Kosuge, K. Underactuated robotic hand for a fully automatic dishwasher based on grasp stability analysis. *Adv. Robot.* **2022**, *36*, 167–181. [CrossRef]
14. Xiong, Y.; Ge, Y.; Grimstad, L.; From, P.J. An autonomous strawberry-harvesting robot: Design, development, integration, and field evaluation. *J. Field Robot.* **2020**, *37*, 202–224. [CrossRef]
15. Arad, B.; Balendonck, J.; Barth, R.; Ben-Shahar, O.; Edan, Y.; Hellström, T.; Hemming, J.; Kurtser, P.; Ringdahl, O.; Tielen, T.; et al. Development of a sweet pepper harvesting robot. *J. Field Robot.* **2020**, *37*, 1027–1039. [CrossRef]
16. Zhao, Y.W.; Geng, D.X.; Liu, X.M.; Sun, G.D. Kinematics Analysis and Experiment on Pneumatic Flexible Fruit and Vegetable Picking Manipulator. *Trans. Chin. Soc. Agric. Mach.* **2019**, *50*, 31–42. (In Chinese) [CrossRef]
17. Jia, J.M.; Ye, Y.Z.; Cheng, P.L.; Zhu, Y.P.; Fu, X.P.; Chen, J.N. Design and Experimental Optimization of Hand-held Manipulator for Picking Famous Tea Shoot. *Trans. Chin. Soc. Agric. Mach.* **2022**, *53*, 86–92. (In Chinese) [CrossRef]
18. Cui, Y.J.; Ma, L.; He, Z.; Zhu, Y.T.; Wang, Y.C.; Li, K. Design and Experiment of Dual Manipulators Parallel Harvesting. *Trans. Chin. Soc. Agric. Mach.* **2022**, *53*, 132–143. (In Chinese) [CrossRef]
19. Yin, J.J.; Chen, Y.H.; He, K.; Liu, J.Z. Design and Experiment of Grape-picking Device with Grasping and Rotary- cut Type of Underactuated Double Fingered Hand. *Trans. Chin. Soc. Agric. Mach.* **2017**, *48*, 12–20. (In Chinese) [CrossRef]
20. Ma, T.; Yang, D.; Zhao, H.W.; Li, T.; Ai, N. Grasp Analysis and Optimal Design of a New Underactuated Manipulator. *Robot* **2020**, *42*, 354–364. (In Chinese) [CrossRef]
21. Chen, W.; Xiong, C.; Chen, W.; Yue, S. Mechanical adaptability analysis of underactuated mechanisms. *Robot. Comput. -Integr. Manuf.* **2018**, *49*, 436–447. [CrossRef]
22. Lei, X.P.; Tang, S.F.; Liang, W.; Guo, Z.R. Optimization Design and Grasping Stability Analysis of Under-actuated Multi-finger Manipulator. *J. Mech. Transm.* **2020**, *44*, 53–58. (In Chinese) [CrossRef]
23. Guo, H.S.; Ma, R.; Zhang, Y.X.; Li, Z.Y. Design and Simulation Analysis of Ya Shaped Underactuated Korla Fragrant Pear Picking Manipulator. *J. Agric. Mech. Res.* **2023**, *45*, 110–117. (In Chinese) [CrossRef]
24. El Didamony, M.I.; El Shal, A.M. Fabrication and evaluation of a cabbage harvester prototype. *Agriculture* **2020**, *10*, 631. [CrossRef]
25. Blok, P.M.; Van, E.F.K.; Tielen, A.P.M.; Van Henten, E.J.; Kootstra, G. The effect of data augmentation and network simplification on the image-based detection of broccoli heads with Mask R-CNN. *J. Field Robot.* **2021**, *38*, 85–104. [CrossRef]
26. Calzado, J.; Lindsay, A.; Chen, C.; Samuels, G.; Olszewska, J.I. SAMI: Interactive, Multi-sense Robot Architecture. In Proceedings of the 2018 IEEE 22nd International Conference on Intelligent Engineering Systems, Las Palmas de Gran Canaria, Spain, 21–23 June 2018. [CrossRef]
27. Wang, Y.; Wang; Wen, Z.J.; Lian, Y.; Hu, S.; Li, X.J.; Shao, X.G. Broccoli Harvesting Machine. China Patent 201821455355.X, 2 April 2019. (In Chinese)
28. Du, J.C.; Yu, H.M.; Zhang, J.W.; Chen, Y.T.; Chen, K.J. Design on Broccoli Harvesting Suspension Bucket Based on ANSYS. *Agric. Eng.* **2019**, *9*, 102–107. (In Chinese) [CrossRef]

29. Zhang, X.M.; Chen, X.A.; Hou, X.N.; Liu, S.Y.; Xie, P.F. A Selective Broccoli Harvester. China Patent 202211324085 X, 31 January 2023. (In Chinese)
30. Zhao, Y.F.; Han, F.F.; Tang, Z.; Chen, S.R.; Ding, H.T. A Kind of Harvester for Precise Cutting of Broccoli. China Patent 202210599247.4, 16 August 2022. (In Chinese)
31. Niu, C.Y. Design and Experiment of Broccoli Harvesting Manipulator. Master's Thesis, Zhejiang Sci-Tech University, Hang Zhou, China, 1 April 2022. (In Chinese)
32. Zaicovski, C.B.; Pegoraro, C.; Ferrarezze, J.P.; Cero, J.D.; Lund, D.G.; Rombaldi, C.V. Effects of mechanical injury, temperature decreasing, and 1-MCP on the post-harvest metabolism of Legacy broccoli. *Food Sci. Technol.* **2008**, *28*, 840–845. [CrossRef]

Article

Simulation and Experiment of Compression Molding Behavior of Substate Block Suitable for Mechanical Transplanting Based on Discrete Element Method (DEM)

Jingjing Fu [1], Zhichao Cui [1], Yongsheng Chen [1], Chunsong Guan [1], Mingjiang Chen [1,*] and Biao Ma [2,*]

[1] Nanjing Institute of Agricultural Mechanization, Ministry of Agriculture and Rural Affairs, Nanjing 210014, China; tutujing12@163.com (J.F.); cuizhichaoabc@163.com (Z.C.); cys003@sina.com (Y.C.); cs.guan@163.com (C.G.)

[2] Graduate School, Chinese Academy of Agricultural Sciences, Beijing 100083, China

* Correspondence: chenmingjiang@caas.cn (M.C.); mabiao@caas.cn (B.M.)

Abstract: The compression molding performance of a substrate block has a significant effect on the quality and stability of mechanical transplanting. The physical experiment and DEM simulation were combined to evaluate the compression molding behavior of substrate block in this study. A calibration procedure of DEM parameters of peat particles was proposed at first. Then, the above parameters were brought into the contact model of the compression system–particles, and the effect of the loading speed on the compression behavior of the peat substrate block was investigated. The compressive force–displacement curves of the simulated and measured tests were all contained in the initial linear stage and non-linear stiffing stage. The particle number of central sections was higher than side section, and the variable coefficient was greater at higher loading speed. The substrate blocks all expanded after demolding. The higher the loading speed, the greater the expansion in the height's direction, and the easier it was for cracks to be generated near the bottom. This study will provide a reference for the design of substrate block forming machines.

Keywords: substrate block; compression molding; DEM; parameter calibration; seedling transplanting

Citation: Fu, J.; Cui, Z.; Chen, Y.; Guan, C.; Chen, M.; Ma, B. Simulation and Experiment of Compression Molding Behavior of Substate Block Suitable for Mechanical Transplanting Based on Discrete Element Method (DEM). *Agriculture* **2023**, *13*, 883. https://doi.org/10.3390/agriculture13040883

Academic Editor: Jacopo Bacenetti

Received: 17 March 2023
Revised: 7 April 2023
Accepted: 14 April 2023
Published: 17 April 2023

1. Introduction

Seedling transplanting is an important process in vegetable and economic crop production, and mechanical automation and intelligent production is an imperative portion of its development trend. However, at present, there still exists some problems, such as the rupture and fall of the seedling substrate or the damage of seedling stems and leaves, which will affect the quality, efficiency, and reliability of the transplantation [1]. Except for the structural design and stability of the transplanting machine, the quality of seedlings also has a significant effect on the seedlings' transplanting performance. A strong and unified seedling is expected to reduce management costs, and it is also suitable for mechanical operation. There are three types of seedlings for transplanting: bare root, plug trays, and pot/substrate block seedlings. The latter two are most used for mechanized transplantation [2,3]. The substrate block seedlings have better functional and developed root systems that are capable of withdrawing more water and nutrients, and they have higher survival rates than bare seedlings and plug seedlings [4,5]. Additionally, they also have relatively regular characteristics for the automatic mechanized transplanting than pot seedlings, which would help to expand the fully automated transplanting [6].

Although the substrate block is suitable for seedlings and mechanical operation, it also has stricter requirements than other types due to its compressed structure. In addition to the chemical properties (pH, EC, nutrients, etc.) provided by the components of the substrate materials, the physical characteristics have put forward higher requirements for the compressed substrate block, such as its mechanical strength, density, porosity, water availability, air capacity and expansion rate [7]. Those are mainly dependent on the compression process. Some studies have investigated the moisture content of the raw materials, component formula, and compression parameters (compression stress, temperature) on the quality of the substrate block and the seedling's performance [8–10]. In addition to the above factors, the processing machinery of the substrate block is also one of the key factors affecting its performance. However, there is currently no certain standard for the compression molding of the substrate blocks, and there is no uniform manufacturing standard for molding equipment. Nowadays, the existing equipment used in China still has some problems, such as substrate shedding or sticking to the mold. Hence, studying the working mechanism between the substrate and molding machinery will help to improve the operation performance of the processing equipment and further improve the quality of substrate blocks. The force and movement process between the machine and substrate particles are complex. The combination of a simulation method and a practical experimental method is usually the commonly used and effective method. In the research on agricultural products and equipment, the discrete element method (DEM) is widely employed to capture the dynamic behavior of the particle materials and the interaction mechanisms between the materials and tools [11–14]. For the substrate block seedlings, understanding the microscopic processes of the particle flow and compression is critical for the quality control of the substrate block. The application of DEM can be beneficial to reduce the risk of failure and cost of production.

In this work, the commonly used peat was used as the growth substrate. The calibration of the DEM parameters of the peat was determined by a combination of a physical test and a simulation test. The stacking angle was adopted as a response value, and the optimal parameters were obtained by Plackett–Burman test, the steepest climbing test, and Box–Behnken test methods. Then, those calibrated parameters were applied into the DEM simulation of the compression molding of the substrate block. The effects of the loading speed on the compression behavior of the substrate block were evaluated from the screen of the distribution movement of the pressed particles. Finally, the reliability and accuracy of the simulation parameters were verified through experiments. It was expected to provide a theoretical basis for the simulation of experiment research and the preparation of substrate blocks with good qualities that are suitable for mechanical transplanting. It also provided an improvement direction for the design of a substrate block forming machine.

2. Materials and Methods

2.1. Determination of the Intrinsic Parameters of the Peat

2.1.1. Moisture Content and Density

The moisture content of peat was measured by the low-temperature drying method specified in GB/T 16913-2008. The density of the peat was determined using the pycnometer method according to GB/T 16913-2008. After calculation, the moisture content and density of the peat substate were 26.69% and 1560 kg·m^{-3}, respectively.

2.1.2. Stacking Angle

The injection limited bottom surface method, according to GB/T 11986-1989, was used to measure the stacking angle of the peat. The measuring device was, mainly, composed of a fixed frame, a funnel, and a steel plate. The front-view image of the peat pile was captured by a high-definition camera, and computer image processing technology was used to process the images. The software MATLAB was used to extract the image boundary pixels through edge detection, and the software Origin was used to pick up the boundary points and was further subjected to polynomial fitting. The arc tangent of the slope of the tangent line of the fitted curve was the stacking angle. The measured value of the stacking angle of the peat was 51.15°.

2.2. Calibration of Discrete Element Simulation Parameters of the Peat

In order to establish a proper model for simulating the compressing process of peat substrate blocks by the DEM method, the parameter calibration of the peat was the prerequisite. The related simulation parameters mainly include the intrinsic parameters of the peat and the contact parameters between the peat and peat as well as between the peat and boundary material (impact recovery coefficient, static friction coefficient, rolling friction coefficient, etc.). The stacking angle referred to the maximum slope angle accumulated and formed on the horizontal surface when particles slowly fell from a certain height under the action of gravity. It was usually taken as the response value, which was mainly influenced by the friction mechanisms between the particles and between the boundaries. Hence, the stacking angle test was performed and reproduced by numerical simulation to determine the optimal parameters of the DEM model.

2.2.1. Setting of the Simulations

The 3D model of the test device for the simulation stacking angle of the peat substate was established in the DEM software (EDEM 2020, Altair Engineering, Troy, MI, USA) according to its practical dimension. The material of the test device was set to stainless steel with a Poisson's ratio of 0.3, density of 7850 kg·m^{-3}, and shear modulus of 7×10^{10} Pa. The single spherical particles with a radius of 1 mm with a normal distribution were used to form the peat particles. The Hertz–Mindlin model and the JKR contact model were adopted in this simulation.

2.2.2. The Sensitivity of the DEM Parameters to the Stacking Angle

In order to explore the sensitivity of the external input DEM parameters on the stacking angle, a Plackett–Burman (PB) test with 11 factors and 1 central point was designed using the software Design-Expert version 12 (Stat-Ease, Inc., Minneapolis, MN, USA). There were 9 real variables (A to I), and 2 other virtual variables (J and K) were used to estimate the errors, while each parameter was set to low and high levels, according to the recommended range value. According to related studies [15–18] and a large number of pre-tests in the early stage, the ranged of each parameter were selected, as listed in Table 1.

Table 1. The parameter table of the Plackett–Burman test.

Symbol	Parameters	Low Level (−1)	High Level (+1)
A	Poisson's ratio of the particles	0.2	0.4
B	Shear modulus of the particles (MPa)	1	10
C	Impact recovery coefficient of particle–particle	0.2	0.6
D	Static friction coefficient of particle–particle	0.2	0.8
E	Rolling friction coefficient of particle–particle	0.1	0.5
F	Impact recovery coefficient of particle–steel	0.2	0.6
G	Static friction coefficient of particle–steel	0.2	0.6
H	Rolling friction coefficient of particle–steel	0.05	0.25
J	Surface energy (J·m^{-2})	0.1	1
K, L	Virtual parameter	−1	1

2.2.3. Calibration of DEM Parameters of Peat

Based on the significant parameters selected by the Plackett–Burman test, the steepest climbing test was designed to further narrow the range of significant parameters. In the simulation test, the non-significant parameter was set to the middle value of the range, and the significant parameter was gradually increased, according to the designed step length. The relative error between the simulation and the physical measured value of the stacking angle was taken as the evaluation index. Based on the results of the steepest climbing test, the Box–Behnken test was designed by the response surface methodology (RSM). The significant parameters were set with high, medium, and low levels. The set of non-significant parameters was the same as the steepest climbing test.

2.3. DEM Simulation for the Compression Molding of Peat Substrate Blocks

According to the size requirements of substrate blocks, a cuboid compression system was set in the DEM simulation environment (Figure 1a), which consisted of a top punch, a square tube, and a bottom plate. The compression press was simulated to understand the quantitative relationship between the DEM parameter settings and the compression behavior of simulated particles. The DEM parameters of particles were set based on the obtained optimal parameters from the results of the calibration of the DEM parameters of the peat. Before the compression simulation, a pre-test of the particle generation was carried out first. According to the actual load capacity of the mold and the pre-test of the particle generation, the results showed that 16,500 particles could almost fill the container. Additionally, further compression molding simulations were all based on the model with generated particles. During the simulation of compression molding process, the top punch part started to move down to compress the substrate with a loading displacement of 60 mm. The different loading speeds at 50 mm·min^{-1}, 100 mm·min^{-1}, 200 mm·min^{-1}, and 500 mm·min^{-1} were set in the simulation. The compressive force–displacement curves and particle numbers on the center section, quarter section, and inner wall (Figure 1b) were collected in the post-simulation process.

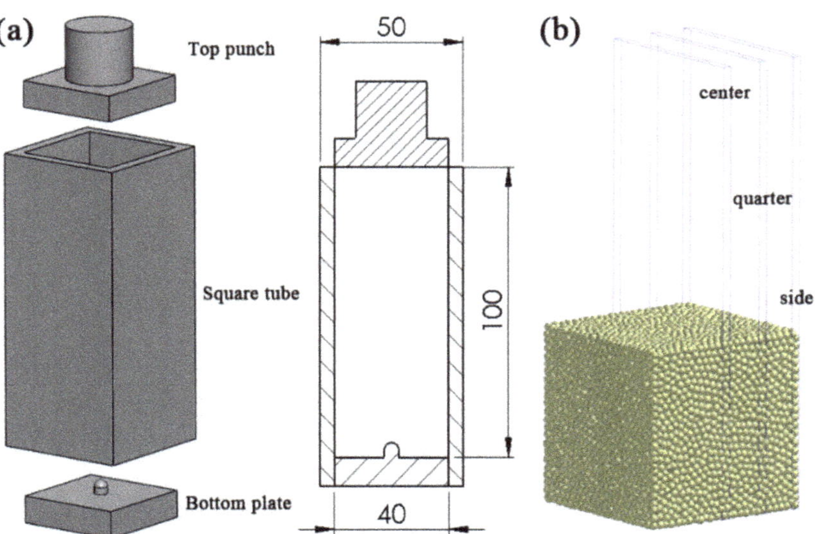

Figure 1. (**a**) Geometry of compression simulation platform and (**b**) data selection surfaces of particle numbers.

2.4. Experiment on the Compression Molding of Peat Substrate Blocks

The compression molding test of peat substrate blocks was carried out on a WDW-20 universal test machine (Jinan Chuanbai Instrument Equipment Co., Ltd., Jinan, China). The compression molding device consisted of a compression system and demolding part (Figure 2). The composition of the compression system was the same as the simulation platform, including a top punch, square tube, and bottom plate. When the compression molding test was carried out, the demolding part was not placed first. Before the test, the square tube was filled in a natural loose state. The compression molding parameters were set similar to the simulation: a loading displacement of 60 mm and different loading speeds of 50 mm·min^{-1}, 100 mm·min^{-1}, 200 mm·min^{-1}, and 500 mm·min^{-1}. When the compression molding process was complete, the compression molding device was placed, as shown in Figure 2. Then, the peat substrate block was demolded by the top punch moving down.

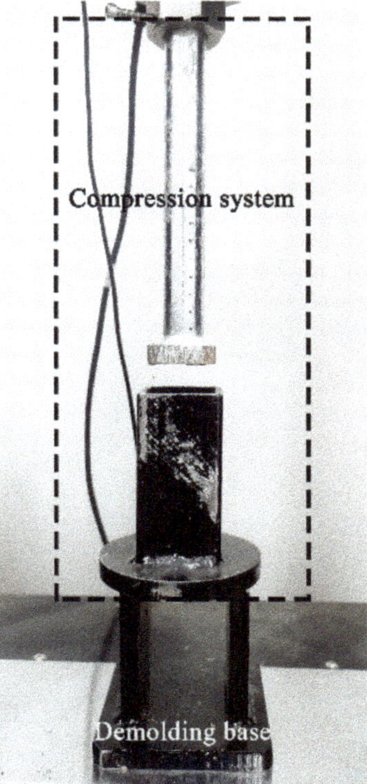

Figure 2. Compression molding device of peat substrate block.

3. Results and Discussion

3.1. Calibration of the DEM Parameters of Peat

3.1.1. Plackett–Burman (PB) Test

The protocol and results of the PB test and the significant results of each parameter are listed in Tables 2 and 3, respectively. The total contribution of B, D, and J was 83%, and they all significantly affected the stacking angle and followed the order: J > B > D. As shown in the Pareto chart (Figure 3), the contribution values of those three parameters were all exceeded the Bonferroni limit, which also indicated that the parameters were significant for the response. The contribution value lower than the t-value limit usually meant there was no effect for the response [19]. The contribution value of E between the Bonferroni limit and t-value limit was considered E and also had some effect on the response. Additionally, the information provided by Pareto chart showed that J and B had a positive effect on the response, while D and E had negative effects. Therefore, those four parameters were identified as the key parameters for the substrate to be calibrated in the steepest climbing test and BBD test below.

Table 2. The protocol and results of PB test.

Run	A	B	C	D	E	F	G	H	J	K	L	Stacking Angle (°)
1	−1	−1	1	−1	1	1	−1	1	1	1	−1	59
2	−1	−1	−1	−1	−1	−1	−1	−1	−1	−1	−1	32.76
3	−1	1	1	1	−1	−1	−1	1	−1	1	1	41.43
4	−1	1	−1	1	1	−1	1	1	1	−1	−1	57.17
5	1	1	−1	1	1	1	−1	−1	−1	1	−1	33.38
6	−1	−1	−1	1	−1	1	1	−1	1	1	1	46.23
7	1	−1	−1	−1	1	−1	1	1	−1	1	1	37.47
8	0	0	0	0	0	0	0	0	0	0	0	61.64
9	1	−1	1	1	1	−1	−1	−1	1	−1	1	44.28
10	−1	1	1	−1	1	1	1	−1	−1	−1	1	54.66
11	1	1	−1	−1	−1	1	−1	1	1	−1	1	84.23
12	1	1	1	−1	−1	−1	1	−1	1	1	−1	83.65
13	1	−1	1	1	−1	1	1	1	−1	−1	−1	43.13

Table 3. Significance analysis of parameters in PB test.

p	Contribution (%)	*p*-Value	Order
A	3.05	0.0822	6
B	21.02	0.0133 *	2
C	3.05	0.0821	5
D	18.56	0.0150 *	3
E	5.17	0.0509	4
F	1.42	0.1549	9
G	1.85	0.1253	8
H	1.88	0.1236	7
J	43.42	0.0065 **	1

Note: * and ** represent the significance at the 0.05 and 0.01 probability level, respectively.

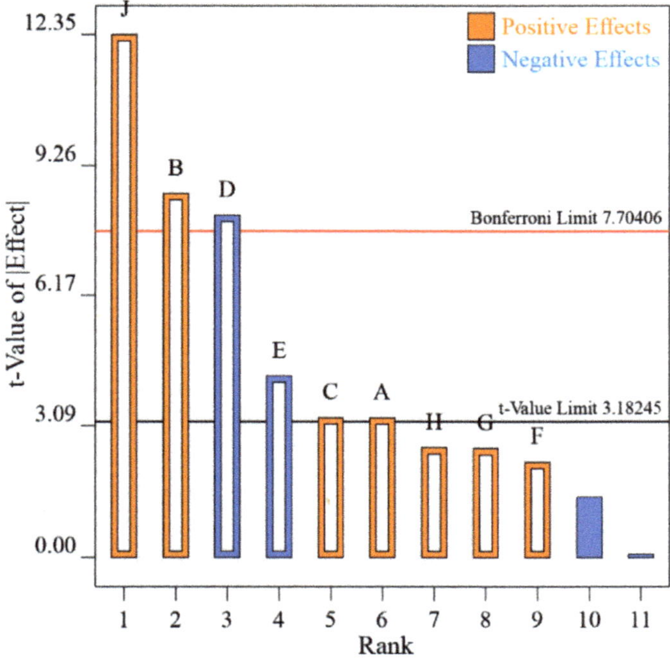

Figure 3. Pareto chart of parameters on response.

3.1.2. The Steepest Climbing Test

The design scheme and results of the steepest climbing test are presented in Table 4. The relative error was the value between the simulated stacking angle and the measured stacking angle (51.15°). The stacking angle gradually increased with the arrangement order of the test groups, and the relative error presented as first decreasing and then increasing. The stacking angle and relative error of the No.2 run were 51.07° and 0.16%, respectively. It indicated that the most optimal parameter level range was around the No.2 test, which was chosen to be the intermediate level (0) in the Box–Behnken test design. Additionally, the parameter levels of No.1 and No. 3 were chosen as the low (−1) and high levels (+1).

Table 4. Design scheme and results of steepest climbing test.

No.	B	D	E	J	Stacking Angle (°)	Relative Error (%)
1	1	0.8	0.5	0.1	32.74	35.99
2	3.25	0.65	0.4	0.325	51.07	0.16
3	5.5	0.5	0.3	0.55	54.81	7.16
4	7.75	0.35	0.2	0.775	70.67	38.16
5	10	0.2	0.1	1	82.75	61.78

Note: positive effects (B and J) were in ascending order, and negative effects (D and E) were in descending order.

3.1.3. Box–Behnken Test

Based on the parameter selection from the PB test and the level selection from the steepest climbing test, a Box–Behnken test design (BBD) was carried out, and the results are shown in Table 5. The regression model describing the relationship between the stacking angle and selected parameters can be described as:

$$\text{Stacking angle} = 53.57 + 2.18B - 1.71D - 0.4179E + 6.81J + 1.25BD + 0.5006BE - 2.76BJ + 1.15DE - 1.01DJ + 1.53EJ + 0.0268B^2 - 0.6149D^2 + 2.86E^2 - 3.41J^2 \quad (1)$$

Table 5. BBD experiment results.

Run	B	D	E	J	Stacking Angle (°)
1	0	1	−1	0	53.0542
2	0	−1	1	0	56.04395
3	0	1	0	1	50.8817
4	1	−1	0	0	55.8771
5	1	0	−1	0	61.63415
6	0	−1	0	−1	46.189
7	0	0	0	0	53.98465
8	0	0	0	0	50.0464
9	0	0	0	0	53.55925
10	1	0	0	−1	44.7601
11	0	0	0	0	53.65835
12	0	0	−1	−1	47.10265
13	0	1	0	−1	41.7636
14	−1	0	1	0	50.258
15	−1	0	0	−1	39.92395
16	0	0	1	1	62.2329
17	0	−1	0	1	59.3303
18	0	0	1	−1	44.90385
19	0	0	−1	1	58.3123
20	−1	0	−1	0	55.354
21	1	0	1	0	58.5406
22	−1	1	0	0	47.8322

Table 5. *Cont.*

Run	B	D	E	J	Stacking Angle (°)
23	−1	−1	0	0	51.86525
24	−1	0	0	1	60.9074
25	1	1	0	0	56.82535
26	0	−1	−1	0	57.6126
27	1	0	0	1	54.70705
28	0	1	1	0	56.0758
29	0	0	0	0	56.5945

The ANOVA results of the quadratic model for BBD are listed in Table 6. The regression model's p-value = 0.0001 demonstrated a significant relationship between the responses and parameters. Consistent with the results of the PB test, B, D and J significantly affected the stacking angle. Moreover, the quadratic terms E^2 and J^2 had significant influence on stacking angle. The lack of fit was insignificant, and the correlation coefficient (R^2 = 0.8962) of the model was high, which demonstrated that the regression model fit well.

Table 6. ANOVA of the quadratic model for BBD.

Source	Sum of Squares	df	Mean Square	F-Value	p-Value
Model	867.18	14	61.94	8.63	0.0001 **
B	57.22	1	57.22	7.98	0.0135 *
D	34.97	1	34.97	4.87	0.0444 *
E	2.10	1	2.10	0.2921	0.5974
J	556.63	1	556.63	77.58	<0.0001 **
BD	6.20	1	6.20	0.8646	0.3682
BE	1.00	1	1.00	0.1397	0.7142
BJ	30.45	1	30.45	4.24	0.0585
DE	5.27	1	5.27	0.7342	0.4059
DJ	4.05	1	4.05	0.5640	0.4651
EJ	9.36	1	9.36	1.30	0.2725
B^2	0.0047	1	0.0047	0.0007	0.9800
D^2	2.45	1	2.45	0.3419	0.5680
E^2	52.94	1	52.94	7.38	0.0167 *
J^2	75.29	1	75.29	10.49	0.0059 **
Residual	100.44	14	7.17		
Lack of fit	78.70	10	7.87	1.45	0.3851
Pure error	21.74	4	5.44		
Cor Total	967.62	28			

Note: * and ** represent the significance at the 0.05 and 0.01 probability level, respectively.

By considering the measured stacking angle (51.15°) as the target value and solving the regression model by using the optimization function of the software Design Expert, the optimal values of the four parameters were obtained as B = 2.43 MPa, D = 0.77, E = 0.41, and J = 0.34 $J \cdot m^{-2}$. The obtained optimal parameters were employed to perform the DEM validation test. The simulated stacking angle was 50.89°, and its relative error with the measured stacking angle was 0.51% (Figure 4), indicating that the DEM models calibrated in this work were found to have good accuracy.

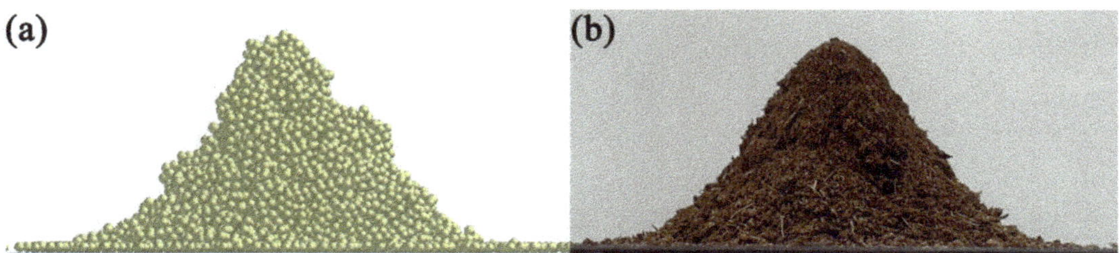

Figure 4. Comparison of the shape of peat pile between the (**a**) simulation test and (**b**) physical test.

3.2. Compression Molding Behavior of Substrate Block

The curves of the force–displacement responses in the compression molding process of the substrate block of measured and simulated experiments under different loading speeds are displayed in Figure 5a,b. The compressive force slightly increased when the displacement was lower than 40 mm and then increased rapidly until it reached the set displacement in both the measured and simulated processes. The first stage (displacement < 40 mm) could be considered as the linear compression process, which was mainly due to the pore reduction caused by the discharge of air and moisture. The second non-linear stage was the result of realignment of substrate particles. In this stage, the pores between the particles were filled, and their contact areas and frictional forces were increased, leading to further compression difficulties and jumpy pressures. This compression behavior was similar with other researchers, including an initial linear stage followed by a non-linear stiffening stage [20]. There was no great difference in the force–displacement curves among different loading speeds in either the measured or simulated processes. Therefore, only one set of the loading speeds (500 mm·min^{-1}) was selected for comparison (Figure 5c). It can be seen that the compression force began to increase after being compressed to 20 mm in the simulated process, while the force increased from the beginning of the compression in the measured process. It was due to the step of particle generation in the simulation. Before compression, the generated particles happened with a certain compression stack under their own gravity. Hence, in the beginning of the compression, the contact force between the top punch and particles was very small, and the two did not even touch. When pressing through this distance, the force of the simulation process increased significantly, and it, essentially, matched the actual process when the punch reached 50 mm. The comparison of the maximum compressive force of the measured and simulated tests is exhibited in Table 7. The maximum compressive force increased with the increasing loading speed both in the measurement and simulation. At higher loading speeds, the compressed displacement of the substrate particles per unit of time was larger, and then they will receive a larger compressive force. The simulated results under different loading speeds were all higher than the measured results, which was due to the unavoidable experimental errors in practical tests. However, the relative error was lower than 15%, especially at a loading speed of 500 mm·min^{-1} (5.45%), which demonstrated the DEM simulations had a good match with experimental measurements, and the DEM model could capture the compression behavior of particles.

The instant particle number at the set displacement and the total accumulated particle number of three selected sections are shown in Table 8 and Figure 6. When the top punch reached the instant of the set displacement, the center and quarter sections possessed similar particle numbers around 850, while those in the side section were only in the range of 650–690 at all different loading speed conditions. Additionally, the variable coefficient was larger at higher loading speeds. This phenomenon also happened in the total accumulated particle number, and the difference between the inside section (center and quarter) and side section was greater. The average and total accumulated particle numbers decreased with increasing loading speed. It was noted that when the displacement of the punch part exceeded 40 mm, there was a trend in the accumulated particle numbers of the quarter section being larger than the center section (Figure 6). As mentioned above, the particles were subjected to the non-linear stage of particle realignment, and the center particles started to spread around, causing the increasing particle number of the quarter section. Hence, at the instant of the set displacement, the quarter section had more particles. However, in total, the center section possessed the most total accumulated particles.

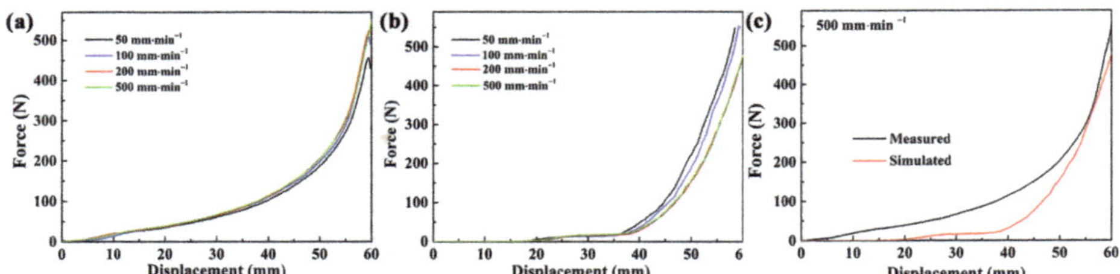

Figure 5. Compressive force–displacement curves of compression process: (**a**) measured, (**b**) simulated, and (**c**) comparison of measured and simulated process at 500 mm·min^{-1} loading speed.

Table 7. The maximum compressive force of measured and simulated tests and their relative error.

Loading Speed (mm·min^{-1})	Maximum Compressive Force (N)		Relative Error (%)
	Measured	Simulated	
50	456.2	520.9	14.18
100	509.6	550.9	8.10
200	526.6	573.0	8.81
500	550.8	580.8	5.45

Note: relative error = | simulated − measured | /measured × 100%.

Table 8. The instant particle number and average particle number of selected sections at set displacement.

Loading Speed (mm·min^{-1})	Particle Number			Average Particle Number	Variable Coefficient (%)
	Center	Quarter	Side		
50	827	861	690	793 (90.52)	11.42
100	836	848	691	792 (87.39)	11.04
200	823	836	660	773 (98.08)	12.69
500	814	841	655	770 (100.50)	13.05

The heights of the peat substrate blocks of different loading speeds after standing for 48 h were collected: 63 (±1) mm, 64 (±1) mm, 65 (±1) mm, and 67 (±1) mm, respectively. According to the length of the tube and the moving displacement of the punch, the theoretical height of a block should be 40 mm, which indicates that the substrate blocks expanded in the height direction after compression molding. Additionally, their surfaces have cracks of varying degrees: as the loading speed increased, the location of the crack was closer to the bottom (Figure 7). At slower loading speeds, the firmness of the block from bottom to top was better, which could be reflected by the particle number. Therefore, the higher the loading speed the easier it was for the blocks to expand after molding to form cracks from the bottom. However, in order to achieve a certain production efficiency in actual production, the compression speed is, generally, fast. Therefore, in order to reduce the expansion problem of the substrate block and ensure its quality, it is recommended to explore the molding quality through some methods, such as adding agglomerants.

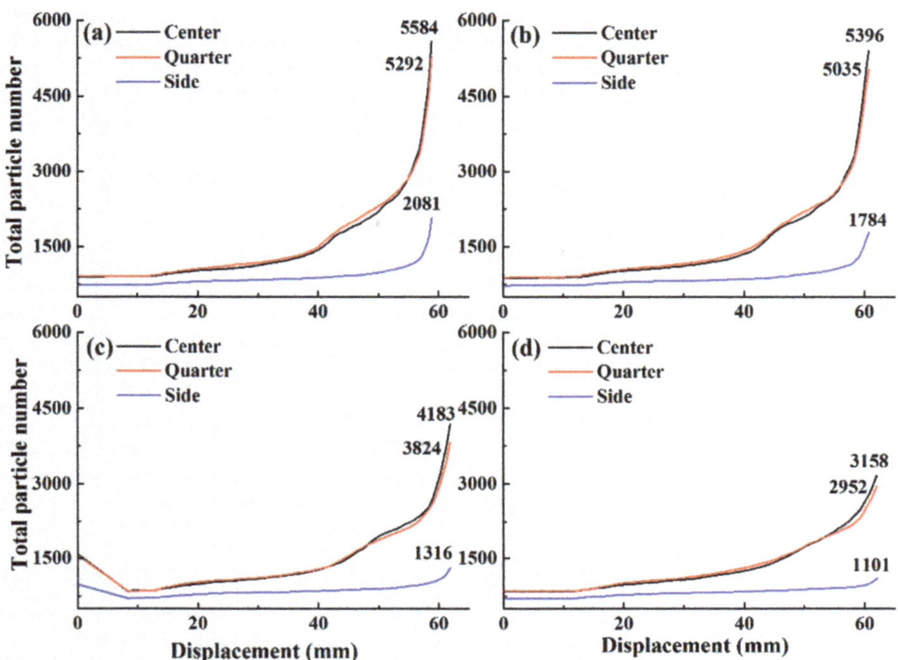

Figure 6. Total accumulated particle number on selected sections of substrate block at different loading speed: (**a**) 50 mm·min^{-1}, (**b**) 100 mm·min^{-1}, (**c**) 200 mm·min^{-1}, and (**d**) 500 mm·min^{-1}.

Figure 7. Image of peat substrate blocks with different loading speed after standing 48 h.

4. Conclusions

The compression molding behavior of peat substrate blocks was evaluated by the combination of a physical experiment and simulation. The simulation parameters of peat particles were calibrated first by the discrete element and response surface methodology. Then, the calibrated parameters were brought into the contact model of the compression system–particles, and the effect of the loading speed on the compression behavior of the substate blocks was investigated. The following conclusions were drawn:

(1) In the proposed calibration method of the DEM parameters of peat particles, the shear modulus of the particles, the static friction coefficient of the particle–particle interactions, the rolling friction coefficient of the particle-particle interactions, and surface energy were found have significant effects on the stacking angle of peat through the Plackett–Burman test. After the range was narrowed by the steepest climbing

test, the significant parameters were optimized by the second-order regression model established in the Box–Behnken test. The result of the validation test of the stacking angle of peat under optimal parameters showed the average relative error between the simulated (50.89°) and the physical stacking angle (51.15°) was 0.51%, indicating the good reliability and accuracy of the DEM model and the simulated parameters of the peat particles.

(2) In the physical and simulated experiments of the compression molding of peat substrate blocks, the compression force of the measured and simulated test all included an initial linear stage followed by a second non-linear stiffening stage in the whole compression process. Additionally, the maximum force increased with increasing loading speed in both the measured and simulated test. The relative errors between measured and simulated maximum force were around 10%, and, especially, the loading speed at 500 mm·min^{-1} was 5.45%. In general, the DEM simulations had a good match with the experimental measurements, and the DEM model could capture the compression behavior of particles.

(3) The peat substrate blocks all exhibited surface cracks under different loading speeds, and the location of the crack was closer to the bottom as the loading speed increased. It was due to the inhomogeneous particle distribution, resulting in varying degrees of compaction along the forming direction at higher loading speeds. While the practical production of substrate blocks mostly involves rapid prototyping to ensure operational efficiency, the compression molding of pure peat usually has lower mechanical properties that do not meet the requirements of mechanical transplanting. Hence, it is suggested to add some agglomerant to enhance the interparticle binding.

In summary, this study established a DEM model that could better simulate the compression molding of peat particles to prepare substrate blocks, which will contribute to the performance improvements of substrate blocks. The finding of this work could provide a reference for the design of the forming machine of substrate blocks suitable for mechanical transplanting.

Author Contributions: Conceptualization, J.F., Y.C. and M.C.; methodology, J.F., Z.C., C.G. and B.M.; software, J.F.; validation, J.F., M.C. and B.M.; formal analysis, J.F.; investigation, J.F., Z.C. and C.G.; data curation, J.F.; writing—original draft preparation, J.F.; writing—review and editing, J.F., M.C. and B.M.; visualization, J.F.; supervision, Y.C. and M.C.; funding acquisition, C.G., Z.C. and Y.C. All authors have read and agreed to the published version of the manuscript.

Funding: This research was funded by the Nanjing Modern Agricultural Machinery Equipment and Technology Innovation Demonstration Project (Grant No. NJ [2022] 03), Jiangsu Agricultural Science and Technology Innovation Fund (Grant No. CX (22) 3092), and Innovative research group of Agricultural Production Waste Resource Utilization Equipment.

Institutional Review Board Statement: Not applicable.

Data Availability Statement: Data are contained within the article.

Acknowledgments: The authors thank the editor and anonymous reviewers for providing helpful suggestions for improving the quality of this manuscript.

Conflicts of Interest: The authors declare no conflict of interest.

References

1. Jin, X.; Tang, L.; Li, R.; Zhao, B.; Ji, J.; Ma, Y. Edge recognition and reduced transplantation loss of leafy vegetable seedlings with Intel RealsSense D415 depth camera. *Comput. Electron. Agric.* **2022**, *198*, 107030. [CrossRef]
2. Nandede, B.M.; Carpenter, G.; Byale, N.A.; Chilur, R.; Jadhav, M.L.; Pagare, V. Manually operated single row vegetable transplanter for vegetable seedlings. *Int. J. Agric. Sci.* **2017**, *9*, 4911–4914.
3. Rahul, K.; Raheman, H.; Paradkar, V. Design and development of a 5R 2DOF parallel robot arm for handling paper pot seedlings in a vegetable transplanter. *Comput. Electron. Agric.* **2019**, *166*, 105014. [CrossRef]
4. Zhao, W.; Cui, Z.; Guan, C.; Chen, Y.; Yang, Y.; Gao, Q. Design and experiment of vegetable seedling substrate block forming machine. *J. Chin. Agric. Mech.* **2021**, *42*, 77–82.

5. Gao, G.; Liang, Y.; Du, Y.; Zhao, W. Optimization design and test of compression molding mechanism. *J. Chin. Agric. Mech.* **2020**, *41*, 43–49.
6. Xu, T.; Cui, Z.; Guan, C.; Chen, Y.; Xiao, T. Design and experiment of sending and taking seedling device of substrate block seedling transplanter. *J. Chin. Agric. Mech.* **2021**, *42*, 50–55.
7. Yang, L.; Cao, H.; Yuan, Q.; Luoa, S.; Liu, Z. Component optimization of dairy manure vermicompost, straw, and peat in seedling compressed substrates using simplex-centroid design. *J. Air Waste Manag.* **2018**, *68*, 215–226. [CrossRef] [PubMed]
8. Yang, L.; Yuan, Q.; Liu, Z.; Cao, H.; Luo, S. Experiment on seedling of compressed substrates with cow dung aerobic composting and earthworm cow dung composting. *Trans. Chin. Soc. Agric. Eng.* **2016**, *32*, 226–233.
9. Xin, M.; Chen, T.; Zhang, Q.; Jiao, J.; Bai, X.; Song, Y.; Zhao, R.; Wei, C. Parameters optimization for molding of vegetable seedling substrate nursery block with rice straw. *Trans. Chin. Soc. Agric. Eng.* **2017**, *33*, 219–225.
10. Qu, P.; Cao, Y.; Wu, G.; Tang, Y.; Xia, L. Preparation and Properties of Coir-Based Substrate Bonded by Modified Urea Formaldehyde Resins for Seedlings. *BioResources* **2018**, *13*, 4332–4345. [CrossRef]
11. Wu, Z.; Wang, X.; Liu, D.; Xie, F.; Ashwehmbom, L.G.; Zhang, Z.; Tang, Q. Calibration of discrete element parameters and experimental verification for modelling subsurface soils. *Biosyst. Eng.* **2021**, *212*, 215–227. [CrossRef]
12. Liu, K.; Su, H.; Li, F.; Jiao, W. Research on parameter calibration of soil discrete element model based on response surface method. *J. Chin. Agric. Mech.* **2021**, *42*, 143–149.
13. Hoshishima, C.; Ohsaki, S.; Nakamura, H.; Watano, S. Parameter calibration of discrete element method modelling for cohesive and non-spherical particles of powder. *Powder Technol.* **2021**, *386*, 199–208. [CrossRef]
14. Wan, D.; Wang, M.; Zhu, Z.; Wang, F.; Zhou, L.; Liu, R.; Gao, W.; Shu, Y.; Xiao, H. Coupled GIMP and CPDI material point method in modelling blast-induced three-dimensional rock fracture. *Int. J. Min. Sci. Technol.* **2022**, *32*, 1097–1114. [CrossRef]
15. Luo, S. Molding Process and Molding Machine Design and Experiment of Vermicompost Substrate. Ph.D. Thesis, Huazhong Agricultural University, Wuhan, China, 2019.
16. Zhao, L.; Zhou, H.; Xu, L.; Song, S.; Zhang, C.; Yu, Q. Parameter calibration of coconut bran substrate simulation model based on discrete element and response surface methodology. *Powder Technol.* **2022**, *395*, 183–194. [CrossRef]
17. Xing, J.; Zhang, R.; Wu, P.; Zhang, X.; Dong, X.; Chen, Y.; Ru, S. Parameter calibration of discrete element simulation model for latosol particles in hot areas of Hainan Province. *Trans. Chin. Soc. Agric. Eng.* **2020**, *36*, 158–166.
18. Liu, H.; Zhang, W.; Ji, Y.; Qi, B.; Li, K. Parameter calibration of soil particles in annual rice-wheat region based on discrete element method. *J. Chin. Agric. Mech.* **2020**, *41*, 153–159.
19. Xia, R.; Li, B.; Wang, X.; Li, T.; Yang, Z. Measurement and calibration of the discrete element parameters of wet bulk coal. *Measurement* **2019**, *142*, 84–95. [CrossRef]
20. Maraldi, M.; Molari, L.; Regazzi, N.; Molari, G. Analysis of the parameters affecting the mechanical behaviour of straw bales under compression. *Biosyst. Eng.* **2017**, *160*, 179–193. [CrossRef]

Article

Design and Testing of a Self-Propelled Dandelion Seed Harvester

Zhe Qu [1], Qi Lu [1], Haihao Shao [1], Long Liu [2], Xiuping Wang [2] and Zhijun Lv [1,*]

[1] College of Mechanical and Electrical Engineering, Henan Agricultural University, Zhengzhou 450002, China; quzhe071171@henau.edu.cn (Z.Q.); luqi@stu.henau.edu.cn (Q.L.); shaohaihao@stu.henau.edu.cn (H.S.)

[2] Changyuan Branch of Henan Academy of Agricultural Sciences, Xinxiang 453004, China; liulong@hnagri.org.cn (L.L.); wangxiuping@hnagri.org.cn (X.W.)

* Correspondence: lvzhijun@henau.edu.cn

Abstract: At present, there are few harvesters for dandelion (*Taraxacum mongolicum*) seeds, which limits the large-area planting of dandelion. Furthermore, manual harvesting is characterized by huge labor intensity, low efficiency, and high costs. Combining the material characteristics of dandelion plants and seeds with agronomic requirements for harvesting dandelion seeds, a self-propelled dandelion seed harvester was designed. The harvester is mainly composed of collection devices, separation devices, transmission devices, and a rack. It can facilitate seed collection from plants, seed transportation, and seed–pappus separation in one operation. The collection and separation processes of dandelion seeds were studied to ascertain the main factors that affect the collection rate. Then, the collection and separation devices were designed, and their parameters were analyzed. Taking the forward speed, wind velocity of blowers, and rate of rotation of the drum as test factors and the collection rate as the evaluation index, quadratic regression orthogonal rotating field tests were performed. In this way, the optimal combination of operation parameters was determined: the collection rate is optimal when the forward speed is 0.8 $m \cdot s^{-1}$, the air velocity from the blowers is 1.63 $m \cdot s^{-1}$, and the rate of rotation of the drum is 419 rpm. Field test results showed that a favorable harvesting effect was achieved after operation of the harvester, and only small amounts of dandelion seeds remained unharvested. Under the optimal parameter combination, the collection rate reached 89.1%, which could meet requirements for practical field harvesting of dandelion seeds. The test results satisfy the design requirement.

Keywords: dandelion seeds; fluent simulation; harvester; design; testing

Citation: Qu, Z.; Lu, Q.; Shao, H.; Liu, L.; Wang, X.; Lv, Z. Design and Testing of a Self-Propelled Dandelion Seed Harvester. *Agriculture* **2023**, *13*, 917. https://doi.org/10.3390/agriculture13040917

Academic Editors: Cheng Shen, Zhong Tang and Maohua Xiao

Received: 13 March 2023
Revised: 11 April 2023
Accepted: 12 April 2023
Published: 21 April 2023

1. Introduction

Dandelion (*Taraxacum mongolicum*) is a perennial herb that is rich in vitamins A and C and nutrient elements such as potassium [1–4] and that has high medicinal value [5,6] and economic value [7]. Dandelion leaves with extremely high medicinal value are commonly picked and sold as an herbal medicine (to considerable economic benefit). The harvesting of dandelion seeds requires suitable timeliness. Dandelion seeds need to be harvested quickly, or mature seeds will detach from the plants and fall with the wind, which brings difficulties to harvesting and reduces the yield of dandelion seeds. In order to meet the needs of food or medicine, people began to try to plant dandelion artificially. However, due to the special structure of dandelion seeds, the existing seed collection machine cannot be applied to the collection of dandelion seeds. After the maturation of dandelion, the seed collection process becomes complex, requiring a lot of manpower and material resources, and the collection efficiency is very low, which is not conducive to the artificial cultivation of dandelion.

At present, dandelion seed harvesters are hardly studied by foreign researchers. Some Chinese researchers have studied the mechanized harvesting of dandelion seeds. Xie [8] designed a handheld dandelion seed harvesting device which can initially achieve the purpose of dandelion seed collection, save labor and material resources to a certain extent, and improve the speed of collection. Zhang [9] designed a cross-walking remote-controlled dandelion seed harvester based on the principle of a vacuum cleaner. The adsorption seed collection device is connected with the walking device and with a car structure, and the height of the car body is adjustable. As the dandelion harvester can operate at different heights, multi-line dandelion seeds can be quickly harvested, and the efficiency of dandelion harvesting operations can be improved. Yin [10] devised a dandelion seed harvester on which the rocker is equipped with a seed-drawing disc, allowing seeds to be drawn into the seed storage tank. The harvester solves the time-consuming and laborious problems of artificial picking, although it still needs operators to bend over for a long time, leading to high labor intensity and low operational efficiency. Zhang [11] designed a dandelion seed harvester that draws seed–pappus mixtures of dandelion into the separation drum through an air-suction pipeline to separate seeds from pappi. Wang [12] designed a dandelion seed collection device. The controller can adjust the width of the feed port according to the actual data through image processing; moreover, the working powers of the servo-motor, vibration motor, and scattering motor can also be adjusted, which allows better harvesting. However, most relevant research on the topic in China is still in its infancy, and since the harvesters have not been verified as reliable in the field, they cannot be applied to practical production.

Aiming to improve the harvesting rate, reduce harvesting loss, decrease labor intensity, and increase the degree of mechanization, a self-propelled dandelion seed harvester was designed according to agronomic requirements for harvesting dandelion seeds and material characteristics of dandelion plants and seeds. By analyzing the collection and separation processes of dandelion seeds, the main factors that affect the collection rate were determined. Then, the collection and separation devices were designed, and their parameters were analyzed. Taking the forward speed, wind velocity of blowers, and rate of rotation of the drum as experimental factors, and taking the collection rate as the evaluation index, quadratic regression orthogonal rotating field tests were conducted. In this way, the optimal combination of operation parameters was determined. By conducting field tests, the harvesting quality of the harvester was assessed and the design was verified as being both reasonable and reliable.

2. The Structure and Working Mechanism

2.1. The Structure

The dandelion seed harvester was mainly composed of the collection devices, central control board, transmission pipeline, separation devices, seed storage tank, transmission system, and a rack. The three-dimensional (3D) structure of the harvester is shown in Figure 1. The collection devices mainly consist of a disturbing roller, a collection cover, negative-pressure blowers, and a height-regulating device. The separation devices mainly include a cap, bafflers, a concave board, and a separation drum. The collection devices draw dandelion seeds with pappi from plants to the devices using negative-pressure blowers. The separation devices separate seeds from pappi using a separation drum so that seeds fall in the seed storage tank. The parameter table of the dandelion seed harvester is shown in Table 1.

Table 1. Parameters table of dandelion seed harvester.

Item	Value
Overall dimension (mm)	2450 × 1200 × 1100
Height of chassis off the ground (mm)	500
Walking system power (w)	1500
Forward speed (m·s^{-1})	0.8~1.2
Wind velocity of blowers (m·s^{-1})	1.0~2.0
Drum rotational speed (rpm)	300~500
Cutting width (mm)	1200

Figure 1. The three-dimensional (3D) structure of the harvester. 1. Collection devices; 2. Central control board; 3. Transmission pipeline; 4. Separation devices; 5. Seed storage tank; 6. Rack; 7. Height-regulating device.

2.2. *Working Mechanism*

The harvester uses a harvesting scheme that collects dandelion seeds using negative-pressure blowers and separates seeds using the separation drum. A schematic representation of the operation of the harvester is shown in Figure 2. Before field operation, the height-adjusting device is adjusted so that the collection devices have an approximate height, and operation parameters of various devices are adjusted to their desired values. Thereafter, the harvester is set to work in the field.

Figure 2. Harvest operation diagram.

(1) Collection. The disturbing roller in the lower part of the collection device is used to disturb plants so that seeds have a lower connection force to the plant and even fall from the plant. As a result, the wind velocity of blowers needed to collect seeds can be decreased, easing their collection. The negative-pressure blowers create a negative-pressure environment near the feed port to collect seeds with pappi in the devices [13–15]. The mixtures are then transported to the separation devices via the transmission pipeline.
(2) Separation. Seeds with pappi are transported to the separation devices through the transmission pipeline. Two types of separation elements, namely, wire-loops and hairbrushes on the separation drum, beat and rub seeds with pappi so that the seeds are uniformly dispersed in the devices, followed by the separation of seeds from pappi.

The separated seeds fall off in the seed storage tank and pass through holes on the concave board for separation. The separated pappi are discharged from the harvester by the impurity removal mechanism on one side of the harvester.

3. Design of Key Components

3.1. Material Characteristics

Material characteristics of dandelion plants and seeds provide an important basis for the design of the harvester [16]. To determine key parameters including the plant height, diameter of spherical seed heads, and moisture content of seeds, 100 dandelion plants were randomly selected in the field and measured. The material characteristics of measured dandelion plants and seeds are summarized in Table 2.

Table 2. Material characteristics of dandelion plants and seeds.

Item	Value
Row spacing (cm)	15
Plant height (cm)	35~45
Moisture content of seeds (%)	9.82
Diameter spherical seed heads (cm)	4~6
Suspension speed of seeds with pappi ($m \cdot s^{-1}$)	0.82
Seed length (cm)	0.30~0.49

3.2. Design of the Collection Devices

Collection is one of the key processes required to finish the harvesting of dandelion seeds. The collection devices mainly include the disturbing roller, collection cover, negative-pressure blowers, and the height-regulating device. The 3D structure of the collection devices is illustrated in Figure 3. Dandelion seeds with pappi detach from plants after being disturbed by the disturbing roller and then collect to the collection cover using negative-pressure blowers, thus completing their collection [17,18]. The design solves problems of difficulty collecting dandelion seeds with pappi, high labor intensity, and low harvesting efficiency.

Figure 3. 3D structure of the collection devices. 1. Fan fixing bracket; 2. Disturbing roller; 3. Electrical machine; 4. Collection cover; 5. Height-regulating device; 6. Fan blades.

3.2.1. Design of the Disturbing Roller

The disturbing roller is one of the core components of the collection devices. The collection quality, to a great extent, is influenced by the device [19,20]. Figure 4 illustrates the 3D structure of the disturbing roller, which mainly consists of a drive motor, a support frame, a rotation shaft, and spring-teeth; it is driven by the motor.

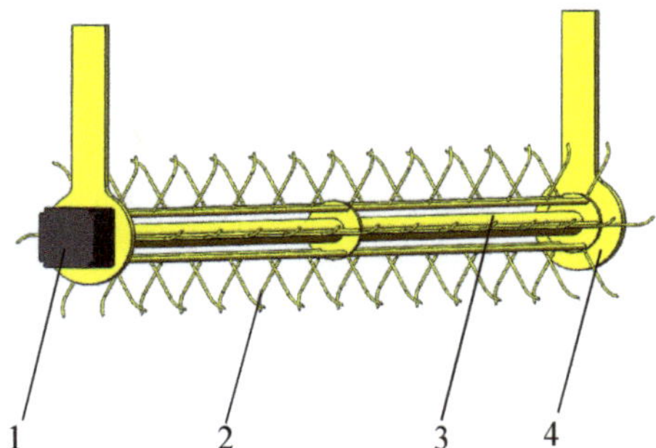

Figure 4. The 3D structure of the disturbing roller. 1. Electrical machine; 2. Spring-teeth; 3. Rotation shaft; 4. Support frame.

In this article, the diameter range of 4~6 cm was obtained by measuring the crown balls of 100 dandelion plants. The maximum diameter of 6 cm was used as the design basis; that is, the minimum horizontal length of the elastic teeth is 6 cm. At the same time, in order to better make the downed seeds move to the lower part of the fan, the head of the spring teeth is bent. The horizontal length of the bent spring teeth is 2 cm, and the horizontal length of the spring teeth is at least 6 cm. In this paper, when designing the horizontal spacing of the spring teeth, the minimum diameter of the crown ball was measured to be 4 cm, so the horizontal distance between the two spring teeth should be set to be less than 4 cm. In order to better facilitate contact between the disturbance roller and the crown ball, we designed the horizontal distance between the spring teeth to be 2 cm, ensuring that at least two spring teeth have contact with the crown ball at the same time and improving the shooting ability of the spring teeth. After analyzing the shape of the spring teeth of a wheat combine harvester and baler, it was determined that they have a cylindrical shape; therefore, the spring teeth designed in this paper are spring steel with a diameter of 6 mm. The spring teeth are evenly arranged on the disturbance roller with a horizontal spacing of 2 cm, and the horizontal length of the spring teeth is 6 cm.

As shown in Figure 5, when collecting dandelion seeds the disturbing roller first disturbs dandelion plants to loosen the connection between the plants and seeds or to cause seeds to detach from the plants. Negative-pressure blowers are then used to finish the collection. It can be seen from Figure 5b that in order to improve the collection efficiency, it is necessary to position the spring-teeth just below the blowers and rotate them by 90°. For this to be effective, the advance time t_1 of the harvester needs to be equal to the period of rotation t_2 of the disturbing roller.

$$t_1 = \frac{R}{V} \tag{1}$$

$$t_2 = \frac{\frac{\pi}{2}}{2\pi N} = \frac{1}{4N} \tag{2}$$

where V, R, and N represent the forward speed of the harvester (m·s^{-1}), radius of rotation of the disturbing roller (m), and rotational speed of the disturbing roller (rpm), respectively.

It is calculated that

$$N = \frac{V}{4R} \tag{3}$$

According to Equation (3), to make spring-teeth appear just below the blowers after the rotation of 90° intended to enhance the collection effect, the rotational speed of the disturbing roller should be related to the radius of rotation of the disturbing roller and the forward speed of the harvester.

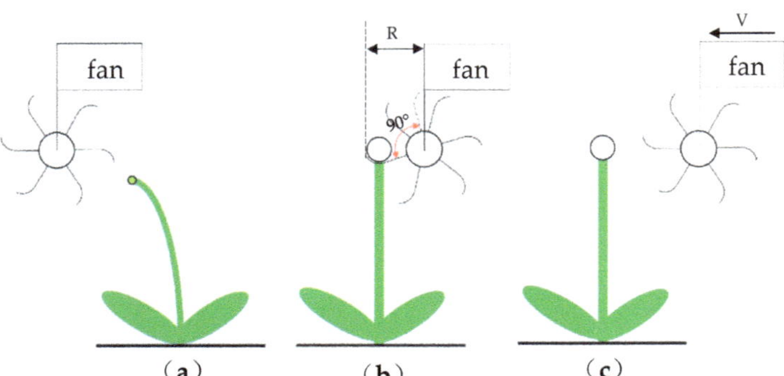

Figure 5. Schematic diagram of perturbation process. (**a**) After the perturbation process. (**b**) During disturbing the process. (**c**) Before the perturbation process. Screen, V and R, represent the forward speed of the harvester (m·s^{-1}) and radius of rotation of the disturbing roller (m).

The radius of rotation of the disturbing roller needs to be designed according to the diameter of the spherical seed heads of dandelion. It can be seen from Table 2 that the diameter of spherical seed heads is 4~6 cm; that is, the radius of rotation of the disturbing roller is

$$R = r + d \tag{4}$$

where R is the radius of rotation of the disturbing roller (m); r is the radius of the rotation shaft of the disturbing roller (m); and d denotes the diameter of the spherical seed heads of dandelion plants (m).

The diameter of the rotation shaft of the disturbing roller is set to be 3 cm by comprehensively considering the feasibility of the entire design and the dimension reliability of the disturbing roller. That is, $R \geq 4.5$ cm, then R = 5 cm, the horizontal length of the spring-teeth is 3.5 cm, and the diameter of the rotating shaft of the disturbing roller is 3 cm. Under these conditions, the rate of rotation of the disturbing roller is only related to the forward speed; that is, the rate of rotation of the disturbing roller N is $5V$ (where V is the forward speed).

At the same time, the selection of materials for producing spring-teeth on the disturbing roller is also critical. At present, the most widely used materials for producing spring-teeth include metals, plastics, and nylons. Compared with other materials, metal spring-teeth are hard and heavy; plastic spring-teeth are light while having poor toughness, which is likely to damage dandelion plants; and nylon spring-teeth are light, which reduces the load on the motor, and they are also highly elastic, tough, and hard to break, showing good performance for their cost. Considering this, nylon was selected to produce spring-teeth on the disturbing roller.

3.2.2. Design of Negative-Pressure Blowers

By measuring the material characteristics of dandelion seeds, the average values of the long and short axes of dandelion seeds were found to be 3.096 and 0.588 mm, respectively, and the suspension speed of dandelion seeds with pappi was found to be about 0.82 m·s^{-1}. Due to their characteristically small volume, light weight, and slow suspension speed, it is appropriate to collect dandelion seeds using negative-pressure blowers.

In this paper, based on the planting agronomic requirements of dandelion plants, the row spacing used was 15 cm, the harvester was designed to harvest the seeds of eight rows of dandelion plants at one time, and the fan was used for negative pressure harvesting. Three schemes of collecting devices were designed. The first scheme consists of a small fan used to harvest two rows of dandelion seeds and four small fans used to harvest eight rows. The second scheme uses a larger fan to harvest four rows at a time and uses two larger fans to harvest eight rows; the third scheme involves installing a small fan in the middle and back of the adjacent two fans of the first scheme and harvesting the dandelion seeds missing from the front fan. Three blower layouts shown in Figure 6 were preliminarily designed, and the optimal one needed to be selected from these layouts. The harvesting breadth was 120 cm, and the maximum diameters of blowers at the small and large feed ports were 30 and 60 cm, respectively. The layout and reliability of the blowers and the specifications of common blowers on the market were considered. Figure 6a shows the single-row layout of four blowers with diameters of 24 cm; Figure 6b illustrates the single-row layout of two blowers with diameters of 50 cm; Figure 6c presents the double-row staggered layout of seven blowers with diameters of 24 cm.

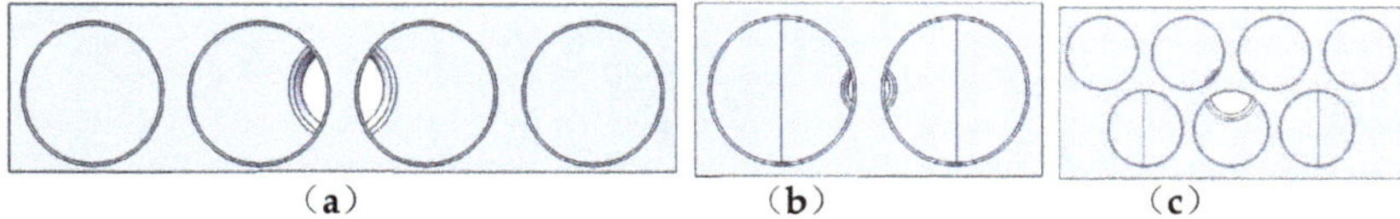

Figure 6. Negative pressure fan layout diagram. (**a**) Single row arrangement of small fans. (**b**) Single row arrangement of large fans. (**c**) Small fan double row layout.

3.3. *Flow-Field Simulation of the Collection Devices*

The collection devices are key devices used to harvest dandelion seeds using the dandelion seed harvester. The objective of improving the collection rate of dandelion seeds can be achieved through the reasonable design and improvement of the collection devices. The size and layout of air inlets greatly influence the collection quality. Through flow-field simulation in the collection covers with the three designed blower layouts using Fluent software, flow fields in the three structures at different wind velocities were compared and analyzed. In this way, the optimal structure of the collection devices was determined [21,22], providing a basis for field harvesting tests.

3.3.1. Mesh Generation

The 3D modelling and mesh generation were performed for collection covers with the three different blower layouts, as illustrated in Figure 7.

As displayed in Figure 7d, when simulating flow fields in the collection covers, a cross-section I passing the axis of air outlet was selected to observe and analyze the velocity nephograms in flow-field simulation. The velocity vector diagrams of the entire collection covers were observed and investigated to better understand the motion of airflows in the collection covers.

3.3.2. Simulation Results and Analysis

Figure 8 shows the velocity nephograms and velocity vector diagrams of the collection devices with the single-row layout of small blowers that rotate at 1.0, 1.5, and 2.0 $m\cdot s^{-1}$. The velocity nephograms demonstrate that the velocity changes slightly in adjacent areas in the collection cover and is uniform while the amount of air at the air outlet is large; thus, the wind velocity is fastest there. As displayed in velocity vector diagrams, the airflow field in the collection cover varies slightly with changes in the wind velocity. Roughly two flow directions are observed at the air inlet: part of the airflow moves towards the outlet while the residue converges with the other part of airflow after being reflected by the wall of the collection cover to flow from the air outlet. This indicates that the intake air is confluent in the collection cover.

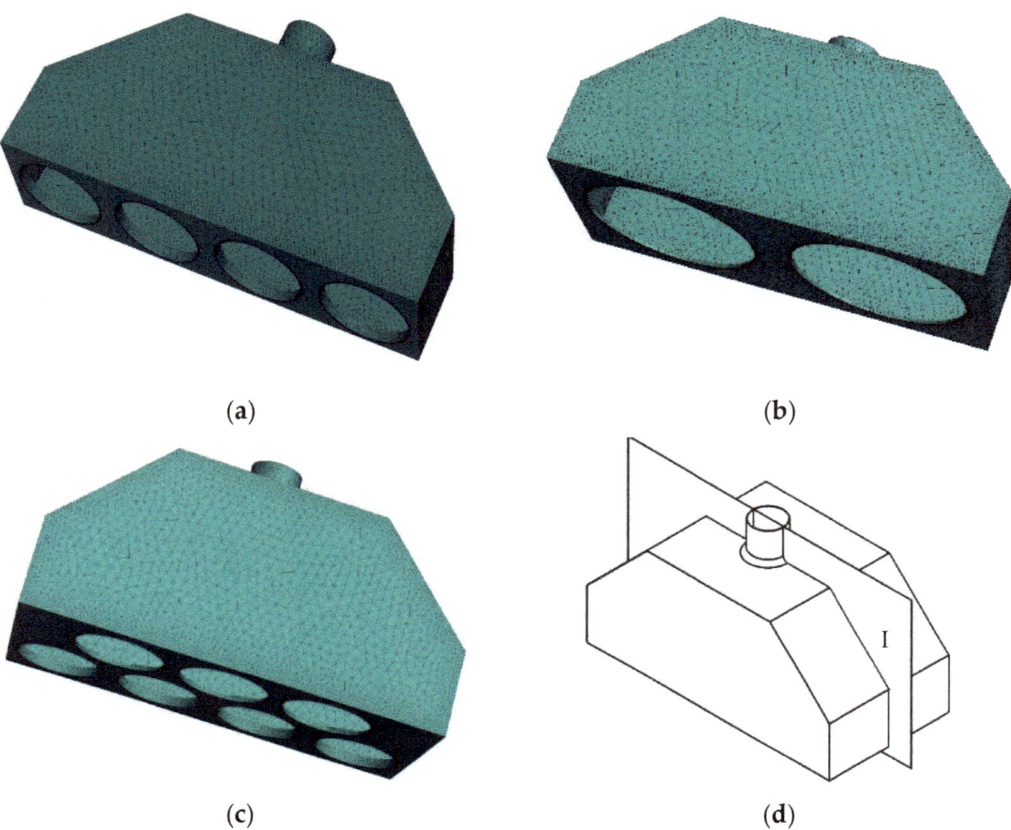

Figure 7. Grid division of collection device model. (**a**) Grid division of the collection hood for the first type of fan layout, (**b**) Grid division of the collection hood for the second type of fan layout, (**c**) Grid division of the collection hood for the third type of fan layout, (**d**) Cross section schematic diagram of velocity cloud map.

Figure 9 illustrates the velocity nephograms and velocity vector diagrams of the collection devices with the single-row layout of large blowers that rotate at 1.0, 1.5, and 2.0 $m{\cdot}s^{-1}$. The velocity vector diagrams show that airflow is poorly confluent in the collection devices with the single-row layout of large blowers, and a large blank area of airflow can be seen in the center. The airflow entering the collection cover from the air inlet does not converge well with the airflow in other directions, which flows towards the air outlet after being reflected at the wall. Instead, some air flows obliquely downward after reflection at the wall, which disturbs the flow field in the collection cover and inhibits airflow at the air inlet, slowing its progress into the collection cover. As a result, the collected dandelion seeds are less readily passed out in the collection process.

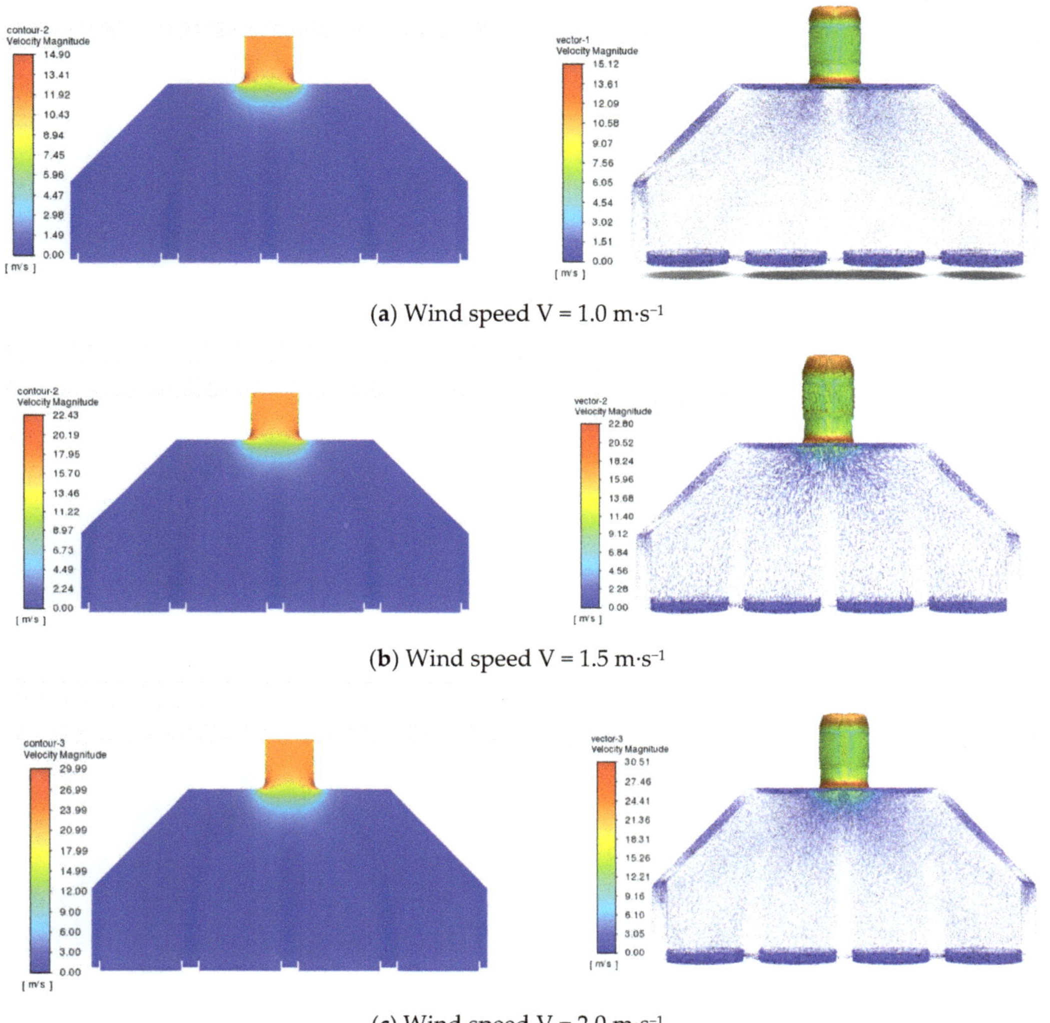

Figure 8. Velocity nephogram and velocity vector diagram under different wind speeds.

Figure 10 shows the velocity nephograms and velocity vector diagrams of the collection devices with the double-row layout of small blowers that rotate at 1.0, 1.5, and 2.0 m·s^{-1}. Analysis of the velocity nephograms reveals that the velocity changes little and is uniform in the collection cover, while there is a significant difference in wind velocities at the inlet and outlet. By observing the velocity vector diagrams, it was found that there is a large blank area of airflow in the collection cover with the double-row layout of small blowers. In addition, the same problem that arose in the collection devices with the single-row layout of large blowers also arises in this layout. That is, some air flows downward after being reflected by the walls, which disturbs the flow field in the collection cover and inhibits the otherwise smoother inflow of air at the air inlet to the collection cover.

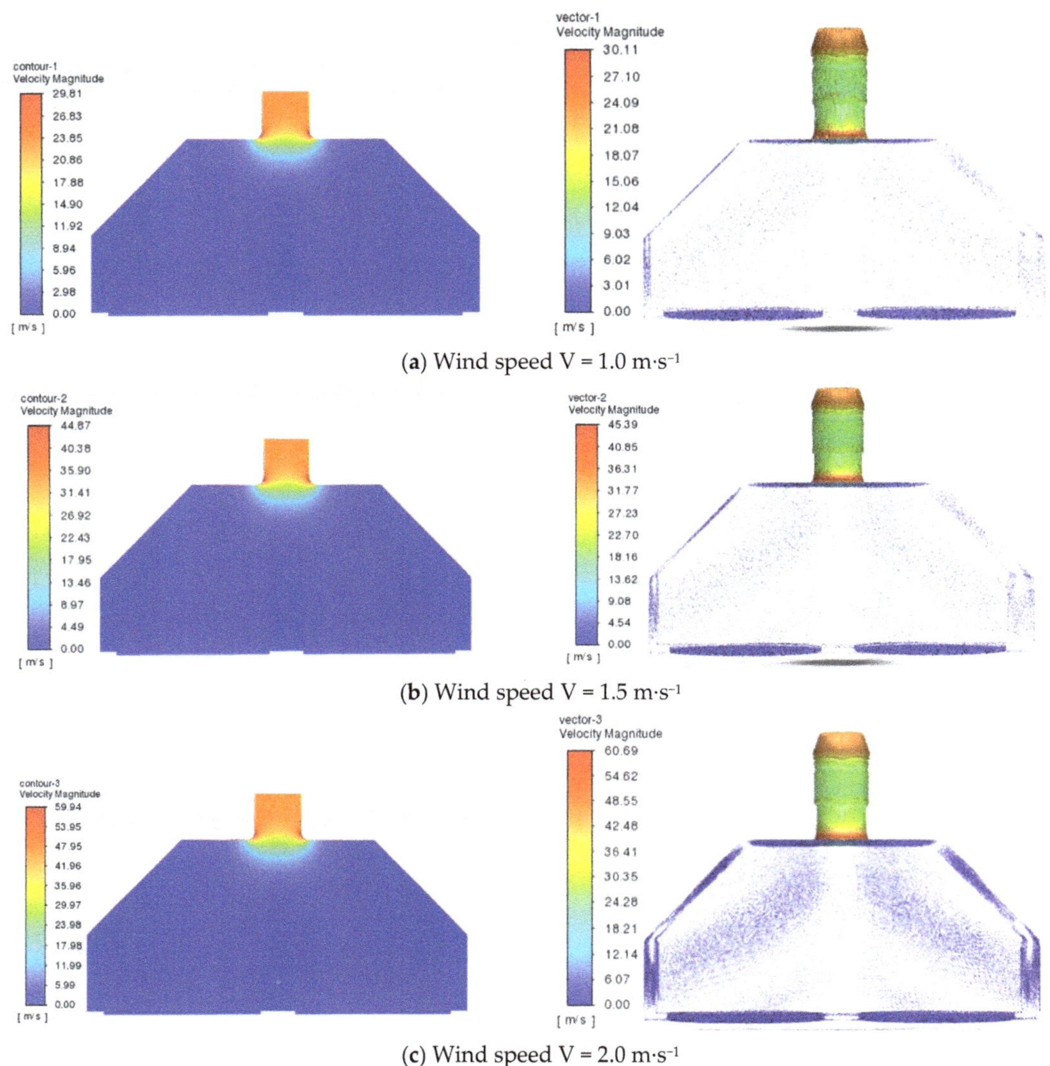

(**a**) Wind speed V = 1.0 $m \cdot s^{-1}$

(**b**) Wind speed V = 1.5 $m \cdot s^{-1}$

(**c**) Wind speed V = 2.0 $m \cdot s^{-1}$

Figure 9. Velocity nephogram and velocity vector diagram under different wind speeds.

Comprehensive analysis of the flow-field simulation results of collection devices with the three blower layouts indicates that the single-row layout of small blowers shows the best effect compared with the other two layouts. The layout enables smooth motion of airflow, meets requirements for collecting and transporting dandelion seeds, saves materials, and is easily machined. The wind field generated by the fan in Figure 7b is large and not concentrated in the lower part of the collection device. The collection rate in the pre-test is small, and it is easy to inhale other impurities into the machine. In the pre-test, the rear fan of the collection device shown in Figure 7b collects fewer seeds, and the overall collection rate is almost the same as that shown in Figure 7a. During the pre-test verification of the collection rate of the three schemes, it was found that the collection device of the style shown in Figure 7a had a higher collection rate than the other two schemes, and that the airflow in the flow field was stable and the cost was low. In summary, we chose the collection device of the style shown in Figure 7a.

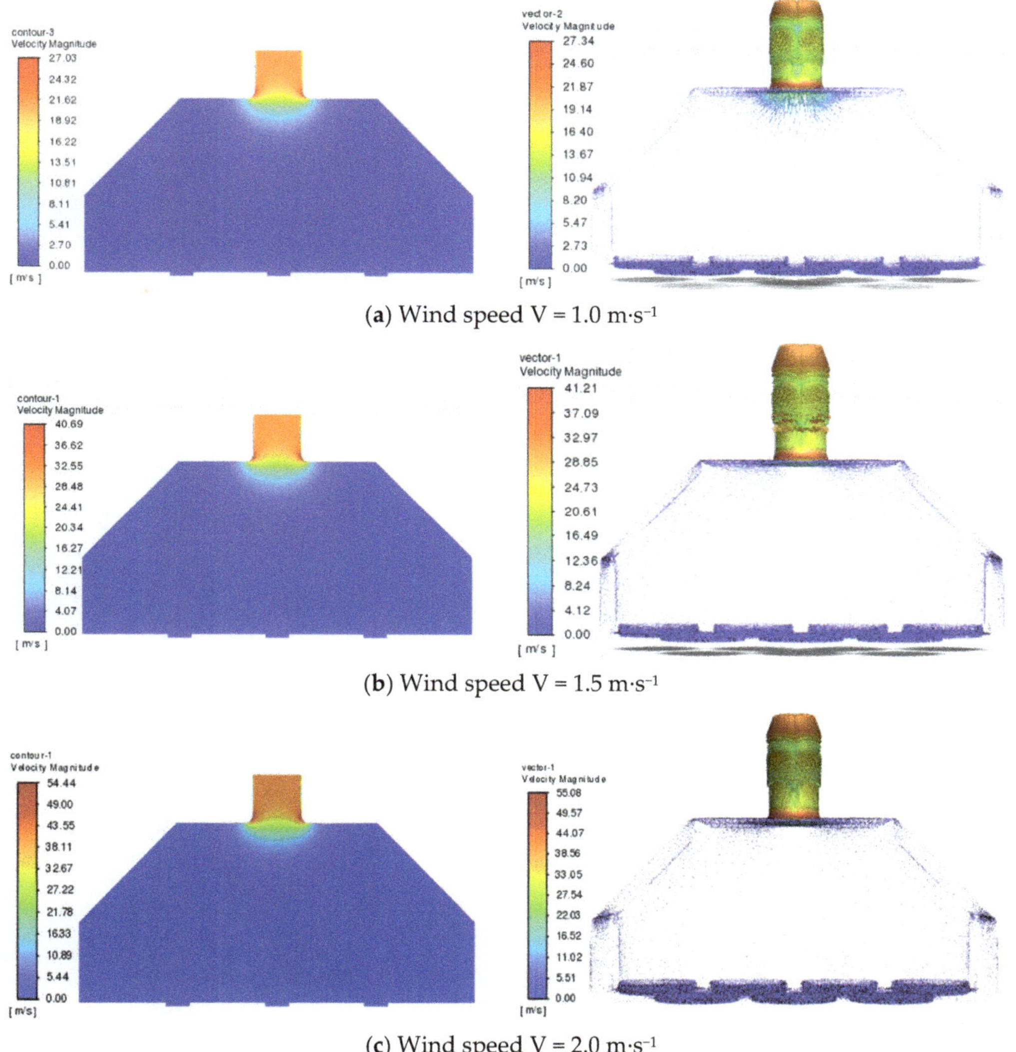

(**a**) Wind speed V = 1.0 m·s^{-1}

(**b**) Wind speed V = 1.5 m·s^{-1}

(**c**) Wind speed V = 2.0 m·s^{-1}

Figure 10. Velocity nephogram and velocity vector diagram under different wind speeds.

3.4. Design of the Separation Devices

Separation is an important process to separate dandelion seeds from pappi so as to obtain dandelion seeds [23]. The separation devices include an impurity removal mechanism, a cap, a concave board for separation, and a separation drum. Figure 11 shows the 3D structure of the separation devices.

After entering the separation devices, dandelion seeds with pappi are separated from pappi under the action of the separation elements [24]. The separated seeds fall in the seed storage tank through the concave board, as shown in Figure 12. Finally, pappi are discharged from the harvester via the impurity removal mechanism, thus completing the seed–pappus separation stage. This overcomes the difficulty in seed–pappus separation, addresses low separation efficiency, and reduces the potential for harm to operators often caused during the manual separation of pappi.

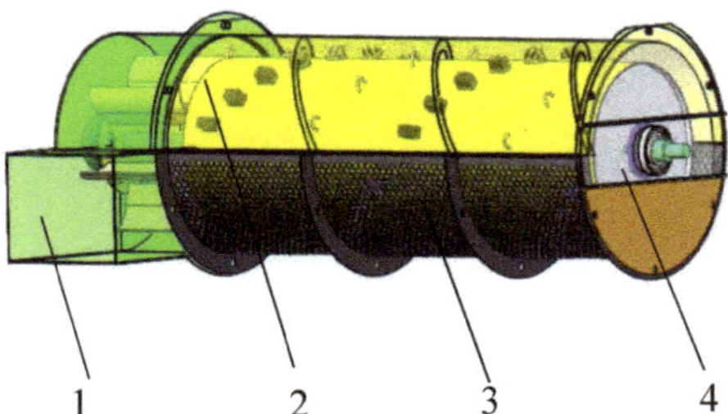

Figure 11. The 3D structure of the separation devices. 1. Impurity removal mechanism; 2. Cap; 3. Concave board for separation; 4. Separation drum.

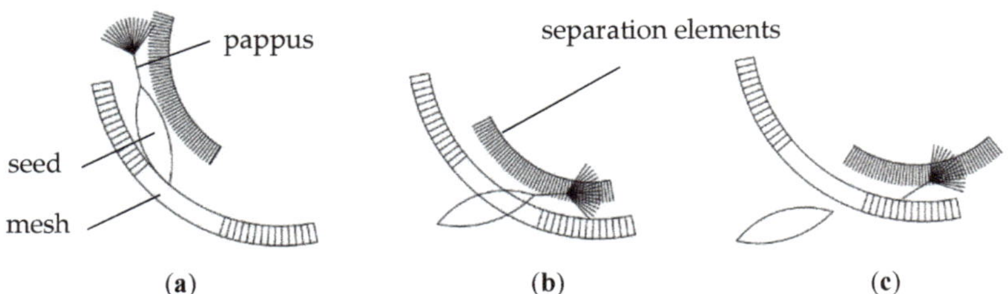

Figure 12. Schematic diagram of separation process of grain and crown hair. (**a**) Before separation. (**b**) During separation. (**c**) After separation.

3.4.1. Design of the Separation Drum

Optimal design of the separation drum can promote efficient seed–pappus separation [25,26]. Seeds with pappi are light and easily affected by the wind as they float. This makes it difficult to bring the seeds into contact with the separation drum to separate seeds from pappi. Therefore, the separation drum is designed as a closed drum (Figure 13), which narrows the space available to the seeds with pappi and to some extent solves the problems of floating, small contact areas with the separation drum and concave board, and low contact probability of seeds [27]. It also improves the separation efficiency of pappi and seeds, increases the separation rate, and reduces impurities in the harvest.

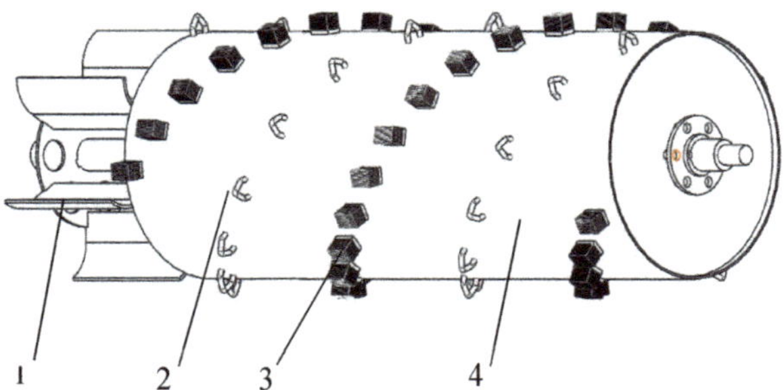

Figure 13. The 3D structure diagram of separation drum. 1. Impurity removal mechanism; 2. Wire-loops; 3. Hairbrush; 4. Closing drum.

Through comparison with separation devices of other harvesters and through combining the special material characteristics of dandelion seeds, the separation elements were designed as hairbrushes and wire-loops. Dandelion seeds with pappi were separated from pappi by the wire-loops and hairbrushes distributed on the drum, thus yielding pure dandelion seeds. The layouts of the separation elements exert a significant influence on the separation effect of pappi and seeds. To explore the optimal layout of separation elements, three element layouts (Figure 14) were designed to select the better combination through pre-testing. Figure 14a–c separately shows the spiral layouts of only wire-loops distribution, only brush distribution, and cross distribution of bow tooth brushes [28,29].

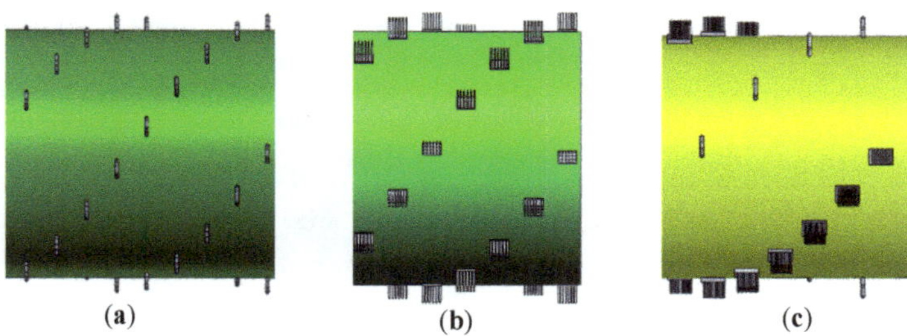

Figure 14. Separation element combination mode. (**a**). Only wire-loops distribution. (**b**). Only brush distribution (**c**). Cross distribution of bow tooth brushes.

Furthermore, the selection of materials for producing the hairbrushes and wire-loops can also heavily influence the separation effect of pappi and seeds. Commonly used materials to produce hairbrushes include polybutylene terephthalate (PBT) fibers, bristles, nylon fibers, polypropylene (PP) fibers, and metal wires. Among them, PBT fibers and bristles are so soft that the separation effect cannot meet the operation requirement; PP fibers show poor elasticity and cannot recover all imposed long-term deformations, hindering the separation of pappi and seeds; metal wires are hard; and nylon fibers exhibit moderate hardness, favorable elasticity, and cost-effective performance, so nylon was also used to fabricate hairbrushes on the separation devices. Wire-loops are commonly made with metal (steel or iron). In the present study, steel was selected to produce the wire-loops.

The length of the separation drum is closely related to its separation capacity: the longer the drum, the longer the separation time and the higher the separation rate [30]. The length of the separation drum in the separation part is calculated using the following formula:

$$L \geq \frac{q}{q_0} \tag{5}$$

where L represents the length of the separation drum (m); q denotes the feed quantity of the separation devices (kg/s) and is set to 0.5 kg/s in the present research; and q_0 is the designed allowable feed quantity per unit length of the separation drum (kg/(s·m)) and is set to 0.5~0.8 kg/(s·m).

From Equation (2), the length L of the separation section of the separation drum is 0.63~1.0 m and it is set to 1.0 m here. Therefore, the length of the separation section of the drum is 1.0 m.

If the diameter of the separation drum is too small, only small amounts of seed–pappus mixtures are allowed to enter the drum, and the contact area between the mixtures and the concave board narrows [31], reducing the time taken for the separation of dandelion seeds from pappi. At present, the diameter of commonly used separation drums is 550~650 mm. The greater the diameter, the greater the feed quantity that can be sustained, although the heavier the load. Compared with the harvesting of wheat and rice, the feed quantity of dandelion is not that large. The diameter D_z of the separation drum satisfies

$$D_z = D_g + 2h_z \tag{6}$$

where D_g is the tooth root diameter of the drum (mm) and h_z is the height of the high separation elements (mm).

Generally, $D_g \geq 300$ mm. Considering the separation effect and the feed quantity of dandelion seed–pappus mixtures, D_g was determined to be 300 mm. To avoid collisions of wire-loops with bafflers on the cap, the heights of wire-loops and hairbrushes were separately set to be 35 and 50 mm. The diameter of the separation drum was then calculated to be 400 mm using Equation (3).

By conducting pre-tests on the separation drum, the separation effects of the separation drums with the three different element layouts were verified. The results show that the combined layout of hairbrushes and wire-loops allows the optimal separation effect of the separation drum and the significant separation effect of seeds and pappi. That is, the combined layout of hairbrushes and wire-loops was selected for separation elements on the separation drum; the length of the drum separation section is 1.0 m, and the diameter of the separation drum is 400 mm.

3.4.2. Design of the Concave Board

The concave board mainly consists of a perforated screen, a connecting plate with the upper cap, and an arcuate side-plate [32,33]. Holes are distributed uniformly on the perforated screen (Figure 15). To enable better contact of seed–pappus mixtures with the drum and convenient separation of seeds from pappi, the wrap angles of the concave board for separation are designed to be 180°; that is, the concave board is hemi-cylindrical. One end of the concave board is connected to the transmission pipeline, while the other end is connected to the impurity-removal device to coordinate with the separation drum to separate pappi from seeds and allow seeds to pass through the holes in the screen. In the meantime, pappi are retained on the concave board and discharged from the harvester via the impurity removal mechanism during the rotation of the separation drum.

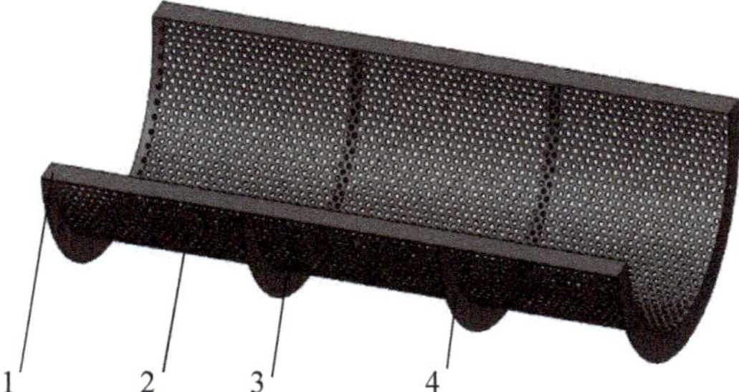

Figure 15. The 3D structure diagram of separation concave plate. 1. Connecting plate with the upper cap; 2. Perforated screen; 3. Arcuate side-plate; 4. Bolt hole.

The screening effect of the perforated screen is closely related to porosity of the screen, which is influenced by the hole shape, size, and layout. Dandelion seeds resemble paddy rice in appearance, both of which are elliptic. Considering the shape and passing performance of seeds through the perforated screen, round screen holes were designed; because dandelion seeds are generally 0.30~0.47 cm long, the screen holes are circular with a diameter of 0.5 cm.

The more porous the screen, the better the screening effect and the more seeds that pass through the screen. Therefore, screen holes are distributed in the form of equilateral triangles that enable high porosity (Figure 16). If the hole diameter d is 0.5 cm and the spacing between two adjacent holes l is 0.1 cm, the center-to-center spacing of two adjacent holes l_1 is 0.6 cm. The porosity K of the screen is

$$K = \frac{\pi d_0^2}{2\sqrt{3} l_1^2} \times 100\% = 63\% \quad (7)$$

where K is the porosity of the screen (%); d is the hole diameter on the screen (cm); and l_1 is the center-to-center spacing between two adjacent holes (cm).

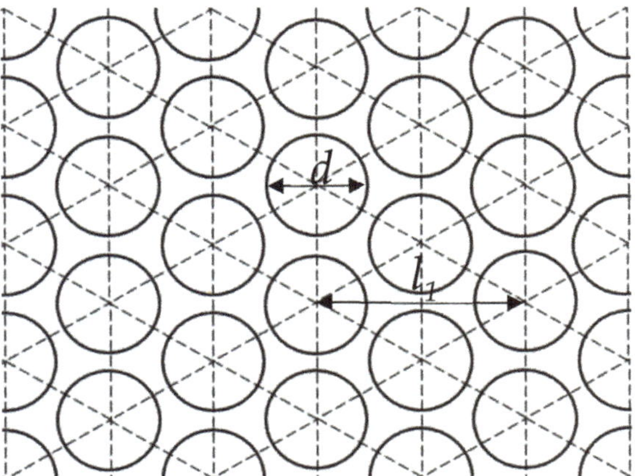

Figure 16. Schematic diagram of sieve arrangement.

4. Field Tests

4.1. Test Conditions

Field tests on dandelion seed harvesting were conducted in a dandelion planting base at the Changyuan branch of the Henan Academy of Agricultural Sciences (Xinxiang City, Henan Province, China) [34]. The planting base was situated on a flat terrain, where dandelion plants grow well. The key parameters of dandelion plants and seeds were measured in the experimental field by measuring instruments. Dandelion plants have an average row spacing of 15 cm, an average height of 40 cm, and a seed moisture content of 9.82%. The growth cycle of dandelion was 3 months. A region some 30 m long and 10 m wide was selected from within the planting base as the harvesting test region. In addition, a harvester starting area and a parking area were also reserved on both sides of the test region (Figure 17).The photos of dandelion plants and dandelion seed crowns are shown in Figure 18.

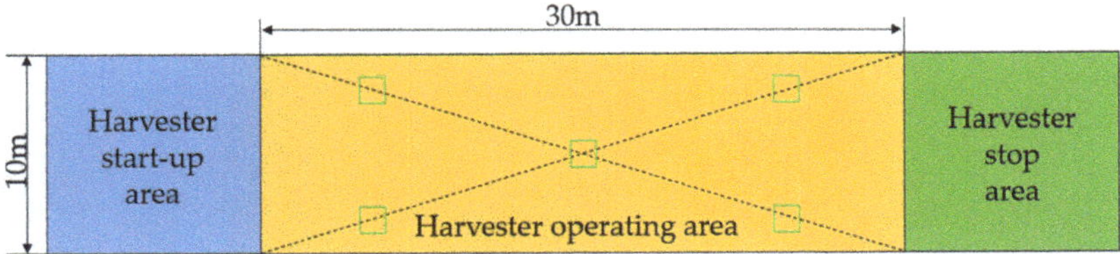

Figure 17. Experimental plot area division diagram.

Figure 18. Photos of dandelion plants and dandelion seed crown balls.

The test instruments and devices include the self-propelled dandelion seed harvester, an anemograph, a tape, an electronic scale, a digital camera, and a rev-counter.

4.2. *Harvesting Tests*

4.2.1. Test Methods

The harvesting tests were conducted to assess the effects of varied test operating parameters of the collection and separation devices to explore the influences of the forward speed, wind velocity at the air inlet, and rate of rotation of the drum on the harvesting effect. Meanwhile, the operating quality of the harvester was verified, and the harvesting quality was reflected by measuring the collection rate. The main performance evaluation index was the rate of collection of dandelion seeds. The performance test of the harvester is shown in Figure 19.

(**a**)

(**b**)

(**c**)

Figure 19. Harvester performance test. (**a**) Comparison diagram before and after harvest. (**b**) Harvester operation diagram. (**c**) Separated seeds and crown hairs.

The collection rate was measured using the following method: in the test region, the harvester was driven to harvest the seeds of all dandelion plants; after harvesting, the dandelion seeds in the seed storage tank were removed and weighed (recorded as m_1); as shown in Figure 17, five sampling points measuring 1 m^2 were selected artificially using the five-point sampling method in the test region to collect unharvested dandelion seeds left by the harvester; these seeds were separated from pappi and weighed (recorded as m_2).

The rate of collection is calculated using the following equation:

$$F_1 = \frac{\frac{m_1}{300} \times 5}{\frac{m_1}{300} \times 5 + m_2} = \frac{m_1}{m_1 + 60m_2} \tag{8}$$

where F_1 is the collection rate of dandelion seeds using the harvester (%); m_1 is the mass of dandelion seeds in the seed storage tank after harvesting (g); and m_2 represents the mass of unharvested dandelion seeds left by the harvester at sampling points (g).

4.2.2. Test Design

Herein, the influences of the forward speed, the rate of rotation of the blowers at the feed port of the collection devices, and the rate of rotation of the drum on the harvesting effect of the harvester were discussed. To this end, taking the forward speed, wind velocity of blowers, and rate of rotation of the drum as test factors, quadratic regression orthogonal rotating combination tests were conducted. Combined with the pre-test data and material characteristics of dandelion seeds, the test ranges of the forward speed, wind velocity at blower inlets, and rate of rotation of the drum were set to 0.8~1.2 $m \cdot s^{-1}$, 1.0~2.0 $m \cdot s^{-1}$, and 300~500 rpm [35], respectively. The test factors are numbered (Table 3) and the simulation test schemes and results are listed in Table 4.

Table 3. Coding of simulation test factors.

Code	Experimental Factors		
	Forward Speed (m·s^{-1})	Wind Velocity of Blowers (m·s^{-1})	Drum Rotational Speed (rpm)
−1	0.8	1.0	300
0	1.0	1.5	400
1	1.2	2.0	500

Table 4. Experiment scheme and results.

No.	Experimental Factors			Experimental Index
	Forward Speed	Wind Velocity of Blowers	Drum Rotational Speed	Harvest Rate
1	−1	−1	0	85.8%
2	1	−1	0	79.4%
3	−1	1	0	88.0%
4	1	1	0	86.9%
5	−1	0	−1	79.1%
6	1	0	−1	75.4%
7	−1	0	1	83.3%
8	1	0	1	78.2%
9	0	−1	−1	73.5%
10	0	1	−1	74.1%
11	0	−1	1	76.1%
12	0	1	1	83.1%
13	0	0	0	87.5%
14	0	0	0	87.0%
15	0	0	0	87.3%
16	0	0	0	87.1%
17	0	0	0	86.9%

4.3. Test Analysis

Software Design-Expert 8.0 was adopted to perform variance analysis of data in Table 4 and results are listed in Table 5 [36]. It can be seen from Table 5 that the test model is extremely significant ($p < 0.0001$), suggesting that the designed tests are reasonable and effective; the coefficient of determination R^2 is 0.9626, indicative of a high degree of fitting of the regression equation. The forward speed, wind velocity of blowers, and rate of rotation of the drum all significantly affect the collection rate ($p < 0.05$). The interaction term between the forward speed and wind velocity of blowers exerts significant influence on the collection rate; the interaction term between the rate of rotation of the blowers and drum also has a significant influence. The response surfaces of the influence of the interaction terms of the test factors on the collection rate are illustrated in Figures 20 and 21. Through a multiple regression of the test results, the regression equation of the collection rate can be obtained as follows:

$$\varphi = +87.16 - 0.24\text{A} + 2.16\text{B} + 2.32\text{C} + 1.32\text{AB} - 0.35\text{AC} + 1.60\text{BC} + 0.083A^2 - 2.22B^2 - 8.24C^2 \quad (9)$$

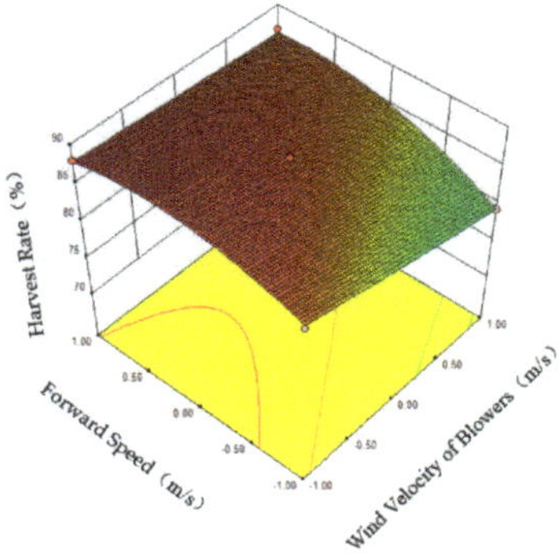

Figure 20. Interaction between forward speed and wind velocity of blowers.

Table 5. Analysis of variance of regression equations.

Source of Variance	Sum of Square	Degree of Freedom	Mean Square	F	p
Model	448.41	9	49.82	95.83	<0.0001
A-A	33.21	1	33.21	63.88	<0.0001
B-B	37.41	1	37.41	71.95	<0.0001
C-C	43.25	1	43.25	83.17	<0.0001
AB	7.02	1	7.02	13.51	0.0079
AC	0.49	1	0.49	0.94	0.3640
BC	10.24	1	10.24	19.7	0.0030
A^2	0.029	1	0.029	0.055	0.8211
B^2	20.7	1	20.7	39.82	0.0004
C^2	286.06	1	286.06	550.19	<0.0001
Residual	3.64	7	0.52		
Lack of Fit	3.41	3	1.14	19.58	0.0075
Pure Error	0.23	4	0.058		
Cor Total	452.05	16			

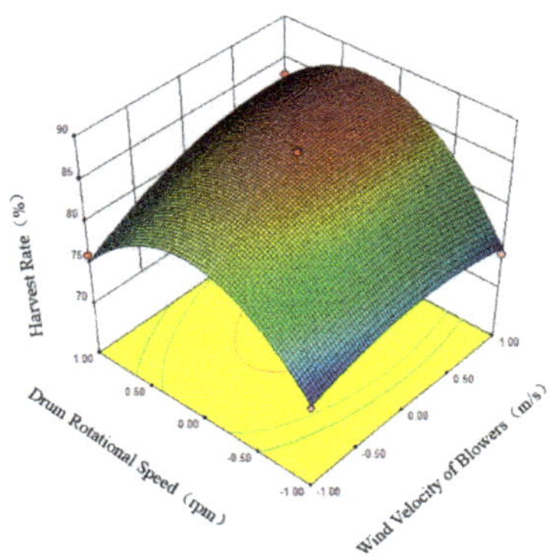

Figure 21. Interaction between wind velocity of blowers and drum rotational speed.

4.4. Optimal Parameters

Taking the maximization of the collection rate as the objective, the optimization module in software Design-Expert 8.0 was used to solve the regression equation, thus determining the optimal solution. The equation set of the objective and constraints is expressed as Equation (10).

$$\left\{\begin{array}{l}\mathrm{MaxY(A,B,C)} \\ 0.8 < A < 1.2 \\ 1.0 < B < 2.0 \\ 300 < C < 400\end{array}\right\} \tag{10}$$

According to the optimized results, the optimal working parameters of the dandelion seed harvester were selected as follows: a forward speed of 0.8 $m \cdot s^{-1}$, a wind velocity of blowers of 1.63 $m \cdot s^{-1}$, and a rate of rotation of the drum of 418.7 rpm (rounded to 419 rpm). After adjusting the relevant parameters of the harvester, harvesting tests were conducted in which the collection rate reached 89.1% (as expected based on experimental evidence).

5. Discussion

Through global analysis of the test process and results of the dandelion seed harvester, the deficiencies in the harvesting quality were investigated:

1. Environmental influences: Under strong external natural wind, the blowers fail to draw all falling seeds into the harvester, which may cause a collection loss. If the ground is rough, seeds falling off under the disturbance of the disturbing roller cannot be collected due to the rugged ground and long distance from the blowers that results in inadequate wind strength, thus inducing the collection loss.

2. Inconsistent maturity: Dandelion seeds have different degrees of maturity: in the maturation period of dandelion seeds, the plants mature quickly and near-simultaneously, whereas the maturation period of the seeds is very short, so the seeds need to be harvested timeously. In the harvesting process, some immature seeds may be entrained, and thus have a high moisture content and a strong connection to the pappi. Therefore, the separation devices cannot timeously and effectively separate the immature dandelion seeds from the pappi, which decreases the separation rate, reducing the yield.
3. Degree of proficiency of drivers: The degree of proficiency of drivers for the dandelion seed harvester exerts certain influences on the harvesting quality. If the driver is less adept, the normal operation of the harvester cannot be guaranteed, thus inducing a lower harvesting quality. Improving the proficiency of drivers for the harvester provides certain assistance in increasing the harvesting efficiency and quality.

6. Conclusions

1. At present, there are few dandelion seed harvesters available, while manual harvesting of dandelion seeds calls for high labor intensity, is inefficient, and may induce a large loss of yield. To overcome these problems, a self-propelled dandelion seed harvester was designed. The harvester is mainly composed of collection devices, separation devices, transmission devices, and a rack. The harvester has the ability to collect and separate dandelion seeds.
2. Important working parts, namely, the collection and separation devices of the harvester, were designed. The collection devices include the disturbing roller, collection cover, height-regulating device, and negative-pressure blowers. They are responsible for collecting dandelion seeds. The operating breadth of the harvester is 1.2 m. The operating height, forward speed, wind velocity of the blowers, and rate of rotation of the drum are adjustable within 0.35~0.45 m, 0.8~1.20 $m \cdot s^{-1}$, 1.0~2.0 $m \cdot s^{-1}$, and 300~500 rpm, respectively. The separation devices include the separation drum, concave board for separation, and impurity removal device, responsible for separating dandelion seeds from pappi. The length, diameter, and rate of rotation of the drum are 1.15 m, 0.4 m, and adjustable within 300~500 rpm, respectively.
3. To explore the optimal working performance of the dandelion seed harvester, multi-factor field tests were performed. Field tests show that the harvesting effect is optimal under the following conditions: a forward speed of 0.8 $m \cdot s^{-1}$, a single-row layout of small blowers, a wind velocity of blowers of 1.63 $m \cdot s^{-1}$, a combined layout of separation elements (hairbrushes and wire-loops), a rate of rotation of the drum of 419 rpm, and the use of a perforated screen with circular holes on the concave board. The rate of collection of dandelion seeds reached 89.1%, which reaches the expected harvesting effect desired in the design of the dandelion seed harvester.

Author Contributions: Conceptualization, Z.Q. and Q.L.; methodology, Z.L. and L.L.; investigation, Z.Q. and L.L.; resources, X.W.; data curation, Q.L. and H.S.; writing—original draft preparation, Z.Q. and Q.L. All authors have read and agreed to the published version of the manuscript.

Funding: This research was funded by the National Key R&D Program of China (2019YFD1002602) and Henan Province Science and Technology Research (212102110217).

Institutional Review Board Statement: Not applicable.

Informed Consent Statement: Not applicable.

Data Availability Statement: The data used to support the findings of this study are available from the corresponding author upon request.

Acknowledgments: The authors would like to thank their college and the laboratory, as well as gratefully appreciate the reviewers who provided helpful suggestions for this manuscript.

Conflicts of Interest: The authors declare no conflict of interest.

References

1. Zhou, R.L.; Lu, F.; Qin, L.L. Nutrition and health care function of dandelion. *Food Nutr. China* **2011**, *17*, 71–72.
2. Gonzalez-Castejon, M.; Visioli, F.; Rodriguez-Casado, A. Diverse biological activities of dandelion. *Nutr. Rev.* **2012**, *70*, 534–547. [CrossRef]
3. Gu, L.W.; Wang, L.; Zhang, H.; Guan, H.B.; Zheng, Y. Nutritional value of dandelion and its development and utilization prospect. *Jilin Veg.* **2013**, *4*, 13–14.
4. Ruiz-Juarez, D.; Melo-Ruiz, V.E.; Gutierrez-Rojas, M.; Sanchez-Herrera, K.; Cuamatzi-Tapia, O. Nutrient Value in Dandelion Flower (*Taraxacum officinale*). *Ann. Nutr. Metabolism.* **2020**, *76*, 102.
5. Hao, W.L.; Mu, C.C. The medicinal value of fresh dandelions. *J. Benef. Readiness Drug Inf. Med. Advices* **2017**, 65–66.
6. Chen, R.J.; Wang, Q.Y.; La, X.J.; Li, J.A.; Liang, G.T.; Cao, H.J.; Wang, Y.Y.; Zhang, M. Advances in medicinal research of dandelion. *Mod. J. Integr. Tradit. Chin. West. Med.* **2021**, *30*, 563–567.
7. Zhao, L.; Yang, Y.J.; Lin, D. Economic value of dandelion. *Liaoning Agric. Sci.* **2006**, 33–35.
8. Inner Mongolia Academy of Agricultural & Animal Husbandry Sciences. A New Handheld Dandelion Seed Harvestingdevice:CN212184213U[P]. Available online: https://patents.google.com/patent/CN212184213U/zh (accessed on 22 December 2020).
9. Zhang, Y. Cross-Walking Remote-Controlled Dandelion Seed Harvester Based on the Principle of Vacuum Cleaner: CN214282208U [P]. Available online: https://patents.google.com/patent/CN214282208U/en?oq=CN214282208U (accessed on 28 September 2021).
10. Yin, G.; Lin, G.F.; Song, J.J.; Li, J.L.; Zhang, Y.B.; Lu, Z.Y.; Xie, S.Q.; Chen, J.; Sun, L. Dandelion Harvester China: CN207476241U [P]. Available online: https://patents.google.com/patent/CN207476241U/en?oq=CN207476241U (accessed on 12 June 2018).
11. Shandong University of Science and Technology. The Utility Model Relates to a Dandelion Seed Harvesting Device China: CN210537495U [P]. Available online: https://patents.google.com/patent/CN210537495U/en?oq=CN210537495U (accessed on 19 May 2018).
12. Zhengzhou Zhifeng Electrical Technology Co., Ltd. The Utility Model Relates to a Dandelion Seed Collecting Device China: CN112121922A [P]. Available online: https://patents.google.com/patent/CN112121922A/en?oq=CN112121922A (accessed on 25 December 2020).
13. Xu, H.Z.; Hua, Y.; He, J.; Chen, Q.L. The Positive and Negative Synergistic Airflow-Type Jujube Fruit Harvester (P-N JH). *Processes* **2022**, *10*, 1486. [CrossRef]
14. Kang, W.S.; Guyer, D. Development of Chestnut Harvesters for Small Farms. *J. Biosyst. Eng.* **2008**, *33*, 384–389. [CrossRef]
15. Bedane, G.M.; Gupta, M.L.; George, D.L. Development and evaluation of a guayule seed harvester. *Ind. Crops Prod.* **2008**, *28*, 177–183. [CrossRef]
16. Zhao, X.Q.; Wang, W.F. Study on ecological characteristics of wild dandelion seed from beihua university. *For. Prospect. Des.* **2020**, *49*, 93–95.
17. Zhang, F.K. *Design and Optimization Test of Air Suction Floor Jujube Picking Machine*; Tarim University: Xinjiang, China, 2020.
18. Hebei University of Technology. An Air Suction Medlar Picking Machine: CN203120470U [P]. Available online: https://patents.google.com/patent/CN203120470U/en?oq=CN203120470U (accessed on 14 August 2013).
19. Xiao, H.R.; Ding, W.Q.; Mei, S.; Song, Z.Y.; Zhang, Z.; Qin, G.M.; Zhao, Y.A. A Vibrating Chinese Wolfberry Picking Mechanism. Chinese Patent CN205491807U, 24 August 2016.
20. Xu, L.M.; Chen, J.W.; Wu, G.; Yuan, Q.C.; Ma, S.; Yu, C.C.; Duan, Z.Z.; Xing, J.J.; Liu, X.D. Design and operating parameter optimization of comb brush vibratory harvesting device for wolfberry. *Trans. Chin. Soc. Agric. Eng.* **2018**, *34*, 75–82.
21. Han, D.D. *Design of Harvesting Device of Safflower Petals Based on Pneumatic and Simulation of Airflow Field*; Shihezi University: Xinjiang, China, 2014.
22. Pang, M.J.; Lu, Y.; Bao, J.; Chen, R. The Design Principle of the Gas Trap Hood and Numerical Simulation of Its Flow Fielo. *Light Ind. Mach.* **2006**, *24*, 45–47.
23. Yang, G.; Chen, Q.M.; Xia, X.F.; Chen, J.N.; Song, Z.Y. Design and optimization of key components of 4DL-5A faba bean combine harvester. *Trans. Chin. Soc. Agric. Eng.* **2021**, *3*, 10–18.
24. Wu, J.; Tang, Q.; Mu, S.L.; Jiang, L.; Hu, Z.C. Test and Optimization of Oilseed Rape (*Brassica napus* L.) Threshing Device Based on DEM. *Agriculture* **2022**, *12*, 1580. [CrossRef]
25. Li, X.P.; Zhang, W.T.; Wang, W.Z.; Huang, Y. Design and Test of Longitudinal Axial Flow Staggered Millet Flexible Threshing Device. *Agriculture* **2022**, *12*, 1179. [CrossRef]
26. Wang, S.S.; Lu, M.Q.; Hu, J.P.; Chen, P.; Ji, J.T.; Wang, F.M. Design and experiment of chinese cabbage seed threshing device combined with elastic short-raspbar tooth. *Trans. Chin. Soc. Agric. Mach.* **2021**, *52*, 86–94.
27. Sudajan, S.; Salokhe, V.M.; Triratanasirichai, K. Effect of type of drum, drum speed and feed rate on sunflower threshing. *Biosyst. Eng.* **2002**, *83*, 13–21.
28. Di, Z.F.; Cui, Z.K.; Zhang, H.; Zhou, J.; Zhang, M.Y.; Piao, L.X. Design and experiment of rasp bar and nail tooth combined axial flow corn threshing cylinder. *Trans. Chin. Soc. Agric. Eng.* **2018**, *34*, 28–34.
29. Koyuncu, T.; Peksen, E.; Sessiz, A.; Pinar, Y. Chickpea threshing efficiency and energy consumption for different beater-contrbeater combinations. *Agric. Mech. Asia Afr. Lat. Am.* **2007**, *38*, 53–57.

30. Meng, F.H.; Jiang, M.; Geng, D.Y.; Lin, J.H.; Xu, H.G. The Design of Longitudinal-axial Cylinder for the Combine. *J. Agric. Mech. Res.* **2019**, *41*, 90–94.
31. Li, H.T.; Wan, X.Y.; Wang, H.; Jiang, Y.J.; Liao, Q.X. Design and Experiment on Integrated Longitudinal Axial Flow Threshing and Separating Device of Rape Combine Harvester. *Trans. Chin. Soc. Agric. Mach.* **2017**, *48*, 108–116.
32. Liao, Q.X.; Xu, Y.; Yuan, J.C.; Wan, X.Y.; Jiang, Y.J. Design and Test of Longitudinal Axial Flexible Hammer-claw Corn Thresher. *Trans. Chin. Soc. Agric. Mach.* **2019**, *50*, 140–150.
33. Sessiz, A.; Koyuncu, T.; Pinar, Y. Soybean threshing efficiency and power consumption for different concave materials. *Ama Agric. Mech. Asia Afr. Lat. Am.* **2007**, *38*, 56–59.
34. Liao, Q.X.; Wan, X.Y.; Li, H.T.; Ji, M.T.; Wang, H. Design and experiment on cyclone separating cleaning system for rape combine harvester. *Trans. Chin. Soc. Agric. Eng.* **2015**, *31*, 24–31.
35. Abbas, A.A.-A. The Effect of Combine Harvester Speed, Threshing Cylinder Speed and Concave Clearance on Threshing Losses of Rice Crop. *J. Eng. Appl. Sci.* **2019**, *14*, 9959–9965.
36. Powar, R.V.; Aware, V.V.; Shahare, P.U. Optimizing operational parameters of finger millet threshing drum using RSM. *J. Food Sci. Technol.* **2019**, *56*, 3481–3491. [CrossRef]

Article

Design of a Spring-Finger Potato Picker and an Experimental Study of Its Picking Performance

Lihe Wang [1,2], Fei Liu [1,*], Qiang Wang [3], Jiaqi Zhou [1], Xiaoyu Fan [1], Junru Li [1], Xuan Zhao [1] and Shengshi Xie [1]

[1] College of Mechanical and Electrical Engineering, Inner Mongolia Agricultural University, Hohhot 010010, China; 18326106301@163.com (X.F.)
[2] Vocational and Technical College, Inner Mongolia Agricultural University, Baotou 014109, China
[3] Agricultural and Animal Husbandry Technology Extension Centre, Hohhot 010010, China
* Correspondence: afei2208@imau.edu.cn

Abstract: To address the problems of low pickup rates and high rates of wounds in the mechanised harvesting of potatoes under sectional harvesting conditions, a spring-finger potato picker was designed and its overall structure and working principles were described. The kinematic principle of the spring-finger was analysed based on the picking constraints, and the kinematic parameters of the picker were determined. A response surface Box–Behnken Design test design was used to carry out a quadratic orthogonal rotational test, with the speed of the spring-finger, the forward speed of the machine, and the embedded depth as test factors, and the loss rate and the wounded potato rate as evaluation indicators. The test results were optimised and analysed. When the spring-finger speed was 19.57 r/min, the forward speed was 0.61 m/s and the embedded depth was 71.31 mm, the loss rate was 1.18%, and the wounded potato rate was 5.71%. The optimised data were verified, and the results showed that the loss rate of the spring-finger potato picker was 1.48% and the wounded potato rate was 4.98%, meeting the potato picking and harvesting requirements. The research can provide a theoretical basis and design reference for the development and application of sectional potato harvesting machinery.

Keywords: agricultural machinery; harvesting; potato; spring-finger; ADAMS-EDEM; loss rate; wounded potato rate

Citation: Wang, L.; Liu, F.; Wang, Q.; Zhou, J.; Fan, X.; Li, J.; Zhao, X.; Xie, S. Design of a Spring-Finger Potato Picker and an Experimental Study of Its Picking Performance. *Agriculture* **2023**, *13*, 945. https://doi.org/10.3390/agriculture13050945

Academic Editors: Cheng Shen, Zhong Tang and Maohua Xiao

Received: 17 March 2023
Revised: 18 April 2023
Accepted: 22 April 2023
Published: 25 April 2023

1. Introduction

The potato is one of the most widely grown crops worldwide due to its high yield and unique nutritional value [1]. In 2015, China categorized the potato as a strategic staple food; under the development process of the national grain growing industry structure adjustment and potato staple food strategy, China's potato planting area and total production have become some of the best in the world. Potatoes are planted all over the country, with both planting area and total production on the increase. Although China's potato planting mechanization level is improving year by year, the potato harvest mechanization level is still very low; the potato machine harvesting rate is less than 50%, which has seriously restricted the development of China's potato industry [2–4]. At present, potato machine harvesting methods can be divided into combined and sectional harvesting according to the differences in the harvesting process. In China, potatoes are generally eaten fresh, for reasons such as dietary structure [5]. Potatoes harvested with combine harvesters have fragile skins, high trash content and are not easily stored for long periods, so potato harvesting in China is still mainly done in sections [6]. It is important to develop efficient and low-loss potato harvesting equipment to meet the customary consumption and harvesting methods of potatoes in China.

In order to solve the problems of manual potato picking with high labour intensity and low production efficiency, researchers have done a lot of research and achieved good application results. Shi et al. [7] designed a small potato picker with a 30 kw tractor, using a picking shovel combined with a spring-fingered bar conveyor chain to achieve potato picking, and a potato block lifting device and a potato block collection device to achieve potato soil separation and potato block collection, but the wounded potato rate and the rate of contamination are higher than the relevant national industry standards. Xiao et al. [8] designed a small potato picking and grading harvester, using a roller-type potato secondary grading device to achieve combined potato picking, grading and potato block collection, but the number of grading levels does not meet the actual grading requirements, and the picking efficiency is low. Hu et al. [9] designed an integrated potato picking and grading harvester, using a cam mechanism and a vibration dampening lever of the wheel mechanism to achieve horizontal and vertical vibration of the screening part, thus achieving the purpose of potato two-monopoly picking and three-stage sorting, but the size of the whole machine increased, causing a little problem for the machine's turning and turning around. Liu et al. [10] designed a potato harvester with drum-type separation, using a rotary-driven digging device arranged in a circle, while using a drum sieve-type separation structure to separate the conveying device, but there were certain problems of broken skins and injured potatoes. Wang [11] designed an active turntable potato picking device and developed a picking mechanism test stand to carry out exploratory trials in a digitised soil trough, but there was still a certain amount of potato leakage and soil congestion. Yang et al. [12] designed the 4UJ-1400 potato picker, which used a forced pushing device to lift potatoes at a large angle, with good soil removal effects, but low picking efficiency. Wei et al. [13] designed a crawler self-propelled sorting potato harvester, using crawler self-propelled technology, automatic row digging technology and a combination of screen surface separation and manual assisted sorting to achieve potato harvesting, but there are certain problems of potato leakage and wounded, which has not yet been transformed into a popular product.

This paper addresses the problems of low picking rates and high wounded potato rate in mechanised potato harvesting under sectional harvesting conditions, and designs a spring-finger potato picker, analyses the theoretical calculations of its key components, establishes the kinetic equations of the spring-finger in the picking process, and determines the optimal operating parameters through laboratory test in the hope of reducing the loss and wounded potato rates during sectional potato picking.

2. Materials and Methods

2.1. Overall Structure and Working Principles

2.1.1. Overall Structure

The spring-finger potato picking device is mainly composed of a spindle, a spring-finger shaft, a crank, a curved cover, a roller, a spring-finger, and a roller disc, as shown in Figure 1. The device has three sets of spring-finger, with each set evenly distributed on the spring-finger shaft. The crank is connected at one end to the roller and at the other end to the spring gear shaft that passes through the drum disc. As the roller rolls in the cam chute, the crank pushes the spring gear shaft to move the spring-finger along a set course. The curved hood is directly connected to the frame at one end and the other end is staggered with the spring-finger to prevent debris from entering the pickup.

2.1.2. Principle of Operation

The motion of the spring-finger in the picking mechanism is a combination of linear motion in the direction of operation of the machine and rotary motion controlled by a cam slide. The cam chute constrains the movement of the rollers and at the same time constrains the different attitudes of the spring-finger during rotation, the different picking attitudes of the spring-finger correspond to different picking phases. For each week of rotation of the roller in the cam slide, the picking spring-finger correspond to four phases, i.e., "pickup", "lifting", "push" and "quick-return". The pickup cycle is shown in Figure 2.

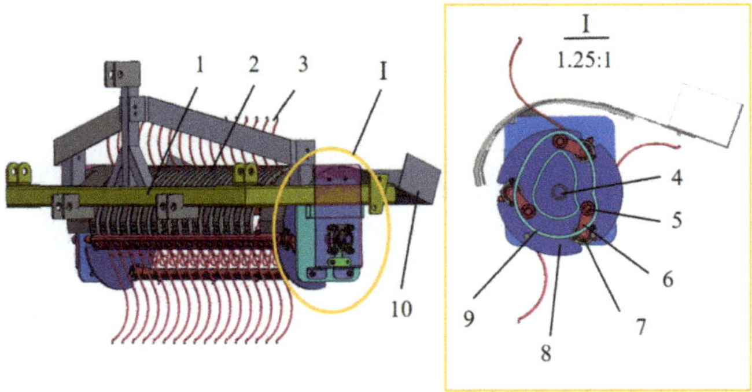

Figure 1. Schematic diagram of the complete assembly of the picking device: (1) frame (2) curved housing (3) popping gear (4) spindle (5) roller (6) crank handle (7) spring-finger shaft (8) roller disc (9) inner and outer cam (10) potato collection box.

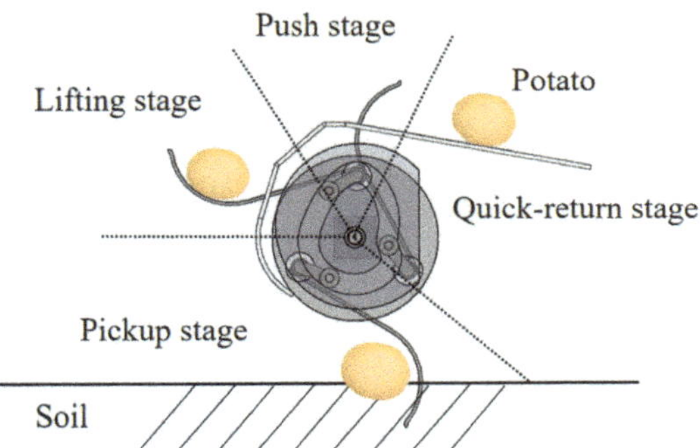

Figure 2. Diagram of the spring-finger picking process.

(1) Pickup stage: at the start of the pickup phase, the drum rotates and at the same time moves the spring-finger underneath the drum, the end of the spring-finger start to protrude through the gap in the curved hood and then are inserted into the ground at a certain angle of entry, then the spring-finger continue to move, as the machine advances and the drum disc rotates, the spring-finger continue to move upwards to pick up the potatoes laid on the ground.

(2) Lifting stage: at the start of the lifting stage, the direction of spring-finger swing is controlled by the rollers in the cam chute and the lifting stage starts in a horizontal position. The drum disc continues to rotate and the end of the spring-finger begin to move backwards and upwards, transporting the pickup potatoes to the top of the curved hood.

(3) Push stage: the phase starts with the pickup spring-finger located at the back front, the relative speed direction is horizontal backwards, pushing the potatoes backwards, in the pushing phase, the pickup spring-finger and the drum disc maintain a large backward angle of inclination between the pickup spring-finger and the curved hood shell to prevent a clamping angle between the pickup spring-finger and the curved hood shell, which is not conducive to backward movement of the potatoes or wounded potato to the potatoes.

(4) Quick-return stage: after the pickup spring-finger have finished picking up, lifting, pushing and unloading potatoes, with the continuous rotation of the drum disc, there is an idling stage before the spring-finger return to the initial position of the pickup stage, this stage does not touch the potatoes and the ground, but the spring-finger rotate with great acceleration in order to quickly return to the starting position of the next pickup stage, shortening the rush back stage and improving the pickup rate.

2.2. Design of Key Components

2.2.1. Spring-Finger

During sectional harvesting, potatoes are excavated and laid out to dry in the field; the distribution is haphazard and irregular, as shown in Figure 3. Potatoes that are excavated and mixed with clods of mud, roots and stems, etc., are exposed on the ground as bright potatoes; meanwhile, others are completely or partially buried in the ground and are known as dark potato [14]. Potatoes are excavated and distributed either on the surface or at a depth of 30–50 mm from the surface [15]. The depth of the pickup tines should be designed to pick up as many potatoes as possible, ensuring that dark potatoes can also be picked up without difficulty, and avoiding soil congestion and severe wear on the tines as a result of going too deep into the soil. Based on the above distribution of potatoes after excavation, the embedded depth should be limited to between 60 and 80 mm to meet the potato pickup requirement. In addition, the mechanical characteristics of the spring-finger should also be taken into account during the design process, as the ground is uneven and mixed with debris in the field due to the constraints of the picking operation.

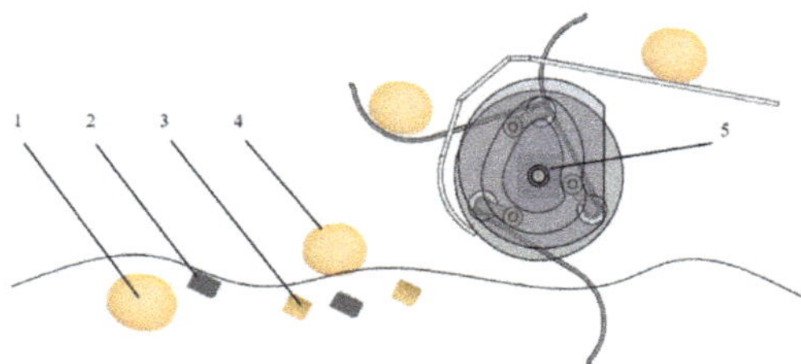

Figure 3. Potato distribution status: (1) dark potato (2) soil block (3) stone block (4) potato (5) pickup device.

The design of the angle of entry of the pickup tines is a function of the pickup effect and is based on the need to pick up potatoes smoothly while reducing wounded potato to the potatoes and not digging up too much soil, in order to reduce the forces on the pickup tines [16]. The potato is analysed for forces, as shown in Figure 4, and the relationship (1) is detailed.

$$\begin{cases} F_N sin\alpha + F_J cos\alpha - F - f cos\alpha = 0 \\ F_N cos\alpha - F_J sin\alpha - \text{mg} + f sin\alpha = 0 \end{cases} \tag{1}$$

where F is the resistance of the potato in the soil, N; m is the mass of the potato, kg; F_N is the support force of the potato with the spring-finger, N; f is the friction force, N; and α is the angle between F and f, °; F_J is the centripetal force on the potato, N; g is the acceleration of gravity, N/kg.

The relevant parameters are brought into the above equation to calculate the pickup spring-finger entry angle $\alpha \geq 53°$. This range ensures that potatoes enter the picker spring-finger smoothly, minimising the wounded potato rate and avoiding the high picking resistance caused by a large angle of entry, thus increasing efficiency and reducing consumption. After determining the lug depth and lug angle, the maximum length that the spring-finger can extend out of the picker is initially determined to be 200 mm, the total length of the spring-finger is 260 mm, the bending angle of the spring-finger δ is 120°, and the material chosen is 65 Mn. The structure of the spring-finger is shown in Figure 5.

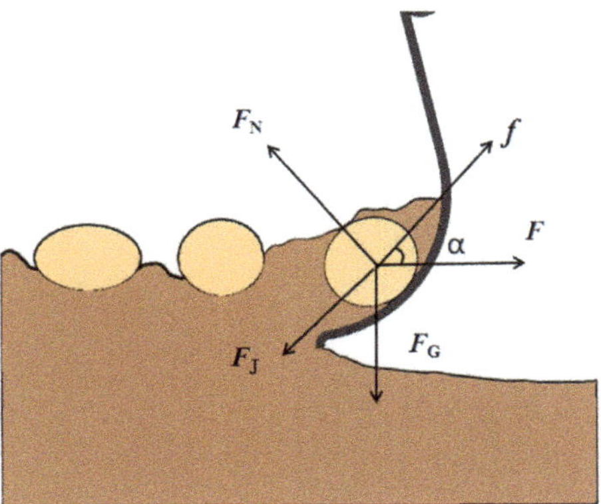

Figure 4. Potato force analysis diagram.

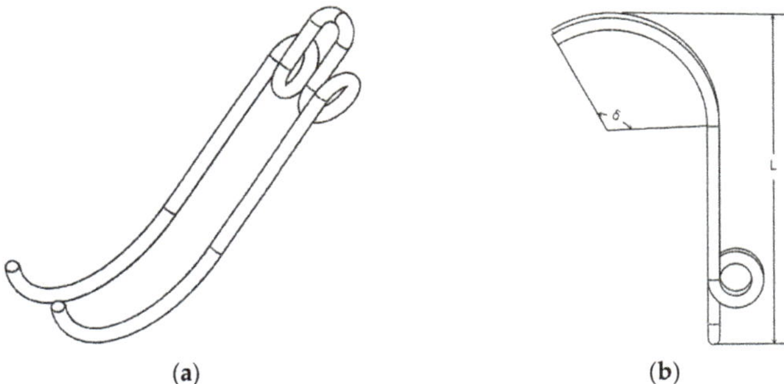

Figure 5. Schematic diagram of the spring-finger structure: (**a**) axonometric drawings. (**b**) side view.

The design of the pickup spacing of the spring-finger should be based on the physical characteristics of the potatoes and their pickup characteristics. The potatoes grown in China are excavated and harvested and then laid on the surface. According to their distribution on the surface, the potatoes are excavated and laid out in an orderly manner, and the grouping state is irregular.

2.2.2. Roller Discs

The picking device is driven by the main drive shaft to rotate the drum disc to rotate the spring-finger shaft, thus completing the picking work. The size of the radius of the drum disc does not directly affect the spring-finger movement, but the size of the drum disc determines the size of the whole machine. Referring to agricultural machinery manuals and other similar picking devices in China, the final diameter of the drum is determined to be 320 mm. The structure of the drum is shown in Figure 6.

2.2.3. Cranks and Rollers

Depending on the movement characteristics of the crank linkage, the crank length should not be designed to be too long or too short. A crank that is too long may result in the rollers becoming stuck in the cam slide and not moving continuously. Meanwhile, a crank that is too short will result in the mechanism running less smoothly and with more impact on the cam slide. The crank is hinged to the roller at one end and pinned to the poppet shaft at the other end. With reference to various models of pickup devices, a crank length of 110 mm was finally chosen.

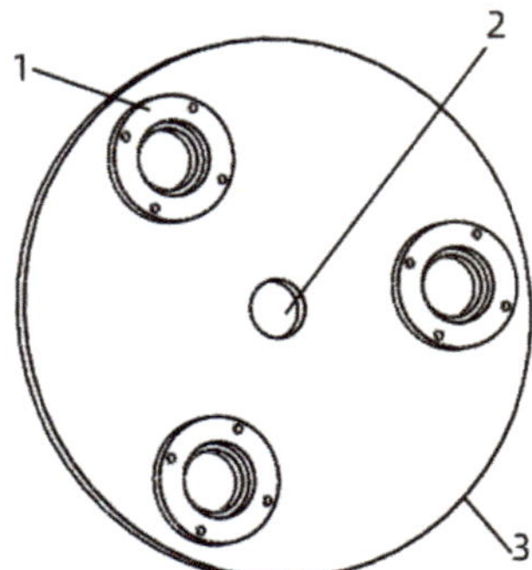

Figure 6. Schematic diagram of the structure of the roller disc: (1) spring gear shaft seat (2) central hole of the drive shaft (3) disc.

The size of the roller radius r_r should take into account the size of the cam slide. As the roller is rolling frictionally and is subjected to large forces when working, a small radius will lead to serious wear on the slide, and the roller is prone to deformation when the radius is too large. Therefore, the size of the rollers should therefore be designed to satisfy Equation (2):

$$r_r = (0.1 - 0.5)\ R_a \quad (2)$$

where R_a is the radius of the base circle, mm; r_r is the radius of the rollers, mm.

The final radius of the rollers was determined to be 45 mm, and the crank poppet shaft-fixing structure is shown in Figure 7.

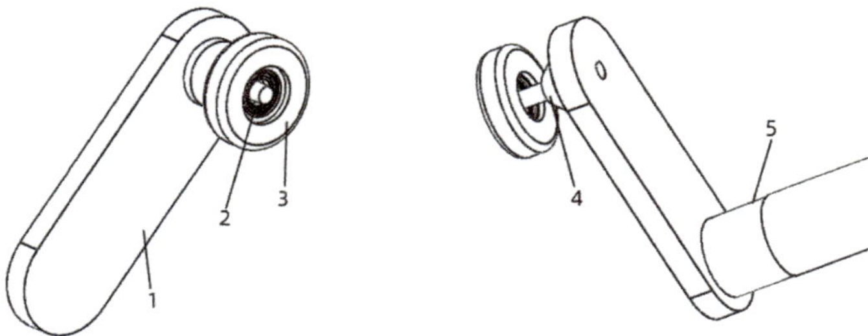

Figure 7. Fixed structure of the crank gear shaft: (1) crank (2) bearing (3) roller (4) connecting rod (5) gear shaft.

2.2.4. Cam Slides

According to the working principle of the picking device, the shape of the cam chute constrains the rotation pattern and attitude of the picking spring-finger during the picking process and is a decisive factor in picking quality. According to the theory of the "reversed pendulum follower disc cam mechanism" by Sheng Kai et al., the design of the cam chute centreline should correspond to the initial position and attitude of the picking spring-finger and the law of motion during the four picking stages [17,18]. The design is based on the polynomial law of motion of the cam slide curve, and a sketch of the picking mechanism based on this design method is shown in Figure 8.

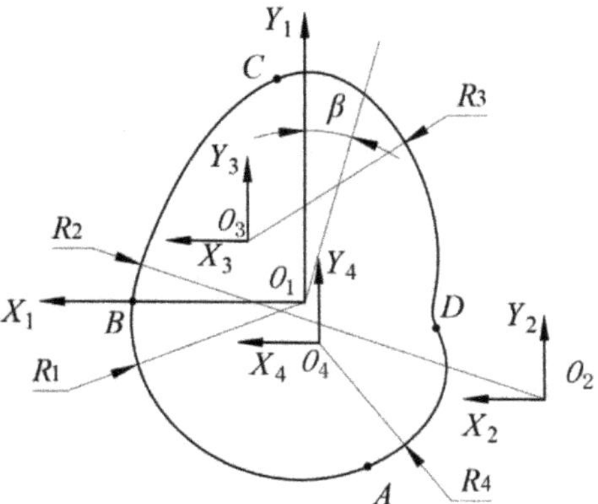

Figure 8. Sketch of the centreline analysis of the cam slide of the poppet pickup mechanism.

Based on the mechanism parameters, a sketch of the pop-up roller pickup mechanism was created in Auto CAD using a graphical method to obtain the phase angles β_1 = 78.36°, β_2 = 89.15°, β_3 = 115.93°, and β_4 = 68.10° for the four stations.

As a result of the above analysis, it can be concluded that the cam slide centreline is composed of multiple circular sliding curves. The mathematical model of the cam slide centreline for each stage is established as follows.

(1) Pickup stage cam chute centreline model.

Curve segment $\overset{\frown}{AB}$ trajectory Equation (3):

$$\begin{cases} X = R_1 cos\beta \\ Y = R_1 sin\beta \end{cases} \tag{3}$$

where R_1 is the $\overset{\frown}{AB}$ radius, mm; β is the angle of the circle corresponding to the centreline of the cam slide, rad.

(2) Model of the centreline of the cam slipway during the lift stage.

Curve segment $\overset{\frown}{BC}$ trajectory Equation (4):

$$\begin{cases} X = R_2 cos\beta_{BC} - X_{O_2} \\ Y = R_2 sin\beta_{BC} - Y_{O_2} \end{cases} \tag{4}$$

where R_2 is the $\overset{\frown}{BC}$ radius, mm; β_{BC} is the angle of circularity of $\overset{\frown}{BC}$, rad; X_{O_2} is the value of the X-axis coordinates of point O_2, mm; Y_{O_2} is the value of the Y-axis coordinates of point O_2, mm.

(3) Model of the centreline of the cam slipway during the pushing stage.

Curve segment $\overset{\frown}{CD}$ trajectory Equation (5):

$$\begin{cases} X = R_3 cos\beta_{CD} + X_{O_3} \\ Y = R_3 sin\beta_{CD} + Y_{O_3} \end{cases} \tag{5}$$

where R_3 is the $\overset{\frown}{CD}$ radius, mm; β_{CD} is the angle of circularity of $\overset{\frown}{CD}$, rad; X_{O_3} is the value of the X-axis coordinates of point O_3, mm; Y_{O_3} is the value of the Y-axis coordinates of point O_3, mm.

(4) Model of the centreline of the cam slipway during the quick-return stage:

Curve segment $\overset{\frown}{DA}$ trajectory Equation (6):

$$\begin{cases} X = R_4 cos\beta_{DA} - X_{O_4} \\ Y = R_4 sin\beta_{DA} + Y_{O_4} \end{cases} \quad (6)$$

where R_4 is the $\overset{\frown}{DA}$ radius, mm; β_{DA} is the angle of circularity of $\overset{\frown}{DA}$, rad; X_{O_4} is the value of the X-axis coordinates of point O_4, mm; Y_{O_4} is the value of the Y-axis coordinates of point O_4, mm.

The final schematic diagram of the cam slide structure is shown in Figure 9.

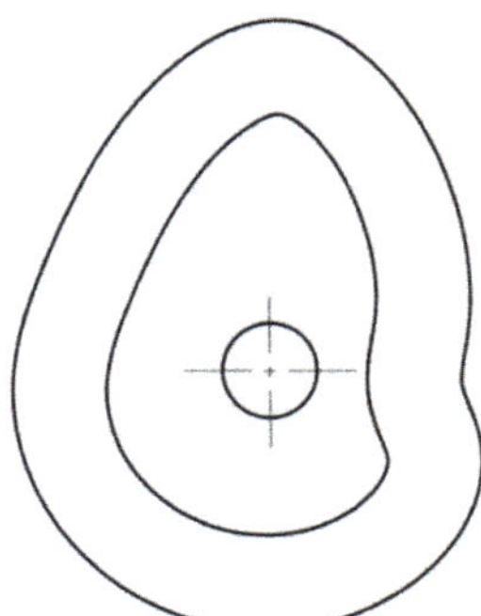

Figure 9. Schematic diagram of the cam slipway structure.

2.3. Analysis of the Principle of Spring-Finger Movement

The law of motion of the spring-finger roller picker is that the cam disk is not moving, the crank and the spring-finger connection point is fixed on the roller, the roller is rotating around the centre of rotation to drive the spring-finger movement. As shown in Figure 10, a coordinate system is established with the centre of the base circle of the cam mechanism as the origin O. The forward direction of the picker is the X-axis direction, and the direction perpendicular to the ground is the Y-axis direction, and the movement of the spring-finger is analysed.

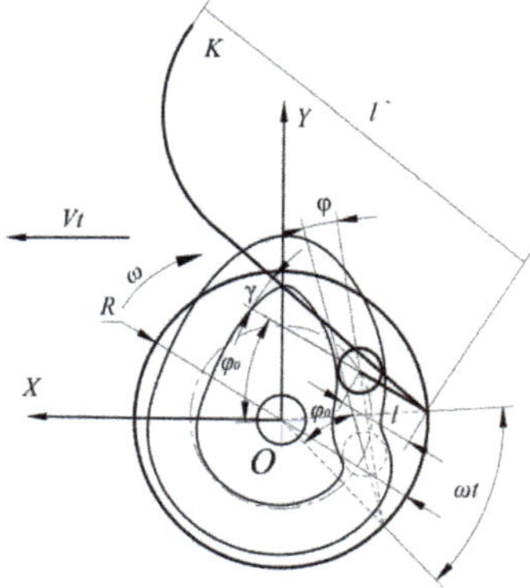

Figure 10. Cam of the spring-finger pickup mechanism: A is the crank and spring-finger connection point; B is the roller centrepoint; K is the spring-finger end point.

The displacement of the K point at the end of the spring-finger is shown in Equation (7):

$$\begin{cases} X = V_t t + Rcos\omega t - Lcos(\omega t + \varphi_0 + \varphi) + L'cos(\omega t + \varphi_0 + \varphi - \gamma) \\ Y = Rsin\omega t - Lsin(\omega t + \varphi_0 + \varphi) + L'sin(\omega t + \varphi_0 + \varphi - \gamma) \end{cases} \quad (7)$$

where X is the horizontal displacement of the end of the spring-finger, m; Y is the vertical displacement of the end of the spring-finger, m; V_t is the forward velocity, m/s; L is the length of the crank, m; L' is the length of the spring-finger, m; γ is the angle between the spring-finger and the crank, rad; φ is the swing angle of the cam mechanism, rad; φ_0 is the initial swing angle of the cam mechanism, rad; t is the time, s; R is the radius of the drum, m; ω picks up the speed of the disc, r/min.

Equation (8) for the radius of gyration of the end of the spring-finger R' is:

$$R' = X^2 + Y^2 \tag{8}$$

where R' is the radius of rotation of the end of the spring-finger, mm.

Without considering the oscillating motion of the spring-finger, the trajectory of the spring-finger is a cycloid, and the shape of the cycloid depends on the size of λ. The equation for the shape of the cycloid (9) is:

$$\lambda = \frac{R'\omega}{V_t} \tag{9}$$

The picking requirements can be met with a value of λ ranging from 0.17 to 0.58. Combined with the forward speed of the potato harvesting machinery in the agricultural machinery design manual and references, the forward speed of the picking device is controlled at 0.4 m/s ~ 0.8 m/s. Upon launch, a disc speed of 15~25 r/min was determined to be the best, a finding that laid the foundation for later tests [19,20].

2.4. Potato Trajectory Simulation Analysis

2.4.1. Simulation Model Building in EDEM

(1) Potato simulation modelling

In this study, the physical parameters of the 'purple-flowered white' potato were tested and analysed at harvest time. The results are shown in Table 1, with the average values for length-width-thickness being 92.59 mm, 63.71 mm and 41.46 mm respectively.

Table 1. Results of measurements of physical parameters of potatoes.

Potato Size	Length (mm)	Width (mm)	Thickness (mm)
Maximum value	149.0	100.7	85.3
Minimum value	71.4	59.2	43.8
Mean value	97.8	77.8	60.4

The spatial coordinates of the 5-sphere potato particles were set in the EDEM software as shown in Table 2, and the final 5-sphere potato particle model was created as shown in Figure 11. The potato particle model was then generated in the particle factory according to the law of normal distribution.

Table 2. Particle model parameters.

Name	Position X (m)	Position Y (m)	Position Z (m)	Physical Radius (m)
sphere 0	−0.02	0	0	0.03
sphere 1	0	−0.008	0	0.032
sphere 2	0	0.008	0	0.032
sphere 3	0.02	0	0	0.03
sphere 4	0	0	0	0.033

(2) Soil simulation modelling

In order to determine the parameters of the sandy soil in the potato growing area, a soil with 6% moisture content was used as the test soil sample and the soil parameters for the simulation tests were determined by a combination of measurements and literature review [21,22] as shown in Table 3.

Figure 11. Potato particle model.

Table 3. Soil parameters.

Parameters	Numerical values
Poisson's ratio	0.2
Modulus of elasticity	13.5 MN/m^2
Density	1.38 g/cm^3
Coefficient of static friction	0.81
Coefficient of rolling friction	0.2095
Resting angle	35.53°
JKR surface energy coefficient	0.356

After practical research, the potato excavator passed through a soil with soft, non-bonded conditions. To facilitate modelling and reduce unnecessary calculations in the simulation, a single sphere soil particle model was established with a particle radius of 2 mm, as shown in Figure 12.

Figure 12. Single sphere particles.

(3) Soil trough simulation modelling

In this experiment, a soil trough with a length, width and height of 2000 × 500 × 300 mm was built with a model number of 50,000 soil particles and a model number of 50 potatoes. The Hertz-Mindlin with JKR model was selected for the discrete element contact model and a simplified pickup model was built as shown in Figure 13.

(4) Determination of contact parameters

In the simulation tests there was contact between soil and soil, soil and potato, potato and potato, pickup device and soil, and pickup device and potato. The material contact parameters can be collated from the literature [23] as shown in Table 4.

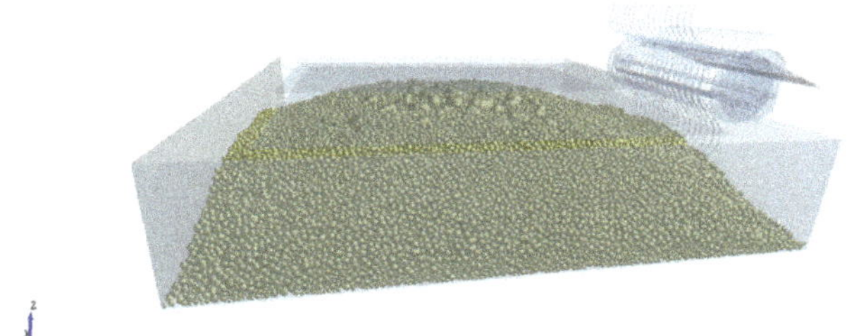

Figure 13. Soil channel model generated by EDEM pre-processing.

Table 4. Physical contact parameters.

Material	Recovery Coefficient	Static Friction Coefficient	Rolling Friction Coefficient
Potato-Potato	0.13	0.2	0.01
Potato-Soil	0.06	0.5	0.01
Potato-Pickup device	0.45	0.5	0.43
Soil-Soil	0.15	0.81	0.15
Soil-Pickup device	0.30	0.5	0.05

2.4.2. Building a Motion Simulation Model in Adams

The 3D model of the potato picking mechanism was simplified as shown in Figure 14. A simulation run was carried out using Adams software to make sure that the settings were correct, then the model building in Adams was ended and subsequent tests were carried out while keeping the machine stable.

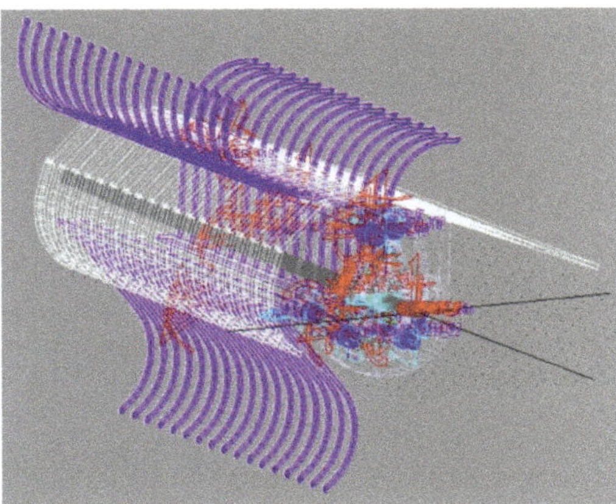

Figure 14. Model of the pickup device in Adams.

2.4.3. ADAMS-EDEM Coupled Analysis Test

A joint ADAMS and EDEM simulation model was established to analyse the potato trajectory and to verify the rationality of the spring-finger potato picker. A forward speed of 0.4 m/s and a spring-finger speed of 25 r/min were chosen. The EDEM software post-processing module can record the movement of the potatoes in real time while simulating the picking process of the picking model. As shown in Figure 15, at 0.600021 s, the spring-finger comes into contact with the potatoes, at 1.30003 s, the potatoes are lifted to the uniformly rising stage, and at 2.00003 s, the potatoes enter the pushing and unloading stage. Then, at 2.40002 s, the potatoes are separated from the spring-finger and continue to move in the opposite direction until fall into the potato collection box, completing the potato picking process. The potatoes are picked up from the surface and lifted to the highest point reached during the process. The trajectory of the potatoes is related to the forward speed of the picker and the speed of the spring-finger.

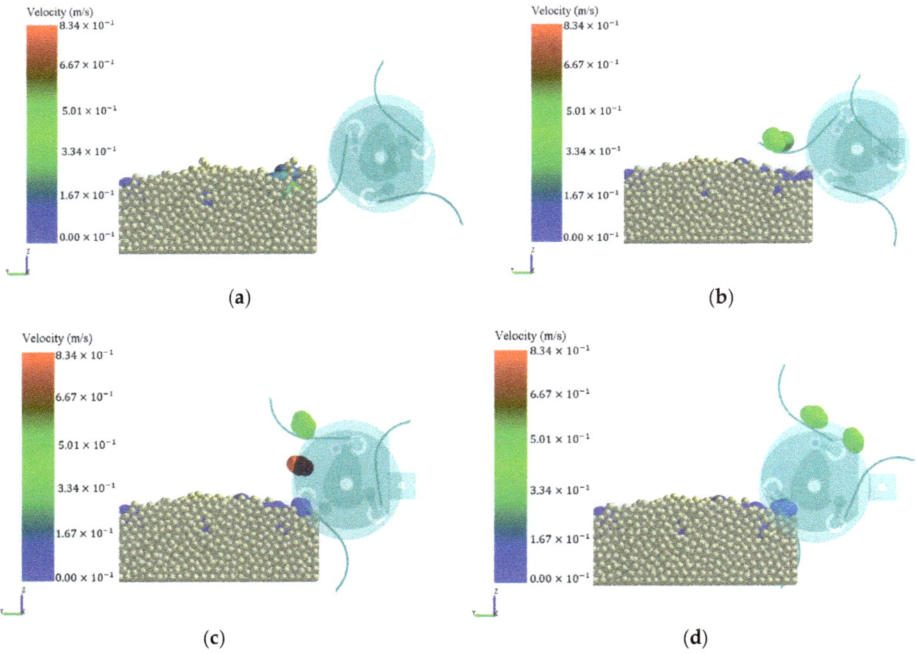

Figure 15. Potato trajectory diagram. (**a**) T = 0.600021 s. (**b**) T = 1.30003 s. (**c**) T = 2.00003 s. (**d**) T = 2.40002 s.

2.5. Laboratory Test

2.5.1. Test Conditions

Performance tests of the spring-finger picker were conducted in May 2022 in the soil tank laboratory of Inner Mongolia Agricultural University. According to the design requirements of the spring-finger picker, the soil should be prepared with a rotary tiller before the picking device works to ensure a consistent environment and ensure that the soil is loose and level. The test procedure is shown in Figure 16.

2.5.2. Evaluation Indicators

Referring to the index requirements in NY/T648-2015 "Technical specification for quality evaluation of potato harvesters" [24] issued by the Ministry of Agriculture of the People's Republic of China, the loss rate and woundedness rate were selected as the evaluation indexes of the spring-finger picking device.

Figure 16. Laboratory test procedure.

(1) Original wounded potato rate: all test potatoes are weighed before the picking operation begins, from which the wounded potatoes are identified and weighed, and the weight of the wounded potatoes in the test area is calculated as a percentage of the total weight of potatoes in the test area, the value of which is the original wounded potato rate calculated according to Equation (10):

$$L_1 = \frac{Q_1}{Q} \times 100\% \quad (10)$$

where L_1 is the original wounded potato rate, %; Q_1 is the original wounded potato quality, kg; and Q is the total potato mass, kg.

(2) Loss rate: after operating the picking device, collect the potatoes that were picked up by the machine and the potatoes missed in the test area and weigh them separately. Calculate the percentage of missed potatoes in the test area to the total potato weight (potatoes picked up by the machine and missed potatoes), and take the value as the loss rate, calculated according to formula (11):

$$L_2 = \frac{Q_2}{Q_2 + Q_3} \times 100\% \quad (11)$$

where L_2 is the loss rate, %; Q_2 is the mass of missed potatoes, kg; and Q_3 is the mass of potatoes picked up by the machine, kg.

(3) Wounded rate: after the device has completed its picking operation, collect the potatoes in the test area (potatoes in potato boxes and missed potatoes) and weigh them. Then, find the wounded potatoes and weigh them. Calculate the percentage of wounded potatoes to the total potato weight, take the value and subtract the original wounded potato rate to obtain the wounded potato rate, calculated according to formula (12):

$$L_3 = \frac{Q_4}{Q_2 + Q_3} \times 100\% - L_1 \quad (12)$$

where L_4 is the percentage of wounded potatoes, %, and Q_4 is the mass of wounded potatoes, kg.

3. Results and Discussion

3.1. Experimental Scheme and Results

Combined with Design-Expert software, the response surface BBD method experimental design [25,26] was carried out, and the spring-finger speed, forward speed and embedded depth were selected as the main influencing factors. The speed of the spring-finger was set at 15–25 r/min, the forward speed at 0.4–0.8 m/s, and the embedded depth at 60–80 mm, and the table of test factor levels is shown in Table 5:

Table 5. Levels and codes of experimental variables.

Level	Test Factors		
	Spring-Finger Speed A (r/min)	Forward Speed B (m/s)	Embedded Depth C (mm)
−1	15	0.4	60
0	20	0.6	70
1	25	0.8	80

The BBD test was chosen for this trial in order to reduce the number of trials. 17 groups of trials were conducted, each group was repeated three times and the final mean was selected in order to take into account the interaction of all factors, with loss rate and wounded potato rate as indicators. The results of the trials are shown in Table 6 below:

Table 6. Design and results of the orthogonal test in the numerical simulation.

Test Number	Test Factors			Test Indicators	
	A r/min	B m/s	C mm	Y_1 %	Y_2 %
1	25	0.8	70	4.34	15.75
2	20	0.4	80	4.56	13.56
3	15	0.4	70	4.72	11.42
4	20	0.8	60	1.77	16.95
5	20	0.6	70	0.47	6.61
6	20	0.8	80	3.04	11.14
7	15	0.8	70	3.16	19.14
8	20	0.6	70	0.36	7.04
9	25	0.6	80	4.32	14.85
10	20	0.6	70	2.04	6.36
11	25	0.4	70	3.39	21.37
12	20	0.6	70	1.06	5.2
13	15	0.6	80	5.07	10.08
14	20	0.4	60	3.01	19.15
15	20	0.6	70	1.34	5.66
16	25	0.6	60	2.66	22.19
17	15	0.6	60	3.52	15.59

3.2. Analysis of Test Results

Using Design Expert12 software multiple regression fitting analysis of the data in Table 6, a quadratic polynomial response surface regression model was established for the three independent variables, loss rate, wounded potato rate, for the spring spring-finger speed A, forward speed B, and embedded depth C. The mathematical model (13) was:

$$\begin{cases} Y_1 = 1.05 - 0.22A - 0.42B + 0.75C + 0.63AB + 0.028AC - 0.070BC + 1.82A^2 + 1.03B^2 + 1.02C^2 \\ Y_2 = 99.77 - 2.02A + 0.74B + 2.28C + 2.71AB + 0.43AC + 0.31BC - 7.43A^2 - 6.16B^2 - 4.91C^2 \end{cases} \quad (13)$$

ANOVA was conducted on the model and the results are shown in Table 7. The p-values of the models for loss rate and wounded potato rate were less than 0.01 and the p-values of the misfit terms were greater than 0.05. The model coefficients of determination R2 were 0.9235 for loss rate and 0.9786 for wounded potato rate respectively, which shows that the optimised regression model was extremely significant and a good fit, and the model was reliable.

Table 7. Analysis variance table of test results.

Source of Variance	Loss Rate Y_1%					
	Sum of Squares	Freedom	Mean Square	F-Value	*p*-Value	Significance
Model	33.09	9	3.68	9.39	0.0037	**
A	0.39	1	0.39	0.99	0.3531	
B	1.42	1	1.42	3.63	0.0986	*
C	4.55	1	4.55	11.61	0.0113	
AB	1.58	1	1.58	4.02	0.0849	*
AC	0.003025	1	0.003025	0.007728	0.9324	
BC	0.020	1	0.020	0.050	0.8293	
A^2	13.99	1	13.99	35.75	0.0006	**
B^2	4.43	1	4.43	11.31	0.0120	
C^2	4.34	1	4.34	11.09	0.0126	
residuals	2.74	7	0.39			
fail to fit	0.86	3	0.29	0.61	0.6415	
error	1.88	4	0.47			
total	35.83	16				
Source of Variance	**Wounded Potato Rate Y_2%**					
	Sum of Squares	**Freedom**	**Mean Square**	**F-Value**	***p*-Value**	**Significance**
Model	502.23	9	55.80	35.55	<0.0001	**
A	40.19	1	40.19	25.60	0.0015	**
B	0.79	1	0.79	0.51	0.5000	
C	73.51	1	73.51	46.83	0.0002	**
AB	44.49	1	44.49	28.34	0.0011	**
AC	0.84	1	0.84	0.53	0.4889	
BC	0.012	1	0.012	0.007709	0.9325	
A^2	132.60	1	132.60	84.48	<0.0001	**
B^2	110.99	1	110.99	70.71	<0.0001	**
C^2	63.77	1	63.77	40.63	0.0004	**
residuals	10.99	7	1.57			
fail to fit	8.80	3	2.93	5.36	0.0692	
error	2.19	4	0.55			
total	513.22	16				

Note: * indicates general significance, $0.05 < p < 0.1$; ** indicates highly significant, $p < 0.01$.

3.3. Response Surface Analysis

Response surface plots were obtained using Design Expert12 software as shown in Figures 17 and 18 to further investigate the effect law of the test factors (spring spring-finger speed A, forward speed B and embedded depth C) and their interaction on the test fingers (loss rate, wounded potato rate).

Figure 17 shows the loss rate Y1 interaction factor response surface analysis. Decreasing the spring-finger speed A and decreasing the forward speed B can reduce the loss rate; however, when the loss rate reaches an extreme point, the loss rate increases as the speed A of the picking gears and the forward speed B of the machine decrease. Decreasing the embedded depth C and increasing the spring-finger speed A can reduce the loss rate; after the loss rate reaches the extreme value, the loss rate will increase as the spring-finger speed A increases and the embedded depth C decreases. Moreover, increasing the forward speed B and decreasing the embedded depth C can help reduce the loss rate; however, when the loss rate reaches a very small value, the loss rate will increase as the forward speed B increases and the embedded depth C decreases.

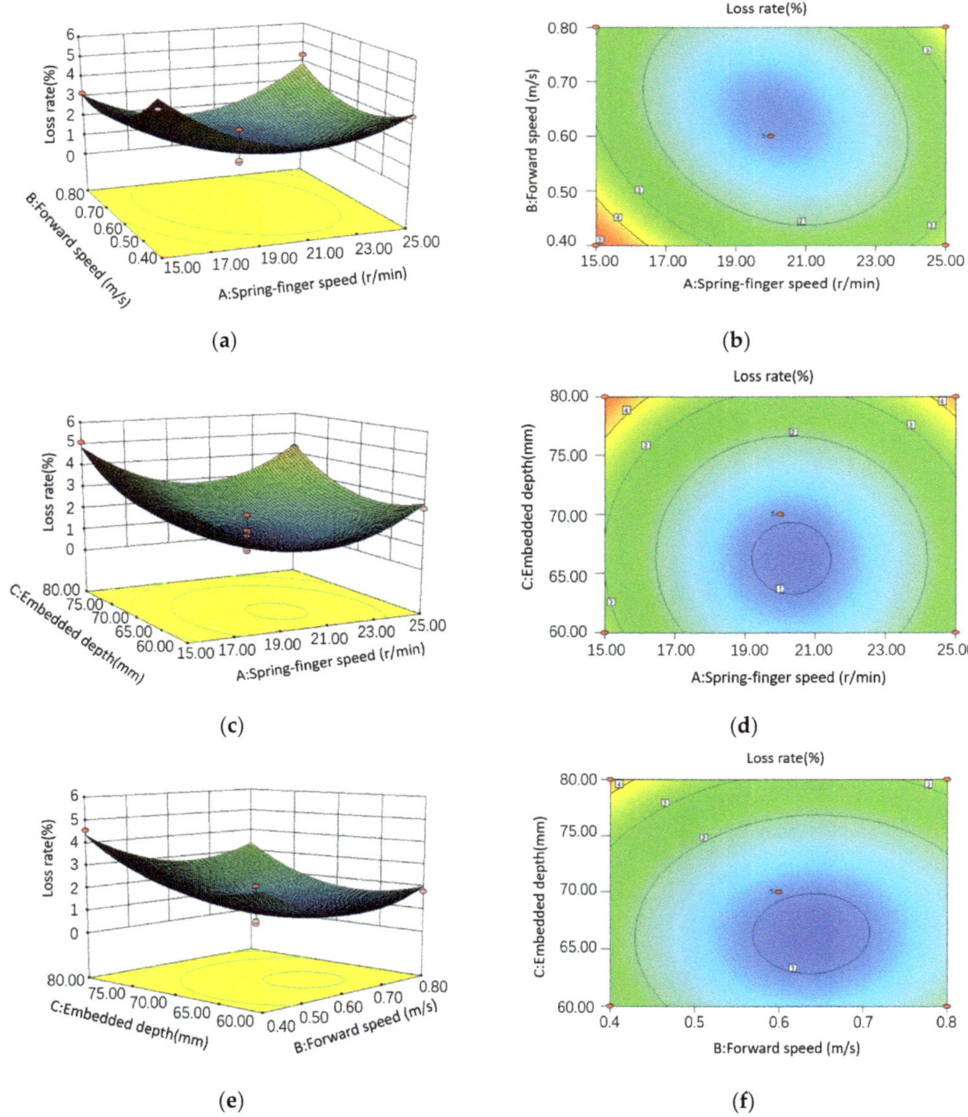

Figure 17. Loss rate interaction factor response surface analysis: (**a**) response surface plot of the effect of spring-finger speed and forward speed on the interaction of the loss rate (**b**) contour plot of the interaction between spring-finger speed and forward speed on the loss rate (**c**) response surface diagram of the interaction between spring-finger speed and embedded depth on the loss rate (**d**) contour plot of the interaction between spring-finger speed and embedded depth on the loss rate (**e**) response surface plot of the interaction between forward speed and embedded depth on the loss rate (**f**) contour plot of the interaction between forward speed and embedded depth on the loss rate.

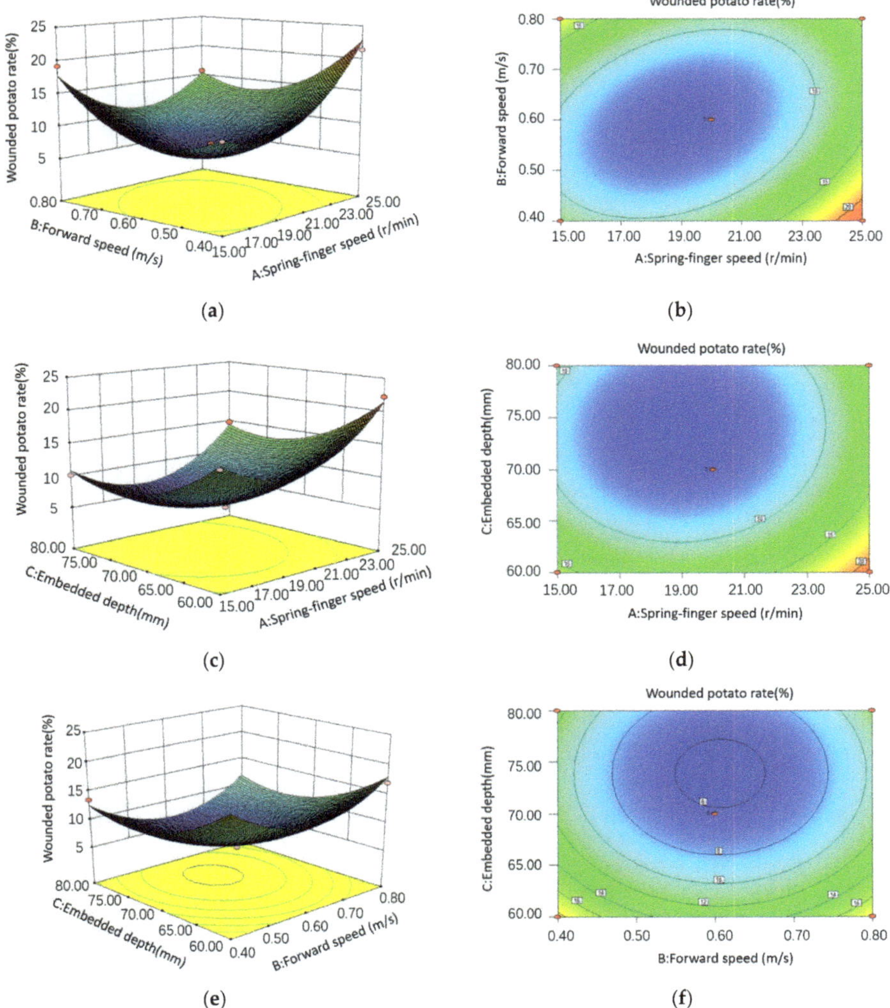

Figure 18. Response surface analysis of interactive factors on the wounded potato rate: (**a**) response surface plot of the effect of spring-finger speed and forward speed on the interaction of the wounded potato rate (**b**) contour plot of the interaction between spring-finger speed and forward speed on the wounded potato rate (**c**) response surface diagram of the interaction between spring-finger speed and embedded depth on the wounded potato rate (**d**) contour plot of the interaction between spring-finger speed and embedded depth on the wounded potato rate (**e**) response surface plot of the interaction between forward speed and embedded depth on the wounded potato rate (**f**) contour plot of the interaction between forward speed and embedded depth on the wounded potato rate.

Figure 18 shows the response surface analysis of the interaction factors for wounded potato rate. It can be seen that reducing the forward speed B and the spring-finger speed A helps to reduce the wounded potato rate. Decreasing the rotational speed of the spring spring-finger A and increasing the embedded depth C reduces the rate of wounded potatoes. Increasing the embedded depth C and decreasing the forward speed B helps to reduce the woundedness rate; however, when the wounded rate reaches a very small value, continuing to decrease the forward speed B and continuing to increase the embedded depth C causes the woundedness rate to gradually increase.

3.4. Optimization Model Analysis and Laboratory Test Validation

Based on the aforementioned results of the regression model of the test indicators and the conditions of picking operation of the potato picking device, the constraints of each test factor were set in the Design Expert12 software, with the objective function shown in Equation (14). The optimum combination of parameters was found to be 19.57 r/min for the spring-finger speed, 0.61 m/s for the forward speed and 71.31 mm for the embedded depth, with corresponding loss rates of 1.18% and 5.71% for the wounded potato rate.

$$\begin{cases} minY_1 = f_1(A, B, C) \\ minY_2 = f_2(A, B, C) \\ S.T. \begin{cases} Y_1 \leq 5\% \\ Y_2 \leq 6\% \\ 5\ \text{r/min} \leq A \leq 45\ \text{r/min} \\ 0.2\ \text{m/s} \leq B \leq 1.6\ \text{m/s} \\ 50\ \text{mm} \leq C \leq 100\ \text{mm} \end{cases} \end{cases} \tag{14}$$

To verify the reliability of the model prediction results, the optimised parameters were rounded and then subjected to a validation test, setting the spring-finger speed at 20 r/min, the forward speed at 0.6 m/s and the embedded depth at 70 mm. The test was carried out three times and the results were averaged, resulting in a loss rate of 1.48% and a wounded potato rate of 4.98%. The test verification results are in general agreement with the optimised analysis and can be used as the optimum working parameters for this picking mechanism.

4. Conclusions

(1) Aiming at the problems of low pickup rate and high wounded rate in potato sectional harvesting, a spring-finger potato picking device was designed, its working principle and key components were introduced, and the main design parameters of its components were determined by analysing key components such as spring-finger, drum disc, crank, roller and cam chute.

(2) The Box-Behnken response surface optimisation test method was used to analyse the effects of spring-finger speed, forward speed and embedded depth on the loss rate and wounded potato rate, to establish a regression model and to analyse their interaction, and the wounded potato rate was 4.98%. The results of the validation tests were basically consistent with the results of the optimisation analysis, indicating that the parameter optimisation regression model is reliable.

(3) Although the spring-finger potato picking device meets the design requirements of the picking process, there is still a certain wounded potato rate, and the subsequent research should improve the picking operation effect by optimising the profile of the cam slide.

Author Contributions: Conceptualization, L.W.; methodology, L.W. and F.L.; software J.Z. and J.L.; validation, L.W. and F.L.; investigation, X.Z. and J.L.; resources, L.W.; data curation, L.W. and Q.W.; writing—original draft, L.W. and X.F.; writing—review and editing, F.L. and S.X.; visualization, L.W. and F.L.; supervision, F.L.; funding acquisition, F.L. All authors have read and agreed to the published version of the manuscript.

Funding: This research was funded by the Science and Technology Program of Inner Mongolia Autonomous Region of China (2020GG0168), Inner Mongolia Autonomous Region Young Scientist Program of China (NJYT20B03), the National Natural Science Found of China (31901409) and the Science and Technology Research Project for Higher Education Institutions in Inner Mongolia Autonomous Region (NJZY21357).

Institutional Review Board Statement: Not applicable.

Data Availability Statement: Not applicable.

Conflicts of Interest: The authors declare no conflict of interest.

References

1. Mystkowska, I.; Zarzecka, K.; Gugała, M.; Ginter, A.; Sikorska, A.; Dmitrowicz, A. Impact of Care and Nutrition Methods on the Content and Uptake of Selected Mineral Elements in Solanum tuberosum. *Agronomy* **2023**, *13*, 690. [CrossRef]
2. Fan, J.; Li, Y.; Luo, W.; Yang, K.; Yu, Z.; Wang, S.; Hu, Z.; Wang, B.; Gu, F.; Wu, F. An Experimental Study of Stem Transported-Posture Adjustment Mechanism in Potato Harvesting. *Agronomy* **2023**, *13*, 234. [CrossRef]
3. Dou, Q.; Sun, Y.; Sun, Y.; Shen, J.; Li, Q. Domestic and foreign potato harvesting machinery status and development. *China Agric. Chem. News* **2019**, *9*, 206–210.
4. Zhao, Q. Domestic and foreign potato harvesting machinery research status and development prospects. *Agric. Eng.* **2020**, *6*, 7–10.
5. Luo, Q.; Lun, L.; Gao, M. Strategic pathways for high-quality development of China's potato industry from 2021 to 2025. *China Agric. Resour. Zoning* **2022**, *3*, 37–45.
6. Wei, Z.; Li, X.; Zhang, Y.; Li, H.; Sun, C. Research progress of potato full mechanized production technology and equipment. *Agric. Mech. Res.* **2017**, *9*, 1–6.
7. Shi, Y.; Yan, S.; Zhu, R.; Li, J.; Huang, S.; Liu, Y. Development and experiment of a small potato picker. *Arid Reg. Agric. Res.* **2016**, *4*, 287–291+298.
8. Xiao, W.; Gao, Y.; Chen, H.; Zhang, Y. Design and experiment of small potato picking and grading machine. *Agric. Mech. Res.* **2019**, *12*, 130–134.
9. Hu, Q.; Xiao, W. Design and test of an integrated potato picking and grading harvester. *Agric. Mech. Res.* **2021**, *11*, 110–114.
10. Liu, J.; Wei, M.; Kang, H.; Wang, Y.; Zhou, J. Design and experiment of potato harvester based on roller separation. *Agric. Mech. Res.* **2021**, *5*, 96–103.
11. Wang, W. Experimental Research on Potato Picking Device Based on Discrete Elements. Master's Thesis, Northwest University of Agriculture and Forestry Science and Technology, Shanxi, China, 2017.
12. Yang, J.; Li, G.; Hao, L.; Chen, W.; Ye, T. Research on the development status and main function structure of potato picker. *Agric. Mach. Use Maint.* **2019**, *3*, 15–16.
13. Wei, Z.; Wang, X.; Li, X.; Wang, F.; Li, Z.; Jin, C. Design and test of a self-propelled sorting potato harvester with crawler. *J. Agric. Mach.* **2023**, *2*, 95–106.
14. Campbell, A.J.; Birt, I.; MacKinnon, B. Modifications of a potato harvester for small plot field research. *Am. Potato J.* **1990**, *11*, 799–803.
15. Zhang, F.; Qiu, Z.; Mao, P. The model and simulation of the potato harvester. *Appl. Mech. Mater.* **2011**, *44*, 900–904. [CrossRef]
16. Yang, F.; Sun, B.; Zheng, S. Research status and development trend of potato harvesting machine. *For. Mach. Woodwork. Equip.* **2021**, *10*, 4–10.
17. Sheng, K.; Zeng, N. Mechanism characteristics and mathematical model of motion of a spring-tooth roller picker. *J. Agric. Mach.* **1991**, *1*, 51–57.
18. Yang, H.; Hu, Z.; Wang, B. Research progress of potato harvesting mechanization technology. *Chin. J. Agric. Mach. Chem.* **2019**, *11*, 27–34.
19. China Agricultural Machinery Research Institute. *Handbook of Agricultural Machinery Design*; Machinery Industry Press: Beijing, China, 2007; p. 1056.
20. Kanafojski, T. *Crop-Harvesting Machines*; China Agricultural Machinery Press: Beijing, China, 1983; pp. 487–512.
21. Sun, Y.; Shao, L.; Fan, Z. Experimental study on Poisson's ratio of non-cohesive soils. *Geotechnics* **2009**, *s1*, 63–68.
22. Song, Z.; Li, H.; Yan, Y. Calibration and testing of a discrete element simulation model for non-equal-sized particles in mulberry soils. *J. Agric. Mach.* **2022**, *6*, 21–33.
23. Wang, Y. Research on the Structure and Loosening Effect of Deep Loosening Shovel Based on Discrete Element Method. Master's Thesis, Jilin Agricultural University, Jilin, China, 2014.
24. *NY/T 648-2015*; Technical Specification for Quality Evaluation of Potato Harvesters. China Standard Press: Beijing, China, 2015.
25. Chen, K. *Experimental Design and Analysis*; Tsinghua University Press: Beijing, China, 2005; pp. 123–134.
26. Pan, L.; Chen, J. *Experimental Design and Data Processing*; Southeast University Press: Nanjing, China, 2008; pp. 78–86.

Article

Calibration of Ramie Stalk Contact Parameters Based on the Discrete Element Method

Yao Hu †, Wei Xiang †, Yiping Duan, Bo Yan, Lan Ma, Jiajie Liu and Jiangnan Lyu *

Institute of Bast Fiber Crops, Chinese Academy of Agricultural Sciences, Changsha 410205, China; 82101202102@caas.cn (Y.H.); xiangwei@caas.cn (W.X.); hunaudyp@stu.hunau.edu.cn (Y.D.); yanbo@caas.cn (B.Y.); malan@caas.cn (L.M.); liujiajie@caas.cn (J.L.)

* Correspondence: yjljn@sina.com

† These authors contributed equally to this work.

Abstract: To obtain the physical parameters and contact parameters of ramie stalk decorticating simulation, the structural dimensions, density, moisture content, elastic modulus, and contact parameters of the ramie stalk were measured in this study based on the phloem and xylem of the ramie stalk. The physical stacking angles of the phloem and xylem were measured by the cylinder lift method and the extraction of the partition method, respectively. The contact parameters between the xylem and phloem of the ramie stalk were directly calibrated. Additionally, the contact parameters of the phloem–phloem, phloem–Q235A steel, xylem–xylem, and xylem–Q235A steel were used as calibration objects, and the simulated stacking angle was used as the evaluation index. Then, the Plackett–Burman test was designed to screen for the parameters which were significantly affecting the simulated stacking angle. Furthermore, the steepest ascent test determined the optimal range of values for two significant parameters of the phloem and three significant parameters of the xylem. Based on the central composite design, the second-order regression equations between the significant parameters of the phloem and xylem and the stacking angle were established, respectively. The physical stacking angles of 37.93° for phloem and 27.17° for xylem were the target values to obtain the optimal parameter group. The results showed that the restitution, static, and rolling friction coefficients between the xylem and phloem were 0.60, 0.53, and 0.021, respectively. The static and rolling friction coefficients between the phloem and phloem were 0.41 and 0.056, respectively. The rolling friction coefficient between the xylem and Q235A steel was 0.033, and the static and rolling friction coefficients between the xylem and xylem were 0.44 and 0.016, respectively. The verification test showed that the relative error values were less than 2.11%, which further indicated that the modeling method and parameter calibration of the ramie stalk phloem and xylem models were accurate and reliable. They can be used for the subsequent calibration simulation tests of ramie stalk bonding parameters and ramie stalk decorticating simulations.

Keywords: ramie stalk; discrete element method (DEM); parameter calibration; stacking angle; contact parameters

Citation: Hu, Y.; Xiang, W.; Duan, Y.; Yan, B.; Ma, L.; Liu, J.; Lyu, J. Calibration of Ramie Stalk Contact Parameters Based on the Discrete Element Method. *Agriculture* **2023**, *13*, 1070. https://doi.org/10.3390/agriculture13051070

Academic Editor: Filipe Neves Dos Santos

Received: 7 April 2023
Revised: 18 April 2023
Accepted: 27 April 2023
Published: 17 May 2023

1. Introduction

Ramie is a culturally significant and traditional economic crop in China. It is widely cultivated in the Yangtze River Basin and southern China, with the planting area and fiber production accounting for over 95% of the global output [1]. Ramie is a perennial herbaceous plant belonging to the family Urticaceae [2], with its entire body being highly valuable. Its fiber exhibits excellent properties, such as strong moisture absorption, great breathability, antibacterial properties, and biodegradability [3–5], which makes it widely useful in textiles, medicine, military, agriculture, and ecofriendly packaging industries. However, the fiber of ramie must be decorticated and processed before use in the textile industry [6]. The ramie decorticator is the main processing equipment for decorticating ramie fiber. The high-speed rotating rollers can crush and eject the xylem through bending and beating the ramie stalk frequently while at the same time keeping the bast fiber intact, thereby obtaining pure raw ramie fiber [7]. However, the movement of each component of the ramie stalk is complex during decorticating, and traditional research methods are unable to simulate and analyze this movement state, which has so far hindered the optimization of decorticating equipment.

Simulation technology is widely used to optimize agricultural machines due to the rapid development of computational speed [8,9]. The EDEM simulation software based on the discrete element method can accurately analyze the movement law of agricultural materials in agricultural machinery. In agricultural machinery simulation operations, agricultural materials can be modeled as particles or clusters. The EDEM software can record the real-time movement trajectory and mechanical behavior of agricultural materials and conduct in-depth research on the interaction mechanism between materials and machinery; this can guide the optimization of the machinery design [10,11]. However, it is crucial to input the accurate physical and contact parameters of the materials to establish a discrete element model, which can faithfully reproduce the characteristic properties of the material and adapt to the real-world operating conditions of the machinery.

The DEM simulation modeling requires the input of intrinsic physical and contact parameters [12]. Due to the individual differences in materials, errors in physical tests, and differences in model constructions, obtaining accurate discrete element parameters through physical tests is difficult. Therefore, the calibration of physical tests and virtual simulations must be conducted to ensure consistency between the simulation and physical results. A stacking angle test can effectively calibrate discrete element parameters [13]. Various stacking angle measurement methods have been developed for different material characteristics, such as the injection, tilted box, cylinder lifting, and extraction of partition methods [14–17]. These methods are widely used to calibrate discrete element simulation parameters in materials such as soil, fertilizer, seeds, and biomass stalks. Xiang et al. [18] built a soil simulation model based on soil stacking tests of physical measurements and EDEM-software-recommended parameters. They used the stacking angle as the response value and completed the calibration and optimization of soil simulation's physical parameters through the Plackett–Burman, steepest ascent, and Box–Behnken tests. Liu et al. [19] obtained the stacking angle of a wheat grain heap through physical and simulation tests under the response value of different parameter combinations, which were finally based on response surface optimization, and they calibrated the discrete element simulation parameters of wheat. Xiao et al. [20] explored the influence of compound fertilizer characteristic parameters on the stacking angle, and they determined the rolling friction coefficients of the three types of granular fertilizers under two particle modeling methods. Shi et al. [21] measured the interval values of the contact parameters of fallen jujube through stacking angle tests. They used EDEM software to establish a simulation test of a fallen jujube stacking angle. They used Plackett–Burman, steepest ascent, and central composite design tests to obtain specific values of simulation parameters from the interval values. Dai et al. [22] used 3D scanning technology to construct a discrete element model of lily bulbs, and they calibrated the contact parameters between the lily bulbs and Q235 steel through bench tests and simulation parameter tests, and then established a regression model for the relative errors of the parameters and optimized the response surface to calibrate the discrete element contact parameters of the lily bulbs. Zhang et al. [17] determined the contact parameters between corn stalks and shredder blades, as well as corn stalks themselves. They used the extraction of the partition method to calibrate the contact parameters for corn stalk discrete element simulation. The research on discrete element simulation modeling and the parameter calibration of agricultural materials mainly focuses on spherical and quasi-spherical particles, such as soil, crop seeds, and fertilizers, and large spherical particles, such as fruits and plant bulbs. Unlike traditional spherical and quasi-spherical particle materials, the ramie stalk has a cylindrical shape in its xylem, and its phloem is strip-shaped after being decorticated from the xylem. Few scholars have studied the calibration of the contact parameters of discrete element models on the ramie stalk.

To establish a discrete element simulation model of ramie stalks, this study established the discrete element models of the phloem and xylem on ramie stalk through the physical and simulation tests. Based on the measurement results of the phloem and xylem of ramie stalks, the stacking angle of the phloem and xylem particle mixture was taken as the response value. This study employed Plackett–Burman, steepest ascent, and response surface tests to complete the calibration and optimization of the contact parameters of the phloem and xylem on ramie stalks. This can provide a basic model and technical support for simulating the decorticating process of ramie fiber.

2. Materials and Methods

2.1. Measurement of the Physical Parameters of Ramie Stalk Phloem and Xylem

We selected the "Zhongzhu No. 1" variety of ramie as the experimental object to measure the structural dimensions, density, shear modulus, and other parameters of the ramie stalk's phloem and xylem. The "Zhongzhu No. 1" variety of ramie was planted at the Shijihu test base of Bast Fiber Crops of the Chinese Academy of Agricultural Sciences Institute in Yuanjiang City, Hunan Province.

2.1.1. Dimensional Measurements

We employed a five-point sampling method to obtain 100 random samples of ramie stalks. The plant height was measured using a tape measure (with an accuracy of 1 mm), and the xylem outer diameter, xylem inner diameter, and phloem thickness of the ramie stalk were measured using a DL91150 Vernier caliper (with an accuracy of 0.01 mm). The results are the average values of the measurements.

The results of the measurements are shown in Table 1. The average values of the ramie plant height, xylem outer diameter, xylem inner diameter, and phloem thickness were 1969.38 mm, 12.56 mm, 7.98 mm, and 0.71 mm, respectively.

Table 1. Structural dimensions of the ramie.

Parameters	Max/mm	Min/mm	Mean/mm	Standard Deviation/mm	Variance/mm
Plant height	2307.00	1457.00	1969.38	217.27	4.77×10^4
Xylem outer diameter	16.11	9.00	12.56	1.58	2.51
Xylem inner diameter	13.07	5.31	7.98	1.26	1.60
Phloem thickness	1.21	0.45	0.71	0.15	0.02

2.1.2. Density and Moisture Content

The density of the phloem and xylem of a fresh ramie stalk was measured using a liquid immersion method [23] by immersing a certain mass of phloem and xylem in water and measuring the drainage volume to obtain the volume of each component of the ramie stalk and then calculating its density. The test was repeated ten times, and the average value was taken. The average densities of the phloem and xylem were 1618.95 kg/m^3 and 751.50 kg/m^3, respectively.

The moisture content of the ramie stalk was determined according to the "Method for determination of the moisture content of wood" [24] using a high-precision balance (with a weighing range of 220 g and an accuracy of 0.001 g) and a DGT-G220 blast drying oven. The moisture content of the ramie stalk was calculated to be approximately 79.72%.

2.1.3. Elastic Modulus and Shear Modulus

Tensile testing is a common method for obtaining the elastic modulus. This method has been used to measure the elastic modulus of wood [25], corn straw [26], rice stalk [27], bamboo stalk [28], and reed stalk [29], providing the most concentrated measurement of the elastic modulus [30]. In this study, based on the standard "Method of sample logs sawing and test specimens selection for physical and mechanical tests of wood" [31] and the relevant literature [32], tensile tests of the phloem and xylem of the ramie stalks were performed using a microcomputer-controlled electronic universal material testing machine (produced by Shanghai Tuofeng Instruments Co., Ltd., Shanghai, China), model TFW-508, with a 500 N transducer, and the accuracies of the force transducer as well as the displacement transducer were within ±0.1%. The test parameters were set to a loading speed of 5 mm/min. The stopping condition was judged to be a drop in the force value of more than 80% of the peak force. The tensile test is shown in Figure 1. The elastic modulus was calculated based on the slope of the linear region of the stress–strain curve. The shear modulus was derived from the elastic modulus using a conversion formula.

(a)

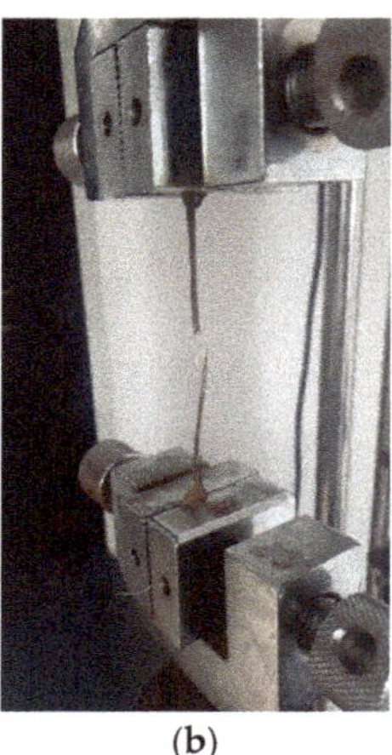

(b)

Figure 1. Tensile test to determine the modulus of elasticity: (**a**) xylem tensile test; (**b**) phloem tensile test.

The results show that the average elastic modulus of the phloem and xylem were 1721.4 MPa and 630.3 MPa, respectively. The average shear modulus values of the phloem and xylem were calculated to be 614.8 MPa and 242.4 MPa, respectively, using Equation (1):

$$\begin{cases} E = \frac{F_1}{\delta A} \\ \varepsilon = \lim_{T_1 \to 0} \frac{\Delta L}{L_1} \\ G = \frac{E}{2(1+v)} \end{cases} \tag{1}$$

where E represents the elastic modulus of ramie, MPa; F_1 represents the axial load on ramie, N; A represents the contact area, mm^2; ε represents the strain value; ΔL represents the deformation of the ramie after tension, mm; L represents the length of the tensile area of the ramie specimen, mm; G represents the shear modulus, MPa; and v is the Poisson's ratio.

2.2. Method for Determining Contact Parameters

The contact parameters of the ramie stalk discrete element simulation model included the coefficient of restitution, coefficient of static friction, coefficient of rolling friction between the phloem and phloem, phloem and Q235A steel, phloem and xylem, xylem and xylem, and xylem and Q235A steel. Among them, the physical tests mainly measured the contact parameters between the phloem and Q235A steel, phloem and xylem, and xylem and Q235A steel. As the phloem–phloem and xylem–xylem contact parameters are difficult to obtain directly from physical tests, physical tests of the stacking angle of phloem and xylem needed to be carried out and calibrated by subsequent simulation tests of the stacking angle.

2.2.1. Coefficient of Restitution

In this study, we mainly measured the coefficient of restitution between phloem and Q235A steel, xylem and Q235A steel, and phloem and xylem. The coefficient of restitution is a parameter that measures the ability of an object to return to its original shape after a collision. It is defined as the ratio of the normal relative separation velocity after a collision to the relative approach velocity before collision (i.e., the ratio of the highest rebound height (h') and the initial drop height (h) during the collision between the test object and the material) [33]. When an object falls freely and collides with the test object, the object bounces freely after the collision, and only gravity works during the falling and rising process. The formula for calculating the coefficient of restitution is as follows (2):

$$e = \left| \frac{v_2' - v_1'}{v_1 - v_2} \right| = \left| \frac{-\sqrt{2gh'}}{\sqrt{2gh}} \right| = \sqrt{\frac{h'}{h}} \tag{2}$$

where v_1 and v_2 are the velocities of the test object and the material before collision, m/s; $v_{1'}$ and $v_{2'}$ are the velocities of the test object and the material after collision, m/s; h and h' are the initial drop height and the highest rebound height, mm; and g is the acceleration due to the fact of gravity, m/s^2.

The maximum rebound height of the coefficient of restitution can be measured using a high-speed camera system. The experimental process was recorded by connecting the high-speed camera, HiSpec5 (Fastec Imaging Inc., San Diego, CA, USA) to a computer. The camera was set to a frequency of 500 Hz, with a resolution of 1280 × 1024 pixels, and a sampling rate of 2 ms. Considering that the ramie stalk has a lower density, the air resistance significantly impacts the experimental results when the falling speed is high. Therefore, the object to be measured was dropped freely from an initial height (h) of 205 mm, collided with the contact material, and the highest rebound height (h′) was captured and recorded. The experimental process is shown in Figure 2.

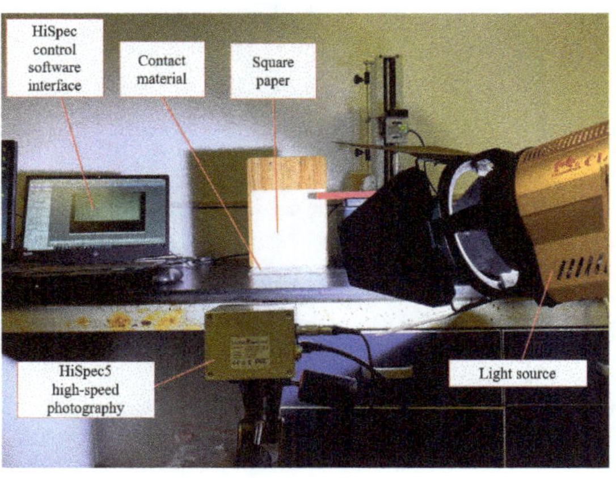

(a)

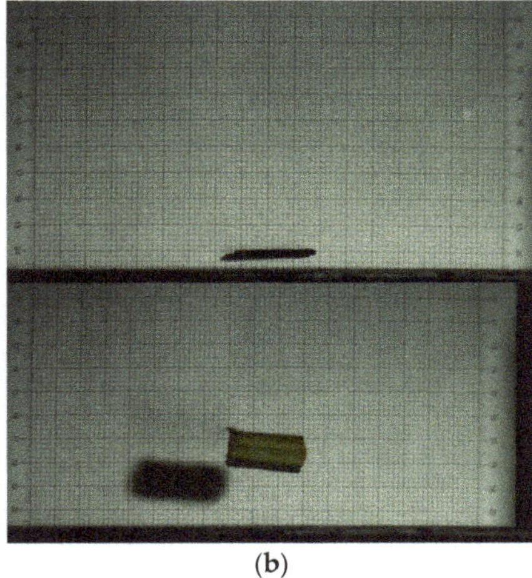

(b)

Figure 2. Measurement of the restitution coefficient: (**a**) test equipment; (**b**) keyframe of the rebound's highest point.

The rebound height (h′) was recorded by capturing the keyframe when the tested object reached the highest position after the collision, from which the coefficient of restitution was calculated. Each group of tests was repeated ten times to analyze the positive collision process between the phloem and Q235A steel, xylem and Q235A steel, and phloem and xylem, the coefficient of restitution was then calculated according to Equation (2), and the average value was taken.

2.2.2. Coefficient of Friction

Measurement of the Static Friction Coefficient

In this study, we mainly measured the static friction coefficients of phloem–Q235A steel, xylem–Q235A steel, and phloem–xylem. The static friction coefficient is the ratio of the maximum static friction force applied to the object to the normal pressure. It is usually measured by the incline plane method [34,35]. The ramie stalk with a high degree of roundness was selected, and the phloem was peeled with a cutter knife to create the test sample. A self-made inclinometer and a digital angle measuring instrument (accuracy of 0.05°) were used to measure and calculate the required static friction coefficient, as shown in Figure 3.

(**a**)

(**b**)

Figure 3. Measurement of the static friction coefficient: (**a**) test equipment; (**b**) sliding angle.

The test object was placed on a steel plate when the static friction coefficients of the ramie stalks for phloem–Q235A steel and the xylem–Q235A steel were measured. The handle of the inclinometer was shaken to raise the incline angle, and the sliding angle was recorded when a sliding trend occurred. When measuring the static friction coefficient of the phloem–xylem, the phloem was attached to the steel plate, and the xylem was axially placed above it for the test. The sliding angle was recorded, and the coefficient of static friction was calculated. The calculation of the static friction coefficient is shown in Equation (3):

$$\mu_1 = \frac{f}{F_2} = \frac{mgsina}{mgcosa} = tan\ \alpha \quad (3)$$

where μ_1 is the static friction coefficient, f is the static friction force between the object and the inclined plane, N; F_2 is the force perpendicular to the inclined plane, N; m is the measured material mass, g; and α is the inclination angle, °.

Measurement of the Rolling Friction Coefficient

The coefficient of rolling friction pertains to the deformation-induced resistance when an object rolls or tends to roll without slipping on another surface [36]. The experimental setup and method for measuring the rolling friction coefficients in this study were the same as those for measuring the static friction coefficients. According to the rolling friction coefficient measurement method provided in the literature [37,38], the test object was placed radially on the test plate. The handle of the inclinometer was shaken to raise the inclination angle of the test plate, and the rolling angle was recorded when the test object showed a rolling trend. The coefficient of rolling friction coefficient was then calculated. When measuring the rolling friction coefficient between phloem–Q235A steel, a high-roundness ramie stalk was selected, and rolling was achieved by external phloem contact. When measuring the rolling trend between the phloem and xylem, the phloem was peeled off in advance and glued to the steel plate. The xylem was placed radially on the phloem plate, the inclination angle was increased, and the rolling friction coefficient was calculated.

2.3. Physical Test of the Stacking Angle

The stacking angle is a microparameter that characterizes the granular materials' flow and friction characteristics. Its numerical value is related to the material type, surface shape, and moisture content, and it is affected by the coefficient of restitution and the coefficient of friction [39]. The stacking angle test is usually used to calibrate discrete element parameters of granular materials. Therefore, we conducted a physical test of the stacking angle. The measurement results can calibrate the contact parameters between phloem and Q235A steel, phloem and phloem, xylem and Q235A steel, and xylem and xylem.

Through preliminary comparative tests, the cylindrical lifting method was found to be suitable for measuring the phloem's stacking angle, and the extraction of the partition method was suitable for measuring the stacking angle for the xylem.

Due to the ramie phloem fibers' lengthy and highly flexible nature, forming a stacking angle is difficult. We referred to the material processing method for calibrating the discrete element parameters of sugarcane leaves [40] and tobacco rods [16], and the phloem was peeled from the ramie stalks with a utility knife and trimmed into 5.6 × 5.6 mm specimens without altering the surface shape of the material. The stacking angle method for the ramie phloem was the cylinder lifting method, where a 20 g sample was placed into a steel cylinder with a diameter of 45 mm and a height of 57 mm. The cylinder was then lifted at a uniform speed of 4 mm/s using a TFW-508 mechanical universal testing machine, and the phloem sample fell onto a 250 mm × 250 mm × 2 mm (length × width × thickness) steel plate from the bottom of the cylinder. After all phloem specimens had come to a complete stop, a stable phloem material pile was formed. The Canon EOS 70D DSLR camera was used to capture the main view of the phloem material pile from 50 cm in front of the pile, as shown in Figure 4a.

(a)

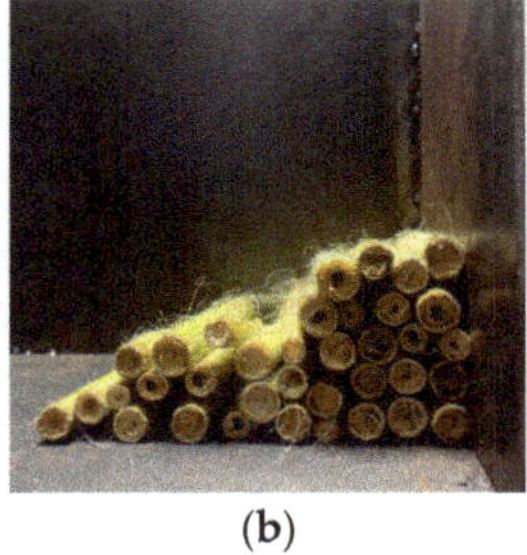

(b)

Figure 4. Physical test of the stacking angle: (**a**) stacking angle of the phloem; (**b**) stacking angle of the xylem.

For the xylem, a ramie stalk xylem radial stacking angle measurement device with removable partitions was created based on reference [17]. The device was made of Q235A steel and had dimensions of 500 mm × 200 mm × 300 mm for the length, width, and height, respectively. To measure the xylem, the ramie stalk sheath was removed, and the phloem was peeled entirely clean. All xylem lengths were controlled at approximately 140 mm. A certain number of xylem stalks were placed into one side of the radial stacking angle measurement device. The entire ramie xylem stalk was moved to the other side after the partition was vertically lifted, forming a radial stacking angle upon collision with the wall. The main view of the radial stacking angle formed after the movement of the xylem stalks was captured using the Canon EOS 70D DSLR camera, as shown in Figure 4b.

2.4. Discrete Element Virtual Simulation Test

2.4.1. Selection of the Contact Model

In this study, Q235A steel was used as the contact material for both the ramie phloem and ramie xylem in the physical tests. Since there was no adhesion between the phloem and xylem and the steel plate during the physical tests, and there was no significant deformation during the accumulation process, the phloem and xylem were idealized as rigid bodies without considering their mechanical properties, only exploring their contact relationship. Therefore, the Hertz–Mindlin (no slip) contact model was selected for the discrete element simulation. The contact parameters included the coefficient of restitution, coefficient of static friction, and coefficient of rolling friction between the phloem and Q235A steel, phloem and phloem, phloem and xylem, xylem and Q235A steel, and xylem and Q235A xylem.

2.4.2. Establishment of the Discrete Element Model for Ramie Stalk

Analysis of the Ramie Stalk Model

An analysis of the ramie stalk structure's composition was conducted to establish a discrete element model for the ramie stalk. The cross-section of the ramie stalk was approximately circular, with the outer to inner cross-sections consisting of the green husk layer, phloem, xylem, and central medulla, as shown in Figure 5. The central medulla is foam-like, with a loose tissue structure and irregular shape. At the same time, the green husk layer was very thin and brittle, both of which have negligible mechanical properties compared to other components.

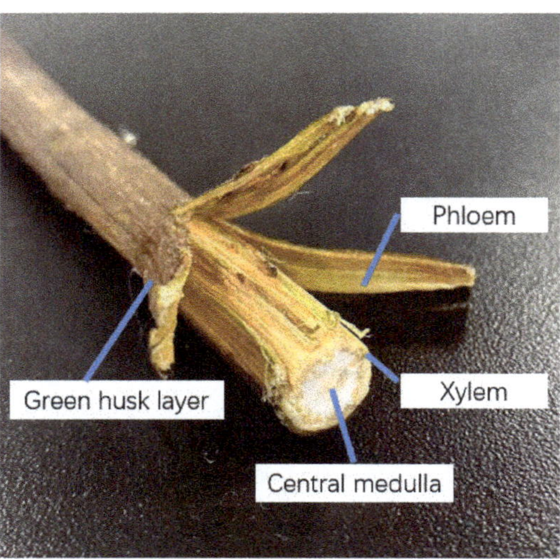

Figure 5. Schematic diagram of the structural composition of the ramie stalks.

Mechanical decortication of ramie stalks is performed by rolling and crushing the xylem with a roller to separate the xylem from the phloem, indicating that the breaking strength of the xylem was less than that of the phloem. From this apparent perspective, it can be assumed that the material properties of the green husk layer and the central medulla can be ignored, and the ramie stalk can be regarded as a combination of phloem and xylem materials. Therefore, only the phloem and xylem should be considered when a discrete element model for the ramie stalk is established. The geometric model is shown in Figure 6. In this paper, the phloem was a composite of the green husk layer and the phloem fibers, and the physical tests did not remove the green husk layer.

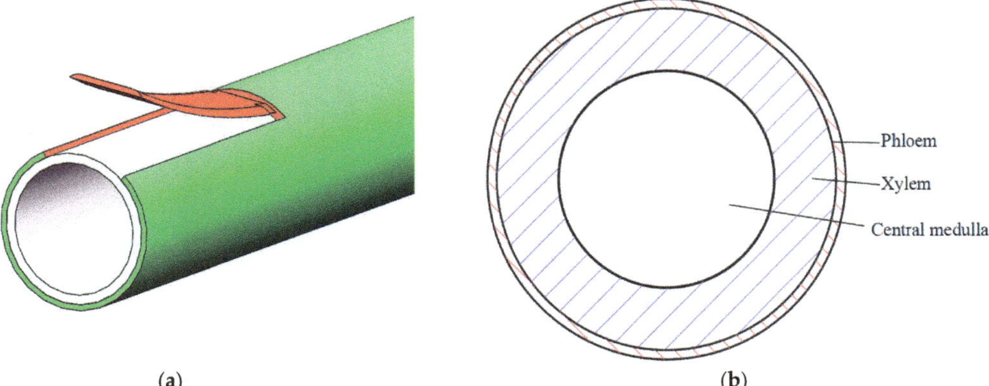

Figure 6. Ramie stalk geometry model: (**a**) ramie stalk structure simulation; (**b**) idealized cross-section of the ramie stalk.

Establishment of a Discrete Element Model for Ramie Stalk Phloem and Xylem

Simulation tests require the first step of building an accurate particle model. Ramie stalk phloem and xylem are nonspherical, and in DEM simulations, the multispherical method is usually used to construct irregular particle models [41–44]. In nonspherical particle discrete element simulations, the multichemical packing model has the advantages of fast calculation rate and simple contact judgment; however, as the number of filling particle elements increases, the simulation time will greatly increase, so the number of filling particle elements needs to be reasonable [45,46].

Since no significant deformation occurs during stacking, this study did not consider introducing bonding. When building the model, the ramie stalk phloem and xylem are first modeled and meshed using the Mesh submodule in the Workbench module of ANSYS 16.0 software. Then, the saved .msh file is imported into Fluent. The coordinate file containing the mesh coordinate information is obtained by reading and compiling the source file CalcRadius.c through the userdefine module, followed by executing CalcRadiusVolume. The coordinate data of the model are then imported into the .xml file saved by EDEM software, and the particle coordinate import is completed by entering EDEM. Based on the size measurement results of the ramie xylem, 732 circular spheres with a radius of 1.15 mm are used to form a cylindrical ramie stalk xylem with an outer diameter and inner diameter of 12.8 mm and 8 mm, respectively, and a length of 140 mm, as shown in Figure 7a.

The related study [40] showed that when calibrating the parameters of the discrete element model if the modeling is performed exactly according to the actual size of the material, it will greatly prolong the simulation time, increase the computational volume, and thus reduce the simulation efficiency. The correct calibration of the discrete element model with a moderately enlarged particle radius can truly reflect the contact parameters of the target material. The ramie phloem was relatively thin, and the filling particles according to the actual size will result in a low simulation efficiency due to the large number of calculations. In the simulation, the thickness of the ramie phloem was doubled, and 16 circular spheres with a radius of 0.7 mm were used to form a rectangular ramie stalk phloem with a length and width of 5.6 × 5.6 mm and a height of 1.4 mm, as shown in Figure 7b. Since the thickness of the ramie phloem is magnified in the simulation, it needs to be redetermined. The mass of a single sphere particle remained unchanged, and the radius was doubled, so the volume increased eight times, and the density was one-eighth of the physical test value of 1618.95 kg/m^3, which is 202.37 kg/m^3.

Figure 7. Ramie stalk discrete element model: (**a**) xylem discrete element model; (**b**) phloem discrete element model.

2.4.3. Calibration of Contact Parameters for Phloem–Xylem Discrete Element Model

Calibration of the Phloem–Xylem Restitution Coefficient

In the physical measurement tests of phloem–xylem restitution coefficients, the coefficient was measured by the rebound height of the phloem and xylem attached to the steel plate. A discrete element simulation test was set up with the same experimental conditions. The collision simulation test was set with an initial collision height of 205 mm, a time step length of 20% of the Rayleigh time, and a save interval of 0.001 s. The rebound height was obtained by reading the data through the analyst module.

Calibration of the Phloem–Xylem Static Friction Coefficient

Based on the physical measurement results of the phloem–xylem static friction coefficient, discrete element simulation tests with the same experimental conditions were established. The friction test set the incline angle of the phloem plate to rise at a speed of 10 deg/s with a time step of 20% of the Rayleigh time and a save interval of 0.001 s. The sliding angle was obtained by reading the data through the discrete element analyst module.

Calibration of the Phloem–Xylem Rolling Friction Coefficient

Based on the physical measurement results of the phloem–xylem rolling friction coefficient, discrete element simulation tests with the same experimental conditions were established. The friction test set the incline angle of the phloem plate to rise at a speed of 10 deg/s with a time step of 20% of the Rayleigh time and a save interval of 0.001 s. The rolling angle was obtained by reading the data through the discrete element analyst module.

2.4.4. Calibration of the Discrete Element Parameters for the Stacking Angle Test

In the simulation of the stacking angle test for the phloem, the inner diameter and height of the cylinder were consistent with those used in the physical test. The interior of the cylinder was set as a particle factory, where the particles were generated freely by the "static" method. This ensured that the phloem was distributed in a relatively dispersed state inside the cylinder, avoiding a situation where an uneven distribution would result in a significant error in the stacking angle after being static. The total mass of the generated particles was 20 g. Subsequently, the particles were allowed to fall freely under gravity, and the simulation model was conducted for 1 s to reach a static equilibrium state. After, the cylinder was lifted vertically at a speed of 4 mm/s. The phloem particles would flow out slowly from the bottom of the cylinder, eventually forming a stable phloem particle heap on the bottom plate, as shown in Figure 8.

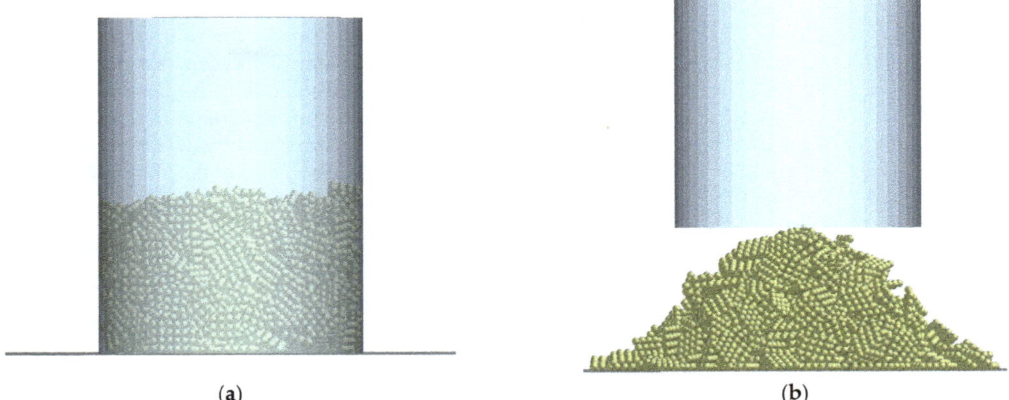

Figure 8. Ramie phloem stacking angle discrete element simulation test: (**a**) before stacking; (**b**) after stacking.

In the simulation of the stacking angle test for the xylem, the dimensions of the device and partition were consistent with those used in the physical test. Thirty-six neat piles of ramie xylem particles were generated at the right wall of the device. After the xylem particle heap and partition became stable, the partition was given a speed of 0.1 m/s to lift upward, and the xylem particle heap began to roll to the left to form the stacking angle. The discrete element simulation of the stacking angle test for the xylem is shown in Figure 9.

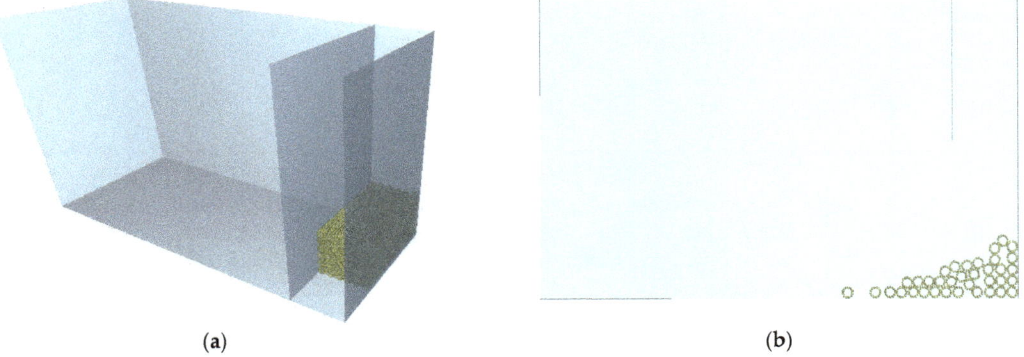

Figure 9. Ramie xylem stacking angle discrete element simulation test: (**a**) before stacking; (**b**) after stacking.

2.5. Discrete Element Simulation Test

In order to achieve a simulation model of the ramie stalk that matched its actual situation and ensured the reliability and authenticity of the model, this study used the Design-Expert 10.0.1 software to carry out tests, such as Plackett–Burman design, steepest ascent test design, and response surface design, to determine the key factors affecting the stacking angle in the simulation parameters of the ramie phloem and xylem, as well as the significant factor levels and parameter optimization of the ramie stalk phloem and xylem. Based on the fitting of the simulation stacking angle and the physical stacking angle of the ramie stalk phloem and xylem, the linear fitting method in MATLAB was used to compare the boundary pixel slope of the simulation model stacking angle and the actual material stacking angle to verify the accuracy of the model.

2.5.1. Plackett–Burman Design

To quickly screen the key factors affecting the response value of the stacking angle in the simulation parameters of ramie stalk phloem and xylem, this study used Design-Expert software to conduct a Plackett–Burman test analysis, taking the stacking angle of the phloem and xylem as the response value, using a 6-factor 2-level test method. The levels were represented in coded form, with a total of 13 groups of tests, each repeated twice, to compare the influence of each factor on the stacking angle of the phloem and xylem. The experimental plan is shown in Table 2.

Table 2. Plackett–Burman test program.

Phloem				Xylem			
Parameters	Symbols	Levels		Parameters	Symbols	Levels	
		Low level(−1)	High level(+1)			Low level(−1)	High level(+1)
Pholem–Q235A steel coefficient of restitution	X_1	0.05	0.45	Xylem–Q235A steel coefficient of restitution	X_1'	0.17	0.57
Pholem–Q235A steel coefficient of static friction	X_2	0.27	0.67	Xylem–Q235A steel coefficient of static friction	X_2'	0.36	0.76
Pholem–Q235A steel coefficient of rolling friction	X_3	0.005	0.100	Xylem–Q235A steel coefficient of rolling friction	X_3'	0.005	0.143
Pholem–Pholem coefficient of restitution	X_4	0.10	0.80	Xylem–Xylem coefficient of restitution	X_4'	0.10	0.80
Pholem–Pholem coefficient of static friction	X_5	0.10	0.60	Xylem–Xylem coefficient of static friction	X_5'	0.10	0.80
Pholem–Pholem coefficient of rolling friction	X_6	0.005	0.100	Xylem–Xylem coefficient of rolling friction	X_6'	0.005	0.050

A first-order polynomial linear model was used for the statistical modeling, as shown in Equation (4). The significance of each factor was obtained through variance analysis, and the significant influencing factors were selected.

$$\Omega = \sigma_0 + \sum_{k=1}^{6} \sigma_k X_k \quad (4)$$

Here, Ω represents the stacking angle, °; σ_0 is the intercept of the model; σ_k is the linear coefficient; and X_k refers to the coded level of the independent variable.

2.5.2. Steepest Ascent Test Design

To screen the optimal parameter range of the significant factors in the phloem and xylem, Design-Expert software was used to conduct the steepest ascent test. During the simulation, based on the results of the Plackett–Burman test, the values of the nonsignificant factors were taken from the physical test values, and the values of the remaining factors were taken from the median values of the Plackett–Burman experimental levels. The significant factors were gradually increased according to the selected step size. The relative error between the measured results and the simulated stacking angle was analyzed until the minimum range of the upper and lower relative errors was selected as the basis for the response surface test value.

2.5.3. Response Surface Optimization Test and Regression Model Establishment

This study used the response surface analysis method to obtain the optimal parameter combination based on the results of the Plackett–Burman test and the steepest ascent test. The phloem and xylem were both using the central composite design method. Three significant levels (high, medium, and low) were taken from the results of the phloem and xylem climbing tests for the experimental design. The interaction effects of the significant factors on the stacking angle were analyzed.

2.5.4. Parameter Optimization and Validation

The Design-Expert software's Optimization module optimized the regression model with the experimental values of the stacking angle as the objective. The steepest ascent test determined the parameter optimization range, and a two-sample *t*-test was performed on the simulation and physical results. The reliability of the optimal combination of parameters was verified by checking whether there was a significant difference between the simulation and physical results.

3. Results and Discussion

3.1. Measurement Results of the Required Parameters for the DEM Simulation

Based on the physical parameter measurement results of the ramie stalk's phloem and xylem, combined with an analysis of the bounce-back process of the phloem and xylem using HiSpec Control Software, the experimental values of the restitution coefficient were obtained. The friction coefficient was calculated by recording the sliding and rolling angle, and the restitution coefficient and friction coefficient results are shown in Table 3.

Table 3. Parameters required for the discrete element simulation.

Parameter Type	DEM Parameter	Parameter Value	Source
Physical Parameters	Density of Q235A steel (kg/m^3)	7850	Literature [47]
	Shear modulus of Q235A steel (Pa)	7.90×10^{10}	Literature [47]
	Poisson's ratio of Q235A steel	0.30	Literature [47]
	Density of phloem (kg/m^3)	202.37	Section 2.1.2
	Shear modulus of phloem (Pa)	6.15×10^{8}	Section 2.1.3
	Poisson's ratio of phloem	0.40	Literature [48]
	Density of xylem (kg/m^3)	751.50	Section 2.1.2
	Shear modulus of xylem (Pa)	2.42×10^{8}	Section 2.1.3
	Poisson's ratio of xylem	0.30	Literature [48]
Contact Parameters	Phloem–Q235A steel coefficient of restitution	0.25	Section 2.2.1
	Phloem–Q235A steel coefficient of static friction	0.47	Section 2.2.2
	Phloem–Q235A steel coefficient of rolling friction	0.05	Section 2.2.2
	Xylem–Q235A steel coefficient of restitution	0.37	Section 2.2.1
	Xylem–Q235A steel coefficient of static friction	0.56	Section 2.2.2
	Xylem–Q235A steel coefficient of restitution	0.074	Section 2.2.2

3.2. Results of the Stacking Angle Measurement

The physical stacking angle images of the phloem and xylem were processed using MATLAB for grayscale processing, binarization, and extraction of the image boundary pixels. The slope of the boundary pixels was linearly fitted using the least squares method, and the angle between the stacking angle tangent and the horizontal line was defined as the stacking angle. The average of the two sides was taken as the stacking angle for the phloem, and the left degree was taken as the stacking angle value for the xylem; the image processing process is shown in Figure 10.

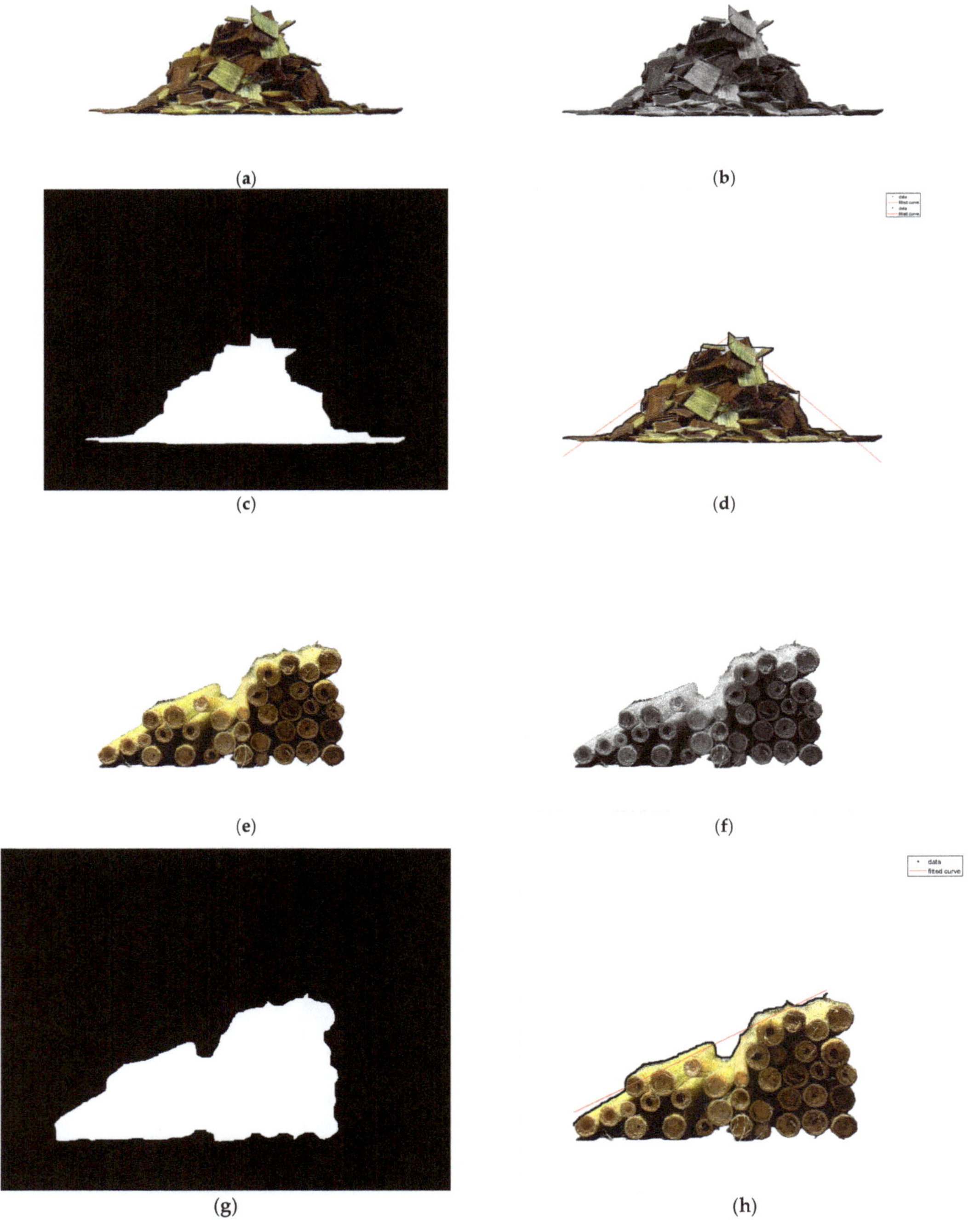

Figure 10. Physical stacking angle image processing: (**a**) original image of the phloem stacking angle; (**b**) grayscale processing of the phloem image; (**c**) binarization of the phloem image; (**d**) fitting phloem image using the least squares method; (**e**) original image of the xylem stacking angle; (**f**) grayscale processing of the xylem image; (**g**) binarization of the xylem image; (**h**) fitting xylem image using the least squares method.

The average value of the stacking angle of the xylem was determined to be 27.17°, with a standard deviation of 1.27°. The average value of the stacking angle of the phloem was 37.93°, with a standard deviation of 2.29°.

3.3. Calibration of the Contact Parameters between Phloem and Xylem

3.3.1. Phloem–Xylem Restitution Coefficient

The physical test result shows that the average maximum rebound height of the ramie xylem on the phloem was 21.3 mm. In the discrete element simulations, to avoid interference, all contact parameters, except the restitution coefficient between the xylem and phloem, were set to 0. After the presimulation tests, the restitution coefficient between the xylem and phloem ranged from 0.4 to 0.7. The simulation test design for the restitution coefficient is shown in Table 4. Three repetitions were conducted for each group, and the mean value was taken.

Table 4. Simulation test results for the restitution coefficient between xylem and phloem.

No.	Coefficient of Restitution, e_1	Maximum Bounce Height of Xylem, h_{max}/mm
1	0.40	9.46
2	0.45	11.26
3	0.50	13.67
4	0.55	16.78
5	0.60	20.80
6	0.65	26.32
7	0.7	34.91

The simulation test results in Table 4 were plotted as a scatter plot and fitted. The fitted curve obtained is shown in Figure 11. The fitting equation for the coefficient of restitution between the ramie xylem–phloem (e_1) and the maximum rebound height (h_{max}) is shown in Equation (5):

$$h_{max} = 45.91 - 187e_1 + 244.5e_1^2 \tag{5}$$

where h_{max} represents the maximum rebound height, in millimeters; e_1 represents the coefficient of restitution between the phloem and the xylem.

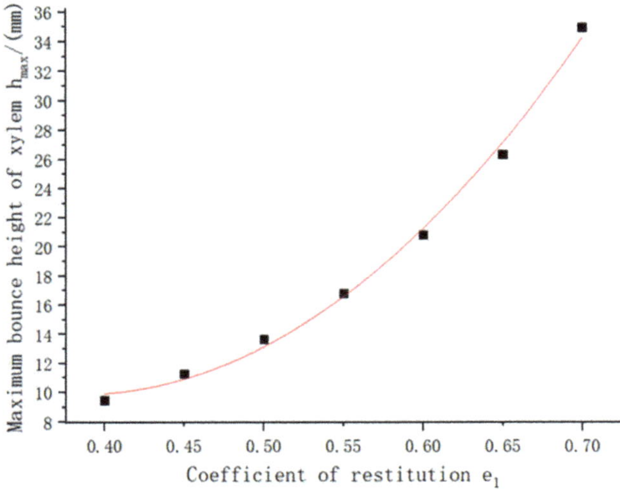

Figure 11. Fitted curve of the restitution coefficient and maximum rebound height.

The determination coefficient (R^2) of the fitting equation is 0.996, indicating the high reliability of the fitting equation. By substituting the measured maximum rebound height into Equation (5), e_1 is calculated as 0.60. Using the coefficient of restitution (e_1) for the simulation tests, repeating five times, and taking the average value, the maximum rebound height is 20.85 with an error of 2.11%. The results show that the simulation results after the calibration are consistent with the physical test results. Therefore, the coefficient of restitution between the ramie phloem and xylem was determined to be 0.60.

3.3.2. Phloem–Xylem Static Friction Coefficient

The coefficient of the static friction (μ_s) between the ramie phloem and xylem was determined by measuring the average sliding angle of the xylem on the phloem's surface during inclined plane sliding. The physical test result shows that average sliding angle between the phloem and xylem was measured to be 31.61°. In the DEM simulation test, the coefficient of restitution was set to the calibrated value, and the range of the coefficient of the static friction was set to 0.1 ~ 0.7 with an interval of 0.1. The remaining contact parameters were all set to 0. The simulation test of the static friction coefficient is shown in Table 5. Each group of tests was repeated three times, and the average value was acquired to obtain the relationship between the sliding angle and the static friction coefficient.

Table 5. Simulation test results for the static friction coefficient between the xylem and phloem.

No.	Coefficient of Static Friction, μ_s	Sliding Angle, α (°)
1	0.10	8.43
2	0.20	14.28
3	0.30	20.75
4	0.40	25.72
5	0.50	30.19
6	0.60	34.26
7	0.70	37.93

The simulation test results were plotted as a scatter plot and fitted, and the fitting curve is shown in Figure 12. The fitting equation for the static friction coefficient between the ramie phloem–xylem (μ_s) and the sliding angle (α) is shown in Equation (6):

$$\alpha = 1.393 + 72.01\mu_s - 28.44{\mu_s}^2 \tag{6}$$

where α is the sliding angle in degrees, °; and μ_s is the static friction coefficient between the phloem and xylem.

The fitting results show that the determination coefficient (R^2) of the fitting equation is 0.996, indicating that the reliability of the fitting equation is high. By substituting the measured sliding angle into Equation (6), μ_s is calculated to be 0.53. A simulated verification test was performed, and the average sliding angle was obtained by repeating the test five times, which was 31.54°. The relative error between the simulated and physical test results was 0.22%, indicating that the calibrated simulation results are consistent with the physical test results. Therefore, the coefficient of the static friction between the ramie phloem and xylem was determined to be 0.53.

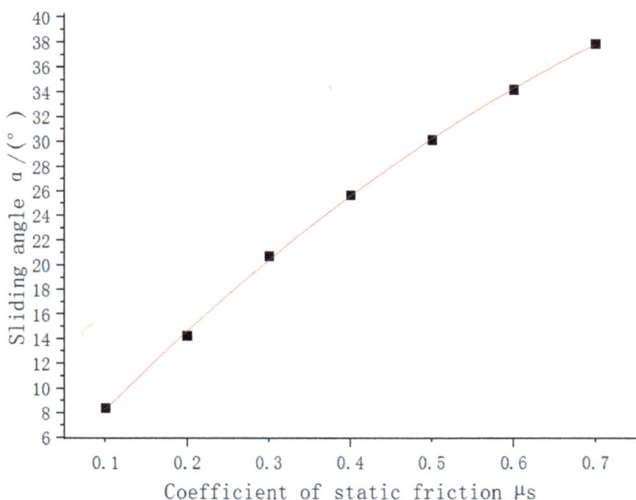

Figure 12. Fitting curve of the static friction coefficient and sliding angle.

3.3.3. Phloem–Xylem Rolling Friction Coefficient

The coefficient of the rolling friction was measured similarly to that of the coefficient of static friction. The xylem was placed radially on the surface of the phloem, and the inclined device was slowly and uniformly raised to gradually increase the phloem plate's inclination angle. When the xylem began to roll, the raising was immediately stopped, and the inclination device was fixed. The rolling angle (β) was measured and recorded by an angle display device. The test was repeated ten times, and the average rolling angle between the phloem and xylem was measured to be 3.91°.

In the DEM simulation test, the values of the restitution coefficient and static friction coefficient that had been calibrated were input. The range of the rolling friction coefficient was set to 0.01–0.04 with an interval of 0.005, and all other contact parameters were set to 0. The simulation test for the rolling friction coefficient was designed as shown in Table 6, and each group of tests was repeated three times to obtain the average value.

Table 6. Simulation test results for the rolling friction coefficient between the xylem and phloem.

No.	Coefficient of Rolling Friction, μ_r	Rolling Angle, β (°)
1	0.010	2.90
2	0.015	3.36
3	0.020	3.83
4	0.025	4.20
5	0.030	4.61
6	0.035	5.10
7	0.040	5.50

The simulation test results in Table 6 were plotted as a scatter plot and fitted, and the fitting curve is shown in Figure 13. The fitting equation for the rolling friction coefficient between the ramie phloem–xylem (μ_r) and the rolling angle (β) is provided by Equation (7):

$$\beta = 2.031 + 88.99\mu_r - 57\mu_r \tag{7}$$

where β represents the rolling angle in degrees, mm; and μ_r represents the rolling friction coefficient between the xylem and phloem.

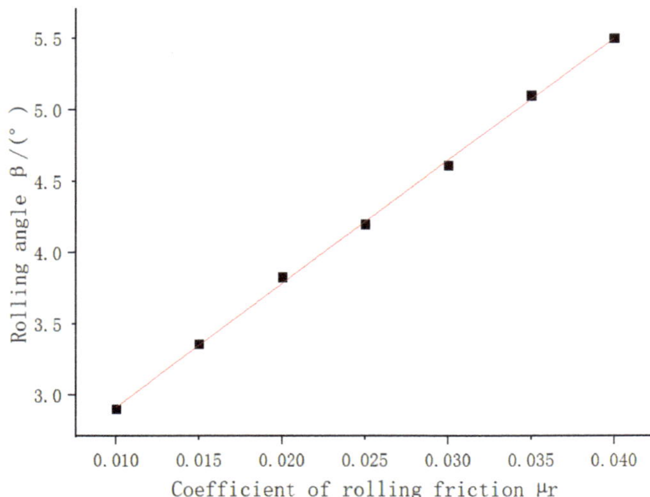

Figure 13. Fitting curve for the rolling friction coefficient and rolling angle.

The fitting result indicates that the determination coefficient (R^2) of the fitting equation is 0.999, thereby indicating a high reliability of the fitting equation. By substituting the measured rolling angle into Equation (7), μ_r was calculated to be 0.021. Through simulation verification tests conducted five times and taking the average value, the rolling angle was found to be 3.90°, with a relative error of 0.3% compared to the physical test results. This shows that the simulated test results after calibration were consistent with the physical test results, thus confirming the rolling friction coefficient between the phloem and xylem as 0.021.

Thus, the calibration of the contact parameters between the phloem and xylem of the ramie stalk by the discrete element method was completed, and the calibrated coefficients of restitution (e_1), coefficient of the static friction (μ_s), and coefficient of the rolling friction (μ_r) between the phloem and xylem were 0.60, 0.53, and 0.021, respectively.

3.4. *Plackett–Burman Parameter Significance Analysis*

3.4.1. Significance Analysis of Phloem Parameters

The Plackett–Burman test results for the phloem are shown in Table 7, and the Design-Expert software was used to perform a significance analysis of the results, as shown in Table 8.

Table 7. Phloem Plackett–Burman test results.

X_1	X_2	X_3	X_4	X_5	X_6	Angle (°)
−1	−1	−1	−1	−1	−1	18.19
−1	1	1	−1	1	−1	38.94
1	−1	1	−1	−1	−1	19.40
−1	1	−1	−1	−1	1	30.67
0	0	0	0	0	0	36.35
1	1	−1	1	−1	−1	20.19
1	1	1	−1	1	1	42.99
1	−1	1	1	−1	1	30.03
1	1	−1	1	1	−1	39.12
−1	−1	−1	1	1	1	41.25
−1	−1	1	1	1	−1	38.64
−1	1	1	1	−1	1	31.28
1	−1	−1	−1	1	1	41.59

Table 8. Parameter significance analysis of the Plackett–Burman test in the phloem.

Parameters	Degree of Freedom	Sum of Squares	*F*-Value	*p*-Value	Effect Value	Significance Ranking
X_1	1	2.67	0.59	0.5239	−0.94	6
X_2	1	16.52	3.65	0.1433	2.35	3
X_3	1	8.82	1.95	0.2647	1.71	4
X_4	1	6.35	1.40	0.3370	1.45	5
X_5	1	717.30	158.39	<0.0001	15.46	1
X_6	1	156.46	34.55	0.0021	7.22	2

The *p*-values of the static friction coefficient between the phloem and phloem (X_5) and the rolling friction coefficient between the phloem and phloem (X_6) were less than 0.05, indicating that X_5 and X_6 have a significant effect on the stacking angle of the phloem. In contrast, the other factors have a relatively small effect. Therefore, only significant factors (i.e., X_5 and X_6) were considered for the subsequent steepest ascent test and response surface design of the phloem.

3.4.2. Significance Analysis of the Xylem Parameters

The Plackett–Burman test results for the xylem are shown in Table 9, and the Design-Expert software was used to perform a significance analysis of the results, as shown in Table 10.

Table 9. Xylem Plackett–Burman test results.

X_1'	X_2'	X_3'	X_4'	X_5'	X_6'	Angle (°)
1	1	1	−1	1	1	38.61
1	1	−1	1	1	−1	27.52
1	−1	1	−1	−1	−1	27.33
−1	1	1	−1	1	−1	31.96
−1	−1	−1	1	1	1	33.84
−1	−1	1	1	1	−1	32.07
1	1	−1	1	−1	−1	22.26
−1	1	1	1	−1	1	34.02
0	0	0	0	0	0	31.10
1	−1	−1	−1	1	1	33.71
−1	1	−1	−1	−1	1	30.55
−1	−1	−1	−1	−1	−1	19.65
1	−1	1	1	−1	1	34.91

Table 10. Parameter significance analysis of the Plackett–Burman test in the xylem.

Parameters	Degree of Freedom	Sum of Squares	*F*-Value	*p*-Value	Effect Value	Significance Ranking
X_1	1	0.43	0.33	0.5620	0.38	6
X_2	1	0.97	0.75	0.3900	0.57	4
X_3	1	82.05	63.22	0.0001	5.23	2
X_4	1	0.66	0.51	0.4741	0.47	5
X_5	1	70.05	53.98	0.0002	4.83	3
X_6	1	167.58	129.14	<0.0001	7.47	1

The results show that the *p*-values of the rolling friction coefficient between the xylem and Q235A steel ($X_{3'}$), the static friction coefficient between the xylem and xylem ($X_{5'}$), and the rolling friction coefficient between the xylem and xylem ($X_{6'}$) were all less than 0.05. Thus, it can be concluded that these three factors are the most critical factors affecting the stacking angle of the xylem, while other factors have relatively small effects. As a result, only the significant factors (i.e., $X_{3'}$, $X_{5'}$, and $X_{6'}$) were considered for the subsequent steepest ascent test and response surface design of the xylem.

3.5. Results of the Steepest Ascent Test

3.5.1. Steepest Ascent Test of the Phloem

Based on the significant factors affecting the stacking angle obtained from the phloem Plackett–Burman test, the phloem–phloem static friction coefficient (X_5) and phloem–phloem rolling friction coefficient (X_6) were taken as the independent variables to carry out the steepest ascent test of the phloem. The test results are shown in Table 11. The results show that, as the values of X_5 and X_6 increased, the relative error between the simulated stacking angle and the physical stacking angle first decreased and then increased. The minimum relative error was obtained when the fifth level was selected, indicating the existence of the optimal value range for the fifth level. Therefore, the stacking angle test result of level 5 was taken as the center point, and the stacking angle test results of levels 4 and 6 were taken as the low and high levels for the subsequent response surface design. The optimization ranges of the phloem–phloem static friction coefficient (X_5) and rolling friction coefficient (X_6) were determined to be 0.35–0.52 and 0.053–0.084, respectively.

Table 11. Results of the steepest ascent test of the phloem.

No.	X_5	X_6	Stacking Angle (°)	Relative Error (%)
1	0.10	0.005	19.50	50.43
2	0.18	0.021	25.49	35.21
3	0.27	0.037	31.35	20.31
4	0.35	0.053	35.56	9.62
5	0.43	0.068	39.57	0.58
6	0.52	0.084	40.45	2.82
7	0.60	0.100	43.13	9.61

3.5.2. Steepest Ascent Test of the Xylem

Based on the Plackett–Burman test for the xylem to derive the significant influences on the stacking angle, the steepest ascent test was carried out with the xylem–Q235A steel rolling friction coefficient ($X_{3'}$), xylem–xylem static friction coefficient ($X_{5'}$), and xylem–xylem rolling friction coefficient ($X_{6'}$) as the independent variables. The test results are shown in Table 12. The results show that as the values of $X_{3'}$, $X_{5'}$, and $X_{6'}$ increased, the relative error between the simulated stacking angle and the physical stacking angle first decreased and then increased. The minimum relative error was obtained when the third level was selected, indicating the existence of the optimal value range for the third level. Therefore, the stacking angle test result of level 3 was taken as the center point, and the stacking angle test results of levels 2 and 4 were taken as the low and high levels for the subsequent response surface design. The optimization ranges of the xylem–Q235A steel rolling friction coefficient ($X_{3'}$), xylem–xylem static friction coefficient ($X_{5'}$), and xylem–xylem rolling friction coefficient ($X_{6'}$) were determined to be 0.028–0.074, 0.22–0.45, and 0.013–0.028, respectively.

Table 12. Results of the steepest ascent test of the xylem.

No.	$X_{3'}$	$X_{5'}$	$X_{6'}$	Stacking Angle (°)	Relative Error (%)
1	0.005	0.10	0.005	16.50	39.27
2	0.028	0.22	0.013	24.60	9.45
3	0.051	0.33	0.020	28.22	3.87
4	0.074	0.45	0.028	29.72	9.40
5	0.097	0.57	0.035	32.41	19.30
6	0.120	0.68	0.043	32.90	21.08
7	0.143	0.80	0.050	33.13	21.93

3.6. Response Surface Design Results

3.6.1. Regression Model Establishment and Experimental Results

Analysis of the Variance of the Regression Model in Phloem

To investigate the impact of the static friction coefficient (X_5) and rolling friction coefficient (X_6) between the ramie phloem and phloem on the phloem stacking angle (Y_1) during the response surface optimization test, we utilized the range of values obtained from the steepest ascent test. The central composite design test was then conducted using Design-Expert to optimize the response surface. Tables 13 and 14 display the factor encoding level values and central composite design experimental results, respectively. A total of thirteen parameter combinations were tested, of which five were center-level repeats.

Table 13. Factor level codes for the phloem central composite design test.

Coding	Phloem–Phloem Coefficient of Static Friction X_5	Phloem–Phloem Coefficient of Rolling Friction X_6
−1.414	0.31	0.047
−1	0.35	0.053
0	0.43	0.068
1	0.52	0.084
1.414	0.56	0.090

Table 14. The phloem central composite design test scheme and results.

No.	X_5	X_6	Stacking Angle Y_1 (°)
1	−1.414	0	35.82
2	−1	−1	35.65
3	−1	1	37.57
4	0	−1.414	37.25
5	0	0	39.69
6	0	0	39.53
7	0	0	39.51
8	0	0	39.56
9	0	0	39.55
10	0	1.414	39.78
11	1	−1	39.43
12	1	1	40.45
13	1.414	0	40.39

By employing the Design-Expert software, a second-order polynomial equation was developed through a multiple regression analysis of the central composite design experimental outcomes. The equation was used to fit the phloem stacking angle and achieve the multivariate nonlinear regression model fitting of the static friction coefficient (X_5) and rolling friction coefficient (X_6) related to the phloem stacking angle (Y_1). Furthermore, the model and coefficients were subjected to a significance test, and the regression equation is shown in Equation (8):

$$Y_1 = 39.54 + 1.64X_5 + 0.81X_6 - 0.23X_5X_6 - 0.74X_5^2 - 0.54X_6^2 \quad (8)$$

Table 15 shows the results of the variance analysis. The significance test of the regression model for the stacking angle of the ramie phloem indicated $p < 0.0001$, a lack of fit of $p = 0.1123$, a determination coefficient of 0.9976, and an adjusted determination coefficient of 0.9958. The regression model was extremely significant, the lack of fit was nonsignificant, and the model was effective. The high value of the determination coefficient (close to 1) indicates a good fit of the regression equation. The coefficient of variation was only 0.45%, and the adequate precision was 68.357, indicating a high correlation between the actual and predicted values and the high reliability of the experimental results. As shown in Table 15, the static friction coefficient (X_5) and rolling friction coefficient (X_6) of the ramie phloem, as well as their quadratic terms ($X_5{}^2$ and $X_6{}^2$), all have an extremely significant effect on the equation. Combined with the linear regression equation, the order of the factors affecting the stacking angle is the static friction coefficient (X_5) > rolling friction coefficient (X_6) of the ramie phloem.

Table 15. ANOVA results of the phloem stacking angle.

Source	Stacking Angle Y_1 (°)				
	Sum of Squares	df	Mean Square	*F*-Value	*p*-Value
Model	32.21	5	6.44	576.29	<0.0001 **
X_5	21.53	1	21.53	1925.51	<0.0001 **
X_6	5.31	1	5.31	475.01	<0.0001 **
X_5X_6	0.20	1	0.20	18.11	0.0038 **
X_5^2	3.81	1	3.81	340.97	<0.0001 **
X_6^2	1.99	1	1.99	178.27	<0.0001 **
Residual	0.078	7	0.011		
Lack of Fit	0.058	3	0.019	3.86	0.1123
Pure Error	0.020	4	5.020×10^{-3}		
Cor Total	32.29	12			

**, Indicate significance at the 0.01 level.

To visually analyze the model's reliability, Design-Expert software's Diagnostics module was used to obtain a quadratic model residual diagnostic plot, as shown in Figure 14. Figure 14a shows a normal plot of the residual, which can be observed to be linearly distributed on both sides of the line for each test group, indicating that the model describes the relationship between the influencing factors and phloem simulation stacking angle with sufficient reliability. Figure 14b is the residual plot of the equation and the predicted values. The random dispersion of the residuals shows an irregular distribution, indicating a good prediction of the equation. Figure 14c shows the distribution of the ratio of the predicted and experimental values of the phloem simulation stacking angle, and the linear distribution indicates a good fit of the model. Overall, these results indicate that the model's reliability is extremely high.

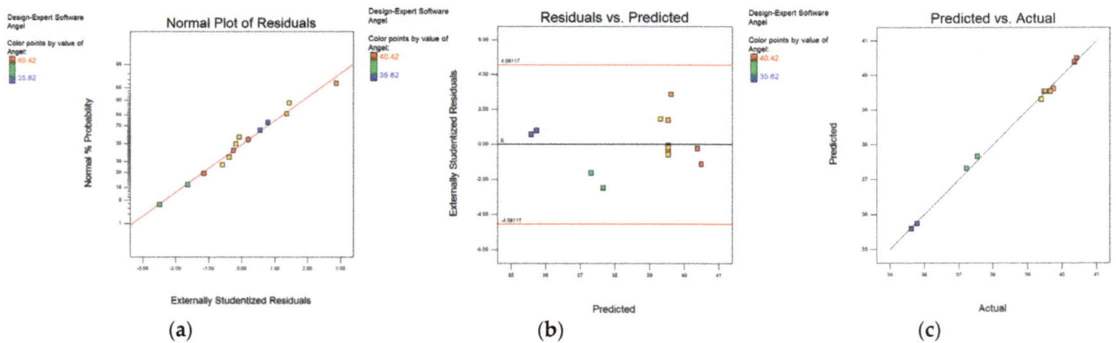

Figure 14. Diagnostic plot of the phloem quadratic model residuals: (**a**) normal plot; (**b**) residual vs. predicted; (**c**) predicted vs. actual.

Analysis of the Variance of the Regression Model in Xylem

Based on the interval of the values of the xylem–Q235A steel rolling friction coefficient ($X_{3'}$), xylem–xylem static friction coefficient ($X_{5'}$), and xylem–xylem rolling friction coefficient ($X_{6'}$) among the ramie xylem obtained from the steepest ascent test in order to investigate the effect of the influencing factors (i.e., $X_{3'}$, $X_{5'}$, and $X_{6'}$) on the xylem stacking angle (Y_2) of the response surface optimization test, the central composite design test was carried out using Design-Expert. The stacking angle simulation test was conducted for 23 sets of parameter combinations of which three sets were repeated at the central level. The results of their factor coding level values and central combination tests are shown in Tables 16 and 17.

Table 16. Factor level codes for the xylem central composite design test.

Coding	Xylem–Q235A Steel Coefficient of Rolling Friction $X_{3'}$	Xylem–Xylem Coefficient of Static Friction $X_{5'}$	Xylem–Xylem Coefficient of Rolling Friction $X_{6'}$
−1.682	0.012	0.14	0.007
−1	0.028	0.22	0.013
0	0.051	0.33	0.020
1	0.074	0.45	0.028
1.682	0.090	0.53	0.033

Table 17. The xylem central composite design test scheme and results.

No.	X_3	X_5	X_6	Stacking Angle Y_2 (°)
1	1	0	0	28.01
2	1	1	−1	27.64
3	1	1	1	29.73
4	0	0	0	27.83
5	−1	−1	1	27.11
6	0	0	0	27.81
7	1	−1	1	28.34
8	0	0	0	28.22
9	0	0	0	28.08
10	0	0	0	28.13
11	−1	1	1	27.98
12	1	−1	−1	26.05
13	−1.682	0	0	26.60
14	1.682	0	0	28.80
15	0	0	0	27.94
16	0	0	−1.682	25.32
17	−1	1	−1	26.28
18	0	1.682	0	28.40
19	0	0	0	28.16
20	0	0	1.682	29.07
21	0	0	0	28.31
22	0	−1.682	0	26.02
23	−1	−1	−1	24.62

A multivariate regression analysis was conducted on the results of the central composite design test using Design-Expert software. After eliminating the insignificant factors while ensuring the model significance and insignificance of the lack-of-fit terms, the second-order regression model was optimized to obtain a new regression equation:

$$Y_2 = 28.06 + 0.69X_3 + 0.70X_5 + 1.09X_6 - 0.12X_5X_6 - 0.15{X_3}^2 - 0.32{X_5}^2 - 0.33{X_6}^2 \quad (9)$$

The results of the variance analysis are shown in Table 18. The p-value ($p < 0.0001$) of the model confirms its significance within the 95% confidence interval. The p-value of the lack-of-fit term was 0.6381, less than 0.05, indicating the effectiveness of the second-order model for the xylem stacking angle. In addition, the determination coefficient and adjusted determination coefficient were 0.9899 and 0.9829, respectively, both close to 1, indicating good agreement between the calculated model and experimental data. The difference between the adjusted determination coefficient and the predictive determination coefficient of 0.9657 was less than 0.2, indicating a good fit, and the adequate precision was 46.594, indicating a high correlation between the actual and predicted values.

Table 18. ANOVA results of the xylem stacking angle.

Source	Stacking Angle Y_2 (°)				
	Sum of Squares	df	Mean Square	*F*-Value	*p*-Value
Model	33.18	9	3.69	141.54	<0.0001 **
X_3	6.57	1	6.57	252.10	<0.0001 **
X_5	6.63	1	6.63	254.38	<0.0001 **
X_6	16.21	1	16.21	622.14	<0.0001 **
X_3X_5	0.025	1	0.025	0.97	0.3422
X_3X_6	4.512×10^{-3}	1	4.512×10^{-3}	0.17	0.6840
X_5X_6	0.12	1	0.12	4.70	0.0492 *
X_3^2	0.35	1	0.35	13.34	0.0029 **
X_5^2	1.64	1	1.64	62.83	<0.0001 **
X_6^2	1.69	1	1.69	64.93	<0.0001 **
Residual	0.34	13	0.026		
Lack of Fit	0.10	5	0.021	0.70	0.6381
Pure Error	0.24	8	0.029		
Cor Total	33.52	22			

**, * Indicate significance at the 0.01 and 0.05 levels, respectively.

Based on the Diagnostics module in the Design-Expert software, the residual diagnostic plots of the quadratic model were obtained, as shown in Figure 15. Figure 15a shows a normal plot of the residual, which can be observed to be linearly distributed on both sides of the line for each test group, indicating that the model describes the relationship between the influencing factors and the xylem simulation stacking angle with sufficient reliability. Figure 15b is the residual plot of the equation and the predicted values. The random dispersion of the residuals shows an irregular distribution, indicating a good prediction of the equation. Figure 15c shows the distribution of the ratio of the predicted and experimental values of the xylem simulation stacking angle. The linear distribution indicates a good fit for the model. Overall, these results suggest that the model's reliability is extremely high.

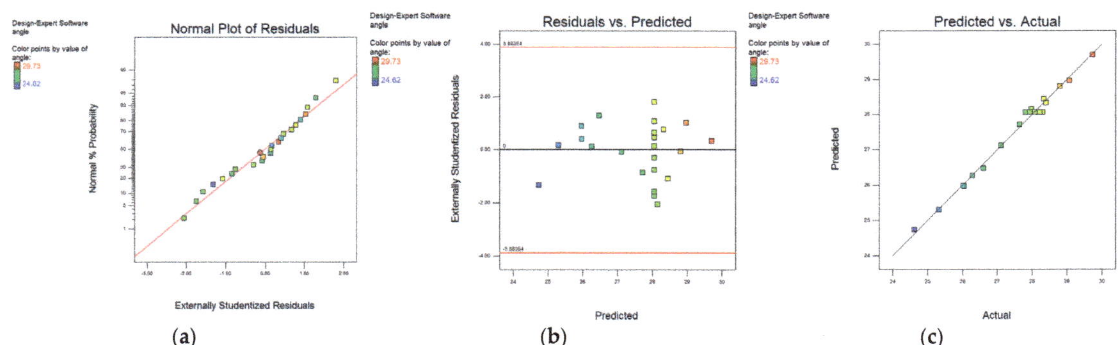

Figure 15. Diagnostic plot of the xylem quadratic model residuals: (**a**) normal plot; (**b**) residual vs. predicted; (**c**) predicted vs. actual.

3.6.2. Analysis of the Interaction Effects among the Factors

Interaction Effects of the Factors in the Phloem on the Simulation Stacking Angle

Based on the variance analysis of the phloem simulation stacking angle, the static friction coefficient (X_5) and rolling friction coefficient (X_6) between the phloem and phloem had an extremely significant impact on the phloem simulation stacking angle. Therefore, the Design-Expert software was used to analyze the nonlinear relationship between X_5, X_6, and the phloem simulation stacking angle. Figure 16 shows the response surface of the interaction between X_5 and X_6 on the phloem simulation stacking angle. When the static friction coefficient among the phloem was constant, the phloem simulation stacking angle increased with the rise in the rolling friction coefficient between the phloem and phloem. When the rolling friction coefficient between the phloem and phloem was constant, the phloem simulation stacking angle increased with the rise in the static friction coefficient between the phloem and phloem. However, the contour slope of X_5 was steeper than X_6, indicating that the static friction coefficient between the phloem and phloem (X_5) had a more significant impact on the phloem simulation stacking angle than the rolling friction coefficient between the phloem and phloem (X_6). Therefore, the order of the effects of each factor on the phloem simulation stacking angle is $X_5 > X_6$, which is consistent with the results of the variance analysis.

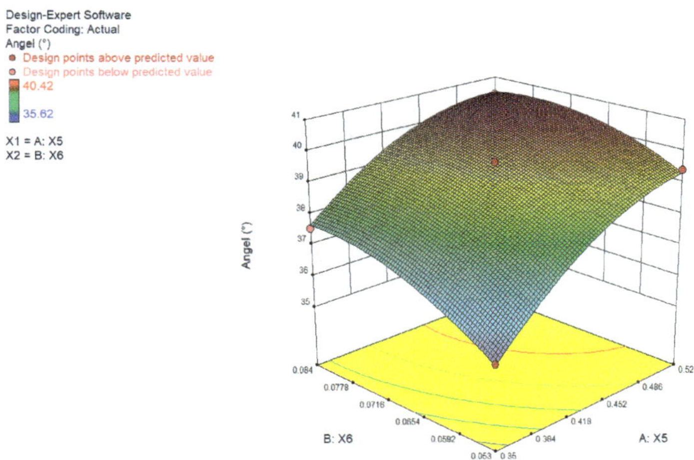

Figure 16. Response surface of the interaction effects of the factors in the phloem on the stacking angle.

Interaction Effects of the Factors in the Xylem on the Simulation Stacking Angle

Based on the analysis of the variance of the xylem simulation stacking angle ($X_3^{2'}$), $X_5^{2'}$ and $X_6^{2'}$ had an extremely significant impact on the simulation stacking angle. $X_{5'}$ and $X_{6'}$ significantly affected the simulation stacking angle, while $X_{3'}$ and $X_{5'}$ and $X_{3'}$ and $X_{6'}$ had no significant effect on the simulation stacking angle. Therefore, Design-Expert software was used to analyze the relationship between the static friction coefficient ($X_{5'}$), rolling friction coefficient ($X_{6'}$), and xylem and the xylem simulation stacking angle. Figure 17 shows the response surface of the interaction between $X_{5'}$ and $X_{6'}$ on the simulation stacking angle. When $X_{5'}$ was constant, the simulation stacking angle increased with the increase in $X_{6'}$. When $X_{6'}$ was constant, the simulation stacking angle increased with the rise in $X_{5'}$. However, the contour slope of $X_{6'}$ was steeper than that of $X_{5'}$, indicating that $X_{6'}$ had a more significant impact on the simulation stacking angle. Therefore, the order of the effects of each factor on the simulation stacking angle is $X_{6'} > X_{5'} > X_{3'}$, which is consistent with the results of the variance analysis.

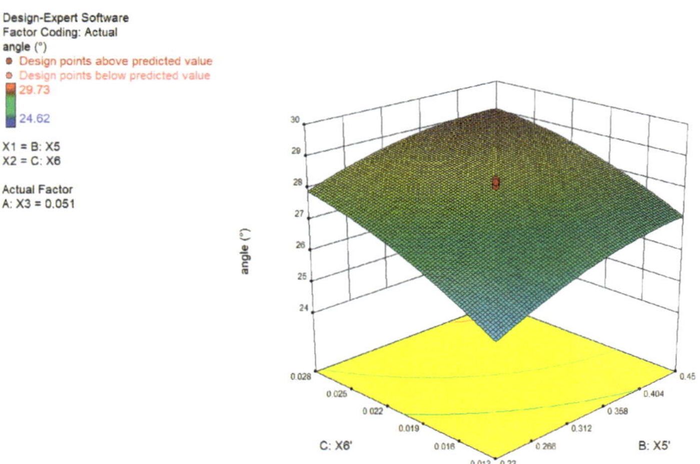

Figure 17. Response surface of the interaction effects of the factors in the xylem on the stacking angle.

3.6.3. Parameter Optimization and Validation

Optimization of the Parameters in the Phloem

Based on the results of the Plackett–Burman test and the steepest ascent test in the phloem, the ranges of X_5 and X_6 were 0.35–0.52 and 0.053–0.084, respectively. Taking the physical test value of the stacking angle of the phloem as the optimization objective, using the parameter Optimization module built into the Design-expert software, the non-significant factors were taken as the physical test values, and the rest were taken as the middle values of the steepest ascent test level to determine the optimal combination of the static friction coefficient (X_5) and rolling friction coefficient (X_6) of the phloem–phloem; the optimization objective function and constraints are shown in Equation (10):

$$\begin{cases} \text{tar}Y_1 = 37.93 \\ 0.35 \leq X_5 \leq 0.52 \\ 0.053 \leq X_6 \leq 0.084 \end{cases} \quad (10)$$

After solving, 44 sets of optimized solutions were obtained. The simulated results of the optimized parameter group were compared with the physical test results. The optimized solution with the most similar shape of the cylindrical lifting physical test stacking angle was found. The static friction coefficient (X_5) between the phloem and phloem particles was determined to be 0.41, and the rolling friction coefficient (X_6) between the phloem and phloem particles was 0.056.

Optimization of the Parameters in the Xylem

Based on the results of the xylem Plackett–Burman and steepest ascent test, the ranges of $X_{3'}$, $X_{5'}$, and $X_{6'}$ were 0.028–0.074, 0.22–0.45, and 0.013–0.028, respectively. Taking the physical test value of the xylem stacking angle as the optimization objective, using the parameter Optimization module built into the Design-expert software, the nonsignificant factors were taken as the physics test values, and the rest were taken as the middle values of the steepest ascent test level to determine the optimal combination of the xylem–Q235A steel rolling friction coefficient ($X_{3'}$), static friction coefficient ($X_{5'}$) of the xylem–xylem, and rolling friction coefficient ($X_{6'}$) of the xylem–xylem; the optimization objective function and constraints are shown in Equation (11):

$$\begin{cases} \text{tar}Y_2 = 27.17 \\ 0.028 \le X_{3'} \le 0.074 \\ 0.22 \le X_{5'} \le 0.45 \\ 0.013 \le X_{6'} \le 0.028 \end{cases} \tag{11}$$

After solving, 100 sets of optimized solutions were obtained. The simulated results of the optimized parameter group were compared with the physical test results. The optimized solution with the most similar shape of the xylem stacking physical test angle was found. The rolling friction coefficient ($X_{3'}$) between the xylem and Q235A steel was determined to be 0.033, the static friction coefficient ($X_{5'}$) between the xylem and xylem was 0.44, and the rolling friction coefficient ($X_{6'}$) between the xylem and xylem was 0.016.

Determination and Validation of the Optimal Parameter Combination

The optimized solutions for the phloem were subjected to simulation tests, and the simulated stacking angles were 38.23°, 38.06°, 37.84°, 37.93°, and 38.12°. The simulated results were close to the cylindrical lifting physical test angle, as shown in Figure 18, with relative errors of 0.79%, 0.34%, 0.24%, 0.11%, and 0.5%, respectively.

Figure 18. Comparison of the physical test and simulation test of the ramie phloem stacking angle: (**a**) physical test; (**b**) simulation test.

To evaluate the difference between the simulation and physical results of the phloem stacking angles, a two-sample *t*-test was conducted. Before conducting the two-sample *t*-test, the existence of significant differences in the variance between the two samples needed to be determined. Therefore, an *F*-test was conducted on the physical stacking angle results to test the simulation results. Table 19 shows the *F*-test results of the phloem.

Table 19. Results of the phloem *F*-test.

	Mean (°)	Variance	Observed Value	df	*F*	p ($F \le f$) One-Tailed	*F* One-Tailed Critical
Simulated value	38.04	0.0219	5	4	0.0041	5.1495×10^{-5}	0.1565
Physical value	37.93	5.2639	5	4			

The phloem samples showed a significant difference between the two variances, with a two-tailed probability of $2p < 0.01$. Therefore, a two-sample heteroskedasticity *t*-test was conducted to assess the significance between the simulation and physical results. Table 20 displays the results of the two-sample heteroskedasticity *t*-test of the phloem.

Table 20. Results of the phloem two-sample heteroskedasticity *t*-test.

	Mean (°)	Variance	Observed Value	Assume Average Difference	df	*t* Stat	p ($T \leq t$) One-Tailed	*t* One-Tailed Critical	p ($T \leq t$) Two-Tailed	*t* Two-Tailed Critical
Simulated value	38.04	0.02193	5	0	4	0.1120	0.4581	2.1318	0.9162	2.7764
Physical value	37.93	5.2639	5							

According to Table 20, $|t|$ < "*t* two-tailed critical" and "*p* two-tailed critical" > 0.05, indicating that there was no significant difference between the phloem simulation and physical results after calibrating the simulation parameters.

The simulated accumulation angles of the xylem were 27.13°, 27.3°, 27.38°, 27.05°, and 27.12°. The simulated results were close to the physical results, as shown in Figure 19, with relative errors of 0.15%, 0.48%, 0.77%, 0.44%, and 0.18%, respectively.

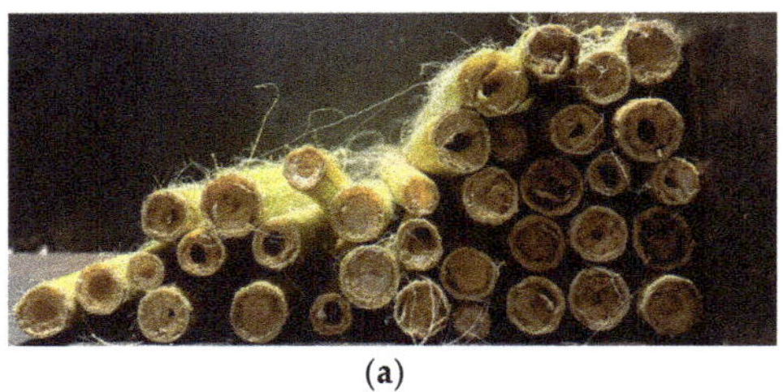

(a)

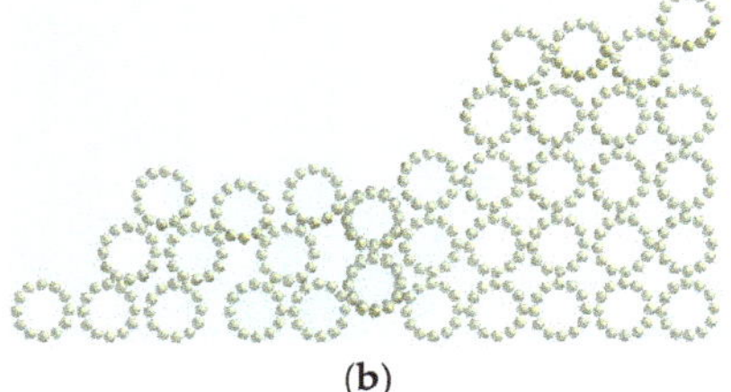

(b)

Figure 19. Comparison of the physical test and simulation test of the ramie xylem stacking angle: (a) physical test; (b) simulation test.

To evaluate the difference between the simulation and physical results of the xylem stacking angles, a two-sample *t*-test was conducted. Before performing the two-sample *t*-test, it was important to assess the variance between the two samples to determine if there was a significant difference. Therefore, an *F*-test was performed on the simulation results based on the physical stacking angle results, and Table 21 shows the *F*-test results for the xylem.

Table 21. Results of the xylem *F*-test.

	Mean (°)	Variance	Observed Value	df	*F*	p ($F \leq f$) One-Tailed	*F* One-Tailed Critical
Simulated value	27.20	0.0190	5	4	0.0119	0.0004	0.1565
Physical value	27.17	1.6004	5	4			

The xylem samples showed a significant difference between the two variances, with a two-tailed probability of $2p < 0.01$. Therefore, a two-sample heteroskedasticity *t*-test was conducted to assess the significance between the simulation and physical results. Table 22 displays the results of the two-sample heteroskedasticity *t*-test of the xylem.

According to Table 22, $|t|$ < "*t* two-tailed critical" value and "*p* two-tailed" > 0.05, indicating no significant difference between the xylem simulated and physical results after calibrating the simulation parameters.

Table 22. Results of the xylem two-sample heteroskedasticity t-test.

	Mean (°)	Variance	Observed Value	Assume Average Difference	df	t Stat	p ($T \leq t$) One-Tailed	t One-Tailed Critical	p ($T \leq t$) Two-Tailed	t Two-Tailed Critical
Simulated value	27.20	0.0190	5	0	4	0.0459	0.4828	2.1318	0.9656	2.7764
Physical value	27.17	1.6004	5							

The results show that after optimizing the simulation parameters, the optimal parameter combination for the stacking angle simulation test and the physical test was when: the static friction coefficient (X_5) between the phloem and phloem was 0.41; rolling friction coefficient between the phloem and phloem (X_6) was 0.056; rolling friction coefficient ($X_{3'}$) between the xylem and Q235A steel was 0.033; the static friction coefficient ($X_{5'}$) between the xylem and xylem was 0.44; and rolling friction coefficient ($X_{6'}$) between the xylem and xylem was 0.016. There was no significant difference between the stacking angle simulation test results and the physical test results. The similarity in the shape and result of the stacking angle between the two indicates that the simulation parameters were accurately set. In addition, the maximum relative error between the simulated and physical results for the phloem was 0.79%, and for the xylem it was 0.77%. The average relative error between the two was only 0.4%. This further verifies the reliability and authenticity of the simulation test. The obtained parameters can be used for subsequent simulation tests on calibrating the ramie stalk's bonding parameters and the ramie stalk's discrete element decorticating simulation test.

4. Conclusions

(1) By comparing the results of the physical and simulation tests, the contact parameters between the ramie stalk phloem and xylem were calibrated. The coefficient of restitution between the ramie stalk phloem and xylem was calibrated to 0.60 by the collision rebound test. The coefficient of the static friction between the ramie stalk phloem and xylem was calibrated to be 0.53 using the sliding test. The coefficient of the rolling friction between the ramie stalk phloem and xylem was calibrated to be 0.021 using the beveled rolling test. The calibration errors of the parameters were 2.11%, 0.22%, and 0.3%, respectively. The results indicate that the calibration results of the contact parameters between the ramie stalk phloem and xylem are reliable.

(2) By using the Plackett–Burman and steepest ascent tests, the significant factors that influence the stacking angle of the ramie stalk phloem and xylem were screened, and their ranges were determined. The Plackett–Burman test results showed that the static friction coefficient (X_5) and rolling friction coefficient (X_6) between the phloem and phloem had a significant influence on the stacking angle of the phloem, while the rolling friction coefficient ($X_{3'}$) between the xylem and Q235A steel, static friction coefficient (X_5), and the rolling friction coefficient ($X_{6'}$) between the xylem and xylem had a significant influence on the stacking angle of the xylem. The results of the steepest ascent test showed that the optimized ranges of the static friction coefficient (X_5) and rolling friction coefficient (X_6) between the phloem and phloem were 0.35–0.52 and 0.053–0.084, respectively, and the optimized ranges of the rolling friction coefficient ($X_{3'}$) between the xylem and Q235A steel, static friction coefficient ($X_{5'}$), and rolling friction coefficient ($X_{6'}$) between the xylem and xylem were 0.028–0.074, 0.22–0.45, and 0.013–0.028, respectively.

(3) Using the response surface methodology, the interaction effects of the various factors in the phloem and xylem on the stacking angle were analyzed. Based on the results of the phloem central composite design test, a regression model was established between the static friction coefficient (X_5) and rolling friction coefficient (X_6) of the phloem–phloem and the stacking angle of the phloem. The results indicate that when the static

friction coefficient (X_5) between the phloem and phloem was constant, the phloem simulation stacking angle increased with the rise in the rolling friction coefficient of the phloem–phloem. When the rolling friction coefficient (X_6) of the phloem–phloem was constant, the phloem simulation stacking angle increased with the rise in the static friction coefficient (X_5) between the phloem and phloem. Factor X_5 had a more significant effect on the stacking angle than X_6. Based on the results of the xylem central composite design test, a regression model was established between the rolling friction coefficient ($X_{3'}$) of the xylem-Q235A steel, the static friction coefficient ($X_{5'}$) of the xylem–xylem, and the rolling friction coefficient ($X_{6'}$) of the xylem–xylem and the stacking angle of the xylem. The results indicated that the order of the influence of each factor on the stacking angle is $X_{6'} > X_{5'} > X_{3'}$.

(4) Utilizing the Optimization module in the Design-Expert software, the optimal parameters for the significant factors of the phloem, xylem, and stacking angle were obtained. Specifically, the static friction coefficient (X_5) between the phloem and phloem was 0.41, and the rolling friction coefficient (X_6) between the phloem and phloem was 0.056. The rolling friction coefficient ($X_{3'}$) between the xylem and Q235A steel was 0.033, the static friction coefficient ($X_{5'}$) between the xylem and xylem was 0.44, and the rolling friction coefficient ($X_{6'}$) between the xylem and xylem was 0.016. A two-sample heteroscedastic *t*-test was performed on the optimized simulation parameter results and the physical stacking angle measurement results. The results showed no significant difference between the simulated and physical values, with a maximum relative error of 0.79% for the phloem and 0.77% for the xylem, and an average relative error of only 0.4%. This verified the reliability and authenticity of the simulation test, which can provide technical support for subsequent simulation tests on ramie stalk bonding parameters to optimize ramie decorticating machines.

Author Contributions: Conceptualization, Y.H. and W.X.; methodology, Y.H. and W.X.; software, Y.H. and Y.D.; validation, Y.H. and L.M.; formal analysis, Y.H. and W.X.; investigation, Y.H. and W.X.; resources, J.L. (Jiangnan Lyu); data curation, J.L. (Jiangnan Lyu) and J.L. (Jiajie Liu); writing—original draft preparation, Y.H. and W.X.; writing—review and editing, W.X. and J.L. (Jiangnan Lyu); visualization, J.L. (Jiangnan Lyu) and B.Y.; supervision, W.X., L.M. and J.L. (Jiajie Liu); project administration, J.L. (Jiangnan Lyu) and W.X.; funding acquisition, J.L. (Jiangnan Lyu) All authors have read and agreed to the published version of the manuscript.

Funding: This research was funded by the CARS—Bast Fiber Crops (No. CARS-16-E21) and the Natural Science Foundation of Hunan Province (No. 2019JJ40333).

Institutional Review Board Statement: Not applicable.

Informed Consent Statement: Not applicable.

Data Availability Statement: All data are presented in this article in the form of figures and tables.

Conflicts of Interest: The authors declare no conflict of interest.

References

1. Cheng, L.; Duan, S.; Feng, X.; Zheng, K.; Yang, Q.; Xu, H.; Luo, W.; Peng, Y. Ramie-degumming methodologies: A short review. *J. Eng. Fibers Fabr.* **2020**, *15*, 1558925020940105. [CrossRef]
2. Zhu, T.; Zhu, A.; Yu, Y.; Sun, K.; Mao, H.; Chen, Q. Research progress of ramie for feedstuff. *Pratac. Sci.* **2016**, *33*, 338–347.
3. Pandey, S. Ramie fibre: Part II. Physical fibre properties. A critical appreciation of recent developments. *Text. Prog.* **2007**, *39*, 189–268. [CrossRef]
4. Du, X.; Zhao, W.; Wang, Y.; Wang, C.; Chen, M.; Qi, T.; Hua, C.; Ma, M. Preparation of activated carbon hollow fibers from ramie at low temperature for electric double-layer capacitor applications. *Bioresour. Technol.* **2013**, *149*, 31–37. [CrossRef] [PubMed]
5. Rehman, M.; Gang, D.; Liu, Q.; Chen, Y.; Wang, B.; Peng, D.; Liu, L. Ramie, a multipurpose crop: Potential applications, constraints and improvement strategies. *Ind. Crops Prod.* **2019**, *137*, 300–307. [CrossRef]
6. Xiang, W.; Ma, L.; Liu, J.; Xiao, L.; Long, C.; Wen, Q.; Lyu, J. Research progress on technology and equipment of ramie fibre stripping and processing in China. *J. Agric. Sci. Technol.* **2019**, *21*, 59–69.
7. Lyu, J.; Long, C.; Zhao, J.; Ma, L.; Lyu, H.; Liu, J.; He, H. Design and experiment of transverse-feeding ramie decorticator. *Trans. Chin. Soc. Agric. Eng.* **2013**, *29*, 16–21.

8. Yan, D.; Yu, J.; Wang, Y.; Zhou, L.; Sun, K.; Tian, Y. A review of the application of discrete element method in agricultural engineering: A case study of soybean. *Processes* **2022**, *10*, 1305. [CrossRef]
9. Zeng, Z.; Ma, X.; Cao, X.; Li, Z.; Wang, X. Critical review of applications of discrete element method in agricultural engineering. *Trans. Chin. Soc. Agric. Mach.* **2021**, *52*, 1–20.
10. Shi, L.; Zhao, W.; Sun, B.; Sun, W. Determination of the coefficient of rolling friction of irregularly shaped maize particles by using discrete element method. *Int. J. Agric. Biol. Eng.* **2020**, *13*, 15–25. [CrossRef]
11. Zhao, H.; Huang, Y.; Liu, Z.; Liu, W.; Zheng, Z. Applications of discrete element method in the research of agricultural machinery: A review. *Agriculture* **2021**, *11*, 425. [CrossRef]
12. Ma, H.; Zhou, L.; Liu, Z.; Chen, M.; Xia, X.; Zhao, Y. A review of recent development for the CFD-DEM investigations of non-spherical particles. *Powder Technol.* **2022**, *412*, 117972. [CrossRef]
13. Roessler, T.; Richter, C.; Katterfeld, A.; Will, F. Development of a standard calibration procedure for the DEM parameters of cohesionless bulk materials—Part I: Solving the problem of ambiguous parameter combinations. *Powder Technol.* **2019**, *343*, 803–812. [CrossRef]
14. Elskamp, F.; Kruggel-Emden, H.; Hennig, M.; Teipel, U. A strategy to determine DEM parameters for spherical and non-spherical particles. *Granul. Matter* **2017**, *19*, 46. [CrossRef]
15. Al-Hashemi, H.M.B.; Al-Amoudi, O.S.B. A review on the angle of repose of granular materials. *Powder Technol.* **2018**, *330*, 397–417. [CrossRef]
16. Jiang, W.; Wang, L.; Tang, J.; Yin, Y.; Zhang, H.; Jia, T.; Qin, J.; Wang, H.; Wei, Q. Calibration and experimental validation of contact parameters in a discrete element model for tobacco strips. *Processes* **2022**, *10*, 998. [CrossRef]
17. Zhang, T.; Liu, F.; Zhao, M.; Ma, Q.; Wang, W.; Fan, Q.; Yan, P. Determination of corn stalk contact parameters and calibration of Discrete Element Method simulation. *J. China Agric. Univ.* **2018**, *23*, 120–127.
18. Xiang, W.; Wu, M.; Lu, J.; Quan, W.; Ma, L.; Liu, J. Calibration of simulation physical parameters of clay loam based on soil accumulation test. *Trans. Chin. Soc. Agric. Eng.* **2019**, *35*, 116–123.
19. Liu, F.; Zhang, J.; Li, B.; Chen, J. Calibration of parameters of wheat required in discrete element method simulation based on repose angle of particle heap. *Trans. Chin. Soc. Agric. Eng.* **2016**, *32*, 247–253.
20. Xiao, W.; Chen, H.; Wan, X.; Li, M.; Liao, Q. Influence of shape and particle size distribution on discrete element simulation flow characteristics of granular compound fertilizers. *Appl. Eng. Agric.* **2021**, *37*, 1169–1179. [CrossRef]
21. Shi, G.; Li, J.; Ding, L.; Zhang, Z.; Ding, H.; Li, N.; Kan, Z. Calibration and tests for the discrete element simulation parameters of fallen jujube fruit. *Agriculture* **2022**, *12*, 38. [CrossRef]
22. Dai, Z.; Wu, M.; Fang, Z.; Qu, Y. Calibration and verification test of lily bulb simulation parameters based on discrete element method. *Appl. Sci.* **2021**, *11*, 10749. [CrossRef]
23. Gupta, M.; Yang, J.; Roy, C. Density of softwood bark and softwood char: Procedural calibration and measurement by water soaking and kerosene immersion method. *Fuel* **2002**, *81*, 1379–1384. [CrossRef]
24. *GB/T 1931-2009*; Method for Determination of the Moisture Content of Wood. Standardization Administration of the P.R.C.: Beijing, China, 2009.
25. Navi, P.; Rastogi, P.K.; Gresse, V.; Tolou, A. Micromechanics of wood subjected to axial tension. *Wood Sci. Technol.* **1995**, *29*, 411–429. [CrossRef]
26. Du, D.; Wang, J. Research on mechanics properties of crop stalks: A review. *Int. J. Agric. Biol. Eng.* **2016**, *9*, 10–19.
27. Xie, H.; Cui, B.; Hao, S.; Li, S.; Jia, X.; Wang, W. Exploring the macroscopic and microscopic characteristics of rice stalk for utilization in bio-composites. *Compos. Sci. Technol.* **2022**, *230*, 109728. [CrossRef]
28. Liu, P.; Xiang, P.; Zhou, Q.; Zhang, H.; Tian, J.; Argaw, M.D. Prediction of Mechanical Properties of Structural Bamboo and Its Relationship with Growth Parameters. *J. Renew. Mater.* **2021**, *9*, 2223–2239. [CrossRef]
29. Tian, K.P.; Shen, C.; Zhang, B.; Li, X.W.; Huang, J.C.; Chen, Q.M. Experimental study on mechanical properties of reed stalk. *IOP Conf. Ser. Earth Environ. Sci.* **2019**, *346*, 012076. [CrossRef]
30. Al-Zube, L.; Sun, W.; Robertson, D.; Cook, D. The elastic modulus for maize stems. *Plant Methods* **2018**, *14*, 1–12. [CrossRef]
31. *GB/T 1929-2009*; Method for Sample Logs Sawing and Test Specimens Selection for Physical and Mechanical Tests of Wood. Standardization Administration of the P.R.C.: Beijing, China, 2009.
32. Guo, Y.; Liang, L.; Zhang, J. *Research on Mechanical Properties of Stalk Crops and Physical Properties of Lemon Sticks and Their Applications*, 1st ed.; Chemical Industry Press: Beijing, China, 2019; pp. 111–118.
33. Sharma, R.K.; Bilanski, W.K. Coefficient of restitution of grains. *Trans. Am. Soc. Agric. Eng.* **1971**, *14*, 216–218.
34. Ma, Y.; Zhang, J.; Wu, Y. *Agricultural Material Science*, 1st ed.; Chemical Industry Press: Beijing, China, 2015; pp. 87–92.
35. Jia, H.; Deng, J.; Deng, Y.; Chen, T.; Wang, G.; Sun, Z.; Guo, H. Contact parameter analysis and calibration in discrete element simulation of rice straw. *Int. J. Agric. Biol. Eng.* **2021**, *14*, 72–81. [CrossRef]
36. *ASTM G194-08*; Standard Test Method for Measuring Rolling Friction Characteristics of a Spherical Shape on a Flat Horizontal Plane. ASTM International: West Conshohocken, PA, USA, 2009.
37. Wolf, D.E. *Friction in Granular Media*; NATO Advanced Study Institute on Physics of Dry Granular Media: Cargese, France, 1997; pp. 441–464.
38. Wen, X.; Yang, W.; Guo, W.; Zeng, B. Parameter determination and validation of discrete element model of segmented sugarcane harvester for impurity removal. *J. Chin. Agric. Mech.* **2020**, *41*, 12–18.

39. Zhao, L.; Zhou, H.; Xu, L.; Song, S.; Zhang, C.; Yu, Q. Parameter calibration of coconut bran substrate simulation model based on discrete element and response surface methodology. *Powder Technol.* **2022**, *395*, 183–194. [CrossRef]
40. Ren, J.; Wu, T.; Mo, W.; Li, K.; Hu, P.; Xu, F.; Liu, Q. Discrete element simulation modeling method and parameters calibration of sugarcane leaves. *Agronomy* **2022**, *12*, 1796. [CrossRef]
41. Liu, F.; Zhang, J.; Chen, J. Modeling of flexible wheat straw by discrete element method and its parameters calibration. *Int. J. Agric. Biol. Eng.* **2018**, *11*, 42–46. [CrossRef]
42. Liao, Y.; Liao, Q.; Zhou, Y.; Wang, Z.; Jiang, Y.; Liang, F. Parameters calibration of discrete element model of fodder rape crop harvest in bolting stage. *Trans. Chin. Soc. Agric. Mach.* **2020**, *51*, 73–82.
43. Zeng, Z.; Chen, Y. Simulation of straw movement by discrete element modelling of straw-sweep-soil interaction. *Biosyst. Eng.* **2019**, *180*, 25–35. [CrossRef]
44. Fan, G.; Wang, S.; Shi, W.; Gong, Z.; Gao, M. Simulation parameter calibration and test of typical pear varieties based on discrete element method. *Agronomy* **2022**, *12*, 1720. [CrossRef]
45. Cabiscol, R.; Finke, J.H.; Kwade, A. Calibration and interpretation of DEM parameters for simulations of cylindrical tablets with multi-sphere approach. *Powder Technol.* **2018**, *327*, 232–245. [CrossRef]
46. Kruggel-Emden, H.; Rickelt, S.; Wirtz, S.; Scherer, V. A study on the validity of the multi-sphere discrete element method. *Powder Technol.* **2008**, *188*, 153–165. [CrossRef]
47. Guan, Z.; Mu, S.; Li, H.; Jiang, T.; Zhang, M.; Wu, C. Flexible DEM model development and parameter calibration for rape stem. *Appl. Sci.* **2022**, *12*, 8394. [CrossRef]
48. Shen, C.; Li, X.; Tian, K.; Zhang, B.; Huang, J.; Chen, Q. Experimental analysis on mechanical model of ramie stalk. *Trans. Chin. Soc. Agric. Eng.* **2015**, *31*, 26–33.

 agriculture

Article

V-Shaped Toothed Roller Cotton Stalk Puller: Numerical Modeling and Field-Test Validation

Zhenwei Wang [1], Weisong Zhao [1], Jingjing Fu [1], Hu Xie [1], Yinping Zhang [2,*] and Mingjiang Chen [1,*]

[1] Nanjing Institute of Agricultural Mechanization, Ministry of Agriculture and Rural Affairs, Nanjing 210014, China; wangzhenwei@caas.cn (Z.W.); wszhao77@163.com (W.Z.); tutujing12@163.com (J.F.); tmjamexh@163.com (H.X.)

[2] School of Agricultural and Food Science, Shandong University of Technology, Zibo 255000, China

* Correspondence: zhangyinping@sdut.edu.cn (Y.Z.); chenmingjiang@caas.cn (M.C.)

Abstract: The V-shaped toothed roller cotton stalk puller has a low removal ratio and weak pulling effect. Hence, we constructed a simplified mathematical model of the V-shaped tooth roller stalk puller based on elastic collision theory and simple beam theory and conducted a mechanical analysis based on this model to explore the causes of pulling errors and fractures. Specifically, the V-shaped tooth plates of the machine were optimized in an orthogonal experiment with the rotational speed, cogging angle, and ground clearance as the influencing factors, and the removal ratio as the evaluation index. This experiment was designed to enable analysis of the physical characteristics of cotton stalks, and the forces applied during the pulling process. Additionally, a V-shaped toothed roller-type stalk-pulling test bench was constructed. The results revealed that, unlike the cogging angle, the ground clearance significantly affected the removal ratio. Furthermore, the highest removal ratio (i.e., 97%) was achieved when the ground clearance was −20 mm, the rotational speed was 300 rpm, and the cogging angle was 32.5°. An L_9 (3^4) orthogonal field experiment was also conducted with the rotational speed, cogging angle, and ground clearance as the influencing factors to investigate their respective influences on the stalk removal ratio. The results revealed that the ground clearance most significantly influenced the ratio, followed by the rotational speed, and cogging angle. The ground clearance and rotational speed of the V-shaped toothed roller were each found to significantly influence the ratio. Furthermore, a ground clearance of −20 mm, rotational speed of 300 r/min, and cogging angle of 25° yielded an average removal ratio of 98.27%. Through this research, the mechanism of toothed roller stalk pulling is further deepened and the toothed series stalk pulling technology provides theoretical support.

Keywords: V-shaped tooth roller; removal ratio; orthogonal experiment; structure design

Citation: Wang, Z.; Zhao, W.; Fu, J.; Xie, H.; Zhang, Y.; Chen, M. V-Shaped Toothed Roller Cotton Stalk Puller: Numerical Modeling and Field-Test Validation. *Agriculture* **2023**, *13*, 1157. https://doi.org/10.3390/agriculture13061157

Academic Editor: Massimo Cecchini

Received: 4 May 2023
Revised: 24 May 2023
Accepted: 25 May 2023
Published: 30 May 2023

1. Introduction

Cotton stalk is a byproduct of cotton that can be utilized as a high-quality renewable resource. China is a large cotton-producing country. In 2020, the national cotton planting area was 3.1699 million square hectares [1,2]. More than 40% of the world's cotton is produced in China, and this area produced a total output of 3.05×10^7 t of cotton stalks [3,4]. However, only approximately one-tenth of this total output was subsequently utilized. The lack of effective cotton stalk harvest machinery is one of the main reasons for this low utilization rate [5,6].

The core component of cotton stalk harvester machinery is the stalk-pulling structure, the performance of which directly influences stalk-pulling efficiency. Thus, to increase the cotton stalk removal ratio, many researchers have investigated the characteristics of cotton stalks, the mechanism of the stalk-pulling structures, and the dynamic characteristics of stalk pulling. Regarding the mechanical properties and pull-out characteristics of cotton stalks, Chen et al. [7,8] examined how the cotton stalk pulling force changed in the field between autumn and the following spring. They obtained the following results: (1) the tensile failure load was 10.9 to 29.8 times larger than the bending failure load and (2) the pulling resistance of the cotton stalk is significantly influenced by the diameter of the cotton stalk root, the soil hardness, and the harvest time. Using a pull-out force measuring device, Li et al. found that the cotton stalk pulling angle significantly affects the pulling force [9,10]. Additionally, the findings reported by Demian T. F. et al. indicate that, within a certain cotton stalk height range, a higher pulling height mandates a larger pulling force [11]. Currently, there are two general types of straw-pulling structures: the row-controlled pulling structure, which includes the tooth-disc type, double-roller type, and chain-clamp type; and the non-row-controlled structure, which includes the knife-roller type and V-shaped toothed roller type. To date, non-row-controlled structures have been preferred over row-controlled structures because they have advantages such as lower operational requirements for the driver and higher working efficiency [12–17].

Among the non-row-controlled stalk-pulling structures, the knife-roller type has a high removal ratio; however, because it penetrates the soil during operation, it is associated with high energy consumption and problems related to residual film and mud, which hinders the follow-up work. The results of theoretical research have shown that, although it is still in an exploratory phase, the V-shaped toothed roller type not only has the advantage of a high removal ratio commonly associated with the knife-roller type but also the advantage of low energy consumption [17]. Tang et al. designed an all-in-one V-shaped toothed roller machine by integrating the functions of cotton stalk pull out, soil clearance, transportation, chopping, and collection; this design yielded a high extraction rate [12]. However, an in-depth theoretical analysis of the V-shaped toothed roller was not performed. The Nanjing Institute of Agricultural Mechanization, Ministry of Agriculture and Rural Affairs, and Binzhou Agricultural Mechanization Research Institute jointly developed a V-shaped toothed roller-type cotton stalk harvester that could perform the functions of stalk pull-out, soil clearance, chopping, and bundling [18–20]. This collaboration resulted in the first application of an integrated elastic collision and simple beam theory to stalk pulling [21]. Thereafter, He et al. and Dai et al. used this theory to study and test other types of stalk-pulling machines [22,23]. However, to date, how the forces applied to V-shaped toothed rollers vary as the stalks are extracted remains to be unknown, and the reasons for cotton stalk fracture and non-pulled stalks lack clear theoretical explanations.

A small-scale test bench is designed in this work, which is driven by hydraulic pressure to pull the stalk roller, and the motor drives the reel and is equipped with torque, speed sensors, and other components that can control the motion parameters very accurately so as to facilitate the research of the experiment. In order to elucidate the operational stress variation profile for V-shaped toothed roller stalk-pulling machines and explore the mechanisms of cotton stalk pull-out, fracture, and missed extraction, elastic collision and simple beam theories were applied to develop a comprehensive stalk-pulling theory. The theoretical results revealed acceleration to be a significant factor affecting the stalk-pulling process. Thus, acceleration was taken into account to increase the accuracy of the mechanical model and yield a more accurate description of the formation mechanism of the three states of stalk pulling. This optimized V-shaped toothed roller model was then applied in a field experiment. Specifically, an orthogonal experiment was carried out to optimize the rotational speed, cogging angle, and ground clearance of V-shaped toothed rollers with respect to the removal ratio. It is believed that the findings of this study can be used as a reference for the development of a highly efficient cotton stalk harvester.

2. Materials and Methods

2.1. V-Shaped Toothed Roller Stalk Pulling Structure and Working Principle

The V-shaped toothed roller stalk-pulling component, which mainly consists of a V-shaped toothed roller, stalk-clearing roller, reel wheel, and land wheel, as shown in Figure 1, is a key component of stalk-pulling machines. The key parameters are listed in Table 1. The V-shaped toothed plates are evenly distributed in three rows along the circumferential direction of the V-shaped toothed roller and will hereafter be referred to as the toothed plates.

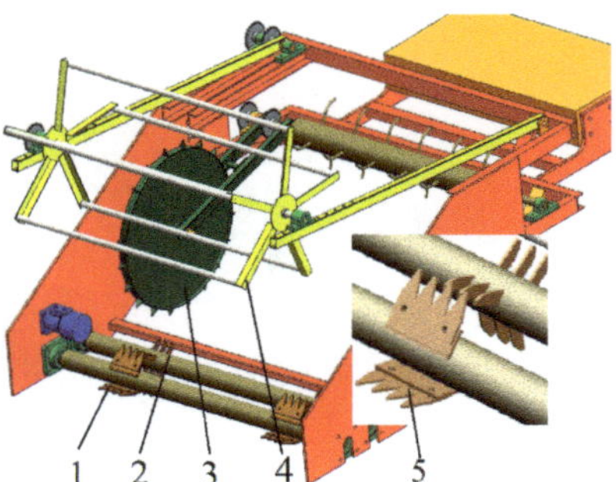

Figure 1. Structural schematic of V-shaped toothed roller. 1. V-shaped toothed roller, 2. Stalk-clearing roller, 3. Land wheel, 4. Reel wheel, 5. V-shaped toothed plate.

Table 1. The key parameters of the V-shaped toothed roller.

Parameters	Value
Length × width × height (mm)	1500 × 1200 × 1100
Number of lines	2
Drive mode	hydraulic
Tooth plate: diameter × thick (mm)	220 × 120
Rotational speed (r/min)	100–600

During the pulling process, the reel wheel rotates to feed the upper part of the cotton stalk into the harvester and the V-shaped toothed roller moves forward and rotates to embed the cotton stalks with lengths within the harvest width into the V-shaped toothed plate through the appropriate space between the V-shaped toothed rollers. The reel wheel works with the V-shaped toothed roller to apply push–pull forces that ultimately extract the cotton stalks embedded in the toothed plates from the soil.

2.2. *Key Component Design*

2.2.1. Determination of V-Shaped Toothed Roller Rotational Speed

The rotational speed of a V-shaped toothed roller should satisfy the following two conditions [10]: (1) the pulled-out, undetached cotton stalk must not hinder the operation of the adjacent toothed plate and (2) when the first toothed plate fails to pull out a cotton stalk, the second or third row of toothed plates must be able to embed and hold the cotton stalk (Figure 2). Considering that the driving speed of a tractor in a field ranges from 0.7 to 1.4 m/s and the plant spacing between cotton stalks typically ranges between approximately 18 and 25 cm, according to Equation (1), the rotational speed of the V-shaped toothed roller should exceed 156 r/min. The results of experimental tests indicate that the rotational speed of a V-shaped toothed roller should not be too high (i.e., it should be less than 500 r/min). This is because an excessively high rotational speed was found to be associated with excessive force being applied to the cotton stalk by the toothed plate, leading to a significantly higher stalk pull-out miss rate and fracture rate.

$$n_0 = \frac{1000 \times 60 \times v_0}{l_0} \times \frac{2}{3} = \frac{4000 v_0}{l_0} \tag{1}$$

n_0—rotational speed of tooth roller;
v_0—driving speed of tractor;
l_0—line spacing of cotton stalk.

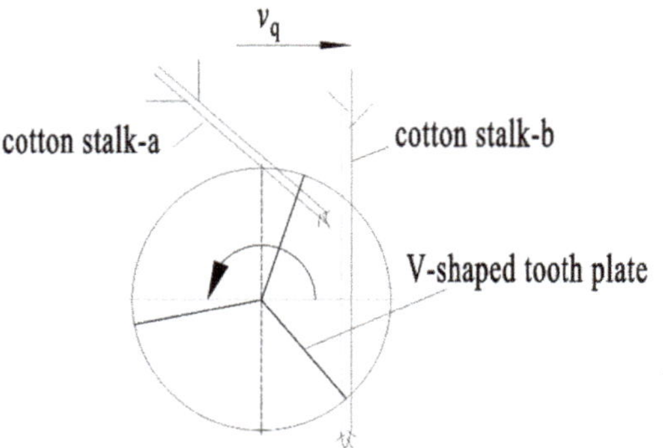

Figure 2. Schematic diagram showing the pull-out process of cotton stalk.

2.2.2. Determination of V-Shaped Toothed Plate Parameters

The maximum diameter of a cotton stalk is 22.75 mm; however, according to the results of multiple field measurements, the diameter of most cotton stalks in Wudi County, Shandong Province ranges from 10 to 18 mm. Thus, we mainly took cotton stalks with a diameter larger than 10 mm into consideration in the proposed V-shaped toothed roller stalk-pulling machine design. Note that the tooth width of the toothed plate was designed to be 30 mm. Additionally, the toothed roller was also designed to satisfy three requirements: (1) to improve the cotton stalk embedding and holding effectiveness, (2) to ensure that each cotton stalk remains embedded and held once it is extracted and swung backward, and (3) to ensure that the stalk-clearing roller can easily clear away each extracted cotton stalk from the toothed roller. The stress analysis results for cotton stalk extraction via the proposed V-shaped toothed roller are shown in Figure 3.

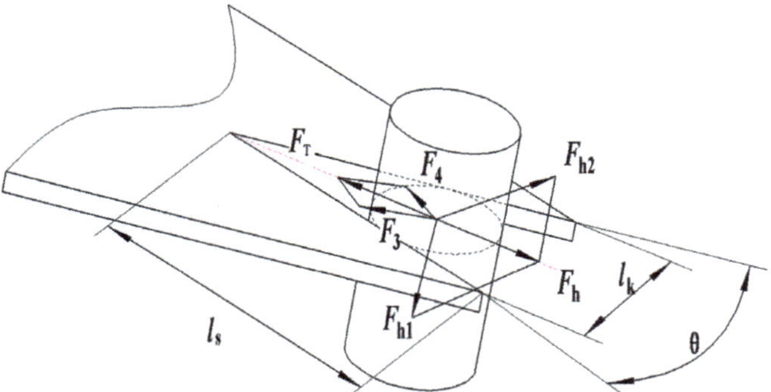

Figure 3. Stress analysis sketch of cotton stalk pulling.

According to the analysis of Figure 3, Formulas (2)–(4) can be obtained:

$$F_h = 2F_{h1}\cos\left(\frac{\pi}{2} - \frac{\theta}{2}\right) = 2F_{h1}\sin\frac{\theta}{2} \quad (2)$$

$$F_T = 2F_3\cos\frac{\theta}{2} = 2F_{h1}\tan\alpha_0\cos\frac{\theta}{2} \quad (3)$$

$$F_h = F_T \quad (4)$$

θ—cogging angle (°);

F_h—combined force of pressure on cotton stalk produced by both sides of tooth cogging (N);

F_3—the friction of cotton stalk under pressure F_{h1} (N);
F_4—the friction of cotton stalk under pressure F_{h2} (N);
F_T—the combined force of friction (N);
α_0—the frictional angle between cotton stalk and tooth plate (°).

The simultaneous Formulas (2)–(4) show that the above formula holds when $\theta = 2\alpha_0$. The frictional angle between the cotton stalk and the tooth plate can be converted by the friction coefficient, which could refer to the friction coefficient of 0.2~0.35 between wood and steel [24]. The converted α_0 value was 11.3°~19.3°, thus the value of θ was 22.6°~38.6°.

The depth of cogging can be calculated from Formula (5):

$$l_s = \frac{l_k}{\frac{2\tan\theta}{2}} = \frac{15}{\frac{\tan\theta}{2}} \tag{5}$$

When the cogging has a smaller angle and a long and sharp tooth tip, the strength of the tooth plate is reduced. Therefore, the cogging angle was suggested to be greater than or equal to 25° after comprehensive consideration (θ = 25°, l_s = 67 mm). To sum up, the range of the cogging angle is set at $25° \leq \theta \leq 40°$.

2.3. Cotton Stalk Pulling Process: Analysis of Toothed Roller Motion Trajectory

During operation, the V-shaped toothed roller, i.e., the main functioning component, makes forward and circular motions. As shown in Figure 4, the process of stalk pulling can be divided into the following four stages according to the motion trajectory: the clamping stage, pulling stage, delivering stage, and detaching stage. The analysis of tooth roller trajectory will provide a theoretical reference of the parameters design in the subsequent stalk removal mechanism. The motion trajectory formula for a certain point on the toothed plate is as follows:

$$\begin{cases} x = v_q t + \frac{R}{1000}\sin(\frac{\pi n t}{30}) \\ y = \frac{R+h}{1000} + \frac{R}{1000}\cos(\frac{\pi n t}{30}) \end{cases} \tag{6}$$

In the formula,
v_q—speed of forward motion (m/s);
R—radius of gyration of V-shaped tooth roller (mm);
h—height of central axis of V-shaped tooth bar above the ground (mm);
n—rotation speed of V-shaped tooth roller (rad/min);
t—time (s).

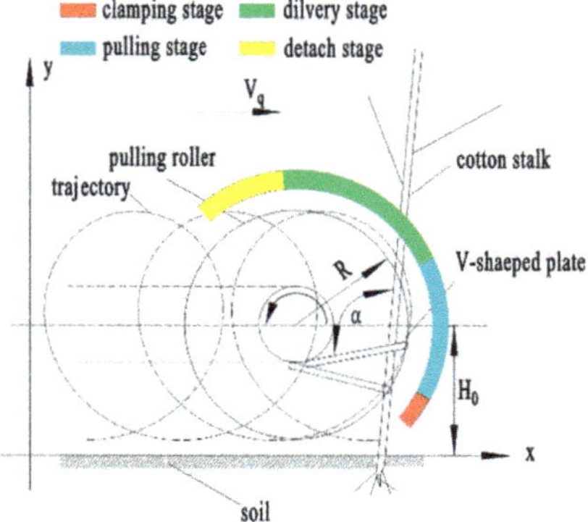

Figure 4. Motion trajectory sketch.

The simulation curves at different speeds and comparison chart of stalk pulling trajectory are exhibited in Figures 5 and 6. It can be clearly seen that the stalk-pulling trajectory is different at different speeds. Assuming that the cotton stalk is contacted and clamped at M, the machine advanced a distance of S and reaches N. It can be obtained from

Figure 6 that for the cotton stalk with different clamping heights, as the speed increases, the stalk pulling distance can be increased faster.

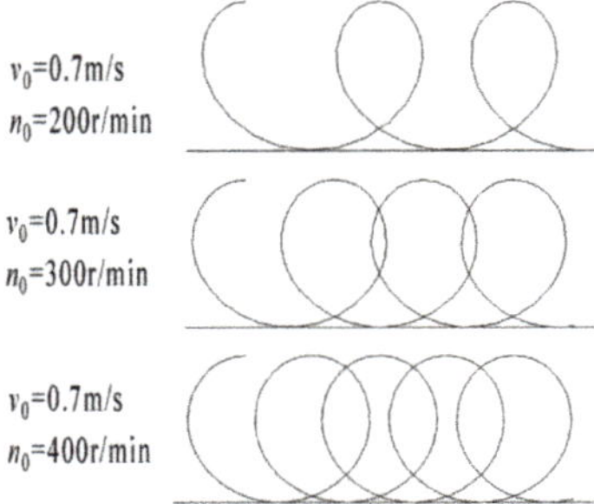

Figure 5. Simulation curves at different speeds.

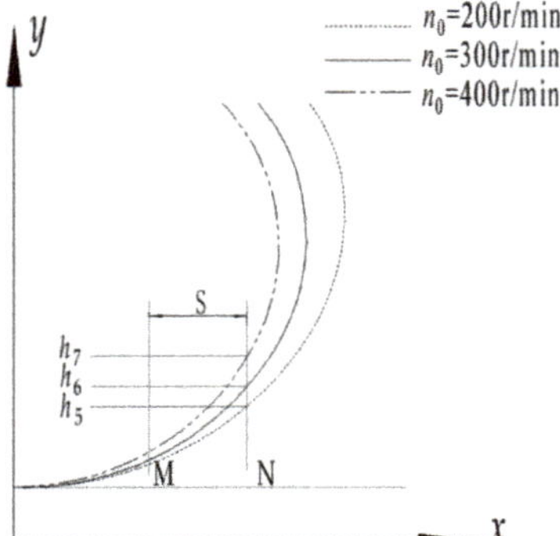

Figure 6. Comparison chart of stalk pulling trajectory.

2.3.1. Cotton Stalk Pulling Process: Collision Analysis

The mechanical changes that occur as the V-shaped toothed roller extracts stalks are very complex; these complex changes occur because of the interaction between the cotton stalk and the toothed plate and between the cotton stalk and soil. Thus, in this study, a simplified mechanical model was established by performing static analysis using data corresponding to a certain moment during the pulling process [21].

The type of collision between the rotating toothed plate and each cotton stalk was assumed to be elastic under ideal conditions [25], meaning that the mechanical properties of each cotton stalk that were temporarily altered as a result of deformation can be recovered without heating, sound, or kinetic energy loss. As such, the deformation of the cotton stalk was set to have a deformation stage and recovery stage. As shown in Figure 7, the force applied by the toothed plate was converted into deformation energy in the deformation stage; in the recovery stage, this deformation energy was defined as the bilateral forces applied to the cotton stalk by the toothed plate, i.e., F_{h1} and F_{h2}. Under the action of F_{h1} and F_{h2}, the cotton stalk and toothed plate moved relative to each other, generating frictional forces, $F_{1'}$ and $F_{2'}$, between them along the direction of the axis, as well as the corresponding resultant force, F_{12}.

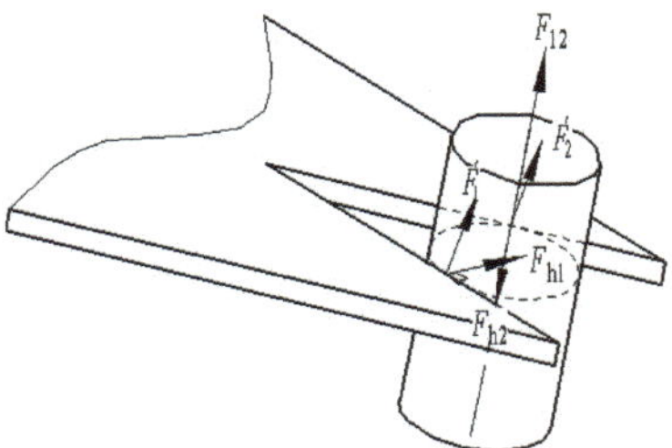

Figure 7. Sketch of the forces applied to a cotton stalk by the toothed plate.

2.3.2. Cotton Stalk Pulling Process: Mechanical Analysis of Cotton Stalk Push and Pull Forces

The toothed plate was designed to clamp around the cotton stalk after the initial collision to enable extraction by applying push and pull forces that can overcome the soil resistance. This stage included the following two processes: the clamping process and the push and pull process. The forces considered in the clamping process stress analysis for a single cotton stalk are illustrated in Figure 8. In general, the toothed plate initially makes contact with the phloem of the cotton stalk before extruding it. Then, the phloem applies bilateral forces, i.e., F_{h1} and F_{h2}, on the toothed plate during the deformation recovery stage. As the machine moves forward, the toothed plate pushes the cotton stalk forward. During this time, the cotton stalk xylem begins undergoing flexible deformation and the push and pull process is initiated. The bilateral forces applied to the toothed roller by the xylem were set as F_{h3} and F_{h4}.

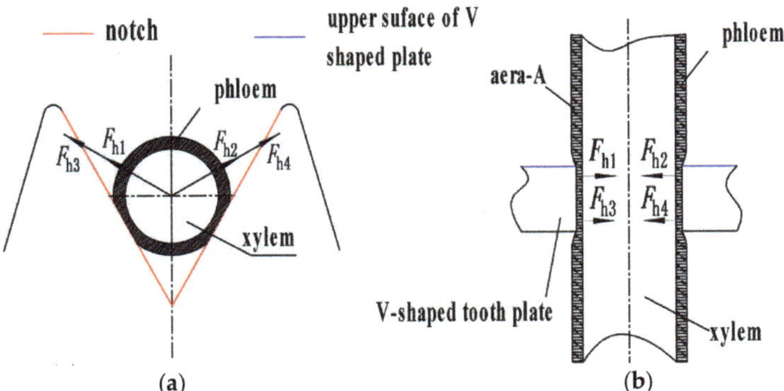

Figure 8. Sketches of cross section and longitudinal plane of cotton stalk clamped by the toothed plate: (**a**) schematic diagram of cross section under clamping state and (**b**) schematic diagram of longitudinal section in a clamped state.

During the push and pull process, the toothed plate continuously exerts force on the cotton stalk, consequently significantly deforming the phloem and creating the expanded zone A. At this moment, the cotton stalk is in an extrusion state, as shown in Figure 9. Thus, the pulling force applied to a single cotton stalk is the resultant force of (1) the upward force F_s exerted on the cotton stalk by the toothed plate, which is perpendicular to the surface of the toothed plate; and (2) the bilateral frictional forces F_{12} and F_{34} applied to the cotton stalk by the clamping tooth, which are directed upward along the length of the cotton stalk.

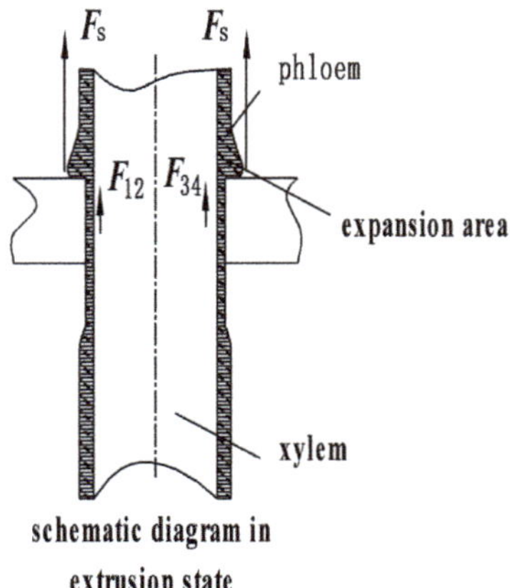

Figure 9. Sketch of longitudinal plane of cotton stalk extruded by the toothed plate.

2.3.3. Cotton Stalk Pulling Process: Analysis of the Bilateral Forces Acting on the Clamping Tooth

Dry friction can be defined as the force that acts to resist the relative motion of two solid objects. In the proposed system, there are two points at which friction occurs between the clamping teeth and the cotton stalk. One is the point at which the bilateral forces F_{h1} and F_{h2} are applied to the clamping tooth by the cotton stalk during the deformation recovery stage. The other is the point at which the cotton stalk undergoing flexible deformation applies bilateral forces to the clamping tooth because of being pushed and pulled. These bilateral forces have been defined as F_{h3} and F_{h4}. The respective interactions between the cotton stalk and soil and the toothed plate and reel could abstract the cotton stalk into a simple beam, wherein the force-exerting points correspond to the contact points between the cotton stalk and the toothed plate. Thus, assuming that the contact points between the cotton stalk and soil, reel wheel and cotton stalk, and toothed plate and cotton stalk are A, B, and C, respectively, the force exerted by the toothed plate on the cotton stalk is P. A schematic showing the application of the simple beam theory to illustrate these forces and corresponding contact points is presented in Figure 10. It can be seen that within a certain range, the greater the impact intensity, the greater the deformation of the cotton stalk, and the greater the F_{h1} and F_{h2} produced by the elastic recovery force of the cotton stalk.

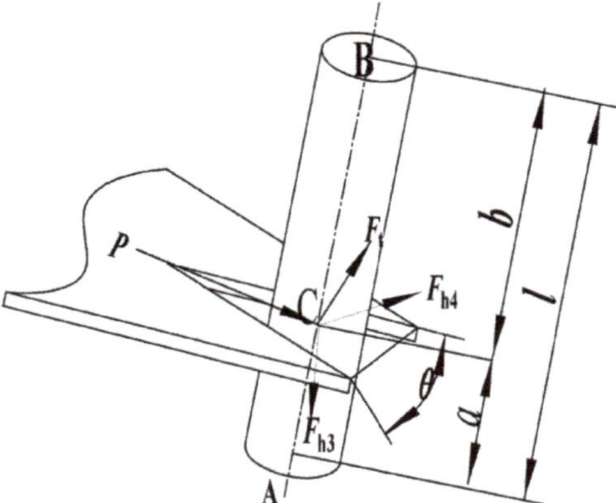

Figure 10. Schematic illustrating simple beam theory application.

According to the stress analysis in Figure 8 and the theory of simple beam, the flexible deformation Δy of cotton stalk was calculated as follows:

$$\Delta y = \frac{Pab}{6lEJ}(a^2 + b^2 - l^2) \tag{7}$$

Through the transformation of the above formula, the force P of the tooth plate exerted on cotton stalk was obtained as:

$$P = \frac{6\Delta ylEJ}{ab(a^2 + b^2 - l^2)} \tag{8}$$

In the formula,
Δy—the flexible deformation (mm);
E—elastic modulus (MPa);
J—moment of inertia (kg·m^2);
l—contact point height of reel wheel (mm);
a—ground clearance of the contact point between tooth plate and cotton stalk (mm);
b—space of contact points among reel wheel, tooth plate, and cotton stalk (mm);
P—the resultant force on the cotton stalk exerted by the tooth plate (N).

According to the parallelogram rule of force, the forces F_{h3} and F_{h4} of the resultant force P on the sides of the tooth plate can be deduced as shown in Formula (9):

$$F_{h3} = F_{h4} = \frac{P\sin(\frac{\pi}{2} - \frac{\theta}{2})\sin\theta}{\sin\theta} = \frac{p\cos\frac{\theta}{2}}{2\sin\frac{\theta}{2}\cos\frac{\theta}{2}} = \frac{p}{2\sin\frac{\theta}{2}} \tag{9}$$

Under the action of P, the maximum friction exerted by the tooth plate on the cotton stalk was $2F_{h3}$, which was along the axis of the cotton stalk. To sum up, the friction F_t between the bilateral sides of the cogging and the cotton stalk was shown in Formula (10):

$$F_t = (F_{12} + F_{34})\,f = 2(F_{h1} + F_{h3})f = 2F_{h1}f + \frac{P}{\sin\frac{\theta}{2}}f \tag{10}$$

In the formula,
F_t—friction between the bilateral sides of the cogging and the cotton stalk, N;
θ—angel of cogging, °;
f—static friction coefficient between the cotton stalk and tooth plate;
F_{h3}—force exerted by cotton stalk under deformation on bilateral sides of cogging, N;
F_{h4}—force exerted by cotton stalk under deformation on bilateral sides of cogging, N.

From Formulas (7)–(10), it can be concluded that under a certain value of Δy, the smaller the value of a is, the larger the value of P is. In the case of the same deformation of the cotton stalk, the lower the height of the clamped cotton stalk, the greater the force between the tooth plate and the cotton stalk, and the greater the friction that can be provided.

2.3.4. Cotton Stalk Pulling Process: Establishment of Mechanical Model

Generally, the forces exerted on the cotton stalk root by the soil act to resist the stalk-pulling process; this force system is quite complex. Therefore, for the purpose of simplification, the soil force system was modeled to have a resultant force, which is referred to as the cotton stalk soil resistance and denoted as F_b. F_b was determined to occur along the length of the cotton stalk.

As the cotton stalk is being pulled, its initial state of static equilibrium is disrupted by sudden movement, which involves an acceleration component. This acceleration was set as a_1, the force producing the acceleration, which is equivalent to the resultant force P, was set as F_a, the frictional force was expressed as F_t, the extrusion force was set as F_s, and the soil resistance was set as F_b. Additionally, the angle between the force vector F_a

and the surface of the toothed plate was set as γ. A schematic illustrating the force system associated with this state of transition is presented in Figure 11.

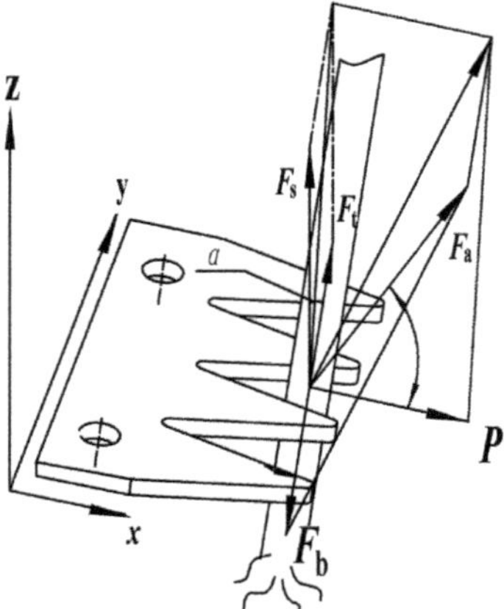

Figure 11. Schematic of forces acting on a cotton stalk transitioning from its initial static equilibrium to a dynamic state.

From the stress analysis of the cotton stalk in Figure 11, Formulas (11) and (12) can be obtained according to the dynamic knowledge such as D'Alembert's principle while taking the acceleration into consideration:

$$F_a \sin\gamma + F_b = F_s \cos\alpha + F_t \tag{11}$$

$$F_a = a_1 m = \frac{v_2 - v_1}{t_1} m = \frac{v_2}{t_1} m \tag{12}$$

By (10) and (11), (13) can be obtained:

$$m a_1 \sin\gamma + F_b = F_s \cos\alpha + 2F_{h1} + \frac{P}{\sin\frac{\theta}{2}} f \tag{13}$$

By (8), (12)–(14) can be obtained:

$$m\frac{v_2}{t_1} \sin\gamma + F_b = F_s \cos\alpha + 2F_{h1} f + \frac{6\Delta y J e l f}{ab(a^2 + b^2 - l^2)\sin\frac{\theta}{2}} \tag{14}$$

In the formula,

a_1—the acceleration (m/s^2) (in actual conditions, the acceleration of cotton stalk a1 is a variable, which is related to the position and angle of the cotton stalk clamping. It is simplified for analysis here);

m—weight of cotton stalk (kg);

v_2—speed of the cotton stalk when being pulled out (m/s);

v_1—speed of cotton stalk at the static state (m/s, set as 0 m/s here);

t_1—time for the pull-out of the cotton stalk (s);

F_a—force-producing acceleration, equal to the combined force of the tooth plate to the cotton stalk and the soil to the cotton stalk (N).

Formula (14) was used to model and analyze the extraction, missed extraction, and fracture states of the cotton stalk pulling process. The following conclusions were made:

F_{h1} generation is mainly dependent on the collision strength between the toothed plate and cotton stalk; more specifically, the elastic recovery of the cotton stalk is dependent on the speed of the collision. Within a certain range, a higher speed corresponds to larger deformation of the cotton stalk. Additionally, the forces associated with deformation recovery are strong and serve to facilitate the cotton stalk clamping process. Upon making contact, when the effects of the collision between the toothed plate and cotton stalk cannot be endured by the cotton stalk, phloem rupture, a missed extraction, or even cotton stalk fracture can occur, as shown in Figure 12c. Under these conditions, the cotton stalk is broken, and the root remains in the soil. Thus, the rotational speed of the V-shaped roller should not be too low or too high.

Figure 12. Conditions of the cotton stalk epidermis following a stalk extraction attempt. (**a**) Example of the epidermis following successful cotton stalk extraction. (**b**) Example of the epidermis of a cotton stalk that failed to be extracted. (**c**) Example of the epidermis of broken cotton stalks.

F_s is the upward force of the toothed plate that acts against the cotton stalk. The effects of F_s are mainly concentrated on the phloem of the cotton stalk. Thus, the magnitude of F_s is primarily dependent on the characteristics of the cotton stalk phloem; this may be a significant reason for considerable differences in the stalk-pulling effects between autumn and spring.

Conclusions ① and ② above could be used to explain the phenomenon depicted in Figure 10. Figure 12a shows an example of the state of the cotton stalk epidermis at the V-shaped toothed plate clamping position after a cotton stalk was successfully extracted in autumn. The epidermis was obviously ruptured; however, in terms of length, the rupture was not remarkably substantial. (Most ruptures were less than 25 mm in length according to the results of rough statistics.) Figure 12b shows the state of the epidermis of a cotton stalk that was unsuccessfully extracted. It can be seen that a large area of the epidermis was destroyed, and some side branches were broken. This state is referred to as the missed extraction state.

In Formula (14), a represents the ground clearance of the contact point between the toothed plate and cotton stalk. Theoretically, a lower value of a should correspond to a better cotton stalk-pulling outcome. θ is the angle between the toothed plates. When P is constant, a lower value of θ corresponds to better gripping force.

Following the analysis of these results, successful cotton stalk extraction has been determined to be associated with a relative displacement between the V-shaped toothed plate and cotton stalk that does not exceed 25 mm. This was determined based on whether the extrusion force F_s exceeded the endurable limit of the cotton stalk. Thus, the extrusion force F_s should be maintained at an appropriate value during the stalk-pulling process. Furthermore, according to Formula (13), under the condition that the soil resistance F_b is constant, the cotton stalk pulling outcome can be improved by reducing the speed v_2, the stalk-pulling height a, or the cogging angle θ.

$$m\frac{v_2\downarrow}{t_1}\sin\gamma + F_b = F_s\cos\alpha\downarrow + 2F_{h1}f\uparrow + \frac{6\Delta yJelf}{ab(a^2\downarrow + b^2 - l^2)\sin\frac{\theta\downarrow}{2}}\uparrow \tag{15}$$

The left side of Formula (15) describes the forces of the soil acting on the cotton stalk root, as well as the force associated with cotton stalk acceleration, which has been set as a passive load. The right side of the equation describes the force exerted by the toothed plate on the cotton stalk; it has been set as the active force. Among all the parameters influencing the active force, the bilateral forces F_{h1} and F_{h2} applied to the clamping tooth by the cotton stalk during the deformation recovery stage of the phloem, the ground clearance α, and the cogging angle θ are controllable. With the exception of the speed of the toothed roller n that influences F_{h1} and F_{h2} that is adjustable, the remaining parameters, such as the friction coefficient f, elastic modulus E, cotton stalk height l, and extrusion force F_s cannot be controlled. Thus, in this study, the speed of the toothed roller n, ground clearance α, and cogging angle θ were determined to be test factors.

2.4. Field Tests

2.4.1. Test Equipment and Materials

To determine the working parameters for toothed rollers, a V-shaped toothed roller stalk-pulling device bench was developed, and field tests were conducted (Figure 13). The test site was the Demonstration Base of New Cotton Variety K836 for Simplified and High-Yield Cultivation in Binzhou City, Shandong Province. The test was conducted in spring. The cotton variety is China Cotton Institute—50 (CCRI50). The cotton stalk rows were spaced at 76 cm, and the target plant spacing was 20 cm, with an average of 30 cm. The height of the cotton stalks typically ranged between 95 and 105 cm, and the diameter of the cotton stalks ranged from 10 to 22 cm. The average soil firmness is 672.3 KPa and the average soil moisture content is 27.4%.

Figure 13. Field test photo: 1. Reel wheel 2. V-shaped tooth roller 3. Torque transducer 4. Cotton stalks.

2.4.2. Test Factors and Levels

The test factors were the speed n of the toothed roller, ground clearance α, and cogging angle θ. According to the results of calculations, the speed of the V-shaped toothed roller varied between 176 and 500 r/min; thus, the test speeds were preliminarily determined to be 200, 300, and 400 r/min. The results of the theoretical analysis indicated that operating the toothed roller at a higher ground clearance yields better performance; thus, α was assigned values of −20 mm (i.e., 20 mm into the soil), 20 mm, and 60 mm in this study. Based on the optimized above-described structural design, the cogging angle θ should range between 25° and 38.6°. However, to account for the coefficient of friction between the cotton stalk and steel plate and to minimize processing difficulty, the maximum cogging angle was set to 40° for the field test. In the case of a certain thrust, the smaller the cogging angle, the greater the clamping force on the cotton stalk. However, there are two uncertainties: 1. Too small an angle leads to too long tooth height, which easily causes the insufficient strength of tooth plate structure; 2. Excessive clamping force will easily cause the cotton stalks to fall off easily. Thus, the cogging angles implemented in the field test were 25°, 32.5°, and 40°. The test levels are summarized in Table 2. The L_9 (3^4) orthogonal experiment was designed to have nine test groups. The experiment was repeated three times for each group, and the mean value was calculated.

Table 2. Factors and levels of orthogonal experiment.

Levels	Factors		
	A. Ground Clearance/(mm)	**B. Rotation Speed/(r/min)**	**C. Cogging Angle/(°)**
1	−20	200	25
2	20	300	32.5
3	60	400	40

2.4.3. Test Methods

The test was carried out in accordance with the standard GB/T8097-2008 [26]. To ensure stable operation of the equipment and minimize error, the puller machine was allowed to adjust its operating posture by traveling a distance of 20 m before data were collected. To facilitate the process of data collection, two rows of cotton stalks were harvested each time. The data collection region was 30 m long, corresponding to approximately 200 cotton stalks. The evaluation index for the test was the removal rate; it was determined as follows:

$$y = \frac{M_b}{M_z} \times 100\% \tag{16}$$

In the formula, y—the removal rate of cotton stalks, %;
M_b—the number of cotton stalks pulled out, plant;
M_z—the sum of cotton stalks, plants.

3. Results and Discussion

The test plan and results were shown in Table 3 (A, B, and C are the ground clearance, the rotation speed, and the cogging angle, respectively).

Table 3. Test results and analysis.

Experiment No.		Factors				y
		A	B	C	D (Emptyrow)	Removal Rate/%
1		1	1	1	1	94.76
2		1	2	2	2	97.34
3		1	3	3	3	93.29
4		2	1	2	3	89.58
5		2	2	3	1	92.77
6		2	3	1	2	90.9
7		3	1	3	2	82.53
8		3	2	1	3	85.71
9		3	3	2	1	80.93
Average removal rate/%	k_1	95.13	88.96	90.46	89.49	
	k_2	91.08	91.94	89.28	90.26	
	k_3	83.06	88.37	89.00	89.53	
	R	12.07	3.57	1.46	0.77	
Influence order: $A > B > C$						

According to the results of numerical analysis for the range R for each factor in Table 2, it can be seen that the order of significance in terms of the influence of each factor on the evaluation index (i.e., the removal rate) was $A > B > C$. The ground clearance most significantly influenced the removal rate, followed by the rotational speed, and the cogging angle. To optimize the combination of these factors to obtain the highest cotton stalk removal rate, the removal rate was plotted as a function of the averaged test factor results presented in Table 2; the resulting graph is shown in Figure 14. It can be ascertained from

Figure 12 that the combination of factors that yielded the highest removal rate was $A_1B_2C_1$, i.e., the combination of a ground clearance of −20 mm, a rotational speed of 300 r/min, and cogging angle of 25°.

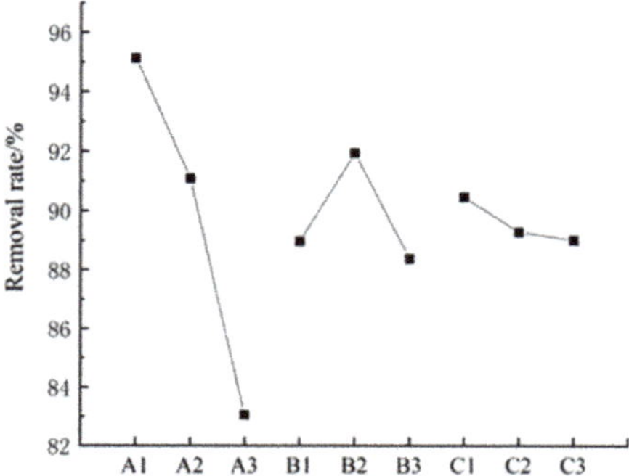

Figure 14. Relationship between factor level and removal rate.

Minitab software was used as a platform to analyze the variance of the results of the orthogonal experiment and quantify the respective influences of the three factors (i.e., cogging angle, rotational speed, and ground clearance) on the removal rate; the results are summarized in Table 4.

Table 4. Analysis of variance for orthogonal experiment results.

Source	Sum of Squares	Degree of Freedom	Sum of Mean Squares	F Value	Significant Level *p*
A	226.568	2	113.284	200.97	0.005 **
B	21.962	2	10.981	19.48	0.049 *
C	2.296	2	1.148	2.04	0.329
Pure error	1.127	2	0.564		
Cor total	251.954	8			

Note: $p < 0.01$ (highly significant, **); $p < 0.05$ (significant, *). Based on the analysis of variance results presented in Table 3, it can be ascertained on the basis of FA > FB > FC that A had an extremely significant influence on the removal rate, B had a significant influence, and C had minimal influence. Such results are consistent with the range analysis results. In addition, it can be also ascertained from Table 3 that the sum of squares error was much less than that of factor A, factor B, or factor C, indicating that the correlations between test factors did not considerably affect the evaluation index.

Compared with the knife-roller type stalk pulling operation equipment, it has the advantages of not entering the soil and reducing energy consumption on the basis of non-alignment operation. Tests have proved that in the Binzhou area, cotton stalks with larger diameters can be effectively pulled out. However, since this type of stalk-pulling device has many V-shaped structures on its stalk-pulling rollers, once cotton stalks are entangled, the cotton stalks will be damaged. The damaged cotton stalk cannot fall off, which leads to its working failure; secondly, this device has been tested in a densely planted cotton area. When the average root diameter of cotton stalks is about 10 mm, the stalk-pulling effect is very poor, so this device still needs further in-depth research. The most foreign equipment introduced in the literature is the double-roller type stalk pulling equipment. Our team has used the domestically produced double-roller equipment to conduct experiments in the Binzhou area, and the effect is not ideal. The variety and planting mode of cotton stalks have a great influence on the mechanical properties of cotton stalks. Up to now, there is no stalk-pulling equipment that can adapt to cotton stalks in different regions. Therefore, further research is needed to solve the issue of stalk-pulling technology.

4. Conclusions

1. Combining parameters such as cotton diameter, plant distance, V-shaped tooth speed, and friction coefficient between cotton and tooth plate, it is determined that the suitable angle range of the groove was 25.6° to 38.6°.
2. On the basis of the cotton stalk pulling model, acceleration was further introduced. A detailed theoretical explanation was given for the state of missing or breaking stalks during pulling.
3. The orthogonal test results demonstrated that the ground clearance of the V-shaped toothed roller had an extremely significant effect on the removal rate. The lower the ground clearance, the better the soil penetration effect. The tooth groove angle has no significant effect on the extraction rate. The effect test showed that the primary and secondary order of the influence of various factors on the cleaning rate was ground clearance > speed > cogging angle. Additionally, under the condition of not considering the extraction energy consumption, the more suitable combination of mechanism parameters was ground clearance −20 mm. The speed of the V-shaped toothed roller was 300 r/min, and the angle of the tooth groove was optional between 25° and 40°.

Author Contributions: Conceptualization, Z.W. and W.Z.; methodology, Z.W., W.Z. and J.F.; software, Z.W. and W.Z.; validation, Z.W., W.Z., J.F. and H.X.; formal analysis, Z.W. and W.Z.; investigation, Z.W., W.Z., J.F. and H.X.; data curation, Z.W., W.Z. and J.F.; writing—original draft preparation, Z.W., W.Z. and J.F.; writing—review and editing, J.F, Y.Z. and M.C.; visualization, Z.W., Y.Z. and M.C.; supervision, Y.Z. and M.C.; funding acquisition, M.C. All authors have read and agreed to the published version of the manuscript.

Funding: This work was financially supported by the National Natural Science Foundation of China (Grant No. 51505242) and the Innovative Research Group of Agricultural Production Waste Resource Utilization Equipment.

Institutional Review Board Statement: Not applicable.

Data Availability Statement: Data are contained within the article.

Acknowledgments: The authors thank the editor and anonymous reviewers for providing helpful suggestions for improving the quality of this manuscript.

Conflicts of Interest: The authors declare no conflict of interest.

References

1. Zhang, Y.; Ding, Y.; Li, J.; Yang, Y. Current Situation and Pest Problems of Cotton Stalk Returning Machinery in China. *Farm Mach.* **2023**, *3*, 79–82.
2. Wang, Y.; Liu, Z.; Cheng, Y. Research on the low-carbon transformation of Xinjiang green and renewable energy cotton stalks. *Mod. Bus. Trade Ind.* **2023**, *44*, 18–20.
3. Zuo, X.; Bi, Y.; Wang, H.; Gao, C.; Wang, L.; Wang, Y. Estimation and Suitability Evaluation of Cotton Stalk Resources in China. *China Popul. Resour. Environ.* **2015**, *25*, 159–166.
4. Zhang, Z.; Qin, C.; Wang, L.; Zhou, L. Cotton stalk Resources Research Status. *Xinjiang Agric. Mech.* **2014**, *5*, 21–23.
5. Shen, M.; Zhang, G.; Zhou, Y.; Zhou, M. The Review of the Mechanization Technology for Cotton Stalk Harvesting in our Country. *J. Agric. Mech. Res.* **2009**, *31*, 7–11.
6. Dong, S.; Wang, F.; Qiu, Z.; Sun, Y. Design and Experiment of Self-propelled Cotton-stalk Combine Harvester. *Trans. Chin. Soc. Agric. Mach.* **2010**, *41*, 99–102.
7. Chen, M.; Wang, Z.; Qu, H.; Chen, Y.; Liu, K.; Song, D. Bending and tensile properties tests of the cotton-stalk. *J. Chin. Agric. Mech.* **2015**, *36*, 29–32. [CrossRef]
8. Chen, M.; Song, D.; Wang, Z.; Wang, R.; Liu, K.; Chen, Y. Research on the Cotton-stalk Uprooting Resistance. *J. Agric. Mech. Res.* **2016**, *38*, 64–68.
9. Li, Y.; Zhang, G.; Zhou, Y.; Ji, W.; Li, Z.; Zhang, Y.; Zai, K. Design and field experiment of drawing resistance measurement system for cotton stalk. *Trans. Chin. Soc. Agric. Eng.* **2013**, *29*, 43–50.
10. Zhang, G.; Li, Y.; Li, Z.; Zhang, Y.; Zhai, K. Measuring System of Cotton Stalk Real-Time Pull Force in The Field Based on LabVIEW. In Proceedings of the 2014 ASABE International Meeting, Montreal, QC, Canada, 13–16 July 2014; ASABE: St. Joseph, MI, USA, 2014; p. 1.

11. Demian, T.F. The Pull and Lift Required to Remove Cotton Stalks in the Sudan. *Exp. Agric.* **1978**, *14*, 129–135. [CrossRef]
12. Tang, Z.; Han, Z.; Gan, B.; Bao, C.; Hao, F. Design and Experiment on Cotton Stalk Pulling Head with Regardless of Row. *Trans. Chin. Soc. Agric. Mach.* **2010**, *41*, 80–85.
13. Chen, M.; Zhao, W.; Wang, Z.; Liu, K.; Chen, Y.; Hu, Z. Operation Process Analysis and Parameter Optimization of Dentate Disc Cotton-stalk Uprooting Mechanism. *Trans. Chin. Soc. Agric. Mach.* **2019**, *50*, 109–120.
14. Ma, J.; Jian, S.; Zhou, H. Design of Zigzag Disk Harvester of Cotton Stem. *Agric. Equip. Veh. Eng.* **2010**, *8*, 3–5.
15. Ying, Y.; Cao, S.; Wang, M.; Lu, Y. Design and test of roller cotton stalk plucking device. *Xinjiang Farm Res. Sci. Technol.* **2018**, *41*, 22–24.
16. Wang, S.; Liu, N. Research on the tray type cotton-wood harvesting machine. *Cereals Oils Process.* **1994**, *6*, 23–25.
17. Guo, Z.; Shi, J.; Kang, X. The Resistance Analysis of Roller-cotton Stalks. *J. Agric. Mech. Res.* **2009**, *31*, 37–39.
18. Zhang, A.; Wang, Z.; Liu, K.; Liu, Y.; Wang, H.; Zhu, D. Design and experiment of the key parts with cotton-stalk on baling combine harvester. *J. Chin. Agric. Mech.* **2016**, *37*, 8–13. [CrossRef]
19. Liao, P.; Liu, K.; Chen, M.; Wang, Z.; Zhang, A. Design and experiment of a roller type cotton stalk cutting mechanism. *J. China Agric. Univ.* **2018**, *23*, 131–138.
20. Zhang, A.; Liu, K.; Wang, Z.; Liu, Y.; Wang, H. Design and Trial on Cotton Stalk Pulling Head with Multi-roller. *J. Agric. Mech. Res.* **2016**, *38*, 91–95.
21. Wang, Z.; Geng, D.; Meng, P.; Jiang, C.; Yan, B. Cutter Installation Performance Impact on the Harvest of Cotton Stalk and Experiment Research. *J. Agric. Mech. Res.* **2015**, *37*, 22–26.
22. He, X. Design and Experiment Research of Cotton Stalk Pulling-Out Mechanism. Master's Thesis, Hunan Agricultural University, Changsha, China, 2016.
23. Dai, Z. Design and Research on Cotton Stalk Drawing Institutions of Pull-out Cotton Stalk Mehanism. Master's Thesis, Hunan Agricultural University, Changsha, China, 2015.
24. Yin, S. The Design Improvement of Cotton Stalk Crushing and Rubbing Machine. Master's Thesis, Northwest A&F University, Xianyang, China, 2016.
25. Li, X.; Qian, L.; Wang, C. Elastic collision is understood from the point of view of force and motion. *J. Phys. Teach.* **2018**, *36*, 60–61.
26. *GB/T 8097*; Equipment for Harvesting-Combine Harvesters-Test Procedure. Standardization Administration of China: Beijing, China, 2008.

agriculture

Article

Detection of Famous Tea Buds Based on Improved YOLOv7 Network

Yongwei Wang, Maohua Xiao, Shu Wang, Qing Jiang, Xiaochan Wang and Yongnian Zhang *

Engineering College, Nanjing Agricultural University, Nanjing 210031, China; 9203010615@stu.njau.edu.cn (Y.W.)
* Correspondence: hczyn@njau.edu.cn

Abstract: Aiming at the problems of dense distribution, similar color and easy occlusion of famous and excellent tea tender leaves, an improved YOLOv7 (you only look once v7) model based on attention mechanism was proposed in this paper. The attention mechanism modules were added to the front and back positions of the enhanced feature extraction network (FPN), and the detection effects of YOLOv7+SE network, YOLOv7+ECA network, YOLOv7+CBAM network and YOLOv7+CA network were compared. It was found that the YOLOv7+CBAM Block model had the highest recognition accuracy with an accuracy of 93.71% and a recall rate of 89.23%. It was found that the model had the advantages of high accuracy and missing rate in small target detection, multi-target detection, occluded target detection and densely distributed target detection. Moreover, the model had good real-time performance and had a good application prospect in intelligent management and automatic harvesting of famous and excellent tea.

Keywords: famous and excellent green tea; bud detection; improved YOLOv7 algorithm; attention mechanics

1. Introduction

China is the most important tea producer in the world, with a tea plantation area of 305.9 million hectares and annual output of 260.9 million tons, accounting for 62.6% and 50.1% of the global tea plantation area and output, respectively [1]. Famous and excellent tea is favored by the people because of its high drinking value and economic value [2]. High-quality tea usually has strict requirements on tenderness and number of leaves, and different grades of high-quality tea often have different requirements on the number of leaves and buds, usually famous and excellent tea picking only pick a bud and a leaf [3]. Famous and excellent tea picking has strong seasonality, short picking cycle and high labor intensity, which is a labor-intensive operation. With the rapid development of the tea industry, the contradiction between the timeliness of famous and excellent tea picking and the shortage of labor force of manual picking is increasingly prominent [4]. In recent years, some famous and excellent tea picking equipment was used for picking tea gardens, but it has some shortcomings, such as imprecise mechanical picking technology and poor quality mechanical tea [5]. In the complex environment of tea gardens, the rapid and accurate detection of young leaves of famous and excellent tea based on vision is the key task to realize automatic picking of famous and excellent tea.

Research on the bud detection of famous and excellent tea is mainly divided into two methods. The first method is the segmentation method based on the physical characteristics of famous and excellent tea [6–10], which mainly takes the shape, color, texture and other physical characteristics of famous and excellent tea as the basis for identifying and segmenting young leaves. Then, traditional methods such as threshold segmentation and watershed segmentation are used to separate and extract the tender leaves from the complex environment. This method is greatly affected by the environment and has a small scope of application. The other is the detection method based on neural network [11–14]. By training the marked famous and excellent tea dataset, the weight model is obtained and then used

Citation: Wang, Y.; Xiao, M.; Wang, S.; Jiang, Q.; Wang, X.; Zhang, Y. Detection of Famous Tea Buds Based on Improved YOLOv7 Network. *Agriculture* **2023**, *13*, 1190. https://doi.org/10.3390/agriculture13061190

Academic Editors: Hongbin Pu and Filipe Neves Dos Santos

Received: 21 April 2023
Revised: 19 May 2023
Accepted: 2 June 2023
Published: 3 June 2023

to detect the tender buds. At present, it is widely used in related agriculture and agronomy fields. YOLO (you only look once) is a target detection algorithm, which has high precision and high efficiency, and it can directly predict the location and attribute of the target in the whole image. Marco Sozzi et al. [15] applied target detection to yield prediction of white grape, compared detection effects of YOLOv3, YOLOv3-Tiny, YOLOv4, YOLOv4-Tiny, YOLOv5x and YOLOv5s, and finally found that the YOLOv5x model, considering bunch occlusion, was able to estimate the number of bunches per plant with an average error of 13.3% per vine. The YOLOv4-tiny model has a better combination of accuracy and speed, which should be considered for real-time grape yield estimation. YOLOv3 model is affected by a false positive–false negative compensation, which decreases the RMSE. Angelo Cardellicchio et al. [16] used the YOLOv5 model to test the phenotypic traits of tomato plants. The train used a challenging dataset acquired during a stress experiment conducted on multiple tomato genotypes, considering the particular challenges of the input images in terms of object size, similarity between objects and their color. The results demonstrated that the models achieve relatively high scores in identifying nodes, fruit and flowers. Dandan Wang et al. developed an accurate apple fruitlet detection method with small model size based on a channel pruned YOLOv5s deep learning algorithm [17]. The experimental results showed that the channel pruned YOLOv5s model provided an effective method to detect apple fruitlets under different conditions. The recall, precision, F1 score and false detection rate were 87.6%, 95.8%, 91.5% and 4.2%, respectively; the average detection time was 8 ms per image; and the model size was only 1.4 MB. It can be used to help growers optimize their orchard management. Compared with the traditional physical method, the deep learning algorithms have the advantages of high identification accuracy, strong robustness and less influence by environmental factors, so it is appropriate for the detection task of famous and excellent tea.

However, with the increase in researchers' attention, Wu et al. [18] found that YOLO has the disadvantage of insufficient frame positioning and difficult to distinguish overlapping detection objects. Famous and excellent tea has a small bud shape and high density, which also have the same problems in detection, and the emergence of attention mechanism can effectively settle the above problems. The attention mechanism can obtain a weight through module calculation and multiply it with input information to achieve the purpose of focusing on important information with high weight and ignoring irrelevant information with low weight. It directly establishes the dependency relationship between input and output without cycling, making the parallelization degree enhanced, the running speed greatly improved and the weight automatically adjusted. So that important information can be selected in different situations, it has higher scalability and robustness. It achieved good results in the detection of famous and excellent tea and other agricultural fields, and it was widely used in the optimization of the model. Liu Tianzhen et al. [19] added SE Block to the YOLOv3 network, and compared with the YOLOv3 model, the F1 score increased by 2.38 percentage points and mAP increased by 4.78 percentage points. Yang et al. applied CBAM Block to wheat detection, and the results showed that the model could effectively overcome the field environmental noise and achieve the accurate detection and counting of wheat ears with different density distributions [20]. The average accuracy of wheat ears detection increased to 94%, 96.04% and 93.11%, respectively. To compare the effect of SE, CBAM and ECA attention modules on the network in the YOLO v5 network model for the posture detection of meat geese, Liu Yingying et al. [21] proved that YOLOv5+ECA had better stability and was more suitable for the posture detection of meat geese in complex scenarios in farms. Fang Mengrui et al. added the CBAM module to YOLOv4-tiny adopted bidirectional feature pyramid network (BiFPN) to integrate feature information of different scales. It was found that the F1 score of the improved Yolov4-Tiny-tea model was 12.11, 11.66 and 6.76 percentage points higher than that of the YOLOv3, YOLOv4 and YOLOv5l network models, respectively [22]. Fu et al. introduced the channel attention-asymmetric spatial pyramid pool (CA-ASPP) module to improve the detection of weak and weak pod targets [23]. The precision of the improved YOLOv5 model increased by about 6%, and the precision of POD number in the 200 soybeans population reached 88.14%. Bao et al. [11] proposed an improved AX-RetinaNet target detection and recognition network for automatic detection and recognition of tea diseases in natural scene images. AX-RetinaNet took the improved X-module multi-scale feature fusion module and added SE Block in the network. Compared with the original network, the mAP, recall rate and recognition accuracy increased by nearly 4%, 4% and nearly 1.5%, respectively. However, it was also found that adding the attention mechanism had the opposite effect for some networks, such as SSD and EfficientNet.

Through research and experiments, it can be found that SE Block [24], CBAM Block [25], ECA Block [26] and CA Block [27] have different degrees of improvement in the detection of different crops, and the improvement effect is related to the position in the model. However, there was no research to compare the effects of four kinds of attention mechanism modules in different positions in the YOLOv7 network on parameters such as the recognition accuracy rate and recall rate of famous and excellent tea.

Therefore, this study focused on the influence of SE Block, ECA Block, CBAM Block and CA Block on the recognition accuracy, recall rate and F1 score in different positions of YOLOv7 network for famous and excellent tea detection. The purpose of this study was to select the most suited network for the detection of famous and excellent tea by comparison.

2. Materials and Methods

2.1. Data Acquisition

Longjing 43 Tea Garden in Juyuanchun, Yangzhou, has a large planting area, standard tea garden and relatively flat shed surface, which meet the experimental requirements. Therefore, the object of our study was Longjing 43 in famous and excellent green tea. In this research, we collected images of famous tea buds in Juyuanchun tea garden in Yangzhou. In order to improve the fitness of the model and the environment, we chose to face the tender buds at a total of 5 shooting angles of $\pm 30^{\circ}$ and $\pm 60^{\circ}$ with the tender buds, respectively, during the shooting, as shown in Figure 1.

Figure 1. Multi-view bud image. (**a**) front view, (**b**) rotate counterclockwise by 30 degrees, (**c**) rotate counterclockwise by 60 degrees, (**d**) turn it clockwise by 30 degrees, (**e**) turn it clockwise by 60 degrees.

Due to the strong seasonality of the picking of famous and excellent tea, the quality of the dataset was the best in a few days before Tomb-Sweeping Day. Therefore, the shooting time of this time was 2 April 2023, and considering the influence of light on the dataset, a total of 1049 images were taken from 9 to 11 a.m. and 3 to 5 p.m., respectively. The resolution of the photos was 3904×2928. Figure 2 shows the overall layout of the tea garden, and Figure 3 shows the images of the dataset taken.

Figure 2. Overall layout of tea garden.

Figure 3. Dataset image.

2.2. Data Enhancement

In order to improve the robustness of the trained model, we used different processing methods to make data enhance for the famous and excellent tea image. In the research, we adopted the following methods:

(1) First, we adjusted the brightness of the image. To be specific, we raised the brightness of the image to 1.3 times and decreased it to 0.7 times compared with the original image, respectively. During our shooting time, the way of dealing with brightness could reflect the change from brightest to darkest during mechanical picking. Through this operation method, our model will be more suitable for the complex tea garden environment with changeable light;

(2) Then, we adjusted the contrast of the image. To be specific, the contrast of the images was increased by 1.2 times and weakened by 0.8 times, so that the sharpness, gray level and texture details of the famous and excellent tea images could be better expressed [20];

(3) Finally, we rotated the image taken. We thought rotation of 30 degrees can better reflect the detection of famous and excellent tea when the machine picks. This operation can enhance the adaptability of the detection model to shoot from different angles.

In the dataset, 200 images were randomly selected each time for brightness enhancement, brightness reduction, contrast enhancement, contrast reduction, horizontal flip and random angle flip. After processing, the number of datasets reached 2049. Figure 4 reflects the results of data enhancement.

Figure 4. Data enhancement. (**a**) Original image, (**b**) brightness enhanced by 30%, (**c**) reduce brightness by 30%, (**d**) contrast enhancement 20%, (**e**) contrast reduction by 20%, (**f**) rotate counterclockwise by 30 degrees.

2.3. Data Annotation

There are mainly two labeling methods for famous and excellent tea. One is to label the front view of famous and excellent tea squarely, as shown in Figure 5, and the other is to label famous and excellent tea images with the front view, side view and top view, respectively, as shown in Figure 6. Both methods can only be picked when the front view images are detected in the picking process. In addition, multi-angle labeling will increase the difficulty of control in mechanical picking and reduce the detection rate, which is unnecessary in the picking process of famous and excellent tea. Therefore, we chose the first labeling method, which only labeled the front view images. We annotated the dataset using the software of labeling. Specifically, the tender bud in the image was selected with a square frame and the label of tea was added. The format of the annotation file was selected in PascalVOC format, and the corresponding xml file was generated after saving. The software will save the image size, label name, target location and other information, as shown in Figure 5.

Figure 5. Front view of labeling.

Figure 6. Multi-view of labeling. (**a**) Red reflects the annotation of side view, cyan reflects the annotation of front view. (**b**) Yellow reflects the annotation of top view.

2.4. Excellent Tea Detection Algorithm

2.4.1. Yolov7 Algorithm

Figure 7 shows the structure diagram of YOLOv7 network. It was mainly composed of input, backbone feature extraction network, enhanced feature extraction network and head. The input module scales the input image to a unified pixel size to reduce the amount of computation [28]. The backbone module is composed of CBS, ELAN and MP modules. The CBS module is composed of batch normalization layer (BN), conv and silu activation function, and it is used to extract multi-scale information of images. ELAN module is composed of multi-branch convolution. It improves the learning ability of the network without destroying the original gradient path. MP module integrates MaxPool and convolution two dimensions of information, improving the feature extraction ability of the network.

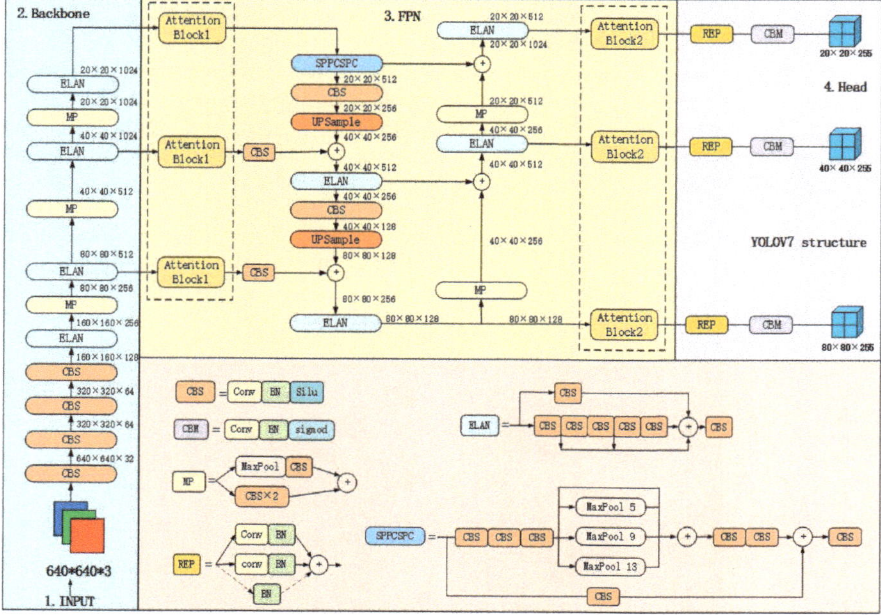

Figure 7. Yolov7 network structure diagram.

The enhanced feature extraction module is composed of path aggregation feature pyramid network (PAFPN) structure. By introducing a bottom-up path, it was easier for the information from the bottom layer to be transferred to the top layer, thus realizing the efficient fusion of features at different levels, improving the accuracy of positioning information. In the SPPCSPC structure, SPP module, composed of CBS and 4 different sizes of maximum pooling, can better distinguish different sizes of the target through different sizes of maximum pooling; the CSP module, composed of two parts, was used for conventional processing and the above part was used for SPP module processing. Finally, the two modules merged together, which can reduce half of the amount of computation, making the processing speed faster, obtaining higher precision.

The head module was composed of REP structure and CBM structure. The REP block (RepVGG Block) was composed of two parallel convolutional layers (Conv), batch normalization layers (BN) and one batch normalization layer (BN) path. The CBM structure was composed of the convolutional layer (Conv), batch normalization layer (BN) and sigmoid activation function. The REP (RepVGG Block) structure adjusted the number of image channels for 3 features of different scales output by PAFPN, including P3, P4 and P5, and it was then used to predict the confidence, category and anchor frame through CBM structure.

2.4.2. Introduction of Attention Mechanism

The attention mechanism can make the network pay more attention to the bud target. It was found that the attention mechanism modules, such as SE Block, ECA Block, CBAM Block and CA Block, play a significant role in improving the model recognition effect. For the input features, CBAM module first learns the weight information of each channel through a shared multilayer perceptron (MLP) and sigmoid function, and then, through a hollow convolution [20] with convolution kernel of 3 × 3 and expansion coefficient of 2 and sigmoid function, learns the weight information of each point in the space. SE Block obtains the channel weight information after passing through the full connection layer twice and sigmoid function. The ECA block changes the two fully connected layers into one-dimensional convolution, and obtains the channel weight information after passing the sigmoid function, which has a good ability to obtain cross-channel information. The CA block divides channel attention into two 1-dimensional feature coding processes, which aggregates features along two spatial directions, respectively. In this way, remote dependencies can be captured along one spatial direction. At the same time, accurate location information can be retained along the other spatial direction, and then, the generated feature map was encoded into a pair of direction-aware and position-sensitive attention maps, which can be applied complementary to the input feature map to enhance the representation of the object of attention. The schematic diagram of the network structure of the four modules is shown in Figure 8.

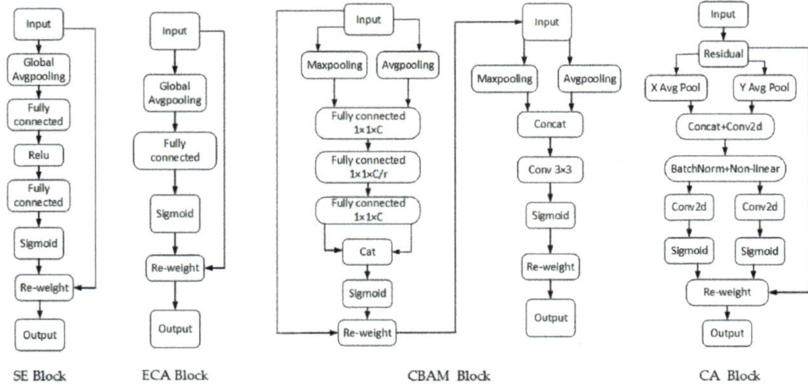

Figure 8. Attention mechanism module.

In order to better adapt to the complex scene of the tea garden, four attention mechanism modules were added to the front and back positions of the enhanced feature extraction network, named Attention Block1 (AB1) and Attention Block2 (AB2), as shown in Figure 7. In the experiment, we compared the recognition effects of adding four attention modules in the above position.

2.5. Training Environment and Parameter Configuration

2.5.1. Experimental Platform and Environment Configuration

This research was based on Pytorch deep learning framework and used cloud server AI-GCI-02PM3 for training. The operating system was Linux, Ubuntu18.04, CPU was Intel 8358P@2.6/10 core, and graphics card (GPU) was NVIDIA GeForce RTX 3090. The video memory was 24 GB, and the deep learning Conda environment was configured as pytorch-1.8.0+python3.8+cuda11.1.

2.5.2. Training Parameter Settings

The number of images in the pre-training stage was 631, and the labels number of buds was 1723 during the annotation. The recognition effect of YOLOv7 model was compared with 4 different attention mechanisms, which were added at different positions. According to the ratio of training set: verification set = 9:1 and training verification set: test set = 9:1, the dataset was randomly divided into training set, verification set and test set. In the improved YOLOv7 network training process, we proceeded with 50 generation freeze training, 300 generation thaw training in turn. In parameter, batch (batch size) was set to 8, the initial learning rate set to 0.001, the minimum learning rate set to 0.00001, using the sgd optimizer, and momentum parameter was set to 0.937. We used the cosine annealing function to dynamically reduce the learning rate and turned off the Mosaic data enhancement method. We set the confidence level to 0.5 and the intersection ratio size used for non-maximum suppression to 0.3. The loss function consisted of three parts: Reg (rectangular box regression prediction) part, 0bj (confidence prediction) part and Cls (classification prediction) part. The Reg part uses CIOU Loss, and the 0bj part and Cls part uses BCE Loss (cross entropy loss).

The number of images in the formal training stage was 2049 and the labels number of buds was 6055 during the annotation. The parameter settings were the same as those in the pre-training stage. We selected the YOLOv7+CBAM model and the YOLOv7+ECA model, which had the preferable recognition effect in the pre-training process, and compared them with the recognition effect of YOLOv7 model in the formal training stage.

2.6. Evaluation Index

In this study, precision (P) was used to represent the percentage of buds correctly identified by the model; recall (R) was used to represent the coverage of bud targets identified in the images; mean average precision (mAP) represents the sum of all classes divided by all classes; F1 score was used to evaluate the performance of the method by balancing the weights of precision and recall; the frames per second (FPS) about the detection time of a single image was used to evaluate the actual bud recognition speed of the model. These parameters were used as evaluation indicators to evaluate the trained model. The relevant calculation formulas are as follows:

$$\mathrm{P} = \frac{\mathrm{TP}}{\mathrm{TP} + \mathrm{FP}} \times 100\% \tag{1}$$

$$\mathrm{R} = \frac{\mathrm{TP}}{\mathrm{TP} + \mathrm{FN}} \times 100\% \tag{2}$$

$$\mathrm{F1} = \frac{2\mathrm{P} \times \mathrm{R}}{\mathrm{P} + \mathrm{R}} \times 100\% \tag{3}$$

$$\text{mAP} = \frac{1}{C}\sum_{k=i}^{N} P(k)\Delta R(k) \quad (4)$$

In the formula: true positives (TP) means that both the detection result and the true value are a famous tea bud; in other words, the number of famous tea buds is detected correctly. False positives (FP) indicates the detection result is famous tea buds, and the true value is the background; in other words, the number of famous tea buds is counted incorrectly. False negatives (FN) means that the detection result is the background, and the true value is the famous tea buds; in other words, the number of famous tea buds are not counted.

"TP + FP" refers to the total number of famous tea buds detected, and "TP + FN" refers to the total number of famous tea buds in an image. C is the number of categories, N represents the number of all pictures in the test set, $P(k)$ represents the precision when k pictures can be recognized, and $\Delta R(k)$ represents the change of the recall value when the number of recognized pictures changes from $k - 1$ to k [20].

2.7. Results and Analysis

(1) We randomly selected 631 images from the captured images to form the pre-training dataset. We used the pre-training dataset to conduct training under the same environment, and compared the recognition effect parameters of the networks with different attention mechanisms added at different positions of the YOLOv7 network. The recognition effect parameters of the model were shown in Table 1. The changes of loss value and mAP of YOLOv7+CBAM network in the training process are shown in Figure 9.

Table 1. Recognition effect parameters of different networks.

Model	P/%	Recall/%	F1 Score/%	Detection Speed/FPS
AB1=SE	85.75	77.32	0.82	54.92
AB2=SE	86.97	81.56	0.83	49.74
AB1=ECA	89.04	84.43	0.84	58.37
AB2=ECA	86.48	81.97	0.84	55.09
AB1=CBAM	89.06	84.70	0.87	60.03
AB2=CBAM	87.11	81.97	0.84	58.79
AB1=CA	87.16	82.38	0.85	56.26
AB2=CA	85.36	81.23	0.83	54.23
No additions	87.33	83.12	0.84	57.37

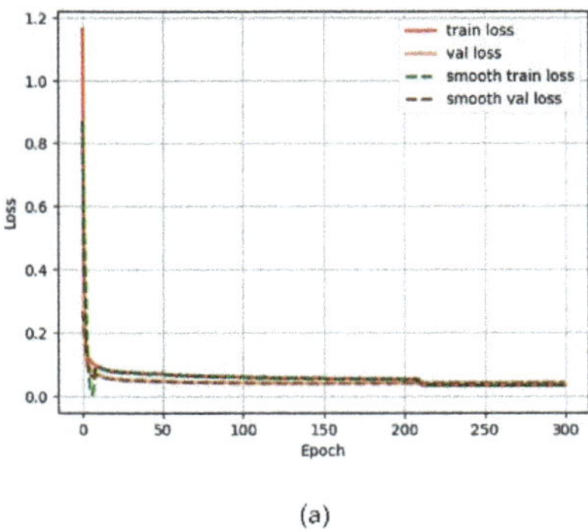

(a)

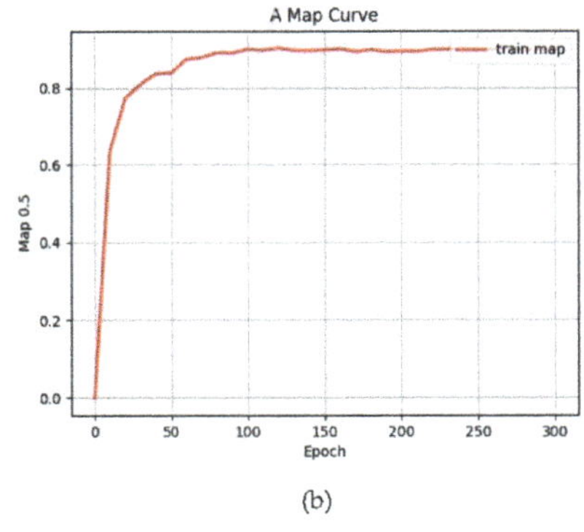

(b)

Figure 9. YOLOv7+CBAM network diagram. (**a**) Loss value change curve during training. (**b**) mAP curve change during training.

By comparing the recognition effects of different attention modules placed at different positions in the YOLOv7 network, it can be seen that after adding CBAM Block and ECA Block, YOLOv7 can achieve certain improvement in the parameters of P, Recall, F1 score and detection speed. Overall, it had a better recognition effect.

(2) In order to compare the influence of the number of images used for training on the network recognition effect, we used 1049 images taken and 1000 images after data enhancement during the formal training phase. The total images number of the dataset was 2049, and the training parameters remained unchanged compared with the pre-training phase. We selected the network of AB1=CBAM and AB1=ECA. The recognition effect of YOLOv7+CBAM network, YOLOv7+ECA network and YOLOv7 network were compared. The recognition effect parameters of the three networks are shown in the following Table 2.

Table 2. Identification effect parameters of different models.

Model	P/%	Recall/%	F1 Score/%	Detection Speed/FPS
YOLOv7	89.23	85.34	0.87	58.21
YOLOv7+ECA	91.83	87.16	0.89	59.64
YOLOv7+CBAM	93.71	89.23	0.91	61.23

(3) The bud images with occlusion, small targets and dense distribution were selected, respectively. The detection effects of the improved attention mechanism network based on the CBAM network, the ECA network and the YOLOv7 network without added attention mechanism were compared, as shown in Figure 10. It was found that the improved YOLOv7 network based on CBAM Block had higher recognition accuracy. Additionally, it had a lower rate of missed detection.

Figure 10. Comparison of recognition effects of different networks.

2.8. Visual Recognition of Heatmap

In order to visually explain the detection process of YOLOv7+CBAM network model, the visualization method of gradient-weighted class activation mapping (Grad-CAM) [29] was adopted in this paper. The recognition effect of YOLOv7 network model, YOLOv7+ECA network model and YOLOv7+CBAM network model was compared, respectively. In the Grad-CAM visualization method, the fusion weights of target feature maps are expressed as gradients, and the global average of gradients is used to calculate the weights. After the weights of all feature maps of each category are obtained, the weights are weighted and heatmap is obtained.

The heatmap can intuitively show the focus of attention of the model when extracting features. The warmer the color, the more attention of the model, and the red part (the warmest part) represents the focus of the model. Leafy, occluded and densely distributed buds were plotted using Grad-CAM, respectively, as shown in Figure 11. As can be seen in Figure 11, the YOLOv7+CBAM network model can accurately focus on different types of images and was little affected by background factors, which further proves that the network proposed in this study had a better effect on improving the detection effect of famous and excellent tea.

Figure 11. Activation graphs of famous and excellent tea image classes of different models.

3. Conclusions

(1) This study compared the attention mechanisms' effects of the SE, ECA, CBAM and CA blocks on the bud detection of famous and excellent green tea in different positions of YOLOv7 network. It was found that the YOLOv7+CBAM network model had the best recognition effect with the recognition accuracy of 93.71%, recall rate of 89.23% and F1 score of 0.91.

(2) The improved YOLOv7 networks basing CBAM, ECA attention mechanism were compared with the YOLOv7 network for the bud images with occluding, small targets and dense distribution, and it was found that the YOLOv7+CBAM network had better recognition effect on various tender leaves.

(3) Multi-leaf, occluding and densely distributed images containing young tea leaves were drawn by Grad-CAM, respectively. It could be seen that the YOLOv7+CBAM network model could accurately focus on different types of images and was little affected by background factors, which further proved that the network proposed in this study had a better effect on improving the recognition effect of famous and excellent green tea.

Author Contributions: Conceptualization, M.X. and X.W.; data curation, S.W. and Q.J.; writing—original draft, Y.W. and Y.Z. All authors have read and agreed to the published version of the manuscript.

Funding: This research was supported by Jiangsu Agricultural Science and Technology Innovation Fund Project (CX (21) 3148), the Fundamental Research Funds for the Central Universities (YDZX2023007) and Key R&D Program of Jiangsu Province (BE2021016).

Data Availability Statement: The data that were used are confidential.

Acknowledgments: The authors would like to thank the help of Juyuanchun Tea Garden for providing us with the experimental site and the funding support above all. Yongwei Wang wants to thank the standing support from the teacher Yongnian Zhang and the standing patience, company and encouragement from Shu Wang, particularly.

Conflicts of Interest: The authors declare no conflict of interest.

References

1. Liang, Y.; Shin, Y.H.; Zhang, L.-J.; Wang, K.-R. Advances in tea Plant Breeding in China. *Agric. Food* **2019**, *7*, 1–10.
2. Hicks, A. Review of global tea production and the impact on industry of the Asian economic situation. *AU J. Technol.* **2001**, *5*, 227–231.
3. Yang, H.; Chen, L.; Ma, Z.; Chen, M.; Zhong, Y.; Deng, F.; Li, M. Computer vision-based high-quality tea automatic plucking robot using Delta parallel manipulator. *Comput. Electron. Agric.* **2021**, *181*, 105946. [CrossRef]
4. Lu, Y.J. Debiao, Significance and realization of mechanized picking of famous green tea in China. *Chin. Tea* **2018**, *40*, 1–4.
5. Fan, W.Y.H.; Xin, Y.-Y.; Fei, L.; Ting, Z.; Li, C.-H. Chinese tea mechanization picking technology research status and development trend. *Jiangsu Agric. Sci.* **2019**, *47*, 48–51.
6. Fuzeng, Y.L.Y.; Yana, T.; Qing, Y. Tea bud recognition method based on color and shape characteristics. *Trans. Chin. Soc. Agric. Mach.* **2009**, *40*, 119–123.
7. Jian, W. Research on tea image segmentation Algorithm Combining color and region growth Wang Jian. *Tea Sci.* **2011**, *31*, 72–77.
8. Miaoting, C. Recognition and Localization of Famous Tea bud Based on Computer Vision. Master's Thesis, Qingdao University of Science and Technology, Qingdao, China, 2019.
9. Wu, X.; Zhang, F.; Lv, J. Research on tea leaf recognition method based on image color information. *Tea Sci.* **2013**, *33*, 584–589.
10. Tang, Y.; Han, W.; Hu, A.; Wang, W. Design and Experiment of Intelligentized Tea-plucking Machine for Human Riding Based on Machine Vision. *Nongye Jixie Xuebao/Trans. Chin. Soc. Agric. Mach.* **2016**, *47*, 15–20. [CrossRef]
11. Bao, W.; Fan, T.; Hu, G.; Liang, D.; Li, H. Detection and identification of tea leaf diseases based on AX-RetinaNet. *Sci. Rep.* **2022**, *12*, 2183. [CrossRef]
12. Yang, H.; Chen, L.; Chen, M.; Ma, Z.; Deng, F.; Li, M.; Li, X. Tender tea shoots recognition and positioning for picking robot using improved YOLO-V3 model. *IEEE Access* **2019**, *7*, 180998–181011. [CrossRef]
13. Wang, T.; Zhang, K.; Zhang, W.; Wang, R.; Wan, S.; Rao, Y.; Jiang, Z.; Gu, L. Tea picking point detection and location based on Mask-RCNN. *Inf. Process. Agric.* **2023**, *10*, 267–275. [CrossRef]
14. Chen, Y.-T.; Chen, S.-F. Localizing plucking points of tea leaves using deep convolutional neural networks. *Comput. Electron. Agric.* **2020**, *171*, 105298. [CrossRef]
15. Sozzi, M.; Cantalamessa, S.; Cogato, A.; Kayad, A.; Marinello, F. Automatic Bunch Detection in White Grape Varieties Using YOLOv3, YOLOv4, and YOLOv5 Deep Learning Algorithms. *Agronomy* **2022**, *12*, 319. [CrossRef]
16. Cardellicchio, F.S.A.; Dimauro, G.; Petrozza, A.; Summerer, S.; Cellini, F.; Renò, V. Detection of tomato plant phenotyping traits using YOLOv5-based single stage detectors. *Comput. Electron. Agric.* **2023**, *207*, 1077757. [CrossRef]
17. Wang, D.; He, D. Channel pruned YOLO V5s-based deep learning approach for rapid and accurate apple fruitlet detection before fruit thinning. *Biosyst. Eng.* **2021**, *210*, 271–281.
18. Wu, D.; Lv, S.; Jiang, M.; Song, H. Using channel pruning-based YOLO v4 deep learning algorithm for the real-time and accurate detection of apple flowers in natural environments. *Comput. Electron. Agric.* **2020**, *178*, 105742. [CrossRef]
19. Liu, T.; Teng, G.; Yuan, Y.; Liu, B.; Liu, Z. Winter jujube fruit recognition method in natural scene based on improved YOLO v3. *Trans. Chin. Soc. Agric. Mach.* **2021**, *52*, 17–25.
20. Yang, B.; Gao, Z.; Gao, Y.; Zhu, Y. Rapid Detection and Counting of Wheat Ears in the Field Using YOLOv4 with Attention Module. *Agronomy* **2021**, *11*, 1202. [CrossRef]
21. Liu, Y.; Cao, X.; Guo, B.; Chen, H.; Dai, Z.; Gong, C. Research on Attitude detection Algorithm of meat goose in complex scene based on improved YOLO v5. *J. Nanjing Agric. Univ.* **2022**, 1–12.
22. Fang, M.; Lü, J.; Ruan, J.; Bian, L.; Wu, C.; Qing, Y. Tea bud detection model based on improved YOLOv4-tiny. *Tea Sci.* **2022**, *42*, 549–560.
23. Fu, X.; Li, A.; Meng, Z.; Yin, X.; Zhang, C.; Zhang, W.; Qi, L. A Dynamic Detection Method for Phenotyping Pods in a Soybean Population Based on an Improved YOLO-v5 Network. *Agronomy* **2022**, *12*, 3209. [CrossRef]
24. Hu, J.; Shen, L.; Sun, G. Squeeze-and-excitation networks. In Proceedings of the IEEE Conference on Computer Vision and Pattern Recognition, Salt Lake City, UT, USA, 18–23 June 2018; pp. 7132–7141.
25. Wang, Q.; Wu, B.; Zhu, P.; Li, P.; Zuo, W.; Hu, Q. Supplementary material for 'ECA-Net: Efficient channel attention for deep convolutional neural networks. In Proceedings of the 2020 IEEE/CVF Conference on Computer Vision and Pattern Recognition, Seattle, WA, USA, 13–19 June 2020; pp. 13–19.
26. Woo, S.; Park, J.; Lee, J.-Y.; Kweon, I.S. Cbam: Convolutional block attention module. In Proceedings of the European Conference on computer Vision (ECCV), Munich, Germany, 8–14 September 2018; pp. 3–19.
27. Hou, Q.; Zhou, D.; Feng, J. Coordinate attention for efficient mobile network design. In Proceedings of the IEEE/CVF Conference on Computer Vision and Pattern Recognition, Nashville, TN, USA, 20–25 June 2021; pp. 13713–13722.

28. Lang, C.; Chao, J. X-ray image rotating object detection based on improved YOLOv7. *J. Graph.* **2023**, *44*, 324–334.
29. Selvaraju, R.R.; Cogswell, M.; Das, A.; Vedantam, R.; Parikh, D.; Batra, D. Grad-cam: Visual explanations from deep networks via gradient-based localization. In Proceedings of the IEEE International Conference On Computer Vision, Venice, Italy, 22–29 October 2017; pp. 618–626.

 agriculture

Article

Design and Experiment of a Low-Loss Harvesting Test Platform for Cabbage

Wenyu Tong [1,2], Jianfei Zhang [1,*], Guangqiao Cao [1], Zhiyu Song [3,*] and Xiaofeng Ning [2]

1 Nanjing Institute of Agricultural Mechanization, Ministry of Agriculture and Rural Affairs, Nanjing 210014, China; tongwenyu@stu.syau.edu.cn (W.T.); caoguangqiao@caas.cn (G.C.)
2 College of Engineering, Shenyang Agricultural University, Shenyang 110866, China; ningxiaofeng@cass.cn
3 Graduate School, Chinese Academy of Agricultural Sciences, Beijing 100083, China
* Correspondence: zhangjianfei@cass.cn (J.Z.); songzhiyu@cass.cn (Z.S.)

Abstract: In order to explore the mechanism and influence mechanism of cabbage harvest damage, a low-loss cabbage harvest test platform was designed on the basis of combining the physical characteristics of cabbage with the mechanical characteristics of mechanical harvest and the cabbage harvest operation process. Through the design of key components of the test platform harvesting, the key parameters of the pulling-out device, the reel device, the flexible clamping and conveying device, and the double-disc cutting device were determined. The movement changes of cabbage during pulling out, conveying, and cutting were analyzed to clarify the process of damage generation and critical conditions of damage in cabbage harvesting operations. The test results showed that when the speed of the pulling out device was controlled at 80–120 r/min, the speed of the clamping and conveying device was controlled at 120–240 r/min, and the speed of the double disc cutter was controlled at 140–180 r/min, the average success rate of pulling on the low-loss harvesting test platform was 92.7%; the average damage rate of the pulling process was 7.32%; the average success rate of clamping and conveying was 88.6%; the average damage rate of the clamping and conveying link was 12%; the average success rate of root cutting was 89.3%; and the average damage rate of the cutting link was 11.34%. The average qualified rate of harvesting in the pulling link was 86.7%, the average qualified rate of harvesting in the clamping and conveying link was 75.3%, and the average qualified rate of harvesting in the cutting link was 77.3%. All the performance indicators meet the design requirements and relevant standards, and the research results can provide a reference for the development and structural improvement of low-loss harvesting equipment for cabbage.

Keywords: cabbage; low-loss harvesting; mechanical harvesting characteristics; design of test platform

Citation: Tong, W.; Zhang, J.; Cao, G.; Song, Z.; Ning, X. Design and Experiment of a Low-Loss Harvesting Test Platform for Cabbage. *Agriculture* **2023**, *13*, 1204. https://doi.org/10.3390/agriculture13061204

Academic Editor: Jin Yuan

Received: 22 May 2023
Revised: 31 May 2023
Accepted: 5 June 2023
Published: 7 June 2023

1. Introduction

Cabbage is one of the staple vegetables in China. It is planted in four seasons in the north and south of China [1]. It reaches 900,000 hm^2, and the total yield accounts for about 50% of the total yield of cabbage in the world [2]. At present, cabbage is still mainly harvested manually, and problems such as increased labor, increased labor intensity, and increased production costs have become increasingly prominent [3].

The soft and easily damaged characteristics of cabbage make their mechanized harvesting quality fluctuate during actual production work. The conveying mechanism structure of cabbage is mostly chain clamping [4], screw conveying [5], and clamping conveying [6]. The CKM-1 [7] and NKH-1 [8] single-row cabbage harvesters developed by the USSR adopt chain clamping and conveying, which have low harvesting efficiency and great damage. Hansen [9] from the United States applied for a patent for a cabbage harvester using a reverse rotation of the double helix conveyor and then transported it to the rear for cutting and packing; Bleinroth [10], Baker [11], and Mori G et al. [12] developed a cabbage harvester using a similar double helix conveying method. Cheng Zhou [13] and Dongdong Du [14] of

China improved and optimized the double helix conveying mechanism and increased the chain clamp mechanism to clamp the root of cabbage and cooperate with the double helix conveying to transport backward synchronously, but the test effect was still not satisfactory and the conveying damage was too large. In order to reduce the damage to cabbage during the conveying process, the conveyor belt conveying structure was invented and applied to the cabbage harvester [15]. Compared with screw conveying, clamping conveying can improve conveying stability and reduce cabbage damage by changing the clamping belt material and tensioning mechanism [16]. Lenker et al. [17] and Wadsworth et al. [18] used rubber as a conveyor belt to complete the transportation of cabbage. Bleinroth et al. [19] designed a pressure-top conveying structure based on the conveyor belt. When the conveying speed of the pressure-top conveyor belt is consistent with the pulling-out speed, the cabbage can be kept upright and pressed during the conveying process, which is conducive to the formation of lotus-type cabbage cutting and improves the accuracy of root cutting. In summary, the current research in the field of low-loss cabbage harvesting technology is relatively weak. The existing cabbage harvesting equipment is not suitable for China's planting agronomy, and there is still a problem of large harvesting damage. Integrating flexible clamping and conveying technology into the design of cabbage harvesting and conveying structures may be a breakthrough to solve the high damage rate in the cabbage harvesting process [20,21].

Therefore, this paper aims to improve the operational performance of the cabbage harvester and reduce harvesting damage. Cabbage was selected as the research object, and on the basis of clarifying the mechanical and physical parameters of cabbage, a low-loss harvesting test platform for cabbage was designed. By obtaining the critical conditions for the damage of cabbage in the harvesting process, the attitude migration law of cabbage in the harvesting mechanism was clarified, which provided a reference for the development and structural improvement of low-loss harvesting equipment for cabbage.

2. Analysis of Mechanical Harvesting Characteristics of Cabbage

The cross-section of the rhizome cutting of cabbage during the harvest period is composed of the central pith, xylem, and phloem fiber layers from inside to outside. The matrix leaves are thicker, the moisture content of the long-term exposed external leaves is lower than that of the internal leaves, and the toughness is stronger. The internal leaves have a high moisture content and are brittle. The basic physical properties and harvesting mechanical properties of cabbage were determined: the rhizome length, diameter, and pulling force were measured for cabbage. The universal testing machine was used to test the crushing, rhizome cutting force, water content, and cutting force of a single plant. The mechanical parameters such as root shear characteristics and ball crushing force under different water content conditions were tested, respectively, which provided a theoretical basis and data support for the study of cabbage harvesting equipment. See Figure 1.

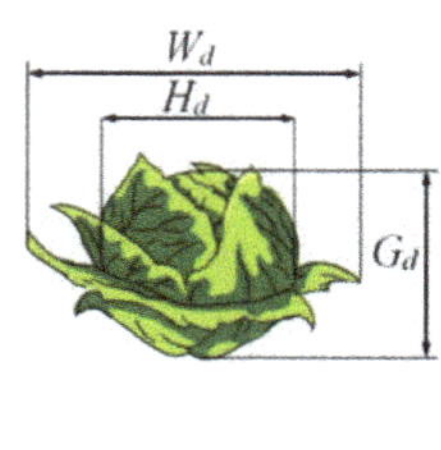

(a)

(b)

(c)

(d)

(e)

Figure 1. Determination of basic physical parameters and harvesting mechanical properties test of cabbage: (**a**) Physical characteristics acquisition: 1: W_d: Expansion degree 2: H_d: transverse diameter 3: G_d: vertical diameter (**b**) Physical characteristics acquisition process. (**c**) Determination of pull-out force. (**d**) Compression test. (**e**) Root cutting force test.

2.1. Measurement Results of Basic Physical Parameters of Cabbage

The determination results of the basic physical parameters of cabbage are shown in Table 1. The vertical diameter of the cabbage ball was (162.3 ± 38.16) mm, the transverse diameter of the sphere was (163.8 ± 25.84) mm, the weight of cabbage was (1.91 ± 0.68) kg, the diameter of the cabbage rhizome was (28.71 ± 4.56) mm, and the expansion degree of cabbage was (325.6 ± 40.89) mm. The measurement results can provide a theoretical basis for the spatial layout of the cabbage harvesting device.

Table 1. Results of basic physical parameters of cabbage.

Statistical Indicators	Vertical Diameter of Cabbage (mm)	Transverse Diameter of Cabbage (mm)	Weight of Cabbage (kg)	Diameter of Rhizome (mm)	Expansion Degree of Cabbage (mm)
Average value	162.3	163.8	1.91	28.71	325.6
Maximum value	190	185	2.265	31.8	360
Minimum value	135	140	1.241	24.6	290
Standard deviation	19.05	12.92	0.34	2.28	20.45
Coefficient of variation	0.12	0.08	0.18	0.08	0.06

2.2. Pulling Force Measurement Results

The determination results of cabbage pull-out force are shown in Table 2: The maximum pullout force of this variety was 228 N, the minimum was 154 N, and the average pullout force was 190.4 N. The measurement results can provide a theoretical basis for the design of the extraction device.

Table 2. Results of pulling force.

Statistical Indicators	Pulling Force (N)
Average value	190.4
Maximum value	228
Minimum value	154
Standard deviation	33.93
Coefficient of variation	0.18

2.3. Test Results of Rhizome Shear Force and Extrusion Force

It can be seen from Figure 2a,b that the curve of force and displacement in the test data is non-linear. During the shear test, the blade first contacts the outer epidermis of the cabbage rhizome, and the shear force gradually increases. When the blade cuts off the outer epidermis and enters the interior of the rhizome, the shear force will decrease, and the change range is small. When the blade cuts to the lower outer epidermis, the shear force gradually increases until the cabbage rhizome is completely broken. The reason for this is that the material of the outer epidermis of the cabbage is fiber. Compared with the internal matrix, its flexibility is much stronger, so the shear force is at its maximum when it just enters the fiber layer. The maximum compressional force is $F_{\max}$ = 1198.3 N, and the maximum root cutting force is F_C = 137.138 N. This value can provide the necessary parameter support for the cutting device and the clamping and conveying device of the cabbage harvesting device.

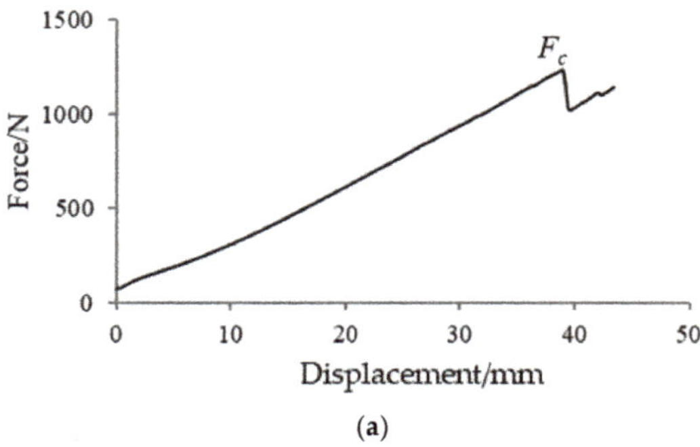

(a)

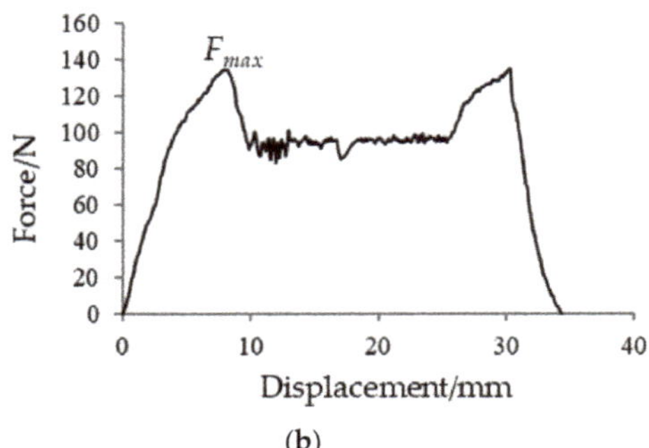

(b)

Figure 2. Harvest mechanical properties test: (**a**) compressional force–displacement curve. (**b**) Root cutting force–displacement curve.

3. Machine Structure and Working Principles

3.1. Structure of the Machine

The low-loss harvesting test platform for cabbage is mainly composed of a cabbage conveying system, cabbage harvesting system, test data acquisition system, and servo motor frequency conversion control box. The conveying system simulates the actual walking state of the cabbage harvesting machine in the field and relies on the cabbage harvesting system to simulate the real harvesting process. The specific structure of the prototype is shown in Figure 3. The conveying system for cabbage mainly consists of a cabbage rootstock clamping cup and conveying chain. The conveying speed (0~0.5 m/s) can be adjusted by the servo motor frequency control box. The cabbage harvesting system is mainly composed of a pulling device, reeling device, flexible clamping conveying device, and double disc cutting device. The speed of the pulling device (0~200 r/min), the speed of the reeling device (0~140 r/min), the speed of the flexible clamping conveyor belt (0~300 r/min), and the speed of the double disc cutting device (0~400 r/min) can also be adjusted by the servo motor frequency control box. Through the torque and pressure sensors installed in the harvesting device, the experimental data acquisition system can collect the motion parameters of the cabbage in the extraction, transportation, cutting, and other links, obtain the motion changes of the cabbage on the test header, find out the damage law, and determine the range of motion combination parameters between different key components. At the same time, the motion trajectory and speed-time curve of a single cabbage plant during transportation are calibrated and tracked. The damage to cabbage after transportation is recorded and saved, and performance indexes such as epidermal damage and cracking are measured.

3.2. Principles of Test Platform

The operation process of the low-loss harvesting test platform for cabbage is shown in Figure 4. The test platform has preliminarily designed a "vertical clamping + flexible conveying" mechanism suitable for cabbage harvesting, which is mainly composed of a reel, a pulling roller, a flexible clamping belt, a cutterhead, a rack, and a transmission shaft. When the test platform is working, the pulling roller is placed under the outer leaf of the cabbages. The external rotation forces the root of the cabbages to be removed from the fixture to complete the pulling. The flexible conveyor belt clamps the cabbages and transports them backward. The cutter head cuts off the root of the cabbages and completes one test. By changing the material and structural parameters and motion parameters of the key components of the conveying mechanism, the damage to the cabbage after the conveying operation was recorded and saved, and the operation performance indexes such as epidermal damage and cracking ball were measured to find out the mechanism of harvesting damage to cabbage.

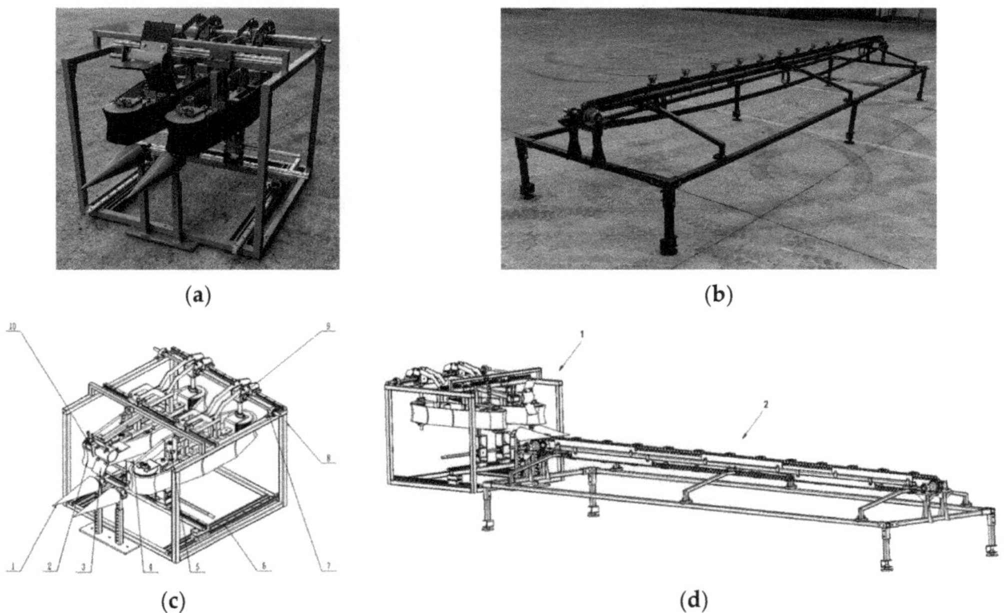

Figure 3. Structure diagram of low loss harvest test platform for cabbage: (**a**) Harvest cutting table. (**b**) Conveying test bench. (**c**) Harvest header structure diagram: 1: pulling device; 2: flexible clamping device; 3: pulling device; 4: disc cutter; 5: tensioning wheel; 6: track; 7: active shaft; 8: stand; 9: conveyor gear box; 10: pulling device motor. (**d**) Structure diagram of test platform: 1: harvest cutting table; 2: conveying test bench.

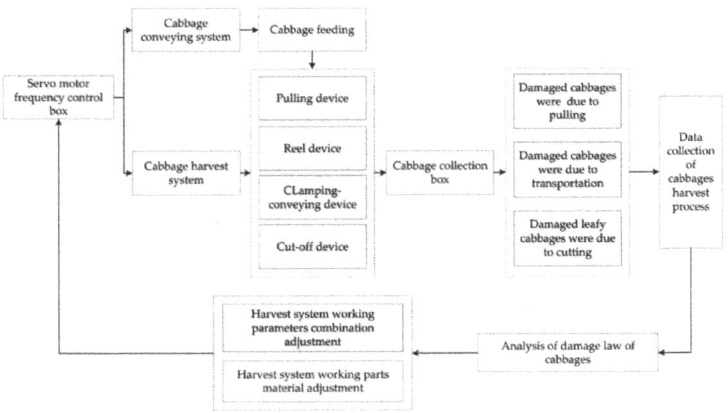

Figure 4. Test platform operation process.

When the low-loss harvesting test platform for cabbage works, firstly, the servo motor frequency conversion control box controls the working parts of the conveying test bench: the pulling device and the reeling device to run at a lower speed while keeping the clamping conveying device and the cutting device in a static state. At this time, the low-loss harvesting test platform for cabbage can only complete the pulling operation. Observe whether the conveying, pulling, and reeling links of the cabbage are smooth. Combined with the damage to the cabbages in the collection box and the data collection results of the harvesting process, the working parameters of the pulling device or the material of the working parts are adjusted according to the test results until the conveying and pulling links are carried out smoothly and the cabbages are not damaged. Recording the working parameters (motion parameters and structural parameters) of the pulling device at this time is a critical condition for preventing damage to the cabbage in the pulling process. Similarly, according to this method, the servo motor frequency conversion control box controls the speed of the clamping and conveying device and the cutting device in turn, makes the low-loss harvesting test platform of the cabbage, completes the pulling + reeling + clamping and conveying, pulling + reeling + clamping and conveying + cutting links in turn, and records the critical conditions for damage in the clamping and conveying and cutting links, respectively.

3.3. The Main Technical Parameters

The main technical parameters of the low-loss harvesting test platform for cabbage are shown in Table 3.

Table 3. The main technical parameters.

Parameters	Value
Harvest header size (length × width × height) (mm × mm × mm)	1560 × 1300 × 975
Conveying test bench size (length × width × height) (mm × mm × mm) (mm × mm × mm)	5000 × 1300 × 780
Conveying speed of conveying system/(m/s)	0–0.5
Pulling device speed/(r/min)	0–200
Reel speed/(r/min)	0–140
Clamping conveying device speed/(r/min)	0–300
Cutting device speed/(r/min)	0–400
Servo motor operating frequency/(Hz)	0–50

4. Analysis of the Working Process and the Selection of Key Parameters

4.1. Design of Pulling Device

As shown in Figure 5a,b, the pulling device of the test platform is mainly composed of a reel and a pulling roller. When the pulling device works, the pulling roller is located below the cabbage. Through their own continuous external rotation, the cabbages are subjected to an upward pulling force. After the root of the cabbage is completely separated from the conveying system, it enters the clamping conveying mechanism through the right position of the reel above the pulling roller.

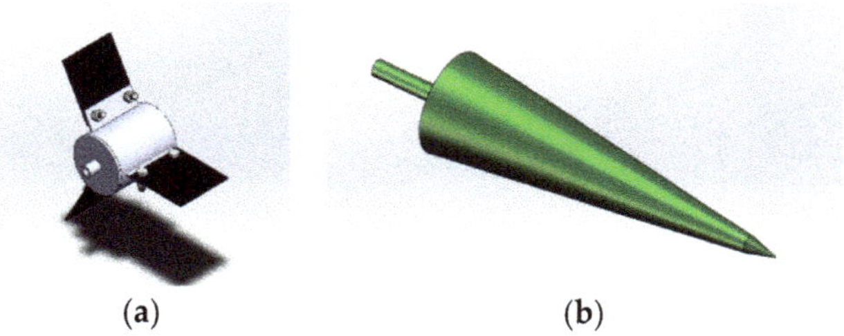

Figure 5. The pulling device: (**a**) reel; (**b**) pulling roller.

As shown in Formula (1), the ratio of the linear velocity at the outer edge of the reel to the forward speed of the machine is called the reeling speed ratio λ. When $\lambda \leq 1$, the cycloid amplitude of the working trajectory of the reel is small, and the function of supporting and guiding the cabbage cannot be realized. As shown in Figure 6, when $\lambda > 1$, the working trajectory of the reel is cycloidal. At this time, the reel can work normally, and the reel effect works well.

$$\frac{V_0}{V_x} = \lambda \quad (1)$$

where V_0 is the speed along the outer line when the wheel is working, m/s, and V_x is the conveying speed of cabbage.

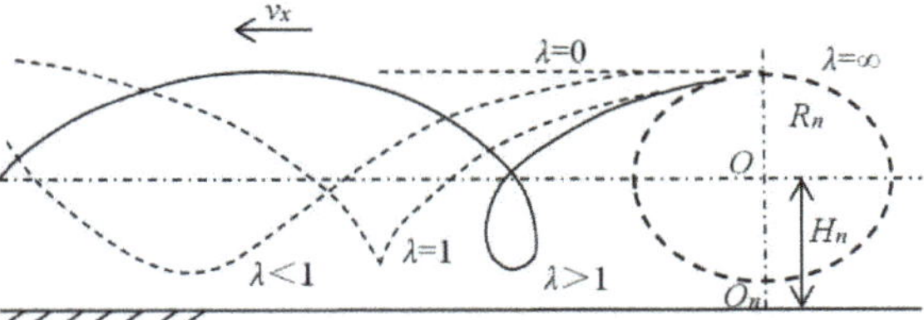

Figure 6. The movement track of the reel.

The displacement equation for the reel:

$$x = V_x + R_n \cos W_n t \tag{2}$$

$$y = H_n - R_n \sin W_n t \tag{3}$$

where R_n is the radius of the reel, mm, and W_n is the rotation speed of the reel, r/min. H_n is the height from the center of the reel to the ground, mm.

It is assumed that the reel has "m" reel leaves. When a reel leaf rotates one circle, the forward distance of the harvester is:

$$L = V_x \frac{60}{mV_n} \tag{4}$$

where L is the forward distance of the harvester, m, and V_n is the angular velocity of the reel, rad/s.

When the reel works normally, the size of the reel should meet:

$$\frac{2\pi R_n}{m\lambda} > D. \tag{5}$$

where D is the diameter of the cabbage.

In order to achieve continuous harvesting, the pitch of the long trochoid of the reel should meet:

$$S_n = \frac{2\pi R_n}{m\lambda} = \frac{S_l}{n} \tag{6}$$

where S_n is the pitch of the long trochoid of the reel; S_l is the distance between two adjacent cabbages; and n is the reel leaf spacing, generally taking 1, 2, and 3.

The number of reel leaves on the test platform is 3, the radius of the reel is 240 mm, the distance of cabbage in the conveying system is 350 mm, take 1 for n, and the conveying speed of the conveying system is set to be 0.3 m/s. The rotation speed of the reel is set to 30, 50, or 70 r/min, and the trajectory of the reel is simulated by MATLAB-ADAMS. The simulation trajectory is shown in Figure 7.

4.2. Design of Pulling Roller

In order to analyze the movement of the cabbage on the pulling roller, the cabbage and the pulling roller are formed into a rigid body system for force analysis. In order to simplify the model, the position of the mass center is not considered in the analysis process. See Figure 8.

$$\sum mv_x = \sum F_x^{(e)}, m_j a_j - m(a_e - a_r \cos\delta) = F_N \cos\delta = \mu F_j \cos\delta \tag{7}$$

$$\sum mv_z = \sum F_z^{(e)}, F_m - (m_j + m)g - F_j \sin\delta = ma_r \sin\delta \tag{8}$$

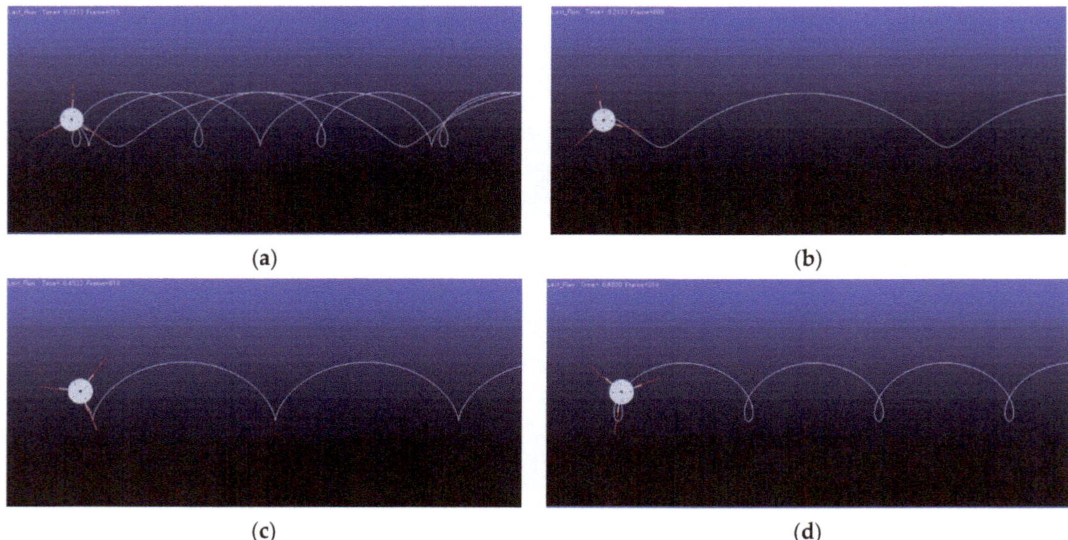

Figure 7. Trajectory diagram simulation of reel movement: (a) motion trajectory integration; (b) rotation speed = 30 r/min; (c) rotation speed = 50 r/min; (d) rotation speed = 70 r/min.

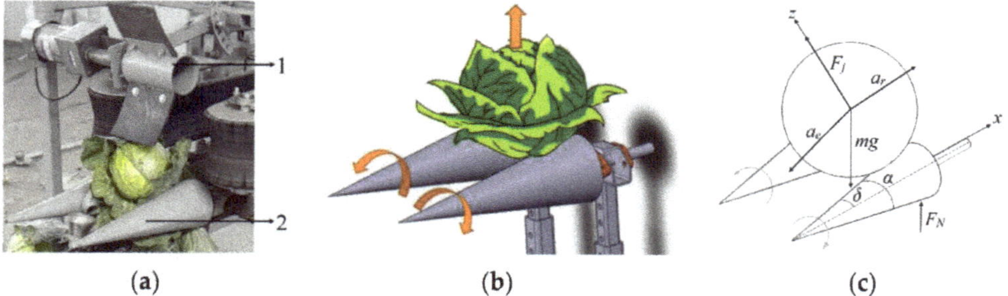

Figure 8. Dynamic analysis of pulling operation: (a) 1. Reel. 2. Pulling roller; (b) pulling roller and cabbage movement direction; (c) rotation speed = 50 r/min.

According to the centroid motion law of rigid body motion, the force balance equations in the x-direction and z-direction are listed:

$$m(a_j \cos\delta - \alpha_r) = mg\sin\delta \tag{9}$$

$$mg\cos\delta = F_j \tag{10}$$

From the above formula, it can be deduced that:

$$a_j = \frac{mg(\mu\ \cos^2\delta + \sin\delta\cos\delta)}{m_j - m + m\ \cos^2\delta} \tag{11}$$

$$a_j = \frac{g(\mu\ \cos^2\delta + \sin\delta\cos\delta)}{K_M - 1 + \cos^2\delta} \tag{12}$$

The conditions that must be met to make the cabbages move upwards without falling are: $\alpha_r > 0$. It can be derived from Formulas (8) and (12):

$$\frac{mg(\mu\ \cos^2\delta + \sin\delta\cos\delta)}{m_j - m + m\ \cos^2\delta} > g\frac{\sin\delta}{\sin\delta} \tag{13}$$

$$\mu > \frac{\tan\delta(K_M - 1 + \cos^2\delta)}{\cos^2\delta} - \tan\delta \tag{14}$$

where μ is the friction coefficient of cabbage; F_m is the overall force of the pulling roller on the cabbage, N; α_r is the relative acceleration of the motion of the cabbage, in m/s^2; α_e is the acceleration of the conveying system, in m/s^2; δ is the angle between the cone on the pulling roller and the horizontal line, (°); F_j is the pull-out force of the pull-out roller on cabbage, N; α_j is the absolute acceleration of the conveying system, in m/s^2; m_j is the weight of a single pulling roller, in kg; F_n is the friction force between cabbage and the pulling roller, N; and K_m is the mass ratio of the pulling roller with cabbage.

To ensure that the cabbages do not fall during harvest, it is necessary to meet Formula (14). Therefore, the material properties should be considered when designing the pulling roller. It can be seen from Formula (12) that the relative motion between the pulling roller and the cabbage has a great influence on transportation, so the rotation speed of the pulling roller plays a vital role in the qualified rate of harvesting.

As shown in Figure 9, the force analysis of cabbage in the pulling process is carried out.

$$f_1\cos\alpha = F_{N1}\sin\alpha \tag{15}$$

$$f_1\sin\alpha + F_{N1}\cos\alpha = mg \tag{16}$$

$$f_1 = \mu F_{N1} \tag{17}$$

where f_1 is the resistance and friction of the cabbage in the pulling process, N; F_{N1} is the supporting force of the working surface of the pulling roller on the cabbage, N; and α is the pulling angle of the pulling roller, (°).

When the cabbage is pulled out, it is subjected to resistance, friction, and support. After overcoming gravity, the pull-out force of cabbage is:

$$F = f_1\sin\alpha + F_{N1}\cos\alpha - mg \tag{18}$$

It can be derived from Formulas (15)–(17):

$$F = \frac{mg}{\tan\alpha\sin\alpha}(\mu\sin\alpha + \cos\alpha) \tag{19}$$

The weight of cabbage is determined by Table 1 at 1.91 kg. According to Formula (19), the pulling force F of the pulling roller on cabbage depends on the pulling angle α. Therefore, a better pulling effect can be obtained by selecting the appropriate pulling angle. The material of the pulling roller designed in this paper is stainless steel, the friction coefficient is 0.3, the distance between the inside of the pulling roller is 50–80 mm, the diameter of the end is 120 mm, the total length of the conical pulling roller is 450 mm, and the angle between the pulling roller (taper) and the horizontal angle is 13.9°. At this time, the pulling force of the two pull-out rollers on cabbage is:

$$2F = \frac{1.91\times 9.8}{\tan 13.9^\circ\sin 13.9^\circ}\times(0.3\sin 13.9^\circ + \cos 13.9^\circ) = 677 \tag{20}$$

At this time, the theoretical pulling force of the pulling roller designed in this paper is greater than the pulling force measured in Table 2, which meets the design requirements.

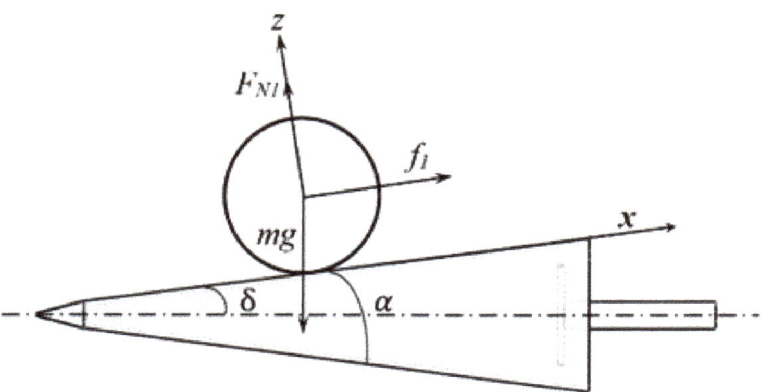

Figure 9. Pulling force analysis.

4.3. Design of Clamping Conveying Device

As shown in Figure 9, the low-loss harvesting test platform for cabbage designed in this paper adopts a new mechanism and new method called "vertical clamping + flexible conveying". By using the flexible feeding and flexible clamping methods, we can improve the adaptability of different ball diameters of cabbage and realize low-loss transportation [22,23].

The analysis of the movement process of the cabbage in the clamping and conveying device is shown in Figure 10, where 1 is the cabbage feeding link, 2 is the clamping and conveying link, and 3 is the harvesting of the finished product link.

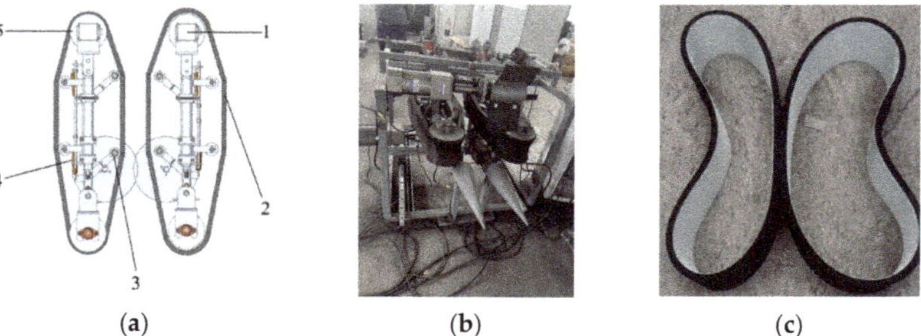

Figure 10. Clamping and conveying device: (**a**) 1: driving motor; 2: CR flexible sponge conveyor belt; 3: tensioner pulley; 4: tightening spring; 5: conveyor belt drive wheel. (**b**) Clamping conveying device physical diagram n. (**c**) High-density CR flexible sponge.

The clamping conveying process of cabbage is shown in Figure 11.

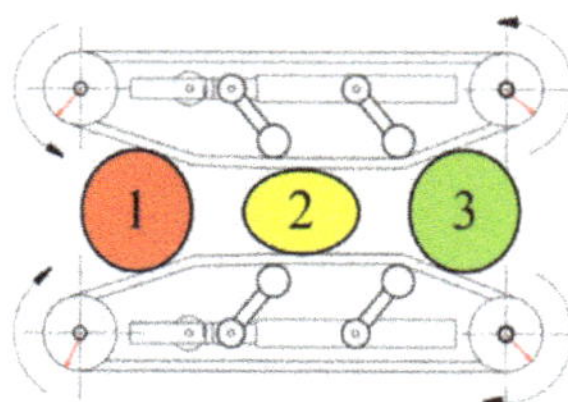

Figure 11. Clamping conveying process of cabbage.

The motion analysis and force analysis of the feeding link of the cabbage are shown in Figure 12a,b.

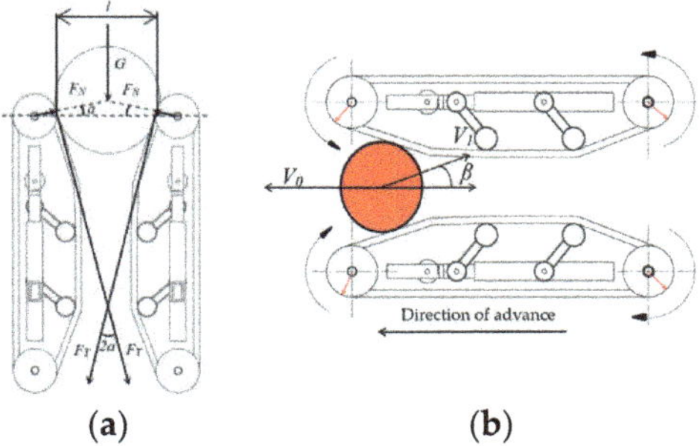

Figure 12. Clamping conveying process of cabbage. (**a**) Force analysis of the feeding link. (**b**) Motion analysis of the feeding link.

If the cabbages are not blocked in the feeding link and smoothly enter the clamping and conveying link, the following formula should be satisfied:

$$\tan\alpha \leq \frac{F_T}{F_N} = \mu \tag{21}$$

$$V_1 \sin\beta > V_0 \tag{22}$$

According to the above formula, it can be calculated by the following formula:

$$\mu \geq \sqrt{\frac{(D+d)^2 - (D-l)^2}{(D+l)^2}} \tag{23}$$

where α is the angle between pressure and the horizontal line of cabbage; V_1 is the linear velocity of the conveyor belt, m/s; β is the lifting angle of the conveyor belt, (°); V_0 is the operating speed of the conveying system, m/s; F_T is the friction force on cabbage, N; F_N is the pressure of the conveyor belt on cabbages, N; μ is the friction coefficient between the conveyor belt and cabbages; D is the diameter of the conveyor belt drive wheel, mm; d is the diameter of cabbage, mm; and l is the distance between the two conveyor belt drive wheels, mm.

The feeding inlet clamping position of the conveyor belt should be at the waist of the cabbage. In this paper, the single weight of "Chun xi" cabbage was 1.2–1.5 kg, the bulb was 180–200 mm, the feeding inlet spacing of the clamping conveying mechanism, which can be adjusted by spring and the minimum spacing, was 120 mm, and the diameter of the conveyor belt drive wheel was 110 mm. Therefore, the maximum value of the friction coefficient between the conveyor belt and the cabbage can be calculated as follows:

$$\mu \geq \sqrt{\frac{(110+200)^2 - (110+120)^2}{(110+120)^2}} = 0.9 \tag{24}$$

The maximum extrusion force of cabbage is:

$$F_{Nmax} = \frac{G}{\mu} \times 2 = 130.67 \tag{25}$$

According to the test of the mechanical harvesting characteristics of the cabbage, the extrusion force is far less than the maximum extrusion crushing force of 1198.4 N. Based on the compressional force test results, the clamping and conveying mechanical structure designed in this paper can avoid the results that the cabbage cannot be clamped due to too small extrusion pressure, and if the extrusion pressure is continuously increased, the cabbage may be blocked, so the structural parameters of the clamping and conveying device are designed reasonably.

By analyzing the force on the base surface of the conveying interval and the deformation of the cabbage [24], the deformation analysis in the extrusion deformation link is shown in Figure 13.

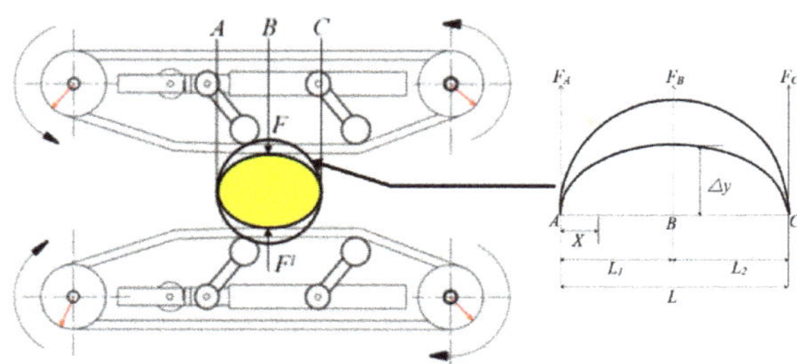

Figure 13. Deformation analysis of cabbage.

It can be seen from Figure 12 that the extrusion force $F = F_1$ at point B and the forces F_A and F_C at points A and C are:

$$F_A = F_C = \frac{FL_1}{L} \tag{26}$$

The reaction force of point B can be obtained from the equilibrium equation:

$$F_B = \frac{FL_2}{L} \tag{27}$$

The deformation bending moment of cabbage is:

$$M_x = \frac{FL_2}{L}x - F(x - L_1) \quad (L_1 \leq x \leq L) \tag{28}$$

The deformation-bending moment is the same on both sides, and the integral on one side can be obtained:

$$EJ\frac{d^2y}{dx^2} = \frac{FL_2}{L}x \tag{29}$$

The deflection curve equation is:

$$y = \frac{F_4L_2x}{6LEJ} = x^2 - L_2^2 + L^2 \tag{30}$$

Because $x = L_1 = L_2$, the clamping deformation Δy of cabbage is:

$$\Delta y = \frac{F_2L_2L_1L}{6EJ} \tag{31}$$

where E is the modulus of elasticity; J is the momentum of inertia; L is the length of AC, mm; L_1 is the length of AB, mm; and L_2 is the length of BC, mm.

According to Formula (15), when the conveyor belt speed is too low, the conveying efficiency will be reduced, which makes it easy to cause conveying blockage, resulting in unsmooth subsequent operations and incomplete root cutting. Through the preliminary test and observation, the main forms of damage in the transportation of cabbages are friction, extrusion, collision, and other forms of damage. The reason is that the deformation of cabbage is too large under the action of rigid parts in the conveying process. If the deformation deflection Δy of cabbage is too large, it is easy to squeeze and break. In Formula (25), the deformation deflection of the cabbage is determined by the elastic modulus. In order to reduce the damage as much as possible, the high-density CR flexible sponge belt is selected to wrap the cabbage for clamping and conveying. While reducing the relative friction, the tensioning mechanism transfers part of the deformation of the cabbage to the flexible clamping conveyor belt. It has a certain anti-deformation effect on the cabbages and can ensure that the cabbages do not slide when they are clamped by the conveyor belt and will not cause ineffective root cutting of the cabbage due to sliding.

In order to adapt to different kinds of cabbage and improve the adaptability of harvesting equipment. The maximum center spacing of the conveyor belt designed in this paper can be adjusted to 280 mm, and the feeding inlet spacing can be adjusted in the range of 200–250 mm. Ensure that different varieties of cabbage can be successfully clamped and transported, even if the ball diameter is different.

4.4. Design of Root Cutting Device

The root-cutting device is one of the key components of the cabbage harvester. It works together with the clamping and conveying device. The main function is to cut off the root of the cabbage. The cutting effect has a great influence on the quality and efficiency of the subsequent harvesting operation. As the rhizomes of cabbage are relatively thick and some kinds of rhizomes are severely fibrotic, when a single disc cutter is used, a higher speed is required to reduce the imbalance of cutting force, which will cause considerable power consumption. Therefore, in order to ensure the stability of the force when cutting the root of the head cabbages and avoid incomplete cutting of the root of the head cabbages, this paper adopts the double disc cutting form. The two cutters keep a certain distance in the direction of the center line, and the two cutters should overlap a little to balance the horizontal force on the rhizome of the cabbage during the root-cutting process, so as to ensure the flatness and integrity of the root cutting. See Figure 14.

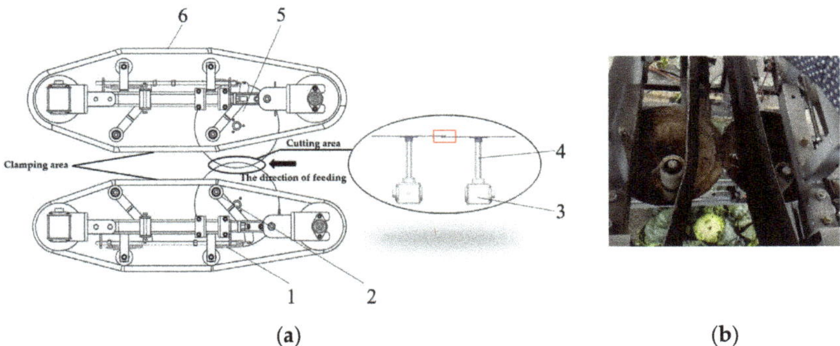

Figure 14. Root cutting device: (**a**) 1: tightening spring; 2: tensioning Wheel; 3: driving motor; 4: transmission shaft; 5: disc cutter; 6: conveyor belt. (**b**) Cutting device physical picture.

As shown in Figure 15, the power of the root-cutting device comes from two servo motors. In the process of root cutting, the roots of cabbage are subjected to cutting forces F_1, F_2, and F_3 in the X, Y, and Z-directions; F_{1xy} and F_{3xy} are the projection components of F_1 and F_3 in the XY plane, respectively; and F_{1x} and F_{1y} are the projection components of F_{1xy} in the X-axis and Y-axis, respectively. F_{3x} and F_{3y} are the projection components of F_{3xy} in the X-axis and Y-axis, respectively. β and φ are the angles between F_{1xy} and F_{3xy} and the X-direction, respectively. Because the cutting positions of the left and right cutters are asymmetric, and the two cutters overlap in the axial direction, there is still some cutting force in the X-axis and Z-axis directions (F_2). Under the combined action of the forces F_{1z} and F_{3z} in the Z-axis direction and the feeding direction and the cutting friction force of the cabbage, the longitudinal component force above the cabbage is larger after receiving, which is helpful to clamp the rhizome of the cabbage [25]. Due to the actual root-cutting process, the vertical plane angle γ, which is the angle between the resultant force F_{3xy} and the force F_3 in the XY plane, is small; therefore, F_{1x} can be approximated as the root main cutting force. In addition, considering the actual installation and use, γ is set to zero, and only the influence of pitch angle θ (the angle between the resultant force F_{1xy} and the force F_1 in the XY plane) is considered in order to determine the best cutting parameters.

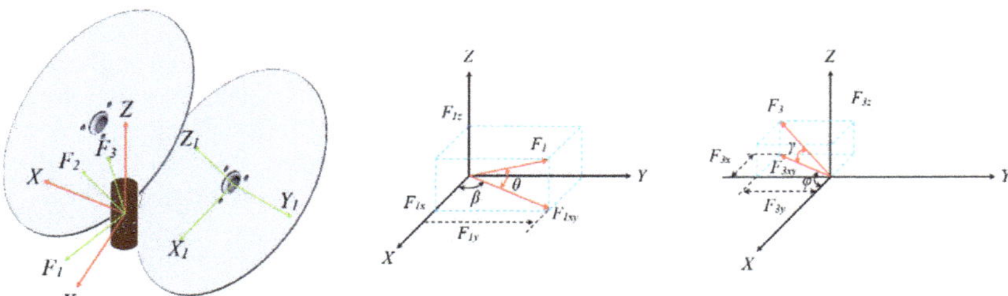

Figure 15. Root cutting force analysis.

As shown in Figure 16, it is assumed that the two disc cutters of the root cutting device are ideal discs, the cabbage is idealized as a round, and the diameter of the rhizome is D_1.

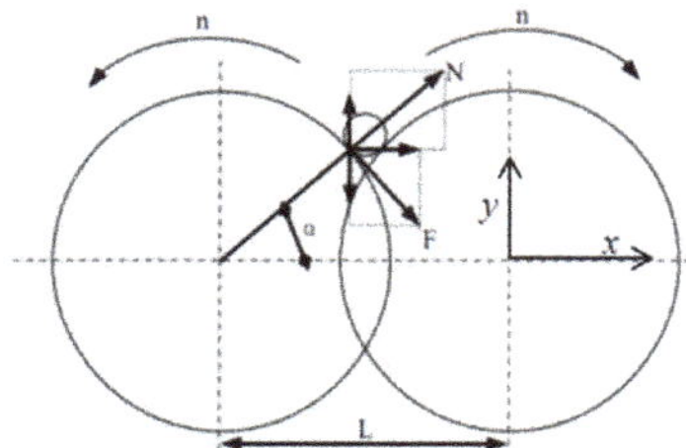

Figure 16. Force on the rootstock of cabbage.

From the force analysis in the diagram, the following equations can be obtained:

$$R_X = N_X + F_X \tag{32}$$

$$T_Y = F_Y + N_Y \tag{33}$$

where R_X is the root cutting force and T_Y is the clamping force of the cutter on the rhizome. N is the normal reaction force of the disc cutter on rhizomes; its horizontal component is N_X and its vertical component is N_Y, N; F is the friction of the circular cutter disc on the friction force of the circular cutter on the root of cabbage; its horizontal component is F_X and its vertical component is F_Y, N.

In order to make the rhizome be held by the disc cutter, the following conditions must be met [26]:

$$T_Y > 0 \tag{34}$$

It can be deduced that: $F_Y > N_Y$, where $F = N \cdot f$, that is:

$$N \cdot f \cos\alpha > N \cdot \sin\alpha \tag{35}$$

Therefore, when $f > tan\alpha$, the disc cutter has better clamping performance:

$$\alpha = \cos^{-1}\frac{L/2}{(D_1 + D_2)/2} = \cos^{-1}\frac{L}{D_1 + D_2} \tag{36}$$

where f is the friction coefficient between the disc cutter and the rhizome of cabbage; α is the angle between the reaction force of the disc cutter on the cabbage rhizome and the

x-axis, (°); L is the distance between two disc cutters, mm; D_1 is the root diameter of the cutting point, mm; and D_2 is the diameter of the disc cutter, mm.

The center distance of the designed double disc cutter is 190 mm, the diameter of the disc cutter is 200 mm, the angle $\alpha = 31.02°$, and the overlap thickness of the two cutters is 5 mm. At this time, $f > tan\alpha$, which can meet the requirements of clamping performance.

4.5. Design of Data Acquisition System

By installing torque and pressure sensors on the test platform to collect the speed and displacement of the cabbage in the harvesting system, the motion trajectory and speed–time curve of a single plant during the harvesting process are calibrated and tracked, and the damage to the cabbage after the harvesting operation is recorded and saved in order to find out the results of the movement of the cabbage in the pulling, conveying, and cutting of roots and determine the range of working parameters between different harvesting components.

The system uses the industrial tablet computer to install the INTOUCH HMI configuration software as the human–machine interface, which can ensure that the test bench starts operation according to the set test parameters. The monitoring acquisition system collects the operating data of each moving part, including torque, tension, speed, acceleration, and other data. The data acquisition cycle can be set, and the data chart curve can be automatically generated according to the recorded data.

The instrument panel graphics and text data list can display the following collected data: Left feeding inlet tension, right feeding inlet tension, conveying torque, left cutter torque, right cutter torque, left clamping conveying tension, right clamping conveying tension. Tension pressure data accuracy static: 1‰, tension pressure data dynamic static: 2‰, torque dynamic accuracy 2‰, each data can set the alarm value; if the data exceed the normal range, the system records and sends an alarm prompt. See Figure 17.

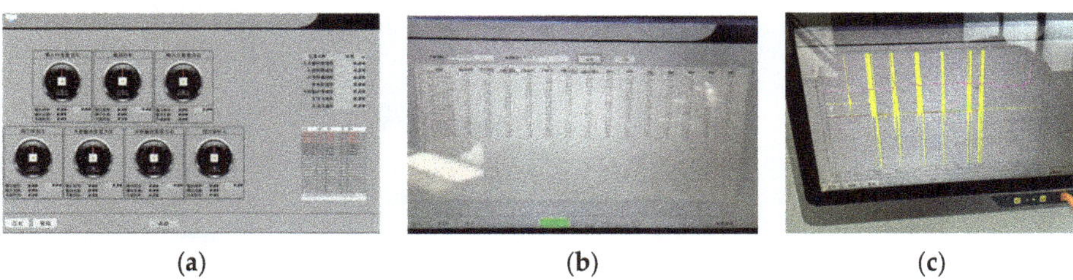

(**a**) (**b**) (**c**)

Figure 17. Data acquisition system: (**a**) test bench working parameters control interface; (**b**) data reports; (**c**) data curves.

According to the data obtained from the data acquisition system and the collection of cabbage in the aggregate box, the structural parameters and working parameters of each harvesting device in the harvesting system are adjusted in real time to reduce the damage rate of cabbage.

5. Single-Factor Test of Low-Loss Harvesting Test Platform

5.1. Test Object and Equipment

As shown in Figure 18, cabbage was selected as the test object, and the cabbage variety was "Chun xi". The expansion degree of cabbage is about 450–500 mm, the height of cabbage is about 140 mm, the width of cabbage is about 149 mm, the outer leaves are about 10–12, and the single cabbage quality range is 1.1–1.4 kg. By observing the working state of the pulling device, the reeling device, the clamping and conveying device, and the cutting device of the test platform. According to the characteristics of the cabbage harvested by the above device, the working frequency (0–50 Hz) of the servo motor driven by the above device is adjusted in real time.

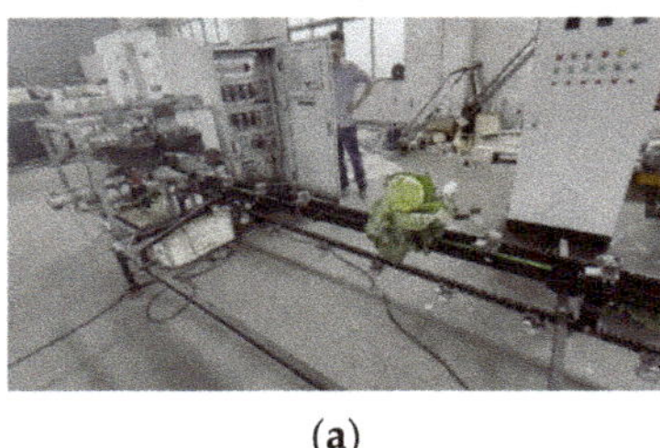

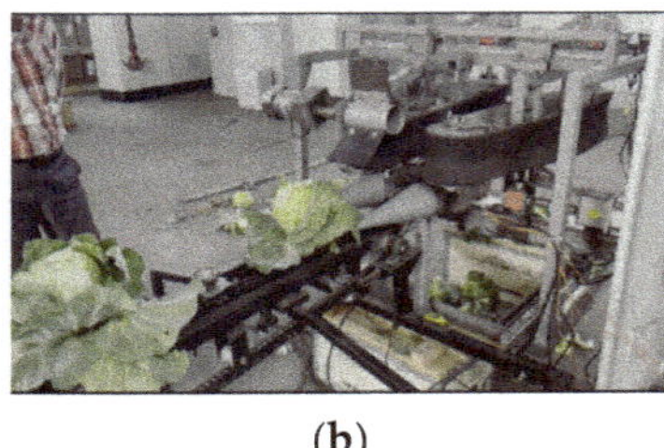

(**a**) (**b**)

Figure 18. Operation performance test: (**a**) conveying system of cabbage; (**b**) harvesting system of cabbage.

5.2. Test Index and Test Results

Since there are no related standards or regulations on mechanized cabbage harvesting. This experiment is based on GB/Z 26582-2011 Production Technical Practice for Cabbage [27] and JB/T 6276-2007 Testing Methods of Beet Harvesting Machinery [28] to select the qualified rate of harvesting as the standard for the harvesting test.

In the single-factor test affecting the qualified rate of cabbage harvest, other influencing factors should be fixed first. By changing one of the factors to obtain the variable relationship between the harvest pass rate and each test factor. Before the test, the machine is idle, and the test is carried out after the machine runs normally. Each group of tests was repeated 5 times, and 30 cabbages were tested each time. After the end of each group of experiments, the total number of test cabbages, the total number of successfully pulled cabbages, the total number of successfully clamped and transported cabbages, the total number of successfully cut cabbages, and the total number of qualified harvests were counted, respectively. Taking the qualified rate of cabbage harvest and the success rate of each link as the evaluation index, the initial level of each factor was set as follows: the rotation speed of the pulling roller was 100 r/min, the rotation speed of the conveyor belt was 160 r/min, and the rotation speed of the cutter head was 140 r/min.

(1) Single-factor test of extraction link: In the test, the speed of the fixed double disc cutter was 140 r/min, and the speed of the fixed flexible clamping conveyor belt was 160 r/min. The rotation speed of the pulling roller started from 60 r/min, and each comparison test increased by 20 r/min in turn. A total of 5 groups of tests were carried out, and the maximum rotation speed of the pulling roller test was 140 r/min.
(2) Single-factor test of flexible clamping conveying link: In the test, the rotation speed of the fixed pulling roller was 100 r/min, the rotation speed of the fixed double disc cutter was 140 r/min, and the rotation speed of the flexible clamping conveyor belt was increased from 80 r/min to 40 r/min each time. Five groups of tests were carried out, and the maximum rotation speed of the flexible clamping conveyor belt test was 240 r/min.
(3) Single-factor test of cutting link: In the test, the rotation speed of the fixed pulling roller was 100 r/min, the rotation speed of the fixed flexible clamping conveyor belt was 160 r/min, and the rotation speed of the cutter was increased from 100 r/min to 20 r/min each time. A total of 5 groups of tests were carried out. The maximum rotation speed of the double disc cutter test was 180 r/min.

The standard of mechanized harvesting of cabbage: At harvest, 2–3 outer leaves (rosette leaves) were retained to protect the leaf balls and ensure that the epidermis was clean and free of cracks.

Qualified rate of harvesting: The harvest qualified rate of cabbage was defined according to the production technical specifications of cabbage and the harvest quality requirements of stem and leaf vegetables: (1) The section of the cutting root must be smooth, and the two sections cannot be broken or cut out. (2) The cutting position should be 10–15 mm above the cabbage outer leaf, and the outer leaf should be cut off at the same time. (3) After harvest, 2–3 outer leaves were retained to protect the leaf head. (4) No crack ball, extrusion damage, cutting damage, and so on caused by the harvesting operation occurred. The formula for the harvest-qualified rate is as follows:

$$N = \frac{N_1}{N_0} \times 100\% \quad (37)$$

where N is the qualified rate of harvest, %; N_1 is the number of qualified cabbages harvested in a single test; and N_0 is the total number of cabbages harvested in a single test.

The test results of the pulling link, clamping and conveying link, and cutting link are shown in Tables 4–6.

Table 4. Analysis of single-factor test results of pulling roller speed.

Rotating Speed of Pulling Roller (r/min)	Total Number of Cabbages Tested	Number of Damaged Cabbages	Number of Successfully Pulled Cabbages	Number of Qualified Cabbages Harvested	The Success Rate of Pulling (%)	Damage Rate (%)	Qualified Rate of Harvesting (%)
60	30	2	28	26	90	6.7	86.7
80	30	1	26	25	86.7	3.3	83.3
100	30	1	29	28	96.7	3.3	93.3
120	30	3	29	26	93.3	10	86.7
140	30	4	29	25	96.7	13.3	83.3
		Average Value			92.7	7.32	86.7

Table 5. Analysis of single-factor test results of clamping conveying speed.

Rotation Speed of Clamping Conveyor Belt Speed (r/min)	Total Number of Cabbages Tested	Number of Damaged Cabbages	Number of Cabbages Successfully Transported by Clamping	Number of Qualified Cabbages Harvested	The Success Rate of Clamping Conveying	Damage Rate (%)	Qualified Rate of Harvesting (%)
80	30	6	25	19	73.3	20	63.3
120	30	4	27	23	86.6	13.3	76.6
160	30	3	29	26	93.3	10	86.6
200	30	3	28	25	96.7	10	83.3
240	30	2	22	20	93.3	6.7	66.7
		Average Value			88.6	12	75.3

Table 6. Analysis of single-factor test results of cutting speed.

Rotation Speed of Double Disc Cutter (r/min)	Total Number of Cabbages Tested	Number of Damaged Cabbages	Number of Successfully Cut Cabbage	Number of Qualified Cabbages Harvested	Success Rate of Cutting	Damage Rate (%)	Qualified Rate of Harvesting (%)
100	30	6	24	18	80	20	60
120	30	5	25	19	83.3	16.7	63.3
140	30	3	28	25	93.3	10	83.3
160	30	2	28	26	93.3	6.7	86.7
180	30	1	29	28	96.7	3.3	93.3
		Average Value			89.3	11.34	77.3

In the single-factor test, the damage types of cabbage and the qualified cabbage are shown in Figure 19.

Figure 19. Damage types of cabbage: (a) broken cabbage; (b) scratched cabbage; (c) cabbage with cut loss; (d) harvest qualified cabbage.

6. Discussion

From the test results, it can be obtained that the average success rate of pulling of low-loss harvesting test platform was 92.7%, the average success rate of clamping and conveying was 88.6%, the average success rate of root cutting was 89.3%, the average qualified rate of harvesting in the pulling link was 86.7%, the average qualified rate of harvesting in the clamping and conveying link was 75.3%, and the average qualified rate of harvesting in the cutting link was 77.3%. The average damage rate of the pulling process was 7.32%, the average damage rate of the clamping and conveying link was 12%, and the average damage rate of the cutting process was 11.34%. During the test, the orderly feeding of cabbage can be realized by controlling the working frequency of the servo motor of the conveying system to avoid blockage and failure of operation performance. The pulling device and the reeling device can successfully pull the cabbage from the conveying bench, straighten up the cabbage to the feeding inlet of the clamping conveying device, and complete an effective harvest with the cutting device. Therefore, the test platform meets the technical requirements for the harvest of cabbage.

According to the test results, it can be inferred that when the pulling device is controlled at 80–120 r/min, the speed of the clamping and conveying device is controlled at 120–240 r/min, and the speed of the double disc cutter is controlled at 140–180 r/min, each link of the harvesting system has a high success rate. By observing the results of the aggregate box of each link, it can be seen that the damage rate of the pulling link in the harvesting system is low, while the damage rate of the clamping, conveying, and cutting links is high. Figure 16 shows the harvesting process and damage types of the low-loss harvesting test platform for cabbage. The reasons for the above phenomena are as follows: (1) The cabbage is not straightened when entering the clamping conveyor belt feeding inlet through the pulling device, which causes the root cutting position to shift. (2) The friction coefficient of the conveyor belt material is large, and the cabbage is scratched during the clamping and conveying process. (3) The too-fast or too-slow speed of the clamping conveyor belt makes the time of the cutter acting on the rhizome shorter in unit time, and the rhizome is not cut off or slipped. From the above causes of damage, it can be determined that the taper of the pulling roller, the material of the clamping conveyor belt, and the ratio of the clamping conveyor belt to the cutter head motion parameters are the key factors affecting the qualified rate of harvesting. Therefore, the experimental optimization will be carried out in the follow-up study in order to find out the critical conditions, the best equipment material, and the optimal working parameter combination for the damage of the cabbage in each link of the harvesting system and provide data support for the low-loss harvesting equipment for the cabbage.

7. Conclusions

(1) The basic physical characteristics and mechanical harvesting characteristics of cabbage, the representative of cabbage, were collected: the maximum shear force of the cabbage rhizome was 137.138 N, and the maximum crushing force was 1198.3 N.

(2) The low-loss harvesting process for cabbage was developed, and the structure of the prototype was determined. Combined with dynamics and kinematics analysis, the working parts of the pulling device, the reeling device, the clamping and conveying device, and the cutting device were designed and selected, and the above devices were integrated to design a low-loss harvesting test platform for the cabbage.

(3) The performance test of the low-loss harvesting test platform for cabbage shows that when the pulling device is controlled at 80–120 r/min, the speed of the clamping and conveying device is controlled at 120–240 r/min, and the speed of the double disc cutter is controlled at 140–180 r/min. Each link in the harvesting system has a high success rate. The success rate of pulling of low-loss harvesting test platform was 86.7–96.7%, the success rate of clamping and conveying was 73.3–96.7%, the success rate of root cutting was 80–96.7%, the qualified rate of harvesting in the pulling link was 76.7–93.3%, the qualified rate of harvesting in the clamping and conveying link was 63.3–86.7%, and the qualified rate of harvesting in the cutting link was 60–93.3%. The damage rate of the pulling process was 3.3–13.3%, the damage rate of the clamping and conveying link was 6.7–20%, and the damage rate of the cutting process was 3.3–20%.

Author Contributions: Conceptualization, W.T. and J.Z.; methodology, W.T. and J.Z.; software, W.T. and Z.S.; validation, W.T., J.Z. and Z.S.; formal analysis, W.T. and X.N.; investigation, W.T.; resources, W.T.; data curation, G.C.; writing—original draft preparation, W.T.; writing—review and editing, W.T.; visualization, Z.S.; supervision, G.C.; project administration, J.Z., G.C. and Z.S. All authors have read and agreed to the published version of the manuscript.

Funding: This research was funded by the National Natural Science Foundation of China (Grant No. 52205273), Jiangsu Agriculture Science and Technology Innovation Fund (Grant No. CX(22)2020), Key R&D Program of Shandong Province in China (Grant No. 2022CXGC010612), Innovation Project of Chinese Academy of Agricultural Sciences (Grant No. CAAS-ASTIP-NIAM).

Institutional Review Board Statement: Not applicable.

Data Availability Statement: The datasets used and/or analyzed during the current study are available from the corresponding author on reasonable request.

Conflicts of Interest: The authors declare no conflict of interest.

References

1. Chunsong, G.; Qing, G.; Xian, L.; Ya, Y.; Zhi, C.; Yong, C. Present situation and development suggestions of cabbage mechanized production in China. *China Veg.* **2019**, *10*, 1–8. (In Chinese)
2. Yang, L.M.; Fang, Z.Y.; Zhang, Y.Y.; Zhuang, M.; Lv, H.H.; Wang, Y.; Ji, J.L.; Liu, Y.M.; Li, Z.S.; Han, F.Q. Research progress on cabbage genetic breeding during 'The Thirteenth Five-year Plan' in China. *China Veg.* **2021**, *1*, 15–21. (In Chinese)
3. El Didamony, M.I.; El Shal, A.M. Fabrication and Evaluation of a Cabbage Harvester Prototype. *Agriculture* **2020**, *10*, 631. [CrossRef]
4. Wang, Z.Y.; Yin, J.; Zhang, K. Design and analysis of vibration device for clamping wheel of vibration conveyor chain of rhizome plant harvester. *Guizhou Sci.* **2018**, *36*, 64–67.
5. Li, H. *Development of Split Combined Header and Directional Conveying Device for Rape Combine Harvester*; Huazhong Agricultural University: Wuhan, China, 2017.
6. Zhao, Q.N.; Wang, J.J.; Wang, Z.; Tian, T.; Zhao, Y.L.; Cui, L.; Hong, M. Application of clamping and conveying technology in vegetable harvesting machinery. *Agric. Eng.* **2020**, *10*, 8–11.
7. Willem, M.; Thomas, P. Cabbage Harvester Machine. Patent WO2002009497A1, 7 February 2002; pp. 38–42.
8. Chattopadhyay, P.S.; Pandey, K.P. Effect of knife and operational parameters on energy requirement in flail forage harvesting. *J. Agric. Eng. Res.* **1999**, *73*, 3–12. [CrossRef]
9. Hansen, C.J. Harvesting Machine for Cabbage, or the Like. U.S. Patent 3827503, 6 August 1974.
10. Bleinroth, H. A Device for Harvesting Headed Cabbages. GB Patent Application No. 1401632, 30 July 1975.
11. Baker, W.M. Cabbage and Lettuce Harvesters. U.S. Patent 3827503, 24 February 1970.

12. Mori, G.; Paterno, F.; Santoro, C. Design and development of multidevice user interfaces through multiple logical descriptions. *IEEE Trans. Softw. Eng.* **2004**, *30*, 507–520. [CrossRef]
13. Zhou, C. *Study on the Cabbage Key Harvesting Technology and Harvester*; Northeast Agricultural University: Harbin, China, 2013.
14. Du, D.D. *Research on Crawler Self-Propelled Cabbage Harvesting Equipment and Development of Its Weighing System*; Zhejiang University: Hangzhou, China, 2017.
15. Zhang, T.; Li, Y.; Song, S.M.; Pang, Y.L.; Shao, W.X.; Tang, X.L. Design and experiment of green cabbage head harvester based on flexible clamping. *Trans. CSAM* **2020**, *51* (Suppl. S2), 162–169. (In Chinese)
16. Cai, J.J.; Zhang, J.X.; Tiemuer, Y.; Gao, Z.M.; Rui, Z.Y.; Liu, X. Design and experiment of clamping belt cotton stalk harvester. *Agric. Mach. J.* **2020**, *51*, 152–160.
17. Lenker, D.H.; Nascimento, D.F.; Adrian, P.A. Apparatus for Harvesting Vegetable Heads. U.S. Patent 4136509, 30 January 1979.
18. Wadsworth, W.F. Feed conveyor apparatus. U.S. Patent 3858660, 7 January 1975.
19. Bleinroth, H. A harvesting Machine for Headed Cabbages. GB Patent Application No. 1402844, 13 July 1975.
20. Horia, M.; El-Sahhar, R.A.; Mostafa, M.M. A developed machine to harvest carrot. *Farm Mach. Power* **2008**, *25*, 1163–1173.
21. Abe, T.; Kobuchi, T.; Tokunaga, H. Harvester for Root Vegetables. U.S. Patent 543123, 11 July 1999.
22. Du, D.D.; Fei, G.Q.; Wang, J.; Huang, J.J.; You, X.R. Development and experiment of self-propelled cabbage harvester. *Trans. CSAE* **2015**, *31*, 16–23. (In Chinese)
23. Zhang, J.F.; Xiao, H.; Yao, S.; Song, Z.; Jin, Y.; Wang, W.; Yang, G. Physical and mechanical properties of cabbage. *Int. Alliance Equest. J.* **2020**, *2*, 1–8.
24. Shi, G.; Li, J.; Kan, Z.; Ding, L.; Ding, H.; Zhou, L.; Wang, L. Design and parameters optimization of a provoke-suction type harvester for ground jujube fruit. *Agriculture* **2022**, *12*, 409. [CrossRef]
25. Zhang, J.; Cao, G.; Jin, Y.; Tong, W.; Zhao, Y.; Song, Z. Parameter Optimization and Testing of a Self-Propelled Combine Cabbage Harvester. *Agriculture* **2022**, *10*, 1610. [CrossRef]
26. Li, X.Q.; Wang, F.E.; Guo, W.J.; Gong, Z.W.; Zhang, J. Analysis of influencing factors on cutting force of cabbage rhizomes. *Trans. CSAE* **2013**, *29*, 42–48. (In Chinese)
27. *GB/Z 26582-2011*; Production Technical Practice for Cabbage. China National GB Standard Research: Shenzhen, China, 2011.
28. *JB/T 6276-2007*; Test Method for Sugar Beet Harvesting Machinery. Ministry of Machinery and Electronics Industry of the People's Republic of China: Beijing, China, 2007.

 *agriculture*

Article

The Random Vibrations of the Active Body of the Cultivators

Petru Cardei [1], Nicolae Constantin [2], Vergil Muraru [1,*], Catalin Persu [1], Raluca Sfiru [1], Nicolae-Valentin Vladut [1], Nicoleta Ungureanu [2,*], Mihai Matache [1], Cornelia Muraru-Ionel [1], Oana-Diana Cristea [1] and Evelin-Anda Laza [1]

[1] National Institute of Research—Development for Machines and Installations Designed for Agriculture and Food Industry—INMA, 013811 Bucharest, Romania; petru_cardei@yahoo.com (P.C.); persu@inma.ro (C.P.); raluca_sfiru@yahoo.com (R.S.); vladut@inma.ro (N.-V.V.); matache@inma.ro (M.M.); cornelia.muraru.ionel@gmail.com (C.M.-I.); diana10.cristea@gmail.com (O.-D.C.); eveline_anda@yahoo.com (E.-A.L.)

[2] Department of Biotechnical Systems, Faculty of Biotechnical Systems Engineering, National University of Science and Technology Politehnica Bucharest, 060042 Bucharest, Romania; nicu.constantin49@yahoo.com

* Correspondence: virgil.muraru@gmail.com (V.M.); nicoleta.ungureanu@upb.ro (N.U.); Tel.: +40-7-4435-7250 (V.M.)

Abstract: The article continues the exposition of the results obtained in researching an agricultural machine for processing soil, designed for research with applications including exploitation. The MCLS (*complex machine for soil tillage*) was designed to research the working processes of the instruments intended for soil processing. The MCLS cultivator is a modulated machine (it can work for three working widths: 1, 2, and 4 m, with tractors of different powers) that is designed to use a wide range of working bodies. The experimental data obtained with the structure with a working width of 1 m and the results of their processing within the framework of the theory of random vibrations are presented in this article. The experimental results are analysed as random vibrations of the supports of the active working bodies. As a result, the main characteristics of random vibrations are exposed: the distribution function, the average value, the autocorrelation, and the frequency spectrum. These general results regarding random vibrations are used for several critical applications in the design, execution, and exploitation of some subassemblies and assemblies of agricultural machines of this type. The main applications include estimating the probability of the occurrence of dangerous load peaks, counting and selecting the load peaks that produce fatigue accumulation in the material of the supports of the working bodies, identifying some design deficiencies or defects in the work regime, and estimating the effects of vibrations on the quality of soil processing. All of the outcomes are composed of applications in MCLS research and exploitation. The applications pursue well-known objectives of modelling the working processes of agricultural machines: safety at work, increasing the quality of work, optimising energy consumption, and increasing productivity, all in a broad context to obtain a compromise situation. The material and the method are based on experimental data acquisition, processing, and interpretation.

Keywords: random; vibrations; tillage; tools; complex; cultivator

Citation: Cardei, P.; Constantin, N.; Muraru, V.; Persu, C.; Sfiru, R.; Vladut, N.-V.; Ungureanu, N.; Matache, M.; Muraru-Ionel, C.; Cristea, O.-D.; et al. The Random Vibrations of the Active Body of the Cultivators. *Agriculture* **2023**, *13*, 1565. https://doi.org/10.3390/agriculture13081565

Academic Editors: Jin He and Caiyun Lu

Received: 25 April 2023
Revised: 26 July 2023
Accepted: 31 July 2023
Published: 4 August 2023

1. Introduction

The phenomena and work processes encountered in agriculture involve, for the most part, important areas of the field of biology (environment, soil, plants, animals, etc.), which is "living". According to all the assessments in the literature, the field of life is a field of unpredictability, uncertainty, randomness, and probability. According to [1], "understanding randomness is essential for modern life, as it underpins decisions under uncertainty".

Random phenomena are treated in physics through theories different from deterministic ones, using the concepts of probability theory and statistics. In [2], it is shown that "in mechanical engineering, random vibration is motion that is non-deterministic, meaning that future behaviour cannot be precisely predicted. The randomness is a characteristic of the excitation or input, not the mode shapes or natural frequencies. Some common examples include an automobile riding on a rough road, wave height on the water, or the load induced on an airplane wing during flight. Structural response to random vibration is usually treated using statistical or probabilistic approaches. Mathematically, random vibration is characterized as an ergodic and stationary process".

In [3], it is explained that an experiment is random when the answer cannot be precisely and unambiguously predicted, as several answers are possible.

According to [2], the main way of approaching random vibrations is through "a measurement of the acceleration spectral density (ASD), which is the usual way to specify random vibration. The root mean square acceleration (Grms) is the square root of the area under the ASD curve in the frequency domain. The Grms value is typically used to express the overall energy of a particular random vibration event and is a statistical value used in mechanical engineering for structural design and analysis purposes." Additionally, in [2], it is stated that "while the term power spectral density (PSD) is commonly used to specify a random vibration event, ASD is more appropriate when acceleration is being measured and used in structural analysis and testing." Such specifications and assessments can be found in [4–9]. In the case of the study whose results are presented in this article, we worked with sequences of specific deformations measured and recorded, which were then converted into loading forces on the supports of the active working bodies.

As the scientific literature shows, work processes in agriculture have random characteristics [10–20]. Much of the specified literature has agro-economic or bio-economic origins, many of which emphasise management. Uncertainties in agricultural work processes are also caused by management system uncertainties [21]. In [22], it is stated that "problems related to agriculture are, in essence, stochastic because of the uncertain nature of their parameters. Many systems in this sector are affected by yield uncertainty caused by factors such as climatic conditions. Uncertainty and imperfect information involved therein challenge decision-making, as decision-makers are led to make decisions before observing the realization of the random factors. Traditional approaches to dealing with agricultural problems do not integrate the risks and uncertainties involved therein, while it is relevant for efficient managerial decision-making to consider uncertainties and respond to opportunities and threats." The authors [23] show that "the second law of thermodynamics states that entropy or randomness in a given system will increase with time". This is shown in science, "where more and more biological processes have been found to be independent" and "randomness is the fundamental and overarching principle that helps to explain how traits are independently passed from parent to offspring". It is the presence of randomness in all biological systems that this paper aims to highlight.

In the exploitation of agricultural machines, the study of vibrations is carried out more and more experimentally. The results are processed using mathematical statistics in the spirit of the theory of random vibrations [24–30]. The classical theory of vibrations cannot effectively study the complex vibrations of agricultural machines due to objective reasons: the difficulty of estimating model constants, nonlinear behaviours, and excitations, most often of a random nature. Apart from these aspects, the influence of games with a functional character, which is impossible to model in a deterministic way, is manifest. Concrete cases are harvesters [24–26], farm tractors in transport or various agricultural works [25,27–33], as well as cultivator subsoilers [34–36].

In the field of agricultural processes that directly involve the soil, the random character of the process is due for the most part to the random properties of the soil. In agricultural machines intended for soil processing, the soil-working body contact excites the analysed structures. The highly random properties of the soil induce the excitation of the process (especially contact forces) to have a deep random character. In general, the dynamic processes produced by the contact between the soil and the working body of agricultural machines intended for soil processing are well included in the category of random vibrations. Sometimes the machines themselves include vibration sources with adjustable, programmable characteristics to achieve a certain type of soil processing. In the latter case, in the experimentally recorded signals, it is natural for the deterministic vibration elements to appear in the spectrum, even if they have been altered to some extent. The random vibrations of agricultural machines intended for soil processing are another example of random vibrations encountered in the technique, apart from those given in [2]. We recall the definition of the agricultural cultivator as given in [37], that is, an agricultural machine that serves to shred and loosen the soil, to destroy weeds from crops of creeping plants, etc. This definition was added because the study material presented in this article has as a main component a working variant of a complex cultivator with variable working width. According to [38], the cultivator can be understood as an agricultural machine used for surface soil work to loosen and destroy weeds without overturning the furrow.

The original aspect of this study is the presentation of a research method based on the theory of random vibrations using the capabilities of a complex modulated cultivator with a variable working width, MCLS. MCLS performance testing is presented simultaneously as a research method for other similar machines or for machines obtained from MCLS by modulating or fitting different active bodies that the structure can use.

Also, the paper tries to show the researchers the benefits produced by the study of vibrations in an experimental framework with the help of the theory of random functions. Most researchers are trained in the scientific spirit of the deterministic theory of vibrations, which is why their first tendency is to look for answers in terms of the deterministic theory. From there, it comes to mathematical modelling and simulation, which are easy to develop on the computer. After the models are run, however, it is found that it is difficult to give answers to the real problems, especially to motivate and validate the hypotheses and mathematical models. Advanced designers in the field of such phenomena have been using the path of experimentation and the theory of random functions for decades. The deterministic approach tries to force physical reality into the model's pattern, while modelling in the experimental framework seeks to take as much as possible from the real conditions and extract a minimum of useful conditions in conception, design, execution, and exploitation. The most difficult problems with this last option are the generalisation (which is low for the time being) and the generalisation costs (which are very high).

2. Material and Methods

Research on the MCLS soil tillage machine (a variant with a working width of 1 m) is the subject of this article. The presented results describe the experiments and the statistical processing of the experimental data. The MCLS complex cultivator was initially conceived, designed, and realised to facilitate research in the field of soil processing [39,40]. The machine was designed for three working widths: 4 m and 2 m (with an 80 HP tractor) and 1 m (for a 45 HP tractor). Additionally, the supporting structure can support the installation of 5–10 types of working bodies to make comparative estimates as complete as possible. After conception, design, and execution, the idea of using the machine in farm operations also appeared. Thus, the machine can be used simultaneously for farm applications and research. Images from the experimental activity with the MCLS machine in all working variants are given in Figures 1–4.

The experiments were carried out in Romania, in the Bucharest-Ilfov area, in an experimental plot owned by the INMA Bucharest Institute, which has a transitional temperate continental climate. The morphological characteristics of the soil were: red-yellow clay loam, in a dry state; medium and small angular agglomerate structures; slightly loosened. The soil moisture at the measurement time was between 20 and 35 percent (Delta-T HH2 digital moisture meter), while the average soil compaction level was 45 KPa.

Traction is provided by agricultural tractors New Holland 80 TD and New Holland 50 TD. The working speeds were between 0.78 and 2.15 m/s. The results presented in this article refer to the experiment carried out at a constant speed of 0.789 m/s.

The experiments were carried out on a plot of 500 square metres, marked with benchmarks. For each test, 10 m were set for the acceleration and 10 m for the deceleration of the tractor, and 30 m were set for the evaluation of the cultivator (see Figure 4).

Figure 1. The MCLS complex cultivator operates in the version with a working width of 4 m.

Figure 2. MCLS in operation in the 2 m working width variant.

Figure 3. MCLS in operation in the 1 m working width variant.

The QuantumX MX1615B data acquisition system was used for the strain gauge measurements, with 32 measurement channels and KFG-6-120-C1-11 N15C2 strain gauge sensors (120 Ω). The strain gauge amplifiers, QuantumX MX1615B, are suitable for precise data acquisition in full-bridge, half-bridge, and quarter-bridge configurations, as well as for strain gauge-based transducers, potentiometers, resistance thermometers (PT100), or normalised voltage (+/−10 V).

The measurement of the forces on each support of the working body was performed by the strain gauge methodology, with the procedure including a calibration stage of the indications of the deformation sensors. Although it is not the most precise measurement method, it is convenient and accessible, especially considering that only values measured by the same procedure are compared and are, therefore, hypothetically, affected by comparable errors.

Figure 4. The location and parcelling of the land on which the experiments were carried out with the MCLS machine.

The experiments took place in the testing ground of INMA Bucharest, where the soil was classified as reddish-brown from the forest and pre-processed with autumn ploughing (coarse). Figure 5 depicts the location of the experimental polygon in detail with reference to the Bucharest-Baneasa airport runway.

Figure 5. Indexing of measurement locations on the structure.

2.1. The Subject of Research

Approaching the soil processing phenomenon using an MCLS-type structure involves the statistical processing of experimental results (descriptive and inferential statistics, random process theory). Descriptive and partial inferential statistics were discussed in a previous article [41]. In this article, the focus will be on approaching the problem within the theory of random processes. In another article [41], all the data strings that are generated by the contact between the ground and the working body have a random character and are comparable to the pseudorandom strings generated by special programmes or to the strings of prime numbers of the same length.

In order to approach it within the theory of random functions, the phenomenon of ground processing by each organ was considered a phenomenon of random vibrations. The main random component is the excitement of the process. Excitation is generated by the interaction of the working body with the ground. In the range of demands considered, the vibrations can be considered linear-elastic.

In this article, the analysis will be made only for the MCLS variant with a working width of 1 m because, in the case of this variant, signals were collected from all the supports of the working bodies throughout the experiment. The analysis carried out on the simple version with a width of 1 m can later be extended to the versions with higher working widths, for which the load on each support cannot yet be measured and other random parameters also appear (for example, the clearances in the truss with a working width of 4 m).

The indexing of the measurement locations and the signal transmission channels from the deformation sensors to the acquisition board are shown in Figure 5. An image of the MCLS variant in experimental work, on the parcelled track of the polygon, with the complete data acquisition equipment, is shown in Figure 6.

Figure 6. MCLS variant with a working width of 1 m in operation with a 45 HP tractor.

2.1.1. The Formulation of the Problem in Terms of the Theory of Random Vibrations

According to [42], physical processes that cannot be characterised by deterministic functions, i.e., vibrations whose instantaneous values cannot be predicted as functions of time, are called random. The random nature of the vibrations of the investigated structure was demonstrated in [41], where even the degree of randomness was quantified. A relatively simple criterion for experimental recognition of the randomness of a phenomenon that is experimental is formulated, for example, in [42]: if, in several experiments organised under identical controllable conditions, the measured quantities differ only by quantities of the order of errors of measurement, the phenomenon can be considered deterministic; it is reproducible, so its development after the moment of measurement is predictable. Otherwise, the phenomenon is random. Apparently, the criterion seems simple, but in the case of soil processing and, in addition, using non-conventional and insufficiently verified structure load measurement systems (strain gauge methodology), this criterion is difficult to use. Given the findings from [41], we opted for the random vibration variant to describe the working process of the 1 m working width version of the MCLS complex cultivator.

The main characteristic of a random variable is the distribution function [43,44]. This is defined according to Formula (1).

$$F_X(x) = P(X < x),\ x \in \mathbb{R} \tag{1}$$

where F_X is *the distribution function of the random variable* x, x is a real number, and $\mathbb{R}$ is the set of real numbers. In Formula (1), some authors use non-strict inequality.

2.1.2. Distribution Functions

In Figure 7 in the column of graphs on the left, the temporal variations of the forces that load the supports of the twelve working bodies of the version with a working width of 1 m of the MCLS are shown. Additionally, in Figure 7, the histogram of the frequencies of the force values is represented in the right column. The polygon of frequencies is built using the histograms. When the number of classes in the histograms becomes very large (theoretically to infinity), the frequency polygon tends to reflect the real distribution density of each force. The frequency histograms are constructed with 64 classes using the formula of Mosteller and Tukey [45]. We took into account the length of the data strings obtained experimentally in calculating the number of classes in the histograms.

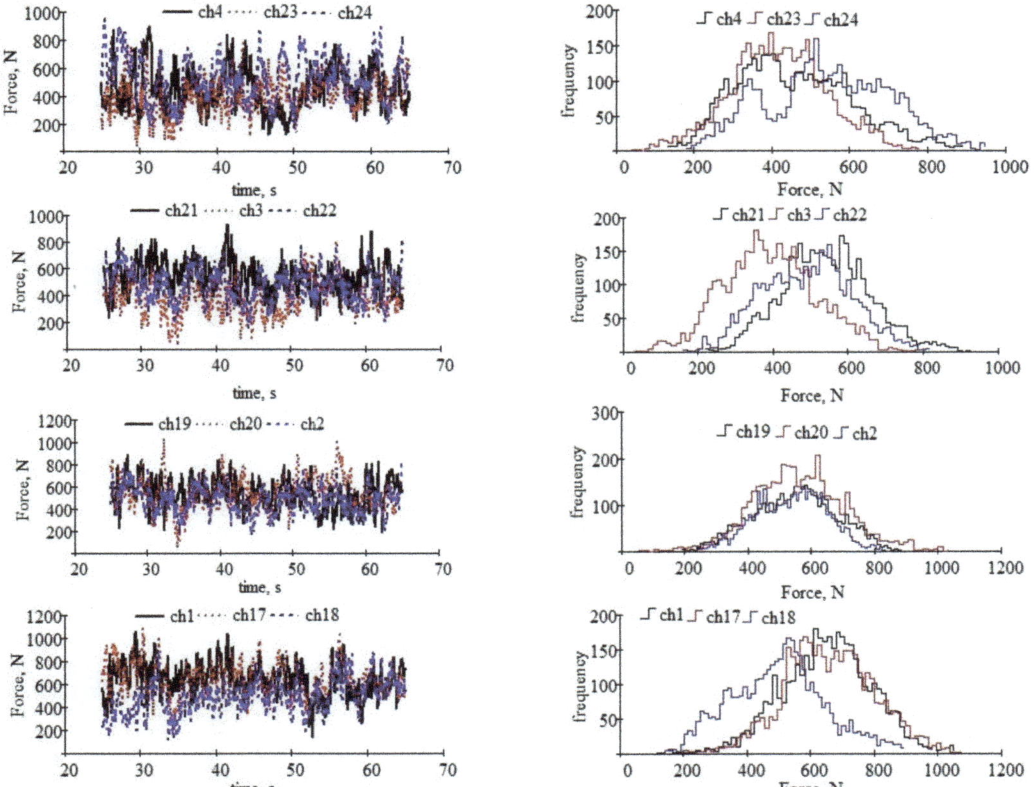

Figure 7. The time dependence of the load in the supports of the active bodies (**left column**) and the frequency distribution of the load values (**right column**). The four rows of figures correspond to the four lines of active organs, starting from the top with the ones closest to the tractor.

The elementary Formula (2) was used to calculate frequency histograms.

$$f_i = \sum_{j=1}^{N} H_i(x_j),\ i = 1, \ldots, n_h,\ j = 1, \ldots, N \tag{2}$$

where:

$$H_i(x_j) = \begin{cases} 1, \; h_i \le x_j < h_{i+1} \\ 0, x_j < h_i \; or \; x_j \ge h_{i+1} \end{cases}, \quad i = 1, \ldots, n_h, \quad j = 1, \ldots, N, \tag{3}$$

$$H_{n_h}(x_j) = \begin{cases} 1, \; h_{n_h - 1} \le x_j \le h_{n_h} \\ 0, x_j \langle h_{n_h - 1} \; or \; x_j \rangle h_{n_h} \end{cases}, \quad j = 1, \ldots, N \tag{4}$$

N is the number of observations from the examined random sequence, and n_h is the number of classes in the frequency histogram. Then the probability density histogram was calculated according to Formula (5):

$$\rho_i = \frac{f_i}{N}, \; i = 1, \ldots, n_h \tag{5}$$

where $\{\rho_i\}_{i=1,\ldots,n_h}$ is the series of probability densities. The series of probabilities (or cumulative probabilities) is calculated using Formula (6).

$$p_i = \sum_{k=1}^{i} \rho_k, \quad i = 1, \ldots, n_h \tag{6}$$

In Figure 8, the probability density and the probability that the random variable (the series of numerical values over time recorded at each of the twelve measurement locations) will take different values in the experimental range are shown.

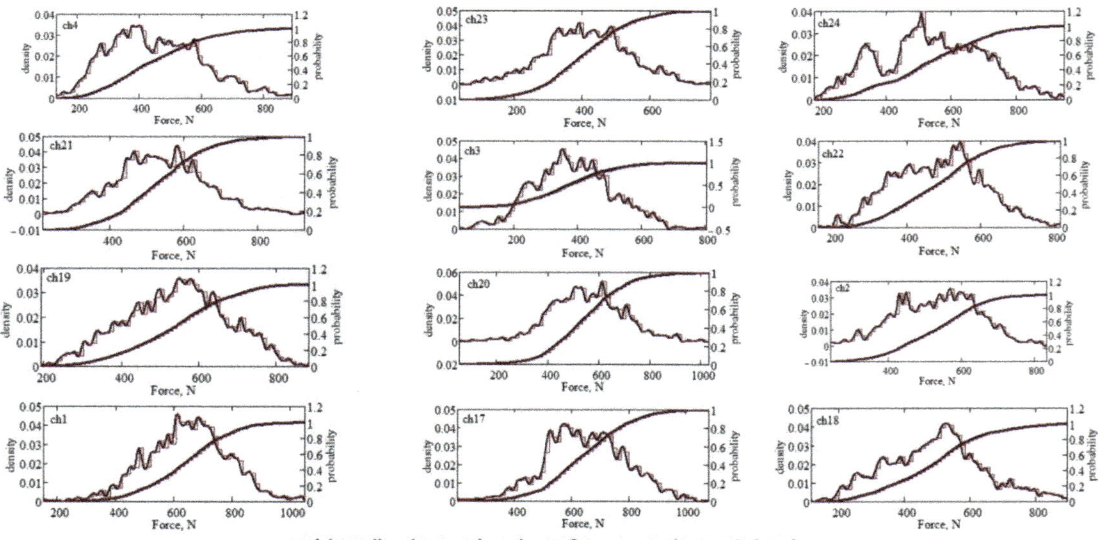

Figure 8. The probability density and the probability of taking certain values from the experimental range for the forces recorded in the twelve measurement locations.

The probability densities calculated as shown above do not resemble those found in statistics books or in articles in which the experimental results are approximated, by hypothesis, with one of the classical statistical distributions (normal, exponential, Student, Fischer, etc.). Probability distributions differ from the statistical densities with which experimental curves are usually approximated. Precision is lost by approximating with the classical statistical distributions of the probability densities, and we cannot precisely estimate the local error in relation to the experimental data. For this reason, we preferred, considering our goals, to interpolate the empirical probability densities through cubic spline functions since we wanted maximum precision and were not interested in using the results in immediate theoretic models. Regarding the probability curves, they have shapes very similar to the classical ones, but for the same reasons, we proceeded with the probabilities by interpolating with spline functions. In Figure 8, the experimental curves interpolated by spline functions are given in each graph for the twelve series of 4000 numerical data points.

The probabilities interpolated with spline functions are used to calculate the probability of the occurrence of a force greater than a fixed force (related to the limit resistance characteristics of the material of the supports of the working body).

The distribution functions of the twelve random variables described by the numerical sequences in Figure 7 are calculated in the same way as the probability densities and probabilities whose graphic representations are given in Figure 8. Therefore, the distribution functions are expressed as interpolations by cubic spline functions, for which the graphic representations are given in Figure 9. The graphic representations of the twelve distribution functions of the sequences of numerical values coming from the measurement points located on the supports of the twelve working bodies are grouped three by one in the graph, corresponding to the three working bodies in each line, with the counting starting from the back of the tractor.

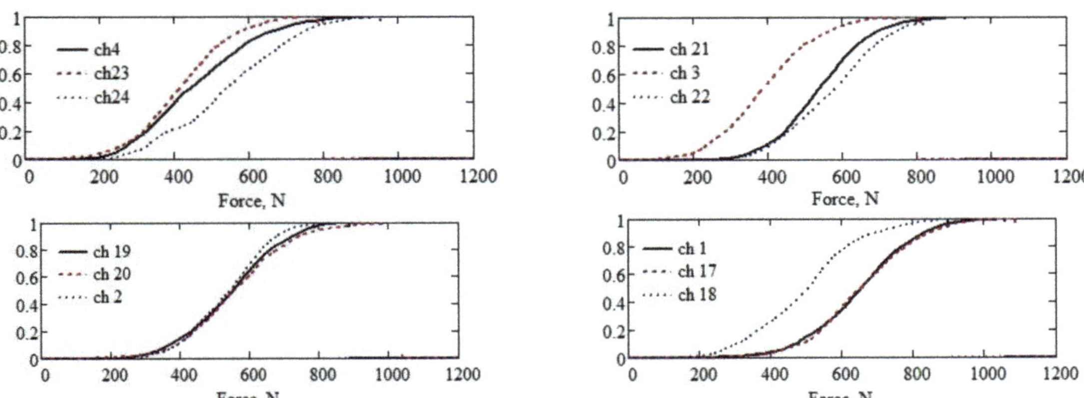

Figure 9. The graphic representation of the distribution functions of the twelve numerical data sequences was obtained by measuring at the locations indicated in Figure 5.

For complete agreement with the theoretical definition, the interpolated distribution functions are extended with the value zero to the left of the minimum value and with the value one to the right of the maximum value of the corresponding random variable.

2.1.3. Characteristic Functions of Random Vibrations

From a theoretical point of view, the first tested properties of random vibrations are *stationarity* and *ergodicity*. The definitions of these notions can be found in all the literature dedicated to random vibrations, for example, in [2,4–9,42,43,46]. In general, the response of structures to random vibrations is studied through experiments that have strictly elaborated procedural standards [47].

A synthetic characterisation of random signals is given in [3]. The non-deterministic signal has specific characteristics: mean, dispersion, global average, global dispersion, histogram, power spectral density, etc. The signal can have a certain degree of predictability in its evolution over time. Depending on specific characteristics, the non-deterministic signal can be:

Stationary, if the mean and dispersion do not depend on time but are constant;
Ergodic, if the mean per portion does not differ from the global mean;
White noise, with a constant power spectral density throughout the frequency band.

Testing the stationarity and ergodicity of the experimental signals (sequences of finite length) cannot be conducted using the definitions of the mean value and the autocorrelation function of the random variables, for example, in [42]. The definitions in [42] assume processes for crossing the limit after the number of samples. We do not have an infinite number of samples, and if we did, we would be constrained by costs to limit the duration of the experiences as much as possible. For this reason, the calculations are made for a finite number of samples, limited by the number of samples in the entire sequence.

In Figure 10, we see the behaviour of the average value of the random variable given by the registration with the ch4 code (coming from the working body in the first line after the tractor, on the extreme left). The tendency to decrease the amplitude is observed with an increase in the number of samples in the sequence considered for calculating the average value. Therefore, one can only suspect asymptotic behaviour, but such behaviour cannot be stated strictly theoretically.

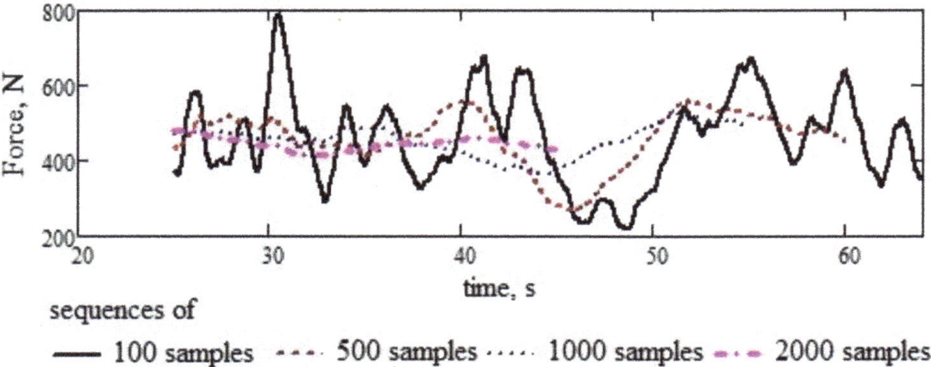

Figure 10. The behaviour of average values increases as the number of samples used in their calculation increases.

In Figure 11, the terms below the limit of the definition of the autocorrelation function are calculated for the subsequence of the ch4 signal, with the graphic representation limited to only five values of the gap. There is a tendency to decrease the amplitude of the autocorrelation function with the gap value and a weak asymptotic tendency, which does not allow us to make statements about the stationarity and/or ergodicity of the signal.

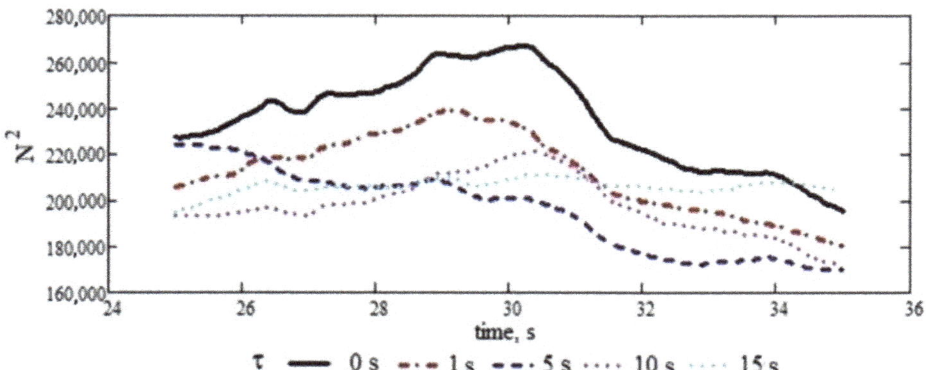

Figure 11. Autocorrelations of the ch4 signal for five offset (τ) values.

In Figure 12, the autocorrelation curves are given for each of the twelve signals collected from the supports of the working bodies of the 1 m working width version of the MCLS. The curves are calculated using the *lcorr* function of the programme [48]. The calculation can also be performed directly by programming the three simple formulas given in [49]. These curves can be considered substitutes in the field of real signals of finite length for the assessment of the stationarity or ergodicity of signals. In Figure 12, it is first observed that all twelve signals tend to asymptote in time towards the value 0. First, this means that the signals are weakly autocorrelated (another argument for their randomness). The fact that they are not constant in time shows that the signals are neither stationary nor ergodic.

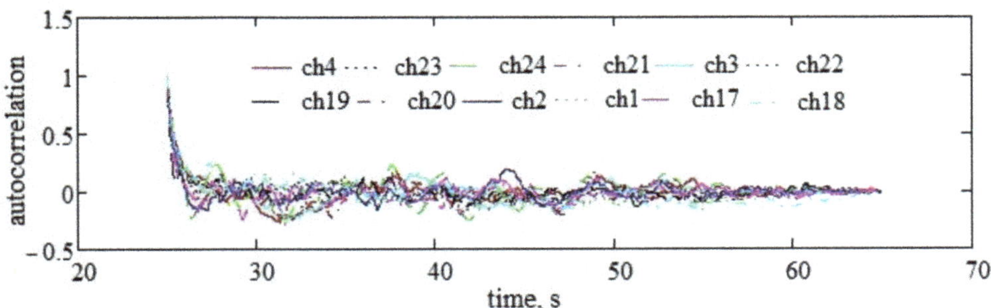

Figure 12. The autocorrelations of the twelve signals recorded in the experiment were calculated using [50].

2.1.4. Spectral Analysis

The frequency spectrum of the signals collected from the twelve working bodies provides interesting information, especially from identifying possible deterministic components that appear in all the examined sequences, which indicate possible deficiencies of the working regime, abnormal clearance, or defects. Deterministic components can also appear in agricultural soil processing machines that produce programmable vibrations, which is not the case for the structure examined in this paper.

Another application of the frequency spectrum of the random sequence is the estimation of the risk of resonant vibrations produced by random components on the elements of the cultivator structure. Components of the excitation that produce resonances are unlikely because the coincidence of a constant excitation with one of the first natural frequencies of the structure (here we refer only to the support of the working organ; in the general case, the calculation is very voluminous) is difficult to manifest. Even if it happens, the damping from the metal material and the ground on which it moves severely diminishes the resonant effect. Therefore, resonance effects are probably due to the improper functioning of some internal organs or a work speed that, combined with the soil profile, can produce them. However, a comparison between the spectrum of the supports of the working body and the spectrum of the excitations can be made to investigate the resonance problem at the supports of the working body or some more complex sub-assemblies. Figure 13 shows an example of a study of the distribution of the natural frequencies of the studied structure (here, the supports of the working body) in relation to the excitation spectrum (signals from the supports), obtained using the Fourier transform. In Figure 13, with solid grey bars, the spectra of the twelve supports of the working bodies are drawn in the 14–16 Hz range, in which the fundamental frequency of the support falls. The red bar with a circle at the upper end marks the position of the fundamental natural frequency of the supports in the chosen frequency range. The fundamental Eigen frequency and the other Eigen frequencies are calculated and shown in Figure 14 (the first five natural frequencies), but they were not experimentally determined as they should have been when calibrating the structure.

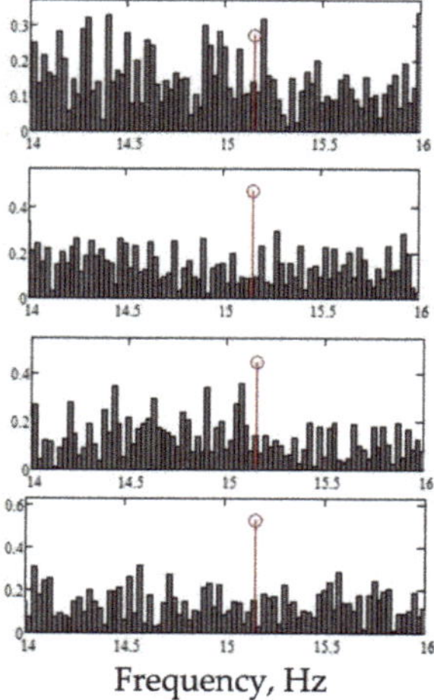

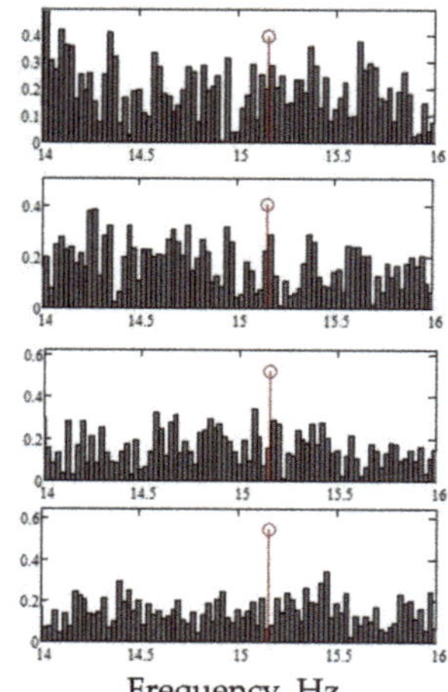

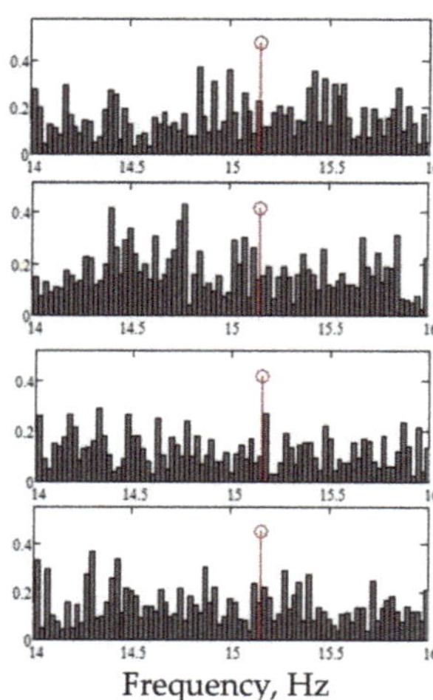

Figure 13. Representation of the frequency spectrum of the twelve signals from the supports of the working bodies, compared to the fundamental frequency of the support, calculated by the finite element method (represented by the thin red bar with a circle at the upper end).

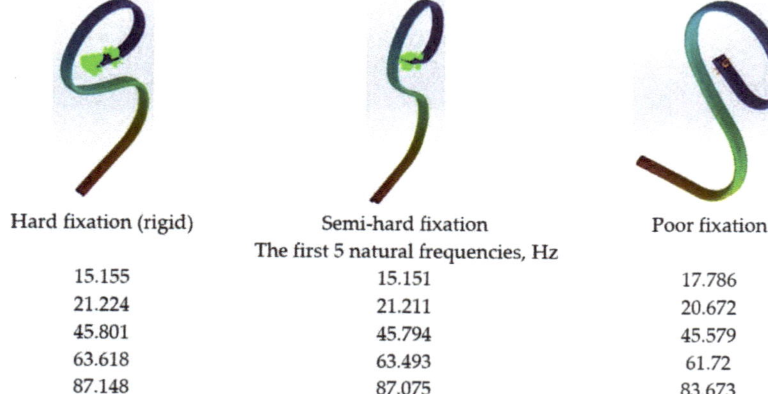

Figure 14. The first five natural frequencies of the supports of the working bodies of the MCLS in three variants of fixation on the load-bearing structure and the deformed forms of the structure in the vibration mode correspond to the fundamental frequency.

Since we had no data about the operation in the non-linear elastic or plastic domains or about the damping capacity of the steel from which the supports are built, we worked in the linear elastic domain, which makes the displacement values exaggerated when calculating the natural frequencies. For this reason, we have not given numerical data relative to the deformed shapes corresponding to the five fundamental natural frequencies.

The value of the fundamental natural frequency represented by the red bar with a circle at the upper end in each of the twelve graphs in Figure 13 corresponds to the support or semi-hard fixation of the support (15.151 Hz). Overlaps of the fundamental frequency over local maxima of the excitation spectrum occur for channels ch4, ch23, ch24, ch3, ch20, ch2, and ch1. However, there are no reasons for concern because the peaks in the spectra of the supports have small values and the damping capacity of the material of the supports, and especially of the soil in which the working body moves, is high.

In addition, when calculating the spectra of all the twelve signals coming from the twelve measurement locations, it is found that they are all highly correlated, with the minimum value of the correlation of two such numerical sequences being 0.979. The most intensely correlated signals correspond to channel pairs: ch2 and ch17, ch2 and ch1, and ch1 and ch21. At the lowest intensities, we find correlations between channels ch4 and ch1, and between ch4 and ch17.

A better understanding of the distribution of the frequency spectrum of all twelve signals coming from the measurement location of the supports of the working body of the cultivator in relation to the natural (calculated) frequencies of the supports is possible using the graphic from Figures 15 and 16.

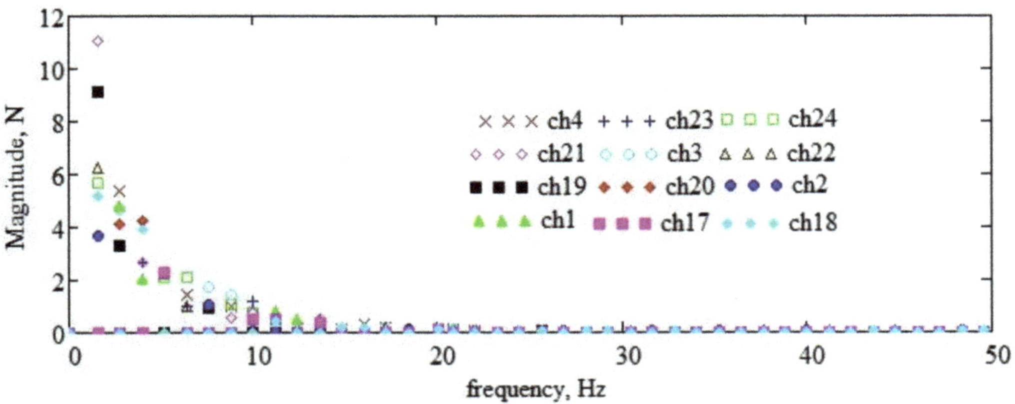

Figure 15. The frequency spectrum for the twelve signals (in the 0–50 Hz window).

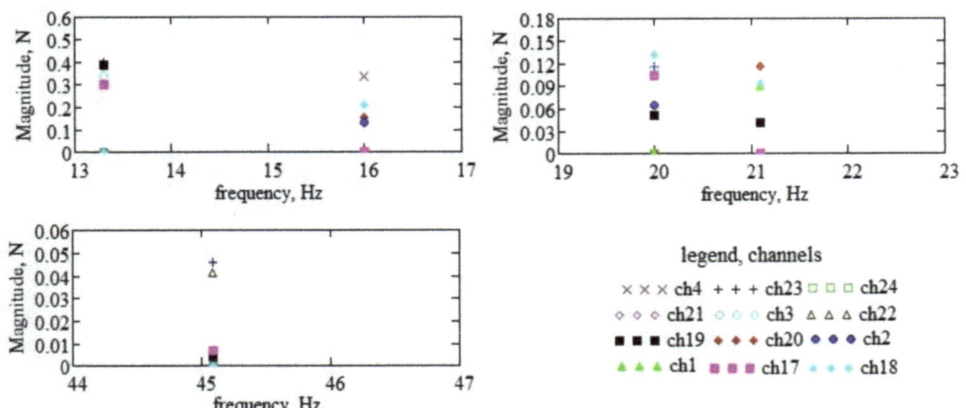

Figure 16. The frequency spectra of all twelve signals in narrow windows around the first three natural frequencies of the support of the working organ (see Figure 14).

In Figure 15, the spectral distribution is plotted for all signals in the range (window) 0–50 Hz, which includes the first three fundamental frequencies. Details in narrow intervals around the first three fundamental frequencies (see Figure 15) are plotted in Figure 16. It is found that, around the first three natural frequencies of the supports of the active organs, there are few frequencies from the signal spectrum, and, in addition, their magnitude is very small (when comparing the magnitudes of the frequencies in Figure 17 with those of the spectrum components in Figure 15).

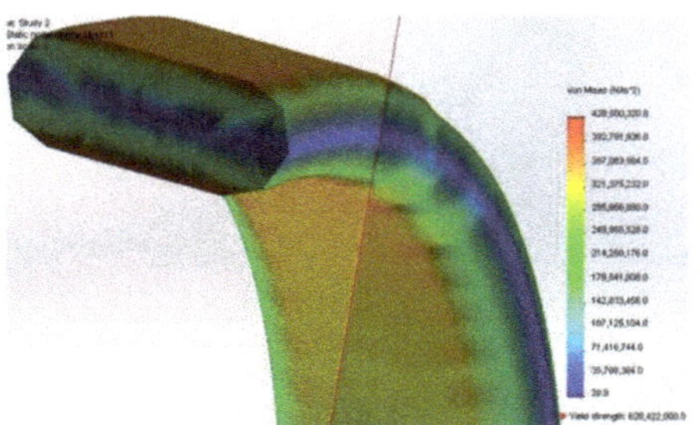
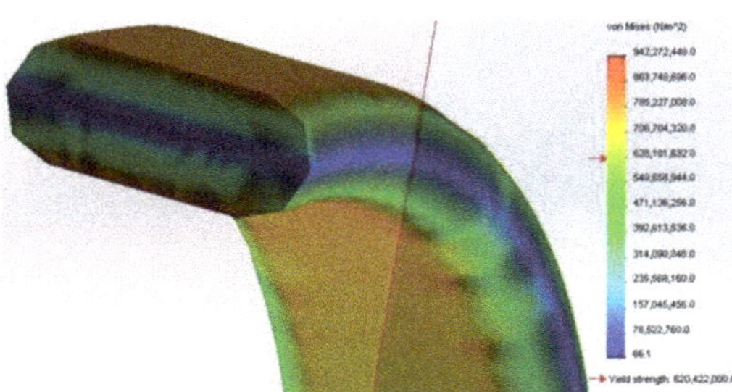

Figure 17. The von Mises stress map on the section with the maximum load on the support of a working body of the MCLS. The load values were 518 N (**left**) and 1086 N (**right**).

2.1.5. RMS and the Average Value

In many studies of random vibrations, RMS (root mean square) is calculated as a measure of amplitude extremes. The frequent use of RMS in the study of vibrations is also due to its direct connection with the energetic content of vibrations and implicitly with their destructive capacity [51]. RMS is related to the vibration amplitude and its average value.

Table 1 gives the RMS values and the average value for each sequence recorded in the examined experiment. It can be seen that the two measures of the force amplitude are close. Moreover, the correlation between the two matrices (that of the RMS and the average values) is very high: 0.999.

Table 1. RMS and average values for the numerical sequences corresponding to the twelve measurement points (see Figure 6), organised in the table according to the order of the physical structure.

RMS, N			Average Values, N		
477.145	**429.029**	**560.354**	453.124	410.584	537.523
549.272	405.641	498.193	536.507	385.602	483.493
557.469	571.771	543.765	541.699	552.885	532.069
659.646	665.703	514.895	644.269	650.431	493.188

3. Results

The main results that will be presented are applications for the design, execution, and operation of agricultural machines such as the MCLS complex cultivator, which is presented in Section 2. The results refer to the calculation of the probability of the occurrence of dangerous peaks, the selection and counting of the peaks of force that produce fatigue accumulation (in the supports of the active organs), the identification of design defects or deficiencies or of the work regime, and effects on the quality of the work.

3.1. *The Probability of the Occurrence of Dangerous Loads*

To identify and count the dangerous peaks of force in support of the working organ, the characteristic resistance limits of the material from which it is constructed will be used: the bending fatigue limit stress, σ_{-1}; the yield stress (plasticisation), σ_Y; and the breaking or yielding limit stress, σ_r. To be able to solve the proposed problem, the finite element method was used to determine the state of equivalent tension in support of a working body as a result of the application of forces with experimentally determined values. The material used has the following limit characteristics: σ_Y = 620.422 MPa, σ_r = 723.826 MPa, modulus of elasticity E = 210,000 MPa, Poisson's ratio ν = 0.28, mass density ρ = 7700 kg/m^3, and the fatigue resistance limit by bending, σ_{-1} = 380 MPa. The structural model used is described in [16,17]. The occurrence of damage phenomena (irreversible deformations, cracks, or breaks in the material) is determined with the help of the structural model by comparing the values of the equivalent stress (von Mises) in the structure with the critical limit stresses of the material. In Figure 18, the colour maps of the state of equivalent tension resulting from the application of the average (518 N) and maximum (1086 N) forces to the support of the working body are shown.

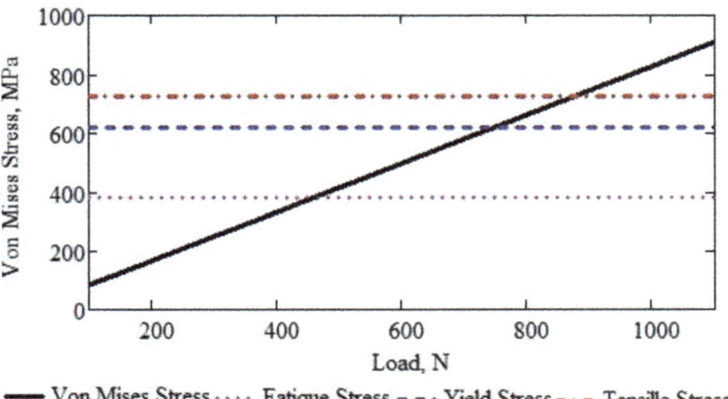

Figure 18. The variation in the maximum von Mises stress in the material supporting the active organs depends on the loading of the organ compared to the critical limits of the material.

It can be observed that the maximum equivalent stress for the load of 518 N has a value of 428.5 MPa (with a maximum resultant relative displacement of 77 mm), and for the load of 1086 N, it has a value of 942.3 MPa. Therefore, the average load of 518 N does not pose damage problems, and the support works in the field of linear elasticity with a safety coefficient of approximately 1.45. In the case of the maximum stress (1086 N), the equivalent stress in the structure reaches maximum values higher than the breaking limit stress of the material. However, the organ will not yield yet because the core of the bar from which the support is built presents an appreciable area that works in the linear elastic domain. For the maximum value of 1086 N, the yield stress is exceeded in an appreciable area (Figure 17), and even the breaking stress is exceeded in narrow areas in strips near the longer sides of the cross-section. However, the section works in linear elastic mode in the central area, so a failure is not directly recorded. However, in this working regime, high peaks lead to the accumulation of fatigue and possibly premature failure.

Using the results provided by the structural analysis of the support of the working body and the probabilities interpolated with cubic spline functions, it is possible to calculate the probability of exceeding the limit forces that cause the critical limits of the material to be exceeded: fatigue, plasticisation, and breaking. First, based on the hypothesis of the linear elastic behaviour of the support material of the working body, the graphic representation is shown in Figure 18.

Note: To calculate the equation of the oblique line in Figure 18, we accepted the hypothesis of linear elastic behaviour of the support material upon bending up to the plasticisation limit. Thus, the fatigue limit is included in the elastic bearing. The equation of the oblique line in Figure 18 is:

$$\sigma = Kx \tag{7}$$

where the elastic constant K is calculated using the result of the structural analysis for the average value (518 N) of the force applied on the twelve sequences of analysed force values: $K = 0.827$ N/Mpa.

The probability that the loading force corresponding to each of the twelve supports of the active organs will exceed the critical values of flow, rupture, and fatigue (characteristics of the material of the supports) is shown in Figure 19.

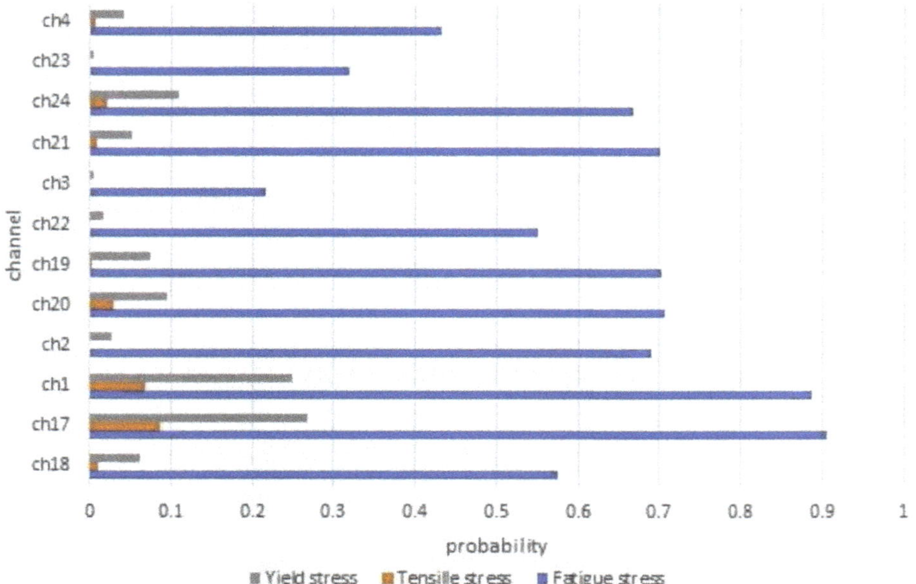

Figure 19. The probability that the loading force will exceed the limits of flow, rupture, and fatigue of the material of the support of the working body.

3.2. Counting the Peaks That Produce Fatigue

One of the immediate applications of this research is the possibility of counting the "cycles" that produce fatigue in the support material of the working body. Using the data and the graph in Figure 18, one can find that the "part" of the random sequence of force values that requires the support of the working body can produce the accumulation of fatigue, cracks, and breaks in the material.

A stress sequence filter is the simplest selection method for the part of the stress that produces fatigue accumulation, plastic yielding, or breaks. The expression of such a filter is given by Formula (8):

$$\varsigma_i = \begin{cases} s_i, & s_i > \sigma_{-1} \\ \sigma_{-1}, & s_i \leq \sigma_{-1} \end{cases}, \; i = 1, \ldots, N \tag{8}$$

where s is the sequence of experimental data in force values converted into values of the equivalent stress (von Mises) in the material, according to Formula (7).

In Figure 20, the load peaks of the support of the active body from the extreme left of the first line of working bodies (ch4) after the back of the tractor are highlighted. Prominent spikes produce fatigue buildup in the support material.

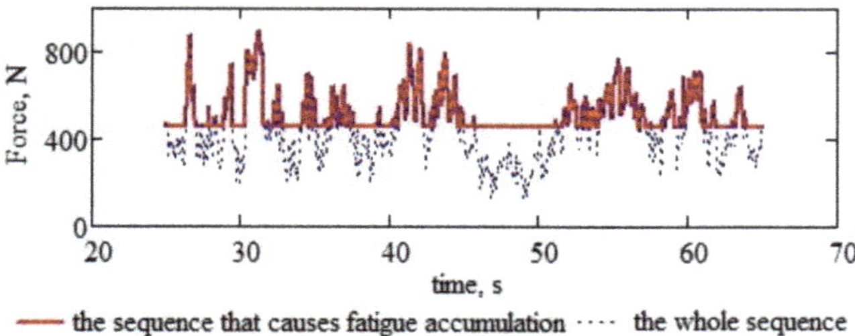

Figure 20. Filtering the sequence of forces that produce the accumulation of fatigue in the signal corresponding to channel ch4.

The wear of the support in 40 s will be calculated by structural engineers using the above data. During this time, the vehicle covers approximately 30 square metres of land. The result allows the estimation of the life span in time (neglecting the wear caused by the action of water or other environmental factors) in hectares of ploughed land or seasons or years of exploitation. In order to determine the lifetime of the supports of the working bodies, any structural analysis programme that has this facility can be used (for example, [52,53]).

3.3. Identification of Defects, Deficiencies, or Operating Errors

Most often, the study of random vibrations is dedicated to or requires the inclusion of the identification of sources of errors, defects, deficiencies, premature wear, etc. in the operating regime. Measuring the random vibrations of structures and improving or even optimising them based on the measurements has become one of the most commonly used methods in the study of phenomena affected by vibrations [54–58].

The research carried out in the MCLS complex cultivator case highlighted work deficiencies, especially the problem of low-quality control of the working depth. It was shown in [41] that using only two working depth control wheels (located in the front of the MCLS) is insufficient. A pair of wheels would also be necessary at the rear. In addition, the suspension-traction system (connection to the tractor) should also be checked to not raise the structure at the front during work. The identification of this problem was made by analysing the distribution of the intensity of the measured force, which indicated appreciably higher forces on the row of organs at the back (ch1, ch17, ch18, see Figure 6) in relation to those measured on the first row of organs behind the tractor (ch4, ch23, and ch24, see Figure 5). The main reason and validation of the suspicion were found by analysing the images in [42]. This deficiency partially generates the non-uniformity of the working depth, treated in Section 2.1.4.

Another deficiency valid only in the case where the MCLS operates in exploitation mode in any of its work variants is unsatisfactory soil fragmentation. According to the analysis in the field and on the photos, it is necessary to couple some shredding elements (discs, rollers, etc.) to the back of the structure (the granulometric analysis of the processed soil was not performed). In the research regime of the pure effects of the working body, the installation of additional shredding structures is prohibited because it alters the desired effects.

Another category of possible deficiencies, which can be avoided partially at the design stage and through systematic experimentation in working mode, are resonances and interferences, also called knocks. Both phenomena are defined in the literature, for example [59]. A search for possible resonances in the supports of the working organs is presented in Section 2.1.4 when calculating the frequency spectra for the twelve supports of the working bodies. The literature includes the results of some studies related to the problems of vibrations in agricultural machines [25,26,34,60–67].

3.4. Effects on the Quality of the Working Depth

The vibration of the supports of the cultivator's working organs involves relative movements horizontally, vertically, and laterally. The vibrations of the cultivator are complex; they are not reduced to the elastic vibrations of the working supports; there are also random rigid vibrations of the supporting structure and the working organs. All these types of vibrations have consequences for the main parameters of the work quality: working depth, initial working width, energy consumption, productivity, and comfort of the tractor driver.

In order to give only an ideal picture of the variations in the working depth due to random vibrations, we performed an elementary and ideal calculation on the deformations of the supports of the working bodies in the work process by applying some forces included in the experimental range. Using the finite element method and the model to calculate the first five natural frequencies, we estimated the relative displacements (deformations) in the three directions for the extremity of the support to which the working body is attached. The results are given in Table 2.

Table 2. The values of the relative displacements of the tip of the support of the working body are calculated using a linear-elastic structural model built by the finite element method.

Force Magnitude, N	Relative Displacement (Deformation), m		
	Ox (Lateral)	*Oy* (Forward)	*Oz* (Vertical)
100	0.001060	0.01329	0.007152
200	0.003340	0.04182	0.023520
500	0.009401	0.01179	0.072160
1000	0.020660	0.25910	0.190700

The results in Table 2 are slightly exaggerated because the support material is assumed to be perfectly linearly elastic, which is not true. However, they suggest a good value as an order of magnitude for the relative displacements. The dominant overshoots are manifested in the forward and vertical directions on the ground. These values are approximately linearly related. At the average value of the force from the recordings related to the analysed experiment (500 N), the relative vertical displacement reaches 7 cm, which seriously affects the programmed working depth (10 cm). These situations are not frequent, but they are facilitated by exceeding a working speed limit, deformations of the working surface (unevenness), and errors in the working depth regulation system.

In such conditions, the cultivator cannot be a tool to process the soil with high precision (a maximum error of the order of 10% of the theoretical depth) at the working depth. A previously levelled and sufficiently crushed work surface, combined with an adapted work speed, are mandatory conditions for precise work in terms of working depth. A general picture of the quality of the work performed in terms of working depth is shown in Figure 21.

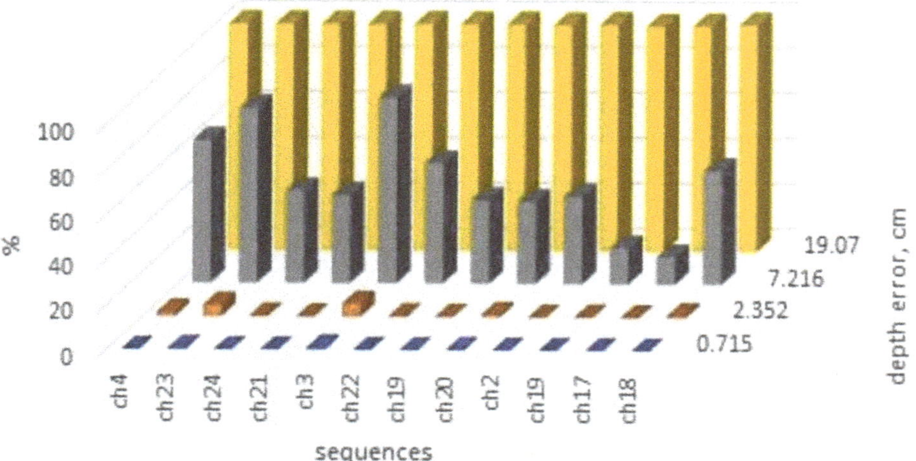

Figure 21. Errors in the working depth are calculated according to the measurements made on each channel.

The calculation from this subsection is based on the graphic representation in Figure 21, which uses the hypothesis of the operation of the supports of the working bodies of the MCLS in a linear-elastic regime without damping, which leads to an overestimation of the relative displacements. Direct measurements regarding the working depth and its monitoring are insufficiently developed in the available experimental techniques; therefore, the estimation method described above was used.

An option to Improve the accuracy of the working depth is to ballast the load-bearing structure by adding some ballast materials. Increasing the precision of the working depth in this way will lead to higher energy consumption and, possibly, a decrease in productivity.

Additionally, a problem that must be solved for this machine is finding the maximum speed at which it can work without manifesting intense vertical oscillations throughout the entire structure, which completely compromises the quality of soil processing. The speed limit, which is determined through tests, is dependent on the characteristics of the soil, so an exact speed limit cannot be indicated a priori.

4. Comments

According to [51], most random vibration analyses are intended to realistically characterise the behaviour of structural systems excited by random inputs. The responses of real systems are known only when measured during their actual physical loading (and even then, only approximately). The effects caused by the use of simplifying assumptions in the process of numerical simulation of physical systems are rarely evaluated. Because of the depth of our knowledge in the analysis of linear systems subject to stationary media, we often idealise real systems as linear and the inputs as stationary. Practically, all real systems are nonlinear and random to a small or large extent. Therefore, the characterisation of the response can only be approximate. Such reasoning also directed our attention to the experimental or theoretical-empirical research of the working processes of agricultural machines.

Starting from these realities, we tried to formulate the problem within the strict framework of random vibrations. We tried to use as few simplifying assumptions or assumptions from the theoretical field of linear vibrations as possible. For this, we constructed the characteristic functions of the empirical sequences exactly as they result from the definition (Section 2). Probability densities and probabilities are obtained directly by using the numerical sequences and interpolating them with spline functions. We did not use probability density modelling with idealised functions (normal distributions, Student, Fischer, etc.), and consequently, the resulting probabilities are also modelled on real data.

Representations of the spectral frequency distributions of the twelve supports of the working bodies (Section 2.1.4, Figures 15 and 16) can be found in [24] for the chassis of a harvester. The orders of magnitude and the numerical values are also comparable, being characteristic of the working processes of agricultural machines. As values, the forces resulting from the experimental study described in this article are comparable to those found for the working bodies and also for a cultivator by the authors [68]. The obtained frequency spectra are similar to those from [69]. Unfortunately, we could not get accelerometer signals from the experimenter, which was very important. Acceleration recording remains a problem to be solved for MCLS. Some of the forces' values and the frequency spectra's frequencies were also obtained in [41]. In the same register of concerns and spectral values, [26] estimates the effects of frequencies and vertical accelerations on several types of tractors using frequency-acceleration diagrams. Such diagrams were not accessible for this paper, so frequency-load force diagrams on the working body were used instead. However, the principle of selecting deficient or dangerous work regimes is the same. The authors [29] use the same principle of spectral selection for the vibration level on the front axle of the Valtra 800 L tractor. As an estimator of the vibration level, the authors [29] use RMS (root mean square) amplitude [70] for the experimentally recorded acceleration sequences in the version indicated by the authors [29]. The RMS maximums are recorded in the frequency range where the maximum oscillation amplitudes of our force sequences were located. The optimisation of the frame of a precision agricultural machine for sowing vegetables by avoiding resonances, based on the same principles as in this article, is described in [61]. The authors proceed similarly [36] for cultivators with working bodies very similar to those with which the MCLS cultivator was equipped in our experiments. Similarly, the authors [62] proceed with the study of the vibration of rapeseed seeds, additionally using high-speed photography and image recognition. All the frequency spectra (in this case normalised) are used in [67] to carry out experimental investigations with vertical damping for block-modular aggregates. As in this article, the authors [71] use spectral analysis and finite element analysis to predict fatigue accumulation in the arm of an excavator. The use of statistical analysis to estimate fatigue life dates back a long time [72], even for an agricultural cultivator; this is one of the main applications of the study of random vibrations. A similar approach for the testing of agricultural machines in an accelerated regime is proposed by the authors [73], in the same manner that we also proposed in Section 2.1.2. It is emphasised that not only the value of the load matters but also its frequency. The authors proceed similarly in [74,75]. A solution to reduce the force of resistance to advance for a cultivator by inducing forced vibrations is presented in [34]. Additionally, the return to some classical solutions, which had been partially abandoned, appeared after studies of random vibrations [76].

5. Conclusions

(C1) The analysis of random vibrations on any of the MCLS complex cultivator's variants is an opportunity and an obligation that must be mentioned in the work methodology that uses this equipment for research. The experiments carried out with the MCLS variants were also its first research applications. The literature shows that this is the most realistic way of approaching work processes in agriculture, having indisputable advantages over mathematical modelling and simulation. Among the most essential advantages is eliminating some assumptions that are often overly simplistic, lacking real motivation, and, above all, rarely verified, at least in the validation process. Moreover, simulation and mathematical modelling must validate their results experimentally; otherwise, they remain only scientific ballast.

(C2) The main results with immediate applications of random vibrations are the improvement of noise and vibration, the determination of dangerous demands for the elements of the structure, the determination of some limit values necessary for the choice of the material of some parts and sub-assemblies in the design, the estimation of the lifetime of the required elements in the regime of accumulation of fatigue, the identification of design or work regime deficiencies, and procedures for improving the quality of soil processing.

(C3) The presented results underline the deficiencies that must be resolved to increase the grower's performance. The recommendations refer to the control of the horizontalisation of the structure during work (with consequences for increasing the quality of soil processing relative to the working depth), the possibilities of ballasting the structure for better control of the working depth, limiting the operating speed, and the requirement of soil processing on some lands with satisfactory flatness.

(C4) A vital conclusion drawn from the results obtained is that to avoid malfunctions or damage due to premature fatigue, it is recommended that the "weakest" design elements, the ones that give way first, be among the cheapest or be simple, easy-to-replace, and cheap elements inserted into the subassemblies that will fail first in the event of an overload. We refer in particular to the fact that it is not rational to take into account the maximum loading for the structure (experimentally recorded). However, it is good to impose provisions for introducing safety elements. Thus, important and expensive subassemblies and parts will be protected. In the design of the 1960s and 1970s, for example, there were safety screws on ploughs that, by giving way first, protected the working bodies, their supports, and the load-bearing structure. Later, the abandonment of some of these safety elements produced the failure of load-bearing structures, for example.

(C5) Since we failed to validate the intuitive assumption that the working organs in the first line behind the tractor would be the most intensively requested, neither in these investigations nor in those described in [41], we deduce that, at least in the experimental conditions, the distribution of the maximum load is random over the twelve working bodies. An important consequence of this finding is that we cannot schedule a rotation of the working bodies on the load-bearing structure positions to equalise the wear and use the full working capacity of all working bodies. Moreover, a statistically higher load was observed for the working body in the last line of organs in the load-bearing structure. First, only the result of the statistical analysis was validated based on the analysis of the images taken during the experiments. It is shown that certain aspects must be fixed in the operation of this machine, which will improve the quality of the work.

(C6) The results of the experimental research agree with those in the literature, presenting significant randomness and revealing average values of the efforts at the level of working bodies expected from a theoretical point of view. The experiments present an original character through a large number of measurement points, the study of the distribution of their results, and the investigation of the links between these measurements.

Obviously, in such a complex problem as soil processing, there will always be future work directions. In the case of the problem whose research is described in this article, there are many open problems, among which are mentioned: the dependence of the descriptive and inferential statistical characteristics on the work speed; the estimation of the soil processing problem with these work organs on all the work variants of the MCLS; solving the problems already listed for all the categories of working bodies that can be mounted on the load-bearing structure of the MCLS; and the study of the processing speed limit depending on the quality of the work performed. All these problems, which do not cover even half of the ways to continue the studies, should be carried out on several types of soil at different humidity levels, aiming, if not to optimise, at least to improve agricultural soil processing. Another interesting problem is the selection of a minimum number of measurements of the soil characteristics that can be used to fix the starting parameters of the soil processing. Among these measurements, we mention humidity, resistance to penetration, plant residue amount or density, and soil resistance's quantified characteristics (apart from resistance to penetration).

Before choosing a way to continue, evaluating the possible benefits of each possibility of continuing the research is required. The evaluation creates the conditions for an optimal choice of development paths, considering the material and human resources, especially the enormous costs of experimental research, which in the end necessarily also includes the theoretical one.

Author Contributions: Conceptualisation, P.C.,N.C. and V.M.; methodology, P.C.,N.C. and V.M.; software, P.C. and C.P.; validation, P.C., N.C., C.P., R.S., C.M.-I., O.-D.C. and E.-A.L.; formal analysis, P.C.,N.C., R.S., N.-V.V., C.M.-I. and E.-A.L.; investigation, P.C., N.C., C.P., V.M., N.-V.V., N.U., M.M., C.M.-I. and O.-D.C.; resources, P.C., R.S., N.-V.V., N.U. and M.M.; data curation, P.C, C.P., V.M., O.-D.C. and E.-A.L.; writing—original draft preparation, P.C. and N.C.; writing—review and editing, P.C., V.M. and N.U.; visualisation, N.U., O.-D.C. and E.-A.L.; supervision, P.C., V.M., N.-V.V., N.U. and M.M.; project administration, P.C., V.M., R.S. and C.M.-I.; funding acquisition, V.M., N.-V.V., N.U. and M.M. All authors have read and agreed to the published version of the manuscript.

Funding: The APC was funded by the National University of Science and Technology Politehnica Bucharest, Romania, within the PubArt Program.

Institutional Review Board Statement: Not applicable.

Conflicts of Interest: The authors declare no conflict of interest.

References

1. Ingram, J. Randomness and probability: Exploring student 'teachers' conceptions. In *Mathematical Thinking and Learning*; Taylor and Francis Group: Oxfordshire, UK, 2022; pp. 1–19.
2. Available online: https://en.wikipedia.org/wiki/Random_vibration (accessed on 9 March 2023).
3. Available online: http://www.mobilindustrial.ro/current_version/online_docs/COMPENDIU/semnale_deterministe_si_nedeterministe__aleatorii_.htm (accessed on 9 March 2023).
4. Crandall, S.H. *Random Vibration*; MIT Press/Wiley: New York, NY, USA, 1958.
5. Crandall, S.H.; Mark, W.D. *Random Vibration in Mechanical Systems*; Academic Press: New York, NY, USA, 1963.
6. Bolotin, V.V. *Random Vibrations of Elastic Systems*; Martinus Nijhoff Publishers: The Hague, The Nederlands, 1994.

7. Elishakoff, I. *Probabilistic Methods in the Theory of Structures: Random Strength of Materials*; Random Vibration, and Buckling; World Scientific: Singapore, 2017.
8. Elishakoff, I. *Solution Manual to Accompany Probabilistic Methods in the Theory of Structures: Problems with Complete*; Worked Through Solutions; World Scientific: Singapore, 2018.
9. Elishakoff, I. *Dramatic Effect of Cross-Correlations in Random Vibrations of Discrete Systems, Beams, Plates, and Shells*; Springer Nature: Basel, Switzerland, 2020.
10. Nyamekye, A.B.; Nyadzi, E.; Dewulf, E.; Werners, S.; Van Slobbe, E.; Biesbroek, R.G.; Termeer, C.J.A.M.; Ludwig, F. Forecast probability, lead time and farmer decision-making in rice farming systems in Northern Ghana. *Clim. Risk Manag.* **2021**, *31*, 100258. [CrossRef]
11. Hamsa, K.R.; Bellundagi, V. Review on Decision-making under Risk and Uncertainty in Agriculture. *Econ. Aff.* **2017**, *62*, 447–453. [CrossRef]
12. Buschena, D.E.; Zilberman, D. What Do We Know About Decision Making Under Risk and Where Do We Go from Here? *J. Agric. Resour. Econ.* **1994**, *19*, 425–445.
13. Kumari, P.L.; Reddy, P.M. Decision Making under Uncertainty in Agriculture-A Case Study. *Int. J. Curr. Microbiol. Appl. Sci.* **2021**, *10*, 188–194.
14. Eisele, M.; Troost, C.H.; Berger, T.H. How Bayesian Are Farmers When Making Climate Adaptation Decisions? A Computer Laboratory Experiment for Parameterising Models of Expectation Formation. *J. Agric. Econ.* **2021**, *72*, 805–828. [CrossRef]
15. Magda, D.; Girard, N.; Angeon, V.; Cholez, C.; Raulet-Croset, N.; Sabbadin, R.; Salliou, N.; Barnaud, C.; Monteil, C.; Peyrard, N. A Plurality of Viewpoints Regarding the Uncertainties of the Agroecological Transition. In *Agroecological Transitions: From Theory to Practice in Local Participatory Design*; Bergez, J.E., Audouin, E., Therond, O., Eds.; Springer: Cham, Switzerland, 2019; pp. 99–120.
16. Spash, C.L. Strong uncertainity, Ignorance and indeterminacy. In *Greenhouse Economics: Value and Ethics*; Routledge: London, UK, 2002; pp. 120–152.
17. Litre, G.; Bursztyn, M. Climatic and socio-economic risks and adaptation strategies among livestock family farmers in the pampa biome. *Ambiente Soc.* **2015**, *18*, 53–78.
18. Coletta, A.; Gianpietri, E.; Santeramo, F.G.; Severini, S.; Triestini, S. A preliminary test on risk and ambiguity attitudes, and time preferences in decisions under uncertainty: Towards a better explanation of participation in crop insurance schemes. *Bio-Based Appl. Econ.* **2018**, *7*, 265–277.
19. Bunyasiri, I.N. Impacts of income uncertainty and diversified agriculture on savings of thai agricultural households. *Bus. Manag. Rev.* **2017**, *4*, 311–322.
20. Burgman, M. *Risks and Decisions for Conservation and Environmental Management*; Cambridge University Press: Cambridge, UK, 2005.
21. Costake, N. Cu privire la conceptul de Management societal informatizat. *Rev. Romana De Stat. Supl.* **2013**, *1*, 153–301.
22. Kawtar El, K.; Driss, M. Stochastic Programming for Agricultural Planning. *J. Posit. Sch. Psychol.* **2022**, *6*, 1938–1944.
23. Raun, W.R.; Dhillon, J.; Aula, L.; Eichoff, E.; Weymeyer, G.; Figueirdeo, B.; Lynch, O.; Nambi, E.; Oyebiyi, F.; Fornah, A. Unpredictible Nature of Environment on Nitrogen Supply and Demand. *Agron. J.* **2019**, *111*, 2633–3403. [CrossRef]
24. Yao, Y.; Li, X.; Yang, Z.; Li, L.; Geng, D.; Huang, P.; Li, Y.; Song, Z. Vibration Characteristics of Corn Combine Harvester with the Time-Varying Mass System under Non-Stationary Random Vibration. *Agriculture* **2022**, *12*, 1963. [CrossRef]
25. Garciano, L.; Torisu, R.; Takeda, J.; Sakai, K. Random, quasi-periodic and chaotic vibrations of farm tractors. *J. Jpn. Soc. Agric. Mach.* **2002**, *64*, 91–98.
26. Vladut, V.; Biris, S.S.; Petre, A.A.; Voicea, I.; Cujbescu, D.; Ungureanu, N.; Matei, G.H.; Popa, D.; Boruz, S.; Constantin, A.M.; et al. Comparative sudy of vibrations in apparatus combinations tangential Threshing. *Ann. Univ. Craiova-Agric. Mont. Cadastre Ser.* **2021**, *51*, 615–627.
27. Matache, M.G.; Voicu, G.h.; Epure, M.; Voicea, I.F.; Gageanu, I.; Ghilvacs, M. Measuring vibrations level during transportation work for electrical tractor. *Acta Tech. Corviniensis-Bull. Eng.* **2020**, *13*, 27–30.
28. Dos Santos, V.C.; Monteiro, L.; De, A.; Macedo, D.X.S.; Costa, E. Whole body vibration in operators using agricultural soil preparation equipment. *Ciência Rural. Santa Maria* **2019**, *49*, 1–9. [CrossRef]
29. Villbor, G.P.; Santos, F.L.; de Queiroz, D.M.; Guedes, D.M. Vibration levels on rear and front axles of a tractor in agricultural operations, Acta Scientarium. *Technology* **2014**, *36*, 7–14.
30. Hosseinpour-Zarnaq, M.; Omid, M.; Biabani-Aghdam, E. Fault diagnosis of tractor auxiliary gearbox using vibration analysis and random forest classifier. *Inf. Process. Agric.* **2022**, *9*, 60–67. [CrossRef]
31. Marin, C. *Vibratiile Structurilor Mecanice*; Editura Impuls: Bucharest, Romania, 2003.
32. Xin, J.; Kaikang, C.h.; Jiangtao, J.; Kaixuan, Z.h.; Xinwu, D.; Hao, M. Intelligent vibration detection and control system of agricultural machinery engine. *Measurement* **2019**, *145*, 503–510.
33. Choi, W.-S.; Pratama, P.S.; Supeno, D.; Woo, J.-H.; Lee, E.-S.; Chung, S.-W.; Park, C.-S. Dynamic vibration characteristics of electric agricultural vehicle based on finite element method. *Globa J. Eng. Sci. Res.* **2017**, *4*, 27–31.
34. Esehaghbeygi, A.; Abedi, M.; Razavi, J.; Hemmat, A. Field evaluation of a vibrating dual bent-share cultivator. *Res. Agric. Eng.* **2020**, *66*, 126–129, 130–139. [CrossRef]

35. Badescu, M.; Croitoru, S.; Petcu, A.; Marin, E.; Boruz, S.; Ivan, G.h.; Vladut, V.; Matache, M.; Cujbescu, D.; Voicea, I. Construction evolution of deep soil loosening equipment in romania in the country specific soil conditions. *Ann. Univ. Craiova-Agric. Mont. Cadastre Ser.* **2014**, *44*, 25–34.
36. Polushkin, O.A.; Ignatenko, V.I.; Ignatenko, I.V.; Vyalikov, I.L.; Bogdanovich, V.P. Dynamic models of cultivator spring tine performance. *MATEC Web Conf.* **2018**, *226*, 01016. [CrossRef]
37. Available online: https://dexonline.ro/definitie/cultivator (accessed on 16 March 2023).
38. Available online: https://www.xn{-}{-}dicionar-qxb.com/Cultivator (accessed on 23 March 2023).
39. Cardei, P.; Constantin, N.; Sfiru, R.; Muraru, V.; Condruz, P. Modular load-bearing structures for the study of interactions between the soil and working parts of agricultural machines. *J. Phys. Conf. Ser.* **2021**, *1781*, 012063. [CrossRef]
40. Cardei, P.; Nicolae, C.; Sfiru, R.; Muraru, V.; Muraru, S.L. Structural analysis of a modulated load-bearing structure designed to investigate the interaction between soil and the working parts of agricultural machines. *J. Phys. Conf. Ser.* **2021**, *1781*, 012064. [CrossRef]
41. Cardei, P.; Constantin, N.; Muraru, V.; Persu, C.; Sfiru, R. Estimation of the random intensity of the soil tillage draft forces in the supports of the working bodies of a cultivator. *Prepr. RG* **2023**, *10*, 49760.
42. Buzdugan, G.H.; Fetcu, I.; Radeş, M. *Vibratii Mecanice*; Editura Didactica si Pedagogic: Bucharest, Romania, 1982.
43. Tiron, M. *Analiza Preciziei de Estimare a Functiilor Aleatoare*; Editura Tehnica: Bucharest, Romania, 1981.
44. Iosifescu, M.; Moineagu, C.; Trebici, V.; Ursianu, E. *Mica Enciclopedie de Statistica*; Editura Stiintifica si Enciclopedica: Bucharest, Romania, 1985.
45. Doğan, N.; Doğan, I. Determination of the number of bins/classes used in histograms and frequency tables: A short bibliography. *J. Stat. Res.* **2010**, *7*, 77–86.
46. Kun, P., II. *Fundamentals of Probability and Stochastic Processes with Applications to Communications*; Springer: Berlin/Heidelberg, Germany, 2018.
47. Tricker, R.; Tricker, S. *Environmental Requirements for Electrochemical and Electronic Equipment*; 8.4.4.5 Other Standards; Elsevier: Amsterdam, The Netherlands, 1999; Available online: https://www.sciencedirect.com/topics/earth-and-planetary-sciences/random-vibration (accessed on 23 March 2023).
48. Mathcad 2001. Available online: https://www.mathcad.com/en (accessed on 23 March 2023).
49. Available online: https://real-statistics.com/time-series-analysis/stochastic-processes/autocorrelation-function/ (accessed on 9 March 2023).
50. Available online: https://blog.endaq.com/vibration-measurements-vibration-analysis-basics (accessed on 16 March 2023).
51. Paez, T.L. *Random Vibrations: Assessment of the State of the Art*; Sandia National Laboratories: Albuquerque, Mexica, 1999.
52. Gheorghe, G.V.; Voicu, G.; Persu, C.; Găgeanu, I. Structural and Fatigue Analysis of Equipment for Cutting Weeds between Plants on the Same Row. *Sci. Bull. Univ. Politeh. Buchar. Ser. D Mech. Eng.* **2022**, *84*, 237–248.
53. SOLIDWORKS. Available online: https://help.solidworks.com/2021/english/SolidWorks/cworks/c_Fatigue_Analysis.htm (accessed on 9 March 2023).
54. Ghazali, M.H.; Rahiman, W. *Vibration Analysis for Machine Monitoring and Diagnosis: A Systematic Review*; Shock and Vibration; Hindawi: London, UK, 2021; pp. 1–25.
55. Li, J.; Liu, Y.; Xing, Z.h.; Zeng, F. Rotating machinery anomaly detection using data reconstruction generative adversarial networks with vibration energy analysis. *AIP Adv.* **2022**, *12*, 035221. [CrossRef]
56. Harris, C.M.; Crede, C.E. *Socuri si Vibratii*; Editura Tehnica: Bucharest, Romania, 1969.
57. Miki, D.; Demachi, K. Bearing fault diagnosis using weakly supervised long short-term memory. *J. Nucl. Sci. Technol.* **2020**, *57*, 1091–1100. [CrossRef]
58. Ebrahimi, E.; Javadikia, P.; Jalili, M.H.; Astan, N.; Haidari, M.; Bavandpour, M. Intelligent Fault Detection of Retainer Clutch Mechanism of Tractor by ANFIS and Vibration Analysis. *Mod. Mech. Eng.* **2013**, *3*, 17–24. [CrossRef]
59. Iacob, C.; Cheorghita, S.I.; Soare, M.; Dragos, L. *Dictionar de Mecanica*; Editura Stiintifica si Enciclopedica: Bucharest, Romania, 1980.
60. Wang, P.; Gao, M.; Sun, Y.; Zhang, H.; Liao, Y.; Xie, S. A vibration-powered self-contained node by profiling mechanism and its application in cleaner agricultural production. *J. Clean. Prod.* **2022**, *366*, 132897. [CrossRef]
61. Wang, J.; Xu, C.h.; Xu, Y.; Wang, J.; Zhou, W.; Wang, Q.; Tang, H. Resonance Analysis and Vibration Reduction Optimization of Agricultural Machinery Frame—Taking Vegetable Precision Seeder as an Example. *Processes* **2021**, *9*, 1979. [CrossRef]
62. Zhan, G.; Ma, L.; Zong, W.; Liu, W.; Deng, D.; Lian, G. Study on the Vibration Characteristics of Rape Plants Based on High-Speed Photography and Image Recognition. *Agriculture* **2022**, *12*, 727. [CrossRef]
63. Sheng, J.; Zeng, Y.; Liu, G.; Liu, R. Determination of Weak Knock Characteristics for Two-Stroke Spark Ignition UAV Engines Based on Mallat Decomposition Algorithm. *Math. Probl. Eng.* **2021**, *2021*, 1250327. [CrossRef]
64. Wang, S.; Lu, B. Detecting the weak damped oscillation signal in the agricultural machinery working environment by vibrational resonance in the duffing system. *J. Mech. Sci. Technol.* **2022**, *36*, 5925–5937. [CrossRef]
65. Jun, L.; Xueping, L.; Zhou, Y.; Tiansheng, H.; Kunpeng, X.; Wei, L. Transversal vibration analysis of resonance condition and frequency for orchard chain ropeway system. *Trans. Chin. Soc. Agric. Eng.* **2014**, *30*, 50–57.
66. Farrokhi Zanganeh, N.; Shahgholi, G.; Agh, S. Studying the effect of balancer on engine vibration of MaSSey Ferguson 285 tractor. *Acta Technol. Agric.* **2021**, *24*, 14–19. [CrossRef]

67. Bulgakov, V.; Kuvachov, V.; Ivanovs, S.; Melnyk, V. Experimental investigations in vertical vibration damping of agricultural aggregate of block-modular type. In *Engineering for Rural Development, Proceedings of the International Scientific Conference, Jelgava, Latvia, 24–26 May 2021*; Tavria State Agrotechnological University: Zaporizhzhia, Ukraine; pp. 635–642.
68. Abbaspour-Gilandeh, Y.; Fazeli, M.; Roshanianfard, A.; Herandez-Hernandez, J.L.; Penna, A.F.; Herrera-Miranda, I. Effect of Different Working and Tool Parameters on Performance of Several Types of Cultivators. *Agriculture* **2020**, *10*, 145. [CrossRef]
69. Gong, Y.; Ren, L.; Han, X.; Gao, A.; Jing, S.h.; Feng, C.h.; Song, Y. Analysis of Operating Condition for Vibration of a Self-Propelled Monorail Branch Chipper. *Agriculture* **2023**, *13*, 101. [CrossRef]
70. Available online: https://vru.vibrationresearch.com/glossary/rms-root-mean-square/ (accessed on 23 March 2023).
71. Zhao, G.; Xiao, J.; Zhou, Q. Fatigue Models Based on Real Load Spectra and Corrected S-N Curve for Estimating the Residual Service Life of the Remanufactured Excavator Beam. *Metals* **2021**, *11*, 365. [CrossRef]
72. Harral, B.B. The application of a statistical fatigue life prediction method to agricultural equipment. *Int. J. Fatigue* **1987**, *9*, 115–118. [CrossRef]
73. Wen, C.h.; Xie, B.; Li, Z.h.; Yin, Y.; Zhao, X.; Song, Z.H. Power density-based fatigue load spectrum editing for accelerated durability testing for tractor front axles. *Biosyst. Eng.* **2020**, *200*, 73–88. [CrossRef]
74. Yeter, B.; Garbatov, Y.; Soares, C.G. Spectral Fatigue Assessment of an Offshore Wind Turbine Structure under Wave and Wind Loading. In *Developments in Maritime Transportation and Exploitation of Sea Resources*; Guedes Soares, C., Lopez Pena, F., Eds.; Francis & Taylor Group: London, UK, 2014.
75. Paraforos, D.S.; Griepentrog, H.W. Surface profiles acquisition for assessing fatigue life of agricultural machinery in test facilities. In Proceedings of the 18th International Conference of the ISTVSAt, Seoul, Republic of Korea, 22–25 September 2014.
76. Cardei, P.; Constantin, N.; Gradinaru, V.; Marin, E.; Manea, D.; Matache, M.; Muraru, V.; Muraru, C.; Pirna, I.; Siru, R.; et al. *Structural Analysis and Materials Focused on Mechanics, Mechatronics, Maintenance and Operation of Machinery for Agriculture and Food Industry*; Editura Terra Nostra: Iasi, Romania, 2012; pp. 166–171.

MDPI
St. Alban-Anlage 66
4052 Basel
Switzerland
www.mdpi.com

Agriculture Editorial Office
E-mail: agriculture@mdpi.com
www.mdpi.com/journal/agriculture

www.ingramcontent.com/pod-product-compliance
Lightning Source LLC
LaVergne TN
LVHW071613170726
843515LV00009B/2388
* 9 7 8 3 0 3 6 5 9 2 1 2 1 *